Diffusions, Markov Processes, and Martingales

Volume 1: Foundations

2nd Edition

Diffusions, Markov Processes, and Martingales

Volume 1: FOUNDATIONS

2nd Edition

L. C. G. ROGERS
School of Mathematical Sciences, University of Bath

and

DAVID WILLIAMS
Department of Mathematics, University of Wales, Swansea

CAMBRIDGE
UNIVERSITY PRESS

CAMBRIDGE
UNIVERSITY PRESS

University Printing House, Cambridge CB2 8BS, United Kingdom

Cambridge University Press is part of the University of Cambridge.

It furthers the University's mission by disseminating knowledge in the pursuit of education, learning and research at the highest international levels of excellence.

www.cambridge.org
Information on this title: www.cambridge.org/9780521775946

First published 1979 by John Wiley & Sons Ltd, Chichester
Second edition 1994 published by John Wiley & Sons Ltd
Reissued by Cambridge University Press 2000
10th printing 2014

A catalogue record for this publication is available from the British Library

ISBN 978-0-521-77594-6 Paperback

For our parents

From the Original (1979) Preface

Long ago (or so it seems today), Chung wrote on page 196 of his book [1]: 'One wonders if the present theory of stochastic processes is not still too difficult for applications.' Advances in the theory since that time have been phenomenal, but these have been accompanied by an increase in the technical difficulty of the subject so bewildering as to give a quaint charm to Chung's use of the word 'still'. Meyer writes in the preface to his definitive account of stochastic integral theory: '*...il faut...un cours de six mois sur les définitions. Que peut on y faire?*'

I have thought up as intuitive a picture of the subject as I can, written it down at speed, and refused to be lured back by piety (or even by wit!) to cancel half a line. 'First' intuition, which is what you need when you are learning the subject, is raw, rough and ready; and, as you have guessed, I make the excuse that it demands a compatible style and lack of polish.

Note that I wrote '*first intuition*'. Consider an example. Meyer's concept of a *right process* is exactly right for Markov process theory, but the concept is the result of a long evolution. To understand it properly, you need a highly developed intuition, and that takes time to acquire. The difficulty with the best advanced literature is that its authors have too *much* intuition; never make the mistake of thinking otherwise.

My aim then is to sharpen your intuition to a point where the advanced abstract literature becomes accessible, enjoyable and 'relevant'. Like my expository article [1], this is a missionary tract not a theological treatise. (Those of you who have read my article [1] will see that this book often follows it very closely, except that now I have the time and the duty to be more obviously appreciative of the abstract theory!)

I believe that, in the end, it is *applications* which justify mathematics. The 'artistic' justification of pure mathematics in terms of intrinsic qualities like elegance and generality rings rather hollow in my ears when I compare the best mathematics with the greatest music. Many applied workers will regard this book as extremely 'pure', but I see it as *one stage in shunting pure theory over towards applications.* The shunting is not always necessary: time and again, one finds 'applied' papers which 'solve' problems long since solved for 'purely theoretical' purposes. Moral: the pure/applied division of probability theory (as of mathematics in general) is a nonsense.

Acknowledgements. This is an appropriate place at which to thank David Kendall and Harry Reuter for teaching me probability theory and for giving me an enthusiasm for the subject which is wearing well. My best way to thank them is to try to share that enthusiasm.

I have to say another huge 'thank you' to David Kendall for the immense amount of work he has done in making editorial comments on the original manuscript. I now see that my determination to convey a sense of adventure did need to be tempered by a greater concern for the reader's sense of security. So I have acceded to many of David Kendall's requests for 'more details'; and as a result, you will learn more techniques of calculation and have a clearer idea of several concepts. (But I still see it as part of my job to keep you on your toes!)

I am very grateful to Ronald Getoor and André Meyer for clearing up some confusions.

I have been extremely fortunate in having been able to rely on the superb typing skills of Sheila Campbell, Eileen Jenkins and Gladys Maddocks; my thanks and best wishes to them.

I thank Springer-Verlag and the authors for granting me permission to quote from Chung [1] in Section III.44, from Getoor [1] in Section III.54, and from Chung [1] and Meyer [1] earlier in ths preface.

Finally, I have to thank James Cameron and Wiley for encouragement and great patience; and subeditors, copy-editors, and printers, whose skills have much impressed me.

David Williams
Swansea, 1978

Preface to the Second Edition

This second edition differs profoundly from the first—and not only in having two authors rather than one. We retain the Gallic tradition of dividing the volume into three massive chapters: Chapter I, which says why the subject is worth studying; Chapter II, which provides background; and Chapter III, which presents an account of Markov processes. Chapter I is now much more extensive and wide-ranging, and covers much work done since the first edition appeared. Chapter II is now a highly systematic account, with detailed proofs, of what every young probabilist must know. It is rather unashamedly a sequel to DW's *Probability with Martingales*, Cambridge University Press having been very generous in allowing us to follow that account closely (but without many proofs, without the examples, etc.). *It is perfectly possible to read Chapter II before Chapter I if you so wish.* We would suggest however that you try things in the order 'heuristics then rigour':

> '*Our doubts are traitors,*
> *And make us lose the good we oft might win,*
> *Through fearing to attempt.*

(W. Shakespeare, *Measure for Measure*.)

Chapter III seems to have been regarded as the most successful part of the original; and it is reproduced here without much modification (except that some of the functional analysis is given fuller treatment). It was always intended as a missionary tract on Markov processes. The full theory may be found in Sharpe [1] and in the final two volumes of the probabilist's bible, Dellacherie and Meyer [1]. All kinds of important developments are ignored in Chapter III: they would require another complete volume, and will be, or are, covered by greater experts. Dawson's eagerly awaited treatment [1] of measure-valued processes has now appeared; Mark Davis has a very nice new book [4] on piecewise-deterministic Markov processes; and so on. You can access the huge literature on measure-valued processes via Dawson's account.

The musical allusions in the first edition have been excised. Apparently many people found them annoying. 'Would David Williams like a book on mathematics filled with references to baseball?', they say. (To which the answer is, of course, 'Yes.') So, this is Mathematics all the way from A to Zzzz—or from Ω on, if you want to be rigorous.

Our thanks to Sue Collins and Wolfgang Stummer, and to other colleagues at Bath, Cambridge, and Queen Mary and Westfield College, London. Our thanks too to Helen Ramsey and other Wiley staff for suggesting this new version; and the copy-editor and printer whose skills have impressed us.

Chris Rogers
David Williams
November 1993

Contents

Some Frequently Used Notation xix

CHAPTER I. BROWNIAN MOTION

1. INTRODUCTION 1

1. What is Brownian motion, and why study it? 1
2. Brownian motion as a martingale 2
3. Brownian motion as a Gaussian process 3
4. Brownian motion as a Markov process 5
5. Brownian motion as a diffusion (and martingale) 7

2. BASICS ABOUT BROWNIAN MOTION 10

6. Existence and uniqueness of Brownian motion 10
7. Skorokhod embedding 13
8. Donsker's Invariance Principle 16
9. Exponential martingales and first-passage distributions 18
10. Some sample-path properties 19
11. Quadratic variation 21
12. The strong Markov property 21
13. Reflection 25
14. Reflecting Brownian motion and local time 27
15. Kolmogorov's test 31
16. Brownian exponential martingales and the Law of the Iterated Logarithm 31

3. BROWNIAN MOTION IN HIGHER DIMENSIONS 36

17. Some martingales for Brownian motion 36
18. Recurrence and transience in higher dimensions 38
19. Some applications of Brownian motion to complex analysis 39
20. Windings of planar Brownian motion 43
21. Multiple points, cone points, cut points 45

22. Potential theory of Brownian motion in $\mathbb{R}^d$ ($d \geqslant 3$) 46
23. Brownian motion and physical diffusion 51

4. GAUSSIAN PROCESSES AND LÉVY PROCESSES 55

Gaussian processes

24. Existence results for Gaussian processes 55
25. Continuity results 59
26. Isotropic random flows 66
27. Dynkin's Isomorphism Theorem 71

Lévy processes

28. Lévy processes 73
29. Fluctuation theory and Wiener–Hopf factorisation 80
30. Local time of Lévy processes 82

CHAPTER II. SOME CLASSICAL THEORY

1. BASIC MEASURE THEORY 85

Measurability and measure

1. Measurable spaces; σ-algebras; π-systems; d-systems 85
2. Measurable functions 88
3. Monotone-Class Theorems 90
4. Measures; the uniqueness lemma; almost everywhere; a.e.(μ, Σ) 91
5. Carathéodory's Extension Theorem 93
6. Inner and outer μ-measures; completion 94

Integration

7. Definition of the integral $\int f \, d\mu$ 95
8. Convergence theorems 96
9. The Radon-Nikodým Theorem; absolute continuity; $\lambda \ll \mu$ notation; equivalent measures 98
10. Inequalities; $\mathscr{L}^p$ and L^p spaces ($p \geqslant 1$) 99

Product structures

11. Product σ-algebras 101
12. Product measure; Fubini's Theorem 102
13. Exercises 104

2. BASIC PROBABILITY THEORY 108

Probability and expectation

14. Probability triple; almost surely (a.s.); a.s.(**P**), a.s.(**P**, $\mathscr{F}$) 108

15. lim sup E_n; First Borel–Cantelli Lemma 109
16. Law of random variable; distribution function; joint law 110
17. Expectation; $\mathbf{E}(X; F)$ 110
18. Inequalities: Markov, Jensen, Schwarz, Tchebychev 111
19. Modes of convergence of random variables 113

Uniform integrability and $\mathcal{L}^1$ convergence

20. Uniform integrability 114
21. $\mathcal{L}^1$ convergence 115

Independence

22. Independence of σ-algebras and of random variables 116
23. Existence of families of independent variables 118
24. Exercises 119

3. STOCHASTIC PROCESSES 119

The Daniell–Kolmogorov Theorem

25. $(E^{\mathrm{T}}, \mathcal{E}^{\mathrm{T}})$; σ-algebras on function space; cylinders and σ-cylinders 119
26. Infinite products of probability triples 121
27. Stochastic process; sample function; law 121
28. Canonical process 122
29. Finite-dimensional distributions; sufficiency; compatibility 123
30. The Daniell–Kolmogorov (DK) Theorem: 'compact metrizable' case 124
31. The Daniell–Kolmogorov (DK) Theorem: general case 126
32. Gaussian processes; pre-Brownian motion 127
33. Pre-Poisson set functions 128

Beyond the DK Theorem

34. Limitations of the DK Theorem 128
35. The role of outer measures 129
36. Modifications; indistinguishability 130
37. Direct construction of Poisson measures and subordinators, and of local time from the zero set; Azéma's martingale 131
38. Exercises 136

4. DISCRETE-PARAMETER MARTINGALE THEORY 137

Conditional expectation

30. Fundamental theorem and definition 137
40. Notation; agreement with elementary usage 138
41. Properties of conditional expectation: a list 139
42. The role of versions; regular conditional probabilities and pdfs 140

43. A counterexample 141
44. A uniform-integrability property of conditional expectations 142

(Discrete-parameter) martingales and supermartingales

45. Filtration; filtered space; adapted process; natural filtration 143
46. Martingale; supermartingale; submartingale 144
47. Previsible process; gambling strategy; a fundamental principle 144
48. Doob's Upcrossing Lemma 145
49. Doob's Supermartingale-Convergence Theorem 146
50. $\mathscr{L}^1$ convergence and the UI property 147
51. The Lévy–Doob Downward Theorem 148
52. Doob's Submartingale and $\mathscr{L}^p$ Inequalities 150
53. Martingales in $\mathscr{L}^2$; orthogonality of increments 152
54. Doob decomposition 153
55. The $\langle M \rangle$ and $[M]$ processes 154

Stopping times, optional stopping and optional sampling

56. Stopping time 155
57. Optional-stopping theorems 156
58. The pre-T σ-algebra $\mathscr{F}_T$ 158
59. Optional sampling 159
60. Exercises 161

5. CONTINUOUS-PARAMETER SUPERMARTINGALES 163

Regularisation: R-supermartingales

61. Orientation 163
62. Some real-variable results 163
63. Filtrations; supermartingales; R-processes, R-supermartingales 166
64. Some important examples 167
65. Doob's Regularity Theorem: Part 1 169
66. Partial augmentation 171
67. Usual conditions; R-filtered space; usual augmentation; R-regularisation 172
68. A necessary pause for thought 174
69. Convergence theorems for R-supermartingales 175
70. Inequalities and $\mathscr{L}^p$ convergence for R-submartingales 177
71. Martingale proof of Wiener's Theorem; canonical Brownian motion 178
72. Brownian motion relative to a filtered space 180

Stopping times

73. Stopping time T; pre-T σ-algebra $\mathscr{G}_T$; progressive process 181
74. First-entrance (début) times; hitting times; first-approach times: the easy cases 183

75. Why 'completion' in the usual conditions has to be introduced 184
76. Début and Section Theorems 186
77. Optional Sampling for R-supermartingales under the usual conditions 188
78. Two important results for Markov-process theory 191
79. Exercises 192

6. PROBABILITY MEASURES ON LUSIN SPACES 200

'Weak convergence'

80. $C(J)$ and $\Pr(J)$ when J is compact Hausdorff 202
81. $C(J)$ and $\Pr(J)$ when J is compact metrizable 203
82. Polish and Lusin spaces 205
83. The $C_b(S)$ topology of $\Pr(S)$ when S is a Lusin space; Prohorov's Theorem 207
84. Some useful convergence results 211
85. Tightness in $\Pr(W)$ when W is the path-space $W := C([0,\infty); \mathbb{R})$ 213
86. The Skorokhod representation of $C_b(S)$ convergence on $\Pr(S)$ 215
87. Weak convergence versus convergence of finite-dimensional distributions 216

Regular conditional probabilities

88. Some preliminaries 217
89. The main existence theorem 218
90. Canonical Brownian Motion CBM($\mathbb{R}^N$); Markov property of $\mathbf{P}^x$ laws 220
91. Exercises 222

CHAPTER III. MARKOV PROCESSES

1. TRANSITION FUNCTIONS AND RESOLVENTS 227

1. What is a (continuous-time) Markov process? 227
2. The finite-state-space Markov chain 228
3. Transition functions and their resolvents 231
4. Contraction semigroups on Banach spaces 234
5. The Hille–Yosida Theorem 237

2. FELLER–DYNKIN PROCESSES 240

6. Feller–Dynkin (FD) semigroups 240
7. The existence theorem: canonical FD processes 243
8. Strong Markov property: preliminary version 247
9. Strong Markov property: full version; Blumenthal's 0–1 Law 249

10. Some fundamental martingales; Dynkin's formula 252
11. Quasi-left-continuity 255
12. Characteristic operator 256
13. Feller–Dynkin diffusions 258
14. Characterisation of continuous real Lévy processes 261
15. Consolidation 262

3. ADDITIVE FUNCTIONALS 263

16. PCHAFs; λ-excessive functions; Brownian local time 263
17. Proof of the Volkonskii–Šur–Meyer Theorem 267
18. Killing 269
19. The Feynmann–Kac formula 272
20. A Ciesielski–Taylor Theorem 275
21. Time-substitution 277
22. Reflecting Brownian motion 278
23. The Feller–McKean chain 281
24. Elastic Brownian motion; the arcsine law 282

4. APPROACH TO RAY PROCESSES: THE MARTIN BOUNDARY 284

25. Ray processes and Markov chains 284
26. Important example: birth process 286
27. Excessive functions, the Martin kernel and Choquet theory 288
28. The Martin compactification 292
29. The Martin representation; Doob–Hunt explanation 295
30. R. S. Martin's boundary 297
31. Doob–Hunt theory for Brownian motion 298
32. Ray processes and right processes 302

5. RAY PROCESSES 303

33. Orientation 303
34. Ray resolvents 304
35. The Ray–Knight compactification 306

Ray's Theorem: analytical part

36. From semigroup to resolvent 309
37. Branch-points 313
38. Choquet representation of 1-excessive probability measures 315

Ray's Theorem: probabilistic part

39. The Ray process associated with a given entrance law 316
40. Strong Markov property of Ray processes 318
41. The role of branch-points 319

6. APPLICATIONS 321

Martin boundary theory in retrospect

42. From discrete to continuous time 321
43. Proof of the Doob–Hunt Convergence Theorem 323
44. The Choquet representation of Π-excessive functions 325
45. Doob h-transforms 327

Time reversal and related topics

46. Nagasawa's formula for chains 328
47. Strong Markov property under time reversal 330
48. Equilibrium charge 331
49. BM($\mathbb{R}$) and BES(3): splitting times 332

A first look at Markov-chain theory

50. Chains as Ray processes 334
51. Significance of q_i 337
52. Taboo probabilities; first-entrance decomposition 337
53. The Q-matrix; DK conditions 339
54. Local-character condition for Q 340
55. Totally instantaneous Q-matrices 342
56. Last exits 343
57. Excursions from b 345
58. Kingman's solution of the 'Markov characterization problem' 347
59. Symmetrisable chains 348
60. An open problem 349

References for Volumes 1 and 2 351

Index to Volumes 1 and 2 375

Some Frequently Used Notation

We use ':=' to mean 'is defined to equal'. This Pascal notation can also be used in reverse. We define

$$\mathbb{Z}^+ := \{0, 1, 2, \ldots\} \supseteq \{1, 2, 3, \ldots\} =: \mathbb{N},$$
$$\mathbb{R}^+ := [0, \infty), \quad \mathbb{R}^{++} := (0, \infty), \quad \mathbb{Q}^+ := \mathbb{Q} \cap \mathbb{R}^+.$$

We neaten layout, and make things easier for our printers, by the use of alternative notations:

$$X(t_1, \omega) \text{ for } X_{t_1}(\omega), \quad f_{n(1)} \text{ for } f_{n_1}, \quad \mathscr{F}(T_1) \text{ for } \mathscr{F}_{T_1}, \quad P_t f(x) \text{ for } (P_t f)(x),$$

etc. Once things are underway, such switches in notation will be made without comment. The composition notation

$$f \circ g(t) := f(g(t))$$

will often be used for tidiness.

If f and g are real numbers or real-valued functions, we define

$$f \vee g := \max(f, g), \quad f \wedge g := \min(f, g), \quad f^+ := f \vee 0, \quad f^- := (-f) \vee 0;$$

hence $f = f^+ - f^-$ and $|f| = f^+ + f^-$.

If $\mathscr{H}$ is a set of real-valued functions, we write

$\mathscr{H}^+$ for the set of non-negative elements of $\mathscr{H}$,

$\mathrm{b}\mathscr{H}^+$ for the set of bounded elements in $\mathscr{H}$.

If Σ is a σ-algebra, we write

$\mathrm{m}\Sigma$ for the set of real-valued (or perhaps $[\infty, \infty]$-valued)

Σ-measurable functions,

$\mathrm{b}\Sigma$ for the space of bounded Σ-measurable functions.

If S is a topological space, we write

$C(S)$ for the space of *all* continuous functions from S to $\mathbb{R}$.
$C_b(S)$ for the space of all *bounded* continuous functions from S to $\mathbb{R}$.

Monotone convergence. We write '$s \uparrow t$' to signify that $s \to t, s \leqslant t$; and '$s \uparrow\uparrow t$' to signify that $s \to t, s < t$. If (s_n) is a sequence then '$s_n \uparrow t$' signifies that $s_n \to t, s_n \leqslant s_{n+1} \leqslant t$; while '$s_n \uparrow\uparrow t$' signifies that $s_n \to t, s_n \leqslant s_{n+1} < t$. If f_n and f are real-valued functions then (for example) $f{:}\uparrow \lim f_n$ signifies that $f_n \uparrow f$ pointwise.

CHAPTER I

Brownian Motion

1. INTRODUCTION

1. What is Brownian motion, and why study it? The first thing is to define Brownian motion. We assume given some probability triple $(\Omega, \mathcal{F}, \mathbf{P})$.

(*1.1*) DEFINITION. *A real-valued stochastic process* $\{B_t : t \in \mathbb{R}^+\}$ *is a* Brownian motion *if it has the properties*

(1.2) (i) $B_0(\omega) = 0$, $\forall \omega$;

(1.2) (ii) *the map* $t \mapsto B_t(\omega)$ *is a continuous function of* $t \in \mathbb{R}^+$ *for all* ω;

(1.2)(iii) *for every* $t, h \geq 0$, $B_{t+h} - B_t$ *is independent of* $\{B_u : 0 \leq u \leq t\}$, *and has a Gaussian distribution with mean* 0 *and variance* h.

The conditions (1.2)(ii) and (1.2)(iii) are the really essential ones; if $B = \{B_t : t \in \mathbb{R}^+\}$ is a Brownian motion, we frequently speak of $\{\xi + B_t : t \in \mathbb{R}^+\}$ as a Brownian motion (started at ξ); the starting point ξ can be a fixed real, or a random variable independent of B.

Now that we know what a Brownian motion is, questions of existence and uniqueness (answered in Section 6) are less important than an answer to the second question of the title, 'Why study it?' There are many answers to this question, but to us there seem to be four main ones:

(i) Virtually every interesting class of processes contains Brownian motion—Brownian motion is a martingale, a Gaussian process, a Markov process, a diffusion, a Lévy process, . . . ;
(ii) Brownian motion is sufficiently concrete that one can do explicit calculations, which are impossible for more general objects;
(iii) Brownian motion can be used as a building block for other processes (indeed, a number of the most important results on Brownian motion state that the most general process in a certain class can be obtained from Brownian motion by some sequence of transformations);
(iv) last but not least, Brownian motion is a rich and beautiful mathematical object in its own right.

The aim of this chapter is to expand on these reasons, and convince you that

Brownian motion is indeed worthy of study; and the rest of this introduction gives a brief outline of some of the main points of the chapter.

2. Brownian motion as a martingale. Let $\{B_t : t \geqslant 0\}$ be a Brownian motion, and define $\mathscr{B}_t = \sigma(\{B_s : s \leqslant t\})$. *Then* $(B_t, \mathscr{B}_t)_{t \geqslant 0}$ *is a martingale.* We shall have a lot more to say about martingales in Chapter II, but for now we need little of the theory developed there. Let us just check that $(B_t, \mathscr{B}_t)_{t \geqslant 0}$ is a martingale (cf. Section II.63); first, $B_t \in L^1$ for all t, because, from (1.2)(i) and (1.2)(iii), $B_t \sim N(0, t)$, and, secondly, for $0 \leqslant s \leqslant t$,

$$\mathbf{E}[B_t - B_s | \mathscr{B}_s] = 0, \qquad \text{equivalently, } \mathbf{E}[B_t | \mathscr{B}_s] = B_s,$$

since $B_t - B_s$ is independent of $\mathscr{B}_s$ by (1.2)(iii). Likewise, since $B_t - B_s \sim N(0, t-s)$ independently of $\mathscr{B}_s$, we have

$$\mathbf{E}[(B_t - B_s)^2 | \mathscr{B}_s] = t - s.$$

But

$$\mathbf{E}[(B_t - B_s)^2 | \mathscr{B}_s] = \mathbf{E}[B_t^2 - 2B_t B_s + B_s^2 | \mathscr{B}_s] = \mathbf{E}[B_t^2 | \mathscr{B}_s] - B_s^2,$$

using properties of conditional expectation (Section II.41), so since we have (almost surely) that $\mathbf{E}[B_t^2 - t | \mathscr{B}_s] = B_s^2 - s$, we conclude that

(2.1) $$B_t^2 - t \quad \textit{is a martingale.}$$

This simple fact is a pointer to the development of stochastic integrals; once that theory is developed, we shall be in a position to prove the following startling converse to (2.1).

(2.2) THEOREM. (Lévy) *Let* $(X_t)_{t \geqslant 0}$ *be a continuous martingale,* $X_0 = 0$, *and suppose that*

$$X_t^2 - t \quad \textit{is a martingale.}$$

Then X is a Brownian motion.

By a continuous martingale, we mean of course one such that $t \mapsto X_t(\omega)$ is a continuous map for all ω. We have not been too specific about the filtration $(\mathscr{F}_t)_{t \geqslant 0}$ with respect to which X is a martingale, but this is not necessary; if X is a martingale with respect to $(\mathscr{F}_t)_{t \geqslant 0}$, and satisfies the hypotheses of Theorem 2.2 then X is an $(\mathscr{F}_t)$ Brownian motion—that is, X satisfies (1.2)(i), (1.2)(ii) and the stronger condition

(1.2)(iii)′ *for any* $t, h \geqslant 0$, $X_{t+h} - X_t$ *is independent of* $\mathscr{F}_t$ *and has a Gaussian distribution with mean zero and variance h.*

The Kunita–Watanabe proof of Theorem 2.2 is given in Section IV.33; a more elementary proof without using stochastic calculus appears in Doob [1].

A remarkable consequence of Theorem 2.2 is that

(2.3) *every continuous martingale is a time-change of Brownian motion.*

For a statement and proof of this, see Section IV.34. One extremely useful consequence is that, since

$$\mathbf{P}\left[\limsup_{t\to\infty} B_t = +\infty,\ \liminf_{t\to\infty} B_t = -\infty\right] = 1$$

(as we shall see in Lemma 3.6), if X is a continuous martingale for which $\mathbf{P}(\liminf X_t = -\infty) > 0$ then we must have $\mathbf{P}(\limsup X_t = +\infty) > 0$. See Section IV.34 for a full discussion.

The elementary arguments that gave (2.1) also show that for any $\theta \in \mathbb{R}$ (or indeed, for $\theta \in \mathbb{C}$)

(2.4) $\exp(\theta B_t - \frac{1}{2}\theta^2 t)$ *is a martingale*;

all one needs is that $\mathbf{E}(\exp[\theta(B_t - B_s)]) = \exp[\frac{1}{2}\theta^2(t-s)]$ for $0 \leqslant s \leqslant t$, which is just the moment-generating function of a Gaussian distribution. These exponential martingales are extremely useful in many ways; in Section 9 we use them to compute the Brownian first-passage distribution to a level, and in Section 16 we derive the Law of the Iterated Logarithm using them.

One small point to note here in connection with the exponential martingales (2.4) is that if we define the Hermite polynomials $H_n(t, x)$ by

$$\exp(\theta x - \tfrac{1}{2}\theta^2 t) := \sum_{n \geqslant 0} \frac{\theta^n}{n!} H_n(t, x),$$

then, for $0 \leqslant s \leqslant t$,

$$\begin{aligned} \mathbf{E}(\exp(\theta B_t - \tfrac{1}{2}\theta^2 t) \mid \mathscr{B}_s) &= \sum_{n \geqslant 0} \frac{\theta^n}{n!} \mathbf{E}(H_n(t, B_t) \mid \mathscr{B}_s) \\ &= \exp(\theta B_s - \tfrac{1}{2}\theta^2 s) \\ &= \sum_{n \geqslant 0} \frac{\theta^n}{n!} H_n(s, B_s), \end{aligned}$$

so, by comparing coefficients of θ^n, we deduce that

$H_n(t, B_t)$ *is a martingale for each* n.

It is easy to check that $H_1(t, x) = x$ and $H_2(t, x) = x^2 - t$, so, in particular, (2.4)$\Rightarrow$(2.1); Lévy's Theorem 2.2 is essentially the converse to this.

(2.5) *Remark.* If $(N_t)_{t \geqslant 0}$ is a standard Poisson process then $X_t := N_t - t$ satisfies all of the hypotheses of Theorem 2.2 except for continuity of the paths.

3. Brownian motion as a Gaussian process. In complete generality, a (real-valued) process $(X_t)_{t \in T}$ indexed by some set T is said to be a *Gaussian process* if, for any $t_1, \dots, t_n \in T$, the law of $(X(t_1), \dots, X(t_n))$ is multivariate Gaussian. Thus the law of the process X is specified by the functions

$$\mu(t) := \mathbf{E}X_t, \quad \rho(s,t) := \operatorname{cov}(X_s, X_t).$$

(By this, we mean no more than that if we were told μ and ρ, we could work out the law of $(X(t_1), \dots, X(t_n))$ for any $t_1, \dots, t_n \in T$.) In the study of Gaussian processes, one usually assumes that $\mu \equiv 0$, to which the general case can be reduced by considering the Gaussian process $X_t - \mu(t)$.

It is obvious that $(B_t)_{t \geq 0}$ is a Gaussian process, with mean zero, and covariance

$$\rho(s,t) = s \wedge t \quad (s, t \geq 0). \tag{3.1}$$

Any continuous real-valued process $(X_t)_{t \geq 0}$ that is a zero-mean Gaussian process with covariance (3.1) *is* a Brownian motion—just check the definition! This simple fact turns out to be an extremely efficient means of checking when a process is a Brownian motion, and the following four simple but extremely important examples serve to illustrate this:

(3.2) the process $(-B_t)_{t \geq 0}$ is a Brownian motion;
(3.3) for any $a \geq 0$, the process $(B_{t+a} - B_a)_{t \geq 0}$ is a Brownian motion;
(3.4) for any $c \neq 0$, $(cB_{t/c^2})_{t \geq 0}$ is a Brownian motion (*Brownian scaling*);
(3.5) the process $(\tilde{B}_t)_{t \geq 0}$ defined by

$$\tilde{B}_0 = 0,$$
$$\tilde{B}_t = tB_{1/t} \quad \text{for } t > 0,$$

is a Brownian motion.

The proofs of these properties are trivial exercises, with the sole exception of the proof of continuity at 0 of $\tilde{B}$. But this is not difficult, because the event that $\tilde{B} \to 0$ at 0 is

$$\tilde{F} = \bigcap_n \bigcup_m \bigcap_{q \in \mathbf{Q} \cap (0, 1/m]} \left\{ |\tilde{B}_q| \leq \frac{1}{n} \right\},$$

since $\tilde{B}$ is certainly continuous in $(0, \infty)$. But the processes $(\tilde{B}_t)_{t > 0}$ and $(B_t)_{t > 0}$ are continuous, and have the same distribution (they are Gaussian processes with the same covariance!), so

$$\mathbf{P}(\tilde{F}) = \mathbf{P}(F) = 1,$$

where F is the event $\bigcap_n \bigcup_m \bigcap_{q \in \mathbf{Q} \cap (0,1/m]} \{|B_q| \leq 1/n\}$ that $B \to 0$ at 0, which, by the definition of B, is certain.

The most important by far of the properties (3.2)–(3.5) is the *Brownian scaling* property (3.4). We shall give here an easy but striking consequence.

(3.6) LEMMA. *We have*

$$\mathbf{P}\left(\sup_t B_t = +\infty, \quad \inf_t B_t = -\infty \right) = 1.$$

Proof. Let $Z := \sup_t B_t$. By Brownian scaling, for any $c > 0$, we have

$$cZ \stackrel{\mathscr{D}}{=} Z,$$

so the law of Z is concentrated on $\{0, +\infty\}$. Let $p = \mathbf{P}(Z = 0)$. Then

$$\mathbf{P}(Z = 0) \leqslant \mathbf{P}[B_1 \leqslant 0 \quad \text{and} \quad B_u \leqslant 0 \quad \text{for all} \quad u \geqslant 1]$$

$$= \mathbf{P}\left[B_1 \leqslant 0 \quad \text{and} \quad \sup_{t \geqslant 0} \{B_{1+t} - B_1\} = 0\right],$$

because $(B_{1+t} - B_1)_{t \geqslant 0}$ is a Brownian motion, whose supremum is therefore 0 or $+\infty$. But $(B_{1+t} - B_1)_{t \geqslant 0}$ is independent of $(B_u)_{u \leqslant 1}$, so we deduce that

$$p = \mathbf{P}(Z = 0) \leqslant \mathbf{P}(B_1 \leqslant 0)\mathbf{P}(Z = 0) = \tfrac{1}{2}p,$$

whence $p = 0$. Combining with (3.2) gives the stated result. □

Lemma 3.6 implies straight away that, almost surely, *for each* $a \in \mathbb{R}$, $\{t : B_t = a\}$ *is not bounded above*. Thus Brownian motion is recurrent—it keeps returning to its starting point.

We shall have more to say about Gaussian processes in Part 4 of this chapter, but point out now that the discussion there is by way of an interesting digression from our main theme; the general setting for Gaussian processes is too general to permit full exploitation of the special features of Brownian motion (notably a completely ordered index set).

4. Brownian motion as a Markov process. Brownian motion is a (time-homogeneous) Markov process; for any bounded Borel $f : \mathbb{R} \to \mathbb{R}$, and $s, t \geqslant 0$,

$$\mathbf{E}[f(B_{t+s}) | \mathscr{B}_s] = P_t f(B_s) \tag{4.1}$$

where the *transition semigroup* $(P_t)_{t \geqslant 0}$ is defined by

$$P_t f(x) := \begin{cases} \displaystyle\int_{-\infty}^{\infty} p_t(x, y) f(y)\, dy & (t > 0), \\ f(x) & (t = 0) \end{cases}$$

where

$$p_t(x, y) := (2\pi t)^{-1/2} \exp\left[-\frac{(x - y)^2}{2t}\right] \tag{4.2}$$

is the Brownian *transition density*. The Markov property (4.1) is immediate from the definition of Brownian motion. It is easy to confirm that $(P_t)_{t \geqslant 0}$ is a semigroup:

$$P_{t+s} = P_t P_s = P_s P_t \quad (s, t \geqslant 0), \tag{4.3}$$

the so-called *Chapman–Kolmogorov equations*. The semigroup property (4.3) suggests that we ought in some sense to have

$$\frac{d}{dt}P_t = \lim_{s\downarrow 0}\frac{1}{s}(P_{t+s} - P_t) = P_t\mathscr{G} = \mathscr{G}P_t, \tag{4.4}$$

where

$$\mathscr{G} := \lim_{s\downarrow 0}\frac{1}{s}(P_s - I) \tag{4.5}$$

is the *(infinitesimal) generator of* $(P_t)_{t\geq 0}$. This is indeed true in complete generality, when suitably interpreted; the suitable interpretation involves us in some fairly careful analysis, because in general $\mathscr{G}$ is not defined for all functions, and much of the classical early work on Markov processes struggled with these technicalities. This functional-analytic viewpoint has many merits, not least that it can suggest quickly what things are likely to be true, but we shall not stress it too much because it is not a very convenient framework in which to prove the conjectures to which it leads. But, for now, let us illustrate the notion by working out the generator of Brownian motion. From (4.5), we should define $\mathscr{G}f$ for suitable f by

$$\mathscr{G}f := \lim_{t\downarrow 0}\frac{1}{t}(P_t f - f),$$

and, indeed, if $f\in C_b^2(\mathbb{R})$ then

$$\begin{aligned}\lim_{t\downarrow 0}\frac{1}{t}(P_t f - f)(x) &= \lim_{t\downarrow 0}\int_{-\infty}^{\infty}\frac{f(x+y\sqrt{t}) - f(x)}{t}\exp(-\tfrac{1}{2}y^2)\frac{dy}{\sqrt{2\pi}}\\ &= \lim_{t\downarrow 0}\int_{-\infty}^{\infty}\frac{1}{t}\{y\sqrt{t}f'(x) + \tfrac{1}{2}y^2 t f''(x+\theta y\sqrt{t})\}\exp(-\tfrac{1}{2}y^2)\frac{dy}{\sqrt{2\pi}}\\ &\qquad\qquad \text{(where } \theta\in(0,1) \text{ depends on } y\sqrt{t})\\ &= \tfrac{1}{2}f''(x).\end{aligned}$$

Thus the infinitesimal generator of Brownian motion is

$$\mathscr{G} = \frac{1}{2}\frac{d^2}{dx^2},$$

at least when applied to $C_b^2(\mathbb{R})$. From (4.4), we find that, for $f\in C_b^2(\mathbb{R})$,

$$\frac{\partial}{\partial t}P_t f(x) = \mathscr{G}P_t f(x) = \tfrac{1}{2}(P_t f)''(x),$$

which leads to *Kolmogorov's backward equation* for the Brownian transition density:

$$\frac{\partial}{\partial t}p_t(x,y) = \frac{1}{2}\frac{\partial^2}{\partial x^2}p_t(x,y), \tag{4.6}$$

since f is arbitrary. Using the other part of (4.4) gives us

$$\frac{\partial}{\partial t} P_t f(x) = P_t \mathscr{G} f(x) = \tfrac{1}{2} P_t f''(x),$$

and an integration by parts now yields *Kolmogorov's forward equation* for the Brownian transition density:

$$\frac{\partial}{\partial t} p_t(x, y) = \frac{1}{2} \frac{\partial^2}{\partial y^2} p_t(x, y). \tag{4.7}$$

This equation is familiar in physics, where it is known as the *heat equation*, or the *diffusion equation*, so called because it determines the physical flow of heat, or the physical diffusion of particles in solution, in a homogeneous medium. Many of the notions of diffusion that probabilists use everyday were known to physicists long ago, and amount to the same things in different language (see, for example, the classic book by Crank [1] for a physicists' exposition—and a broad selection of fascinating and challenging questions). It is however important to stress that we are not simply going to be rederiving results well known in physics; probability provides techniques for the study of *individual* diffusing particles, which are far more flexible and powerful than the classical analysis of the heat equation, which is only a statement about the *average* behaviour of a large number of diffusing particles.

5. Brownian motion as a diffusion (and martingale). Without trying to be too precise, a diffusion (on the real line for now) is a continuous time-homogeneous Markov process X that is 'characterised' in some sense by its local infinitesimal *drift* b and *variance* a: for small h,

(5.1) (i) $$\mathbf{E}[X_{t+h} - X_t | \mathscr{F}_t] \doteqdot hb(X_t),$$

(5.1)(ii) $$\mathbf{E}[\{X_{t+h} - X_t - hb(X_t)\}^2 | \mathscr{F}_t] \doteqdot ha(X_t).$$

If a and b were constant functions then

$$X_t = \sigma B_t + bt, \qquad \sigma := a^{1/2},$$

would satisfy the description (5.1); the more general diffusion is rather similar except that the drift and variance may now depend on position. It is unnecessary to impose conditions on moments of the increment $X_{t+h} - X_t$ beyond the second, which you will certainly accept as plausible if you recall Lévy's Theorem (2.2), which said that Brownian motion is characterised by $b \equiv 0$, $a \equiv 1$.

Broadly speaking, there are three approaches to diffusions: the stochastic differential equation (SDE) approach, the martingale-problem approach, and the partial differential equation (PDE) approach. Each has its merits and peculiar techniques.

The SDE approach constructs the diffusion X with given infinitesimal characteristics a and b by solving

$$X_t = X_0 + \int_0^t \sigma(X_s)\,dB_s + \int_0^t b(X_s)\,ds, \tag{5.2}$$

where $\sigma := a^{1/2}$, an equation that is commonly written in 'differential' form

$$dX_t = \sigma(X_t)\,dB_t + b(X_t)\,dt. \tag{5.3}$$

Thus X has infinitesimal drift b and infinitesimal variance, a, since the increment $X_{t+h} - X_t$ is (approximately)

$$\sigma(X_t)(B_{t+h} - B_t) + hb(X_t).$$

There is a lot of work involved in defining what the second term on the right-hand side of (5.2) means, and in verifying existence and uniqueness of a solution under suitable conditions on σ and b; we shall have almost nothing to say on this until Volume 2.

The martingale problem approach and the PDE approach both begin from the same trivial calculation based on (5.1). For any $f \in C_b^2$,

$$\begin{aligned}(5.4)\quad & \mathbf{E}[f(X_{t+h}) - f(X_t) \mid \mathscr{F}_t] \\ &= \mathbf{E}[f'(X_t)(X_{t+h} - X_t) + \tfrac{1}{2}f''(\theta X_{t+h} + (1-\theta)X_t)(X_{t+h} - X_t)^2 \mid \mathscr{F}_t] \\ &\qquad\qquad\qquad\qquad \text{(where } \theta \in (0,1) \text{ is random)} \\ &\doteqdot f'(X_t)hb(X_t) + \tfrac{1}{2}f''(X_t)[ha(X_t) + h^2 b(X_t)^2] \\ &= h\mathscr{L}f(X_t) + O(h^2),\end{aligned}$$

where $\mathscr{L}$ is the second-order elliptic operator,

$$\mathscr{L}f(x) := \tfrac{1}{2}a(x)\frac{d^2 f}{dx^2}(x) + b(x)\frac{df}{dx}(x). \tag{5.5}$$

The martingale-problem approach takes (5.4) and re-expresses it as

$$\mathbf{E}\left[f(X_{t+h}) - f(X_t) - \int_t^{t+h} \mathscr{L}f(X_s)\,ds \,\middle|\, \mathscr{F}_t \right] = o(h),$$

so that the martingale-problem 'definition' of a diffusion X with drift b and variance a is that X is a continuous process such that, for all $f \in C_b^2$,

$$f(X_t) - \int_0^t \mathscr{L}f(X_s)\,ds \quad \textit{is a martingale.} \tag{5.6}$$

The PDE approach takes expectations on both sides of (5.4) to get

$$P_h f(x) - f(x) = h\mathscr{L}f(x) + O(h^2), \tag{5.7}$$

so that, dividing by h and letting $h \downarrow 0$,

(5.8) *the infinitesimal generator* $\mathscr{G}$ *of* X *is* $\mathscr{L}$.

The PDE approach is now ready to go, with all of the arsenal of PDE techniques at its disposal; for example, one may begin by looking for the fundamental solution $p_t(x, y)$ to

$$\frac{\partial}{\partial t} p_t(x, y) = \mathscr{L}_x p_t(x, y), \qquad p_0(x, y) = \delta_y(x), \qquad p_t \geqslant 0,$$

where $\mathscr{L}_x$ is the operator $\mathscr{L}$ acting on the x-variable, and δ_y is the Dirac delta function at y. This fundamental solution is the transition density of the diffusion, from which one can obtain much information; see Chapter 3 of Stroock and Varadhan [1], which is also the definitive account of the martingale-problem method applied to multidimensional diffusions. There are still real problems of definition, existence and uniqueness for each of the three approaches—least severe for the PDE approach. But the additional price to be paid for using stochastic methods *is* worth it; the conditions imposed on a and b to get a PDE result to work are generally of a *global* nature, whereas the diffusion, being continuous, should only care about *local* behaviour. The stochastic methods are just right for this—once a diffusion leaves a region where everything is nice, we can stop it, and solve in the nice region, thereby giving results under only local conditions. We have great admiration and respect for the PDE approach—the analysts' fine results are not just valid for second-order elliptic operators, which is the case with the probabilistic results. The last word for the moment on the comparison between the three methods must be with Sid Port: 'The one thing probabilists can do which analysts can't is *stop*—and they never forgive us for it.'

You will realise by now that one can perfectly well have diffusions in dimension greater than one, but the one-dimensional diffusion theory is essentially complete, thanks to *Brownian local time*. The existence and properties of Brownian local time form the first non-trivial result in the theory (after the existence of Brownian motion itself).

(5.9) THEOREM (Trotter). *There exists a process* $\{l(t, x): t \geqslant 0, x \in \mathbb{R}\}$ *such that*

(5.10) (i) $(t, x) \mapsto l(t, x)$ *is jointly continuous;*
(5.10)(ii) *for any bounded measurable* f, *and* $t \geqslant 0$,

$$\int_0^t f(B_s)ds = \int_{-\infty}^{\infty} f(x) l(t, x) dx.$$

This is a deep result, whose proof using stochastic calculus we shall finally give in Section IV.44. The key property (5.10)(ii) is the *occupation density formula*; we shall discuss some of the implications for the Brownian sample path in Section 10, but for now we describe the most general regular diffusion (see Sections V.44–54 for the whole story).

(5.11) THEOREM. *A (regular) one-dimensional diffusion X on an interval I can be obtained from Browian motion B as*

$$X_t = s^{-1}(B_{\tau_t})$$

where the scale function $s: I \to \mathbb{R}$ *is continuous and strictly increasing and* $\tau_t = \inf\{u: A_u > t\}$, *where*

$$A_u = \int m(dx)\, l(u, x)$$

for some measure m (the speed measure) *that puts positive finite mass on bounded non-empty open subintervals of I which exclude the endpoints of I.*

Any pair (s, m) gives rise to a regular diffusion, which is uniquely characterised by (s, m).

The theory of diffusions in dimension greater than one is still much less complete, and doubtless will remain so.

2. BASICS ABOUT BROWNIAN MOTION

6. Existence and uniqueness of Brownian motion. The existence proof for Brownian motion that we now give (due to Ciesielski [1]) is the ultimate refinement of Wiener's original idea of representing Brownian motion as a random Fourier series.

(6.1) THEOREM. *There exists a probability space on which it is possible to define a process* $(B_t)_{0 \leqslant t \leqslant 1}$ *with the properties*

(i) $B_0(\omega) = 0$ *for all* ω;
(ii) *the map* $t \mapsto B_t(\omega)$ *is a continuous function of* $t \in [0, 1]$ *for all* ω;
(iii) *for every* $0 \leqslant s \leqslant t \leqslant 1$, $B_t - B_s$ *is independent of* $\{B_u: u \leqslant s\}$ *and has a* $N(0, t-s)$ *distribution.*

Proof. Take some probability space on which there is defined an infinite sequence of independent $N(0, 1)$ random variables. For reasons that will soon be apparent, we assume that they are indexed as $\{Z_{k,n}: n \in \mathbb{Z}^+,\ k \text{ odd},\ k \leqslant 2^n\}$. Now define

$$g_{1,0}(t) = 1$$

$$g_{k,n}(t) = \begin{cases} 2^{(n-1)/2} & ((k-1)2^{-n} < t \leqslant k2^{-n}), \\ -2^{(n-1)/2} & (k2^{-n} < t \leqslant (k+1)2^{-n}), \\ 0 & \text{otherwise}, \end{cases}$$

for $n \geqslant 1$, $k \leqslant 2^n$, k odd. For notational convenience, let $S_n = \{(k, n): k \text{ odd},\ k \leqslant 2^n\}$, $S = \bigcup_{n \geqslant 0} S_n$. The first thing to notice is that $\{g_{k,n}: (k, n) \in S\}$ *is a complete orthonormal system in* $L^2[0,1]$. The orthonormality of the $g_{k,n}$ is easy to check; and, for completeness, if $f \in L^2[0,1]$ were orthogonal to all the $g_{k,n}$ then $F(t) = \int_0^t f(u)\,du$ would vanish at 0 and 1 (since $f \perp g_{1,0}$); and also at $\frac{1}{2}$ (since $f \perp g_{11}$); and also at $\frac{1}{4}, \frac{3}{4}$, (since $f \perp g_{12}, g_{32}$),.... Thus $F = 0$, and $f = 0$.

Now define $f_{k,n}(t) := \int_0^t g_{k,n}(u)\,du$, and the approximations $B_n(\cdot)$ to Brownian motion by

$$B_n(t) = \sum_{m=0}^{n} \sum_{(k,m) \in S_m} Z_{k,m} f_{k,m}(t).$$

Let us describe what these approximations are doing.

The first approximation B_0 is simply $tZ_{1,0}$, a straight line. The next approximation is obtained by adding on a Gaussian multiple of f_{11}, which is a tent-shaped function, vanishing at 0 and 1. The next approximation is obtained by adding on two Gaussian multiples of tent-shaped functions, which both vanish at 0, $\frac{1}{2}$ and 1. The first three approximations are illustrated in Fig. I.1. So what is happening is that the nth approximation is piecewise-linear and continuous, and is equal to the limit value at each point of the form $j2^{-n}$.

The next stage of the proof is to establish that *the* B_n *converge uniformly* almost surely. Indeed, for any positive constant a_n,

$$\begin{aligned}
&\mathbf{P}\left(\sup_{0 \leqslant t \leqslant 1} |B_n(t) - B_{n-1}(t)| > a_n\right) \\
&= \mathbf{P}\left(\sup_k |Z_{k,n}| > 2^{(n+1)/2} a_n\right) \\
&\qquad\qquad\qquad \text{(since the } f_{k,n} \text{ are all at most } 2^{-(n+1)/2}) \\
&\leqslant 2^{n-1} \mathbf{P}(|Z_{1,n}| > 2^{(n+1)/2} a_n) \\
&\leqslant (4\pi)^{-1/2} 2^{n/2} a_n^{-1} \exp(-a_n^2 2^n),
\end{aligned}$$

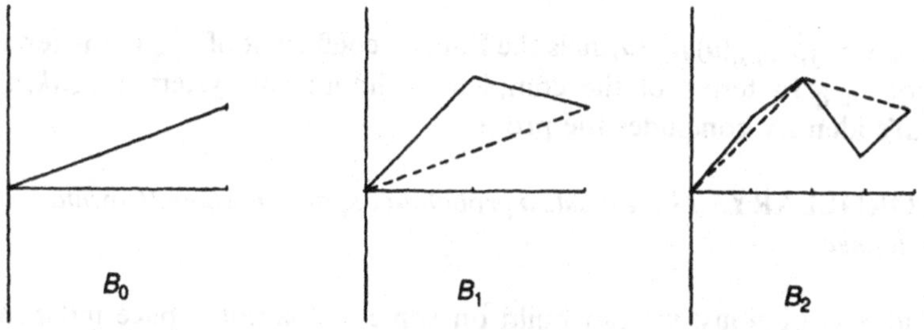

Fig. I.1

by the elementary estimate

$$\int_x^\infty \exp(-\tfrac{1}{2}y^2)\,dy \leqslant x^{-1}\exp(-\tfrac{1}{2}x^2).$$

We now aim to choose the constants a_n in such a way that

$$\sum_n 2^{n/2}a_n^{-1}\exp(-a_n^2 2^n) < \infty,$$

$$\sum_n a_n < \infty.$$

The first of these conditions will ensure that, almost surely,

$$\sup_{0\leqslant t\leqslant 1} |B_n(t) - B_{n-1}(t)| \leqslant a_n \quad \text{for all large enough} \quad n;$$

the second will guarantee that the B_n converge uniformly (almost surely) to a limit B, which is therefore continuous. But these conditions are satisfied by the choice $a_n = (n2^{-n})^{1/2}$, for example.

Thus we have proved that, almost surely, the B_n converge uniformly to some continuous limit B, which we now must show is Brownian motion. As we saw in Section 3, the simplest way to do this is to check that B is a zero-mean Gaussian process with covariance structure $\mathbf{E}(B_s B_t) = s \wedge t$. Obviously, each B_n is a zero-mean Gaussian process: the vector $(B_n(t_1),\ldots,B_n(t_k))$ is multivariate Gaussian. This converges almost surely (and so in distribution) to $(B(t_1),\ldots,B(t_k))$, which also has a zero-mean Gaussian law, and the limit of the covariances of the B_r gives the covariance of B. But

$$\mathbf{E}[B_n(s)B_n(t)] = \sum_{m=0}^{n}\sum_{(k,m)\in S_m} f_{k,m}(s)f_{k,m}(t),$$

by independence of the $Z_{k,m}$, and this converges as $n\uparrow\infty$ to

$$\begin{aligned}\sum_{(k,m)\in S} f_{k,m}(s)f_{k,m}(t) &= \int_0^1 I_{[0,s]}(u)I_{[0,t]}(u)\,du\\ &= s\wedge t,\end{aligned}$$

since $f_{k,m}(s) = \int_0^1 I_{[0,s]}(u)g_{k,m}(u)\,du$ is the Fourier coefficient of $g_{k,m}$ in the representation of $I_{[0,s]}$ in terms of the complete orthonormal system $\{g_{k,n}:(k,n)\in S\}$. Parseval's identity concludes the proof. □

(6.2) COROLLARY. *There exists a probability space on which Brownian motion can be defined.*

The idea is obvious; we can build on some probability space independent copies of the process constructed in Theorem 6.1, and then stick them together to make a process $(B_t)_{t\geq 0}$. We leave the reader to satisfy himself or herself that this can easily be done.

(6.3) *Remark.* The observant reader will notice that in Theorem 6.1 we obtained B as the *almost sure* uniform limit of continuous functions, and may be worrying what happens on the null set—define $B\equiv 0$ there!

Now we turn to the uniqueness of Brownian motion, a much simpler matter, once one decides the *sense* in which Brownian motion is unique. But this can only be in the distributional sense: there is a unique probability measure $\mathbf{P}$ (called *Wiener measure*) on the space $C(\mathbb{R}^+,\mathbb{R})\equiv\{\text{continuous } x:\mathbb{R}^+\to\mathbb{R}\}$ such that under $\mathbf{P}$

(6.4)(ii) for $t_0=0\leqslant t_1\leqslant\cdots\leqslant t_n$, $(x(t_j)-x(t_{j-1}))_{j=1}^n$ are independent zero-mean Gaussian variables with variances $t_j-t_{j-1}(j=1,\ldots,n)$.

If we define $\pi_t:C(\mathbb{R}^+,\mathbb{R})\to\mathbb{R}$ to be the projection map

$$\pi_t(x)=x(t)$$

and $\mathscr{C}$ to be the collection of all cylinder sets

$$\{x:x(t_j)\in A_j \quad \text{for} \quad j=1,\ldots,n\}$$

as n runs through $\mathbb{N}$, $t_1,\ldots,t_n$ run though $\mathbb{R}^+$, and $A_1,\ldots,A_n$ run through $\mathscr{B}(\mathbb{R})$, then $\mathscr{C}$ is a π-system that generates the σ-field $\mathscr{A}\equiv\sigma(\{\pi_t:t\geqslant 0\})$. Any two measures $\mathbf{P}$ and $\mathbf{P}'$ on $C(\mathbb{R}^+,\mathbb{R})$ with properties (6.4)(i) and (ii) must agree on $\mathscr{C}$ and therefore on $\mathscr{A}$ (see Lemma II.4.6). Thus $\mathbf{P}$ is unique.

(6.5) *Remarks.* It is also true that $\mathscr{A}$ is the Borel σ-field on $C(\mathbb{R}^+,\mathbb{R})$ when this space is equipped with the topology of uniform convergence on compacts (see Lemma II.82.3).

7. Skorokhod embedding. In many ways, a zero-mean finite-variance random walk looks like Brownian motion, and we shall shortly see this made precise. The key to understanding this is the celebrated Skorokhod embedding, which allows one to embed *any* random walk with zero mean and finite variance into a given Brownian motion in a very well-controlled way.

We suppose that F is the distribution of the steps of the random walk; thus

$$\int_{-\infty}^{\infty} xF(dx)=0, \qquad \int_{-\infty}^{\infty} x^2F(dx)=\sigma^2<\infty.$$

The aim is to find a stopping time T for Brownian motion B such that

$$B_T\sim F \quad \text{and} \quad \mathbf{E}T=\sigma^2.$$

The requirement that $\mathbf{E}T = \sigma^2$ is essential to our subsequent use of the Skorokhod embedding, and prevents the problem from being a triviality. (As Doob has remarked, we could just take $T := \inf\{t > 1 : B_t = h(B_1)\}$, where h is chosen to make $h(B_1) \sim F$! But for this T, $\mathbf{E}T = \infty$). There are a number of different ways to build the stopping time T based on the path of B alone; we discuss the beautiful Azéma–Yor construction in Section VI.51 to illustrate this by excursion theory. For now, though, we take a simpler though less elegant approach and assume that we are given some further randomisation independent of B, which we use to pick a pair $\alpha < 0 < \beta$ according to the distribution

$$\mu(da, db) = \gamma(b - a)F_+(db)F_-(da), \tag{7.1}$$

where $F_\pm$ are the restrictions of F to $[0, \infty)$, $(-\infty, 0)$ respectively, and

$$\gamma^{-1} = \int_0^\infty bF_+(db) = -\int_{-\infty}^0 aF_-(da) = \tfrac{1}{2}\int |x|F(dx) \tag{7.2}$$

is the appropriate normalisation.

To see why this is a good thing to do, we first need to do a couple of trivial martingale calculations.

(7.3) PROPOSITION. *If* $\tau = \inf\{t : B_t \notin (a, b)\}$, *where* $a < 0 < b$ *are fixed, then* $\tau < \infty$, *a.s., and*

(7.4) (i)
$$\mathbf{P}(B_\tau = b) = \frac{-a}{b - a},$$

(7.4)(ii)
$$\mathbf{E}\tau = |ab|.$$

Proof. Finiteness of τ follows from Lemma 3.6. For (7.4)(i), use the Optional Sampling Theorem on the martingale B:

$$\begin{aligned} 0 &= \mathbf{E}(B_{\tau\wedge n}) \\ &= b\mathbf{P}(B_\tau = b, \tau \leqslant n) + a\mathbf{P}(B_\tau = a, \tau \leqslant n) + \mathbf{E}[B_n : \tau > n] \\ &\to b\mathbf{P}(B_\tau = b) + a\mathbf{P}(B_\tau = a) \end{aligned}$$

as $n \to \infty$, since $B_{\tau\wedge\cdot}$ is bounded.

To obtain (7.4)(ii), consider the martingale

$$M_t = (B_t - a)(b - B_t) + t,$$

and use the Optional Sampling Theorem again:

$$\begin{aligned} \mathbf{E}M_0 = -ab = \mathbf{E}M_{\tau\wedge n} \\ &= \mathbf{E}(\tau \wedge n) + \mathbf{E}(B_{\tau\wedge n} - a)(b - B_{\tau\wedge n}) \\ &\to \mathbf{E}\tau \end{aligned}$$

as $n \to \infty$. □

The Skorokhod embedding is described as follows. Take the two points $\alpha < 0 < \beta$, and run B until it hits one or other:

$$T := \inf\{u : B_u \notin (\alpha, \beta)\}.$$

(7.5) THEOREM (Skorokhod embedding). *The law of B_T is F, and $\mathbf{E}T = \sigma^2$.*

Proof. From (7.1) and (7.4)(i), we see immediately that, for $b \geqslant 0$,

$$\begin{aligned}\mathbf{P}(B_T \in db) &= \int_{(-\infty,0)} \frac{-a}{b-a}\gamma(b-a)F_+(db)F_-(da)\\ &= F_+(db),\end{aligned}$$

using the definition of γ. Thus B_T has law F. Next,

$$\begin{aligned}\mathbf{E}T &= \int_{[0,\infty)}\int_{(-\infty,0)} \mu(da,db)|ab|\\ &= \gamma\int_{[0,\infty)} F_+(db)\int_{(-\infty,0)} F_-(da)(b-a)|ba|\\ &= \int_{-\infty}^{\infty} x^2 F(dx)\\ &= \sigma^2\end{aligned}$$

as required. □

Now we see how to embed the random walk into Brownian motion; we take $T_1 := T$ as just described, and then perform the same construction on the Brownian motion $(B_{T_1+t} - B_{T_1})_{t \geqslant 0}$ to obtain a stopping time T_2' with mean σ^2 and such that $B_{T_1+T_2'} - B_{T_1} \sim F$. Now set $T_2 := T_1 + T_2'$, and proceed to carry out the same construction on the Brownian motion $(B_{T_2+t} - B_{T_2})_{t \geqslant 0}$ (the fact that this *is* a Brownian motion does not follow from (3.3), since T_1 and T_2 are random; we need the *strong* Markov property of Brownian motion—see Section 12—to justify this intuitively obvious fact).

We go on doing this, ending up with a sequence $T_0 \equiv 0 \leqslant T_1 \leqslant T_2 \leqslant \cdots$ of stopping times. To summarise then, we have the following result.

(7.6) THEOREM. *The process $(S_n)_{n \geqslant 0} := (B(T_n))_{n \geqslant 0}$ is a random walk with step distribution F, and $\mathbf{E}T_n = n\sigma^2$.*

The Skorokhod embedding just described reduces many statements about zero-mean finite-variance random walks to trivialities; for example, since

$$\mathbf{P}[\sup B_t = +\infty, \inf B_t = -\infty] = 1,$$

(7.7) $$\mathbf{P}[\sup S_n = +\infty, \inf S_n = -\infty] = 1$$

for *any* such random walk! But we need to be a bit careful here—might the sequence (T_n) not *miss* all those times when B was above 20, say? In principle, yes; but, because we ensured that the $T_n - T_{n-1}$ are independent and identically distributed (IID) with finite mean, that cannot happen. This is the only place where applying the Skorokhod embedding is in the least bit delicate, and in our proof in Sections 8 and 16 of the big classical limit results for IID summands you will see how we deal with this. Generalising to independent summands that have different distributions tends to become more involved; the limit problem for (S_n) gets converted into a limit problem for (T_n)! Nonetheless, if this can be solved, the functional form of the limit theorem often follows very easily. In any case, knowing what Brownian motion does will give a very good idea of what the random walk should do!

8. Donsker's Invariance Principle. It is an important fact that Brownian motion is a weak limit of random walks, both for our intuitive understanding, and for the easy proof it permits of various limit theorems, using the Continuous-Mapping Theorem (see Lemma II.84.2). Theorem 7.6 tells us that a random walk can be embedded in a Brownian motion, so the result cannot be said now to be a big surprise; and, indeed, all that is needed is a little care over the details.

So we shall suppose given any zero-mean step distribution F with unit variance, and use Theorem 7.6 to give us a random walk $S_n := B(T_n)$ with this step distribution. Now for each n define the random function

(8.1) $$S^{(n)}(t) = n^{1/2}\left[\left(t - \frac{k}{n}\right)S_{k+1} + \left(\frac{k+1}{n} - t\right)S_k\right], \quad \frac{k}{n} \leqslant t \leqslant \frac{k+1}{n}.$$

This is a piecewise-linear continuous function, equal to $n^{-1/2}S_k$ at each point of the form $t = k/n$.

(8.2) THEOREM (Donsker's Invariance Principle). *The processes* $(S^{(n)}(t))_{0\leqslant t\leqslant 1}$ *converge weakly to* $(B_t)_{0\leqslant t\leqslant 1}$ *as* $n \to \infty$.

Weak convergence is studied in Chapter II, Part 6.

Proof. We shall prove that, given $\varepsilon > 0$, there is some $n_0 = n_0(\varepsilon)$ so large that, for $n \geqslant n_0$,

(8.3) $$\mathbf{P}\left[\sup_{0\leqslant t\leqslant 1} |S^{(n)}(t) - B^{(n)}(t)| > \varepsilon\right] \leqslant \varepsilon,$$

where $B^{(n)}$ is a Brownian motion; in fact, we define

$$B_t^{(n)} := n^{-1/2}B_{nt}.$$

To begin with, note that we can choose $\delta > 0$ so small that

(8.4) $$\mathbf{P}[|B_s - B_t| > \varepsilon \text{ for some } s, t \in [0, 1], |s - t| \leqslant \delta] \leqslant \tfrac{1}{2}\varepsilon$$

since B is continuous. Note also that, since the embedding times T_n constitute a random walk, by the Strong Law of Large Numbers we have

$$\frac{T_n}{n} \xrightarrow{\text{a.s.}} \mathbf{E}T_1 = 1,$$

so that

$$\frac{1}{n} \sup_{k \leqslant n} |T_k - k| \xrightarrow{\text{a.s.}} 0.$$

Thus, for some n_1, for all $n \geqslant n_1$,

$$\mathbf{P}\left[\frac{1}{n} \sup_{k \leqslant n} |T_k - k| > \tfrac{1}{3}\delta\right] \leqslant \varepsilon/2.$$

The third simple fact we need is that, for any $t \in [k/n, (k+1)/n]$, there is some $nu \in [T_k, T_{k+1}]$ such that

$$S^{(n)}(t) = n^{-1/2} B_{nu} =: B_u^{(n)}. \tag{8.5}$$

Thus we estimate

$$\begin{aligned}
&\mathbf{P}\left[\sup_{0 \leqslant t \leqslant 1} |S^{(n)}(t) - B^{(n)}(t)| > \varepsilon\right] \\
&\quad \leqslant \mathbf{P}\left[\sup_{0 \leqslant t \leqslant 1} |S^{(n)}(t) - B^{(n)}(t)| > \varepsilon, \quad |T_k - k| \leqslant \tfrac{1}{3}n\delta \quad \text{for all } k \leqslant n\right] \\
&\qquad + \mathbf{P}\left[\frac{1}{n} \sup_{k \leqslant n} |T_k - k| > \tfrac{1}{3}\delta\right] \\
&\quad \leqslant \mathbf{P}[|B_u^{(n)} - B_t^{(n)}| > \varepsilon \quad \text{for some} \quad u, t \in [0,1], |u - t| \leqslant \delta] + \varepsilon/2,
\end{aligned}$$

at least for $n \geqslant n_0 = n_1 \vee (3/\delta)$. Indeed, if there is some $t \in [k/n, (k+1)/n]$ such that $|S^{(n)}(t) - B^{(n)}(t)| > \varepsilon$ then, using (8.5), there is some u in $[n^{-1}T_k, n^{-1}T_{k+1}]$ such that $|B_u^{(n)} - B_t^{(n)}| > \varepsilon$; and, since $|n^{-1}T_k - k/n| \leqslant \frac{1}{3}\delta$ for all k (at least off the unlikely event $\{n^{-1} \sup_{k \leqslant n} |T_k - k| > \frac{1}{3}\delta\}$), it follows that $|u - t| \leqslant \delta$. Using (8.4) now yields (8.3):

$$\mathbf{P}\left[\sup_{0 \leqslant t \leqslant 1} |S^{(n)}(t) - B^{(n)}(t)| > \varepsilon\right] \leqslant \varepsilon$$

for all $n \geqslant n_0$. □

Remarks. Notice that Theorem 8.2 implies the Central Limit Theorem for IID zero-mean finite-variance random variables, and the proof makes no use anywhere of characteristic-function methods!

9. Exponential martingales and first-passage distributions. We shall use the Brownian exponential martingales (2.4) to derive the distribution of

$$H_x := \inf\{t > 0 : X_t = x\}$$

for $x > 0$, where $X_t := B_t + ct$ is a Brownian motion with drift, and also to derive the distribution of $H_a \wedge H_b$, where $a < 0 < b$.

Let us fix some $\lambda > 0$. From (2.4),

$$\exp(\theta X_t - \lambda t) = \exp[\theta B_t - (\lambda - \theta c)t]$$

is a martingale provided that

$$\lambda - \theta c = \tfrac{1}{2}\theta^2,$$

that is,

$$\theta = \beta := \sqrt{c^2 + 2\lambda} - c, \quad \text{(or)} \quad \theta = \alpha := -c - \sqrt{c^2 + 2\lambda}.$$

Note that $\alpha < 0 < \beta$. Thus the martingale $\exp(\beta X_t - \lambda t)$ is bounded on $[0, H_x]$, so we can use the Optional Sampling Theorem to conclude that

$$\begin{aligned} 1 &= \mathbf{E}\exp[\beta X(H_x) - \lambda H_x] \\ &= e^{\beta x}\mathbf{E}\, e^{-\lambda H_x}, \end{aligned}$$

from which

(9.1) $$\mathbf{E}\, e^{-\lambda H_x} = \exp\{-x(\sqrt{c^2 + 2\lambda} - c)\}.$$

The Laplace transform can be inverted explicitly to give

(9.2) $$\mathbf{P}(H_x \in dt)/dt = \frac{x}{\sqrt{2\pi t^3}}\exp[-(x - ct)^2/2t].$$

From (9.1), taking the limit as $\lambda \downarrow 0$, we conclude that

$$\mathbf{P}[H_x < \infty] = \begin{cases} 1 & (c \geqslant 0), \\ e^{-2|c|x} & (c < 0). \end{cases}$$

Thus, if the drift is negative, the drifting Brownian motion will with positive probability fail to hit some named positive level. This is not very surprising, since $t^{-1}X_t \to c$, a.s.

Next, if we fix $a < 0 < b$, and let $T := H_a \wedge H_b$, then the process

$$M_t := (e^{\beta b} - e^{\beta a})e^{\alpha X_t - \lambda t} + (e^{\alpha a} - e^{\alpha b})e^{\beta X_t - \lambda t}$$

is a martingale of the form $f(X_t)e^{-\lambda t}$, where f is constructed so that

$$f(a) = f(b) = e^{\beta b + \alpha a} - e^{\beta a + \alpha b}.$$

Another application of the Optional Sampling Theorem gives

$$\mathbf{E}\, e^{-\lambda T} = \frac{e^{\beta b} - e^{\beta a} + e^{\alpha a} - e^{\alpha b}}{e^{\beta b + \alpha a} - e^{\beta a + \alpha b}}. \tag{9.3}$$

This time, there is no explicit inversion of the Laplace transform in any particularly useable form, even in the special case $a = -b$, $c = 0$, when we have the simpler statement

$$\mathbf{E} e^{-\lambda T} = \operatorname{sech}(b\sqrt{2\lambda}). \tag{9.4}$$

Remarks. Everything in this section is entirely classical, but is a good illustration of martingale techniques in conjunction with the Brownian exponential martingales. We shall later see a completely different derivation of (9.2) which is far more illuminating (see Section 13).

10. Some sample-path properties. The fine structure of Brownian sample paths exerts a mesmeric fascination, which you will understand after reading this section; Brownian paths are wilder than we can imagine! As a small aperitif, we give a soft argument to prove that, almost surely, B is not differentiable at zero. Notice first that, by the time-inversion property (3.5) and the oscillation of the Brownian path near infinity (Lemma 3.6),

$$\mathbf{P}[\text{for each } \varepsilon > 0,\ \exists s, t \leqslant \varepsilon \text{ such that } B_s < 0 < B_t] = 1,$$

so the only possible derivative at zero would be zero. But if $B_0' = 0$, we have that, for all small enough t, $|B_t| \leqslant t$. Time inversion again translates this into the statement that, for all large enough s, $|\tilde{B}_s| \leqslant 1$ ($\tilde{B}_s := sB_{1/s}$), contradicting Lemma 3.6.

By Fubini's Theorem, we conclude that, almost surely, B is almost everywhere non-differentiable. What we *really* want, Theorem 10.6, *is* true, but needs another approach. Here is a preliminary result.

(*10.1*) THEOREM. *Almost surely, B is not Lipschitz-continuous anywhere.*

Proof. Fixing some (large) $K > 0$, define, for each $n \geqslant 2$, the event

$$A_n := \{\text{for some } s \in [0,1], |B_t - B_s| \leqslant K|t - s| \text{ whenever } |t - s| \leqslant 2/n\},$$

and let

$$\Delta_{k,n} := |B(k/n) - B((k-1)/n)|, \qquad k = 1, \ldots, n.$$

Then

$$A_n \subseteq \bigcup_{k=2}^{n} \left\{ \Delta_{j,n} \leqslant \frac{4K}{n} \quad \text{for } j = k-1, k, k+1 \right\},$$

and so

$$\mathbf{P}(A_n) \leqslant (n-2)[\mathbf{P}(\Delta_{1,n} \leqslant 4K/n)]^3$$
$$= (n-2)\mathbf{P}(|B_1| \leqslant 4K/\sqrt{n})^3 \leqslant c/\sqrt{n}.$$

But the A_n are increasing, so $\mathbf{P}(A_n) = 0$ for all n. □

The exact modulus of continuity of the Brownian path is an altogether more delicate result; you will find a proof of the following result in Section 1.6 of McKean [1].

(*10.2*) THEOREM (Lévy's Modulus-of-Continuity Theorem)

$$\mathbf{P}\left\{ \limsup_{\delta \downarrow 0} \sup_{0 \leqslant t \leqslant 1} \frac{B_{t+\delta} - B_t}{\left[2\delta \log\left(\frac{1}{\delta}\right)\right]^{1/2}} = 1 \right\} = 1.$$

This exact modulus of continuity should be compared with the celebrated *Law of the Iterated Logarithm*:

$$\mathbf{P}\left\{ \limsup_{t \downarrow 0} \left[2t \log\log\left(\frac{1}{t}\right)\right]^{-1/2} B_t = +1 \right\} = 1. \tag{10.3}$$

We prove this result in Section 16; do note that, although the Law of the Iterated Logarithm (10.3) must hold at almost every point of the path, by Theorem 10.2 there will be places where the oscillation will be wilder than $[2\delta \log\log(1/\delta)]^{1/2}$. Indeed, there will also be places where the oscillation will be like $c\delta^{1/2}$ (if c is small enough); for a fascinating analysis of such *slow points* of Brownian motion, see Greenwood and Perkins [1, 2].

One other sample path property worth mentioning is the following.

(*10.4*) THEOREM. *Almost surely, Brownian motion has no point of increase*:

$$\mathbf{P}[\exists \delta > 0, s > 0 \quad \text{such that} \quad B_{s-h} \leqslant B_s \leqslant B_{s+h} \quad \text{for all} \quad h \in [0, \delta]] = 0.$$

Of the many proofs, we give that of Burdzy [3] in Section 12.

(*10.5*) *Remark.* If $(B_t^1)_{t \geqslant 0}$ and $(B_t^2)_{t \geqslant 0}$ are independent Brownian motions, then $(B_t)_{t \geqslant 0} \equiv (B_t^1, B_t^2)_{t \geqslant 0}$ is Brownian motion in the plane. For any $v \in \mathbf{R}^2$, $|v| = 1$, the process $(v \cdot B_t)$ is a real-valued Brownian motion, and therefore has no points of increase with probability 1. James Taylor has posed the question 'Is there a positive probability that for *some* v the Brownian motion $v \cdot B$ has a point of increase?' As far as we are aware, this problem is unresolved. If true, it would say that, with positive probability, the trace $B[0, 1] := \{B_t : 0 \leqslant t \leqslant 1\}$ of two-dimensional Brownian motion could be cut with a straight line; the best that is known so far is Burdzy's result that $B[0, 1]$ can be cut with a Lipschitz curve.

(*10.6*) THEOREM. *Almost surely, the Brownian path is nowhere differentiable.*

Proof. By Theorem 10.4, symmetry, and the Cameron-Martin-Girsanov Theorem IV.38.9 with $\gamma \equiv 1$, we may (after throwing away a fixed null set of possible paths) assume that for *every* path, the functions $t \mapsto -B(t)$ and $t \mapsto B(t) + t$ have no point of increase. At any point τ at which $B(\cdot)$ is differentiable, we must therefore have both $-B'(\tau) \leqslant 0$ and $B'(\tau) + 1 \leqslant 0$, which is a contradiction. □

Geman and Horowitz [1] give a nice 'local-time' proof.

11. Quadratic variation. While the sole result of this section could have been included in the previous section, we separate it out because of its fundamental importance in the construction of stochastic integrals. Define $t_k^n := t \wedge (k2^{-n})$, and

$$[B]_t^n := \sum_{k \geqslant 1} [B(t_k^n) - B(t_{k-1}^n)]^2.$$

(*11.1*) LEMMA (Lévy). *With probability* 1, *as* $n \to \infty$

$$[B]_t^n \to t \quad \text{uniformly on compact } t\text{-intervals.}$$

The proof is given in Section IV.2, so we defer the details. Since they are not hard, the reader may wish to have a go at proving Lemma 11.1 now.

(*11.2*) *Remarks*

(i) If we take the quadratic variation of B down the dyadic partitions, we get the answer t. If, on the other hand, we take

$$\sup\left\{ \sum_{k=1}^{N} |B(t_k) - B(t_{k-1})|^2 : 0 = t_0 < t_1 < \cdots < t_N = 1 \right\},$$

we get the answer $+\infty$; the *true* quadratic variation of B is infinite almost surely. See Lévy [2, p. 190] or Freedman [1, p. 48].

(ii) Of course, it follows immediately from Lemma 11.1 that the variation of almost every Brownian path is infinite on every interval. See (IV.2.16).

12. The strong Markov property. We give here a quick proof of the strong Markov property for Brownian motion, exploiting heavily the special features of Brownian motion, and standard results on martingales. The strong Markov property holds for a much wider class of processes (we shall see in Chapter III that it holds for all Feller–Dynkin processes, and for all Ray processes), and is proved by approximating a stopping time by a dyadic-rational stopping time. Since this approximation procedure is used in the martingale results to which we appeal, the result we now give contains all the same ingredients. For the definition of Brownian motion relative to a filtered probability space, see Section II.72, or look back to (1.2)(iii)′.

(*12.1*) THEOREM. Let $(B_t)_{t \geqslant 0}$ *be Brownian motion on some filtered probability space* $(\Omega, \mathscr{F}, (\mathscr{F}_t), \mathbf{P})$, *and let* T *be a finite-valued* $(\mathscr{F}_t)$ *stopping time (see*

Section II.73.1). Then the process

$$B_t^{(T)} := B_{T+t} - B_T, \quad t \geqslant 0,$$

is a Brownian motion independent of $\mathscr{F}_T$.

Proof. Fix $0 = t_0 \leqslant t_1 \leqslant \cdots \leqslant t_n$, and reals $\theta_1, \ldots, \theta_m$, and take some $\xi \in b\mathscr{F}_T$. Let $s_j := t_j - t_{j-1}$ $(j = 1, \ldots, n)$. Let S be a stopping time. By applying the Optional Sampling Theorem to the martingale

$$M_t := \exp(i\theta B_t + \tfrac{1}{2}\theta^2 t)$$

and the stopping time $S \wedge N$, we obtain

$$\begin{aligned} &\mathbf{E}[\exp\{i\theta B((S \wedge N) + t) + \tfrac{1}{2}\theta^2((S \wedge N) + t)\} \,|\, \mathscr{F}(S \wedge N)] \\ &\quad = \exp\{i\theta B(S \wedge N) + \tfrac{1}{2}\theta^2(S \wedge N)\}. \end{aligned}$$

Rearranging and letting $N \uparrow \infty$ yields

$$\mathbf{E}[\exp\{i\theta(B_{S+t} - B_S) + \tfrac{1}{2}\theta^2 t\} \,|\, \mathscr{F}_S] = 1. \tag{12.2}$$

Thus, if $\xi \in b\mathscr{F}_T$, we have, with Z_j denoting $B^{(T)}(t_j) - B^{(T)}(t_{j-1})$,

$$\mathbf{E}\left[\xi \exp\left\{\sum_{j=1}^{n} [i\theta_j Z_j + \tfrac{1}{2}\theta_j^2 s_j]\right\}\right] = \mathbf{E}\left[\xi \exp\left\{\sum_{j=1}^{n-1} [i\theta_j Z_j + \tfrac{1}{2}\theta_j^2 s_j]\right\}\right]$$

by taking $S = T + t_{n-1}$, $t = s_n$ in (12.2), and evaluating the expectation by conditioning first on $\mathscr{F}_S$. Conditioning successively on $\mathscr{F}(T + t_k)$ $(k = n-2, \ldots, 0)$ gives

$$\mathbf{E}\left[\xi \exp\left\{\sum_{j=1}^{n} [i\theta_j Z_j + \tfrac{1}{2}\theta_j^2 s_j]\right\}\right] = \mathbf{E}\xi.$$

Hence, conditional on $\mathscr{F}_T$, Z_j $(j = 1, \ldots, n)$ are independent Gaussian variables with mean 0 and variances s_j $(j = 1, \ldots, n)$. □

(12.3) Remarks

(i) The restriction to finite-valued stopping times is not really essential, being included only to save worrying over the definition of $B^{(T)}$.

(ii) If one takes a more general continuous Markov process, the strong Markov property is formulated as follows. Let $\Omega = C(\mathbf{R}^+, \mathbf{R})$, with the canonical process $X_t(\omega) = \omega(t)$, filtration $\mathscr{F}_t^\circ \equiv \sigma(\{X_u : u \leqslant t\})$ and measure $\mathbf{P}^x$ for the process started at x. The *shift maps* $\theta_t : \Omega \to \Omega$ are defined by $(\theta_t\omega)(s) := \omega(t+s)$, and the strong Markov property says that, for any $(\mathscr{F}_{t+}^\circ)_{t \geqslant 0}$ stopping time T, any $\xi \in b\mathscr{F}_{T+}^\circ$, $\eta \in b\mathscr{F}^\circ$ and $x \in \mathbf{R}$,

$$\mathbf{E}^x[\xi\eta \circ \theta_T : T < \infty] = \mathbf{E}^x[\xi \mathbf{E}^{X(T)}(\eta) : T < \infty]. \tag{12.4}$$

This formulation turns out to be ideal for applications. As an example, consider the celebrated Blumenthal 0–1 Law, which says that $\mathscr{F}^\circ_{0+} := \bigcap_{s>0} \mathscr{F}^\circ_s$ *is trivial under* $\mathbf{P}^x$. The proof is not hard either; if $\Lambda \in \mathscr{F}^\circ_{0+}$ then, taking $\xi = \eta = I_\Lambda$ in (12.4), we obtain (with $T \equiv 0$)

$$\mathbf{E}^x[I_\Lambda] = \mathbf{E}^x[I_\Lambda \mathbf{E}^{X(0)} I_\Lambda] = \mathbf{P}^x(\Lambda)^2,$$

so that

$$\mathbf{P}^x(\Lambda) = \mathbf{P}^x(\Lambda)^2 \quad \text{and} \quad \mathbf{P}^x(\Lambda) = 0 \text{ or } 1!$$

The consequences of this result for Brownian motion are far-reaching. In the general setting, though, the measurability questions implicit in (12.4) need to be carefully considered; we return to this in Chapter III.

As a first application of the strong Markov property, we give Burdzy's [1] beautiful proof of Theorem 10.4, that with probability 1 Brownian motion has no point of increase. The aim is to prove that $\mathbf{P}^0(A_0) = 0$, where

$$A_0 = \{\text{for some } 0 < t < u,\ B_s \leqslant B_t \leqslant 1 \text{ for } s \leqslant t, B_s \geqslant B_t \text{ for } t \leqslant s \leqslant u,\ B_u \geqslant B_t + 2\}.$$

This will give the result, since if Brownian motion has a point of increase with positive probability then with positive probability it will have a point of increase before it reaches 1 (scaling), and then with positive probability it will rise at least 2 before returning to the level of the point of increase. For fixed $\varepsilon \in (0, 1)$, define the a.s. finite stopping times T_k, U_k, and non-negative random variables M_k by

$$M_0 = U_0 = 0;$$

$$T_k = \inf\{t > U_k : B_t = M_k - \varepsilon \quad \text{or} \quad B_t = M_k + 2\}, \qquad k \geqslant 0,$$

$$M_{k+1} = \sup\{B_t : t \leqslant T_k\}, \qquad k \geqslant 0,$$

$$U_{k+1} = \inf\{t > T_k : B_t = M_{k+1}\} \qquad k \geqslant 0.$$

Figure I.2 on page 24 illustrates the situation up to the kth stage:

Because of the strong Markov property of Brownian motion, the pieces of path $\{B(U_k + s) - B(U_k) : 0 \leqslant s \leqslant T_k - U_k\}$ are independent, identically-distributed. Thus the random variables

$$X_k = M_{k+1} - M_k, \qquad k \geqslant 0,$$

are also IID, and each is equal in law to the maximum of Brownian motion until it leaves $[-\varepsilon, 2]$. Thus (see Proposition 7.3)

$$\mathbf{P}(X_k > x) = \begin{cases} \varepsilon/(\varepsilon + x) & (0 < x < 2), \\ 0 & (x \geqslant 2). \end{cases}$$

The idea is that the unlikely event that $X_k = 2$ (which has probability $\varepsilon/(\varepsilon + 2)$) is *an approximate point of increase at time* U_{k-1}; convince yourself of the key fact that if A_0 happens then

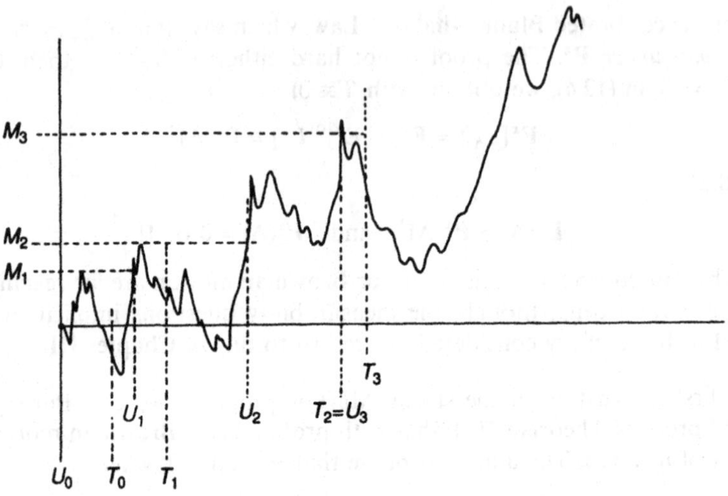

Fig. I.2

$$A_\varepsilon = \{\text{for some } k, X_k = 2, M_k \leqslant 1\}$$

also happens. [*Hint*: If τ is the largest t such that $B_s \leqslant B_t \leqslant 1$ for all $s \leqslant t$, and $B_s \geqslant B_t$ for $t \leqslant s \leqslant u$, $B_u \geqslant B_t + 2$, and if $\xi := B_\tau$ then consider $\inf\{M_k: M_k > \xi\} =: m$. Argue that $m \geqslant 2$.] But now $\mathbf{P}(A_0) \leqslant \mathbf{P}(A_\varepsilon)$ and

$$\mathbf{P}(A_\varepsilon) = \sum_{k \geqslant 0} \mathbf{P}(M_k \leqslant 1, X_k = 2)$$

$$= \sum_{k \geqslant 0} \mathbf{P}(M_k \leqslant 1)\mathbf{P}(X_k = 2)$$

(since X_k is independent of M_k by the strong Markov property)

$$= \frac{\varepsilon}{2+\varepsilon}\mathbf{E}(N+1), \tag{12.5}$$

where N is the number of k for which $M_k \leqslant 1$. But $M_k = \sum_{j=0}^{k-1} X_j$, and $N+1$ is a stopping time for the random walk $(M_k)_{k \geqslant 0}$, so, by Wald's identity (a special case of the Optional Sampling Theorem),

$$\mathbf{E}M_{N+1} = \mathbf{E}(N+1)\mathbf{E}X_0,$$

so that

$$\mathbf{E}(N+1) = \frac{\mathbf{E}(M_{N+1})}{\mathbf{E}X_0} \leqslant \frac{3}{\varepsilon|\log \varepsilon|}.$$

Feeding this into (12.5) gives

$$\mathbf{P}(A_0) \leqslant \mathbf{P}(A_\varepsilon) \leqslant \frac{3}{(2+\varepsilon)|\log \varepsilon|} \to 0 \quad \text{as } \varepsilon \downarrow 0.$$ □

13. Reflection. The basic idea of the reflection principle for Brownian motion is familiar from simple symmetric random walk, and only the technical machinery is any more difficult. For $a \in \mathbb{R}$, define the *hitting time of a*

$$H_a := \inf\{t > 0 : B_t = a\}.$$

(*13.1*) THEOREM. *Fix* $a \in \mathbb{R}$. *The process*

$$\tilde{B}_t := \begin{cases} B_t & (t < H_a), \\ 2a - B_t & (t \geqslant H_a) \end{cases} \tag{13.2}$$

is a Brownian motion.

Proof. Consider the process

$$Y_t := B_t \quad (0 \leqslant t \leqslant H_a), \qquad Z_t := B(t + H_a) - a.$$

By the Strong Markov property, Theorem 12.1, Z is a Brownian motion independent of Y. By (3.2), $-Z$ is also a Brownian motion, also independent of Y. Thus $(Y, Z) \overset{\mathscr{D}}{=} (Y, -Z)$. The map

$$\varphi : (Y, Z) \to (Y_t I_{\{t \leqslant H_a\}} + (a + Z_{t - H_a}) I_{\{t > H_a\}})_{t \geqslant 0}$$

produces a continuous process, which will therefore have the same law as $\varphi(Y, -Z)$. But $\varphi(Y, Z) = B$, and $\varphi(Y, -Z) = \tilde{B}$. □

(*13.3*) COROLLARY. *Define*

$$S_t := \sup\{B_u : u \leqslant t\}.$$

Then, for $a, y \geqslant 0, t \geqslant 0$,

$$\mathbf{P}[S_t \geqslant a, B_t \leqslant a - y] = \mathbf{P}[B_t \geqslant a + y]. \tag{13.4}$$

Proof.

$$\mathbf{P}[S_t \geqslant a, B_t \leqslant a - y] = \mathbf{P}[S_t \geqslant a, \tilde{B}_t \leqslant a - y], \quad \text{with } \tilde{B} \text{ given by (13.2)},$$
$$= \mathbf{P}[B_t \geqslant a + y],$$

by drawing a picture. □

Remarks

(i) Immediately from (13.4), we obtain, for $a > 0$,

$$\begin{aligned} \mathbf{P}[H_a \leqslant t] &= \mathbf{P}[S_t \geqslant a] \\ &= \mathbf{P}[S_t \geqslant a, B_t \leqslant a] + \mathbf{P}[S_t \geqslant a, B_t \geqslant a] \\ &= 2\mathbf{P}[B_t \geqslant a] \\ &= 2\mathbf{P}[B_1 \geqslant a/\sqrt{t}], \end{aligned}$$

by Brownian scaling. Differentiating with respect to t gives

$$(13.5)\qquad \mathbf{P}[H_a \in dt]/dt = \frac{a}{\sqrt{2\pi t^3}} \exp\left(-\frac{a^2}{2t}\right),$$

(ii) The joint distribution of S_t, B_t follows easily from (13.4):

$$(13.6)\qquad \mathbf{P}[S_t \in da, B_t \in dx] = \frac{2(2a-x)}{\sqrt{2\pi t^3}} \exp\left[-\frac{(2a-x)^2}{2t}\right] da\, dx,$$

for $a \geqslant 0$, $x \leqslant a$.

(iii) Let $\mathbf{P}^x$ denote the law of Brownian motion started at x. Then it follows easily from (13.4) that, for $x, y > 0$,

$$\mathbf{P}^x[B_t \geqslant y, H_0 \leqslant t] = \mathbf{P}^x[B_t \leqslant -y],$$

from which

$$(13.7)\qquad \mathbf{P}^x[B_t \in dy, H_0 > t]/dy = p_t(x, y) - p_t(x, -y),$$

where p_t is, of course, the Brownian transition density (4.2). The transition density (13.7) is often called the *taboo transition density*.

(iv) It is possible to obtain the analogue of (13.7) when H_0 is replaced by the first exit of Brownian motion from an interval (a, b); the transition density is

$$(13.8)\qquad \sum_{n \in \mathbb{Z}} \{p_t(x, y + 2n\delta) - p_t(x, 2b - y + 2n\delta)\}, \qquad a < x, y < b,$$

where $\delta := b - a$. We omit the proof of this result, which we shall not be using: see Freedman [1], p. 26 for a proof.

(v) There is a routine way to derive the analogous results for Brownian motion with drift c from the ones we have just calculated. If $\Omega := C(\mathbf{R}^+, \mathbf{R})$, $X_t(\omega) := \omega(t)$, $\mathscr{F}_t^\circ := \sigma(\{X_u : u \leqslant t\})$ is the canonical space, and if $\mathbf{P}^{x,c}$ is the law on $(\Omega, \mathscr{F}^\circ)$ of Brownian motion started at x with drift $c \in \mathbf{R}$, then on $\mathscr{F}_t^\circ$ the laws $(\mathbf{P}^{x,c})_{c \in \mathbf{R}}$ are equivalent with density with respect to Wiener measure given by

$$(13.9)\qquad \left.\frac{d\mathbf{P}^{x,c}}{d\mathbf{P}^{x,0}}\right|_{\mathscr{F}_t} = \exp\left[c(X_t - x) - \tfrac{1}{2}c^2 t\right].$$

This is a special case of the celebrated Cameron–Martin–Girsanov formula, which we discuss in depth in Chapter IV. The reader may like to try proving (13.9) now (*Hint*: consider a cylinder set.) Combining (13.9) and (13.6), we deduce that, for $a \geqslant 0$, $x \leqslant a$,

$$(13.10)\qquad \mathbf{P}^{0,c}[S_t \in da, X_t \in dx] = \frac{2(2a-x)}{\sqrt{2\pi t^3}} \exp\left[-\frac{(2a-x)^2}{2t} + cx - \tfrac{1}{2}c^2 t\right].$$

Can you see how to deduce (9.2) from (13.5) and (13.9)?

14. Reflecting Brownian motion and local time. If $(B_t)_{t\geq 0}$ is a Brownian motion then a *reflecting Brownian motion* is a continuous process identical in law to $(|B_t|)_{t\geq 0}$. It is a continuous non-negative process, and is in fact also a Markov process. The Markov property is not immediately obvious, because we have taken a function of a Markov process ($x \mapsto |x|$ in this case), which will not in general be Markovian. The following simple result covers this situation.

(*14.1*) LEMMA. *Let* $(S, \mathscr{S})$ *and* $(S', \mathscr{S}')$ *be measurable spaces, and suppose that the measurable function* $\Phi: (S, \mathscr{S}) \to (S', \mathscr{S}')$ *is onto,* $(P_t)_{t\geq 0}$ *is a Markov transition semigroup on* $(S, \mathscr{S})$, *and* $(Q_t)_{t\geq 0}$ *is a collection of probability kernels on* S' *such that, for all* $f \in b\mathscr{S}'$,

$$P_t(f \circ \Phi) = (Q_t f) \circ \Phi. \tag{14.2}$$

Then $(Q_t)_{t\geq 0}$ *is a Markov transition semigroup, and if* X *is a Markov process with transition semigroup* (P_t) *then* $\Phi(X)$ *is Markov with transition semigroup* (Q_t).

(A *probability kernel* Q on S' is a map $Q: S' \times \mathscr{S}' \to [0, 1]$ such that $Q(x, \cdot)$ is a probability measure for each $x \in S'$, and $Q(\cdot, A)$ is measurable for each $A \in \mathscr{S}'$.)

Proof. Using the semigroup property of (P_t) and (14.2)

$$P_{t+s}(f \circ \Phi) = (Q_{t+s} f) \circ \Phi = P_t P_s(f \circ \Phi) = P_t((Q_s f) \circ \Phi) = (Q_t Q_s f) \circ \Phi.$$

Since Φ is onto, $Q_{t+s} f = Q_t Q_s f$. To check that $\Phi(X)$ is Markov, just take $0 = t_0 \leq t_1 \cdots \leq t_n$, $s_k \equiv t_k - t_{k-1}$, $f_k \in b\mathscr{S}'$ and compute

$$\begin{aligned} \mathbf{E}^x\left[\prod_{j=1}^n f_j(\Phi(X_{t_j}))\right] &= P_{s_1}(f_1 \circ \Phi) P_{s_2}(f_2 \circ \Phi) \cdots P_{s_n}(f_n \circ \Phi)(x) \\ &= (Q_{s_1} f_1 Q_{s_2} f_2 \cdots Q_{s_n} f_n)(\Phi(x)) \end{aligned}$$

by repeated application of (14.2). □

In our case, the transition semigroup is the Brownian transition semigroup (4.2), and $\Phi(x) = |x|$. If we define, for $f \in C_b(\mathbb{R}^+)$,

$$\begin{aligned} Q_t f(x) &= \int_0^\infty (p_t(x, y) + p_t(x, -y)) f(y)\, dy \\ &= \int_0^\infty (2\pi t)^{-1/2} \exp\left(-\frac{x^2 + y^2}{2t}\right) 2\cosh\left(\frac{xy}{t}\right) f(y)\, dy \end{aligned}$$

then it is trivial to check (14.2). The criterion of Lemma (14.1) for $\Phi(X)$ to be Markov is a very obvious one; see Rogers and Pitman [1] for a discussion of more interesting criteria.

The *strong* Markov property of $|B|$ follows from the strong Markov property of Brownian motion, since any stopping time for the filtration of $|B|$ is a stopping time for the filtration of B.

Now let us consider the process $((S_t, S_t - B_t))_{t \geqslant 0}$, which is a continuous strong Markov process in $(\mathbb{R}^+)^2$. (Here, $S_t := \sup\{B_u : u \leqslant t\}$, as in the previous section). To see the strong Markov property, observe that, for any finite stopping time T,

$$(14.3) \qquad (S_{T+t}, S_{T+t} - B_{T+t}) = (S_T \vee (B_T + \tilde{S}_t), ((S_T - B_T) \vee \tilde{S}_t) - \tilde{B}_t),$$

where $\tilde{B}_t := B_{T+t} - B_T$ and $\tilde{S}_t := \sup\{\tilde{B}_u : u \leqslant t\}$. Since $\tilde{B}$ is independent of $\mathscr{F}_T$, the law of the future of $(S, S - B)$ given $\mathscr{F}_T$ depends only on (S_T, B_T). It is possible to write out explicitly the transition semigroup of $(S, S - B)$, using the results of the last section, and then to confirm, using Lemma 14.1, that $S - B$ is a Markov process, with the same transition semigroup as reflecting Brownian motion $|B|$. We skip the details, because we give a much neater proof in Section V.6. For now, you will not be surprised that $S - B$ is a reflecting Brownian motion even if you do not carry through the calculations we indicated, because $S - B$ is a continuous non-negative process that clearly behaves like Brownian motion in $(0, \infty)$. What is most important for now is the remarkable fact (discovered by Lévy) that *by looking at the path of $S - B$, we can work out what S is*! Let us see how this can be done. Fix $\varepsilon > 0$ and define

$$T := \inf\{u : S_u - B_u > \varepsilon\}.$$

Then it is not hard to see that S_T *must have an exponential distribution*. Indeed, for any $a > 0$,

$$\begin{aligned}
\mathbf{P}[S_T - a > x \mid S_T > a] &= \mathbf{P}[T > H_{a+x} \mid T > H_a] \\
&= \mathbf{P}[S_u - B_u \leqslant \varepsilon \text{ for } H_a \leqslant u \leqslant H_{a+x} \mid S_u - B_u \leqslant \varepsilon \text{ for } u \leqslant H_a] \\
&= \mathbf{P}[\tilde{S}_u - \tilde{B}_u \leqslant \varepsilon \text{ for } 0 \leqslant u \leqslant \tilde{H}_x \mid S_u - B_u \leqslant \varepsilon \text{ for } u \leqslant H_a] \\
&\qquad \text{(using (14.3) with } T = H_a, \tilde{H}_x := \inf\{u : \tilde{B}_u = x\}) \\
&= \mathbf{P}[\tilde{S}_u - \tilde{B}_u \leqslant \varepsilon \text{ for } 0 \leqslant u \leqslant \tilde{H}_x] \text{ (since } \tilde{B} \text{ is independent of } \mathscr{F}(H_a)) \\
&= \mathbf{P}[S_T > x].
\end{aligned}$$

Since S_T has an exponential law, we can compute its mean using the Optional Sampling Theorem:

$$0 = \mathbf{E}B_{T \wedge n} = \mathbf{E}[S_{T \wedge n} - (S_{T \wedge n} - B_{T \wedge n})] = \mathbf{E}S_{T \wedge n} - \mathbf{E}(S_{T \wedge n} - B_{T \wedge n}),$$

whence

$$\mathbf{E}S_{T \wedge n} = \mathbf{E}(S_{T \wedge n} - B_{T \wedge n})$$

and, letting $n \uparrow \infty$, using monotone convergence on the left and dominated convergence on the right, we obtain

$$\mathbf{E}S_T = \varepsilon,$$

so that S_T is an exponential random variable of rate ε^{-1}. Now let us define

$$T'_1(\varepsilon) := 0, \qquad T_n(\varepsilon) := \inf\{u > T'_n(\varepsilon) : S_u - B_u > \varepsilon\},$$
$$T'_{n+1}(\varepsilon) := \inf\{u > T_n(\varepsilon) : S_u - B_u = 0\},$$

so that $T_1(\varepsilon), T_2(\varepsilon), \ldots$ are the successive times at which $S-B$ achieves an upcrossing of $[0,\varepsilon)$ (see Fig. I.3). Define

$$U(t,\varepsilon) := \sup\{k : T_k(\varepsilon) \leqslant t\},$$

the number of upcrossings made by $S-B$ before t. Note that $U(\cdot,\varepsilon)$ is increasing for each $\varepsilon > 0$. Now it does not take too long to realise (using the strong Markov property) that the random variables $S(T_1(\varepsilon)), S(T_2(\varepsilon)) - S(T_1(\varepsilon)), S(T_3(\varepsilon)) - S(T_2(\varepsilon)), \ldots$ are IID exponentials, rate ε^{-1}. Thus

$$U(H_a,\varepsilon) = \sup\{k : S(T_k(\varepsilon)) \leqslant a\}$$

will have a Poisson distribution with mean a/ε. Now hold a fixed, and consider

$$Z_n = 2^{-n}U(H_a, 2^{-n}),$$

which has mean a. If $\mathscr{G}_n := \sigma(\{Z_k : k \geqslant n\})$, we claim that $(Z_n, \mathscr{G}_n)$ *is a reversed martingale*. Indeed, if we consider an upcrossing of $S-B$ from 0 to 2^{-n} then with probability exactly $\frac{1}{2}$ the path of $S-B$ will go on up to 2^{-n+1} before it returns to 0; thus, given $\mathscr{G}_n$, the number of upcrossings to 2^{-n+1}, $U(H_a, 2^{-n+1})$, has a $B(U(H_a, 2^{-n}), \frac{1}{2})$ distribution. Hence

$$\mathbf{E}[U(H_a, 2^{-n+1}) \mid \mathscr{G}_n] = \tfrac{1}{2}U(H_a, 2^{-n}),$$

which implies $\mathbf{E}[Z_{n-1} \mid \mathscr{G}_n] = Z_n$. Thus (see Section II.51)

$$Z_n \to Z_\infty, \quad \text{a.s. and in } L^1.$$

This ensures that $\mathbf{E}Z_\infty = a$; but $\operatorname{var}(Z_n) = a2^{-n} \to 0$, $\mathbf{E}Z_n = a$, $Z_n \xrightarrow{L^2} a$, implying $Z_\infty = a$, a.s. To summarise,

$$\lim_{n\to\infty} 2^{-n}U(H_a, 2^{-n}) = a, \quad \text{a.s.} \tag{14.4}$$

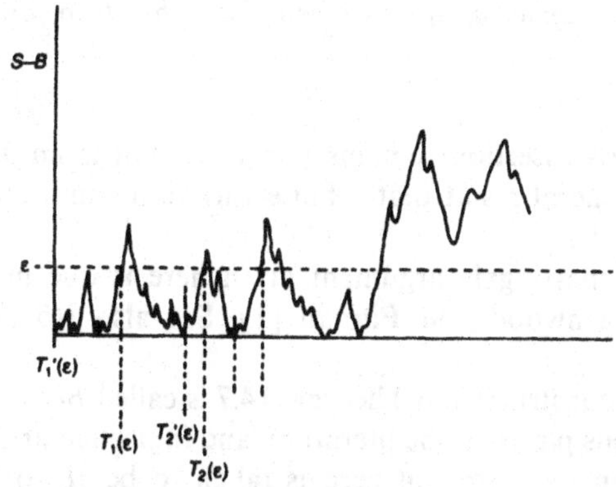

Fig. I.3

Hence we immediately have that

$$\mathbf{P}\left[\lim_{n\to\infty} 2^{-n}U(H_a, 2^{-n}) = a \text{ for all } a \in Q^+\right] = 1,$$

and by the fact that $U(\cdot, \varepsilon)$ is increasing we conclude that

$$(14.5) \qquad \mathbf{P}\left[\lim_{n\to\infty} 2^{-n}U(t, 2^{-n}) = S_t \text{ for all } t \geqslant 0\right] = 1.$$

(For a different martingale argument, see Exercise E79.71c.) To summarize, we can by looking at $S - B$ count the number $U(t, \varepsilon)$ of upcrossings of $[0, \varepsilon]$ by time t, and then, according to (14.5), we can work out S_t simply by the recipe

$$S_t = \lim 2^{-n}U(t, 2^{-n}).$$

But a moment's reflection will show that, since we could reconstruct S just from $S - B$, *we could apply the same construction to the process* $|B|$ (which has the same law as $S - B$) *and thereby obtain some process* l with the properties

(14.6) (i) l is continuous increasing,
(14.6) (ii) l grows only when $|B| = 0$,
(14.6)(iii) $|B_t| - l_t$ is a Brownian motion,

since each of these properties is immediate for the pair $(S - B, S)$, which has the same law as $(|B|, l)$. We have therefore proved the following celebrated result of Lévy.

(*14.7*) THEOREM (Lévy). *There exists a (unique) continuous increasing process l such that $|B_t| - l_t$ is a Brownian motion. The process l grows only when $|B| = 0$, and can be recovered from $|B|$ by the recipe*

$$(14.8) \qquad l_t = \lim_{n\to\infty} 2^{-n}\tilde{U}(t, 2^{-n}),$$

where $\tilde{U}(t, \varepsilon)$ is the number of upcrossings of $[0, \varepsilon]$ by $|B|$ before time t.

(*14.9*) *Remarks*

(i) The uniqueness assertion remains unproved, but is an immediate consequence of a general result that a finite-variation continuous martingale is constant (IV.30.4).
(ii) The reversed martingale argument given here is due in a more general setting to Greenwood and Pitman [1]. See also Itô and McKean [1, Section 2.2].
(iii) The process l constructed in Theorem 14.7 is called *Brownian local time* (at zero). In various places in the literature (and in particular, Itô and McKean [1]), Brownian local time at zero is taken to be $\frac{1}{2}l$, so be careful when moving from one account to another.

(iv) We could evidently repeat the construction (14.8) at other levels by looking at the upcrossings of $[0, 2^{-n}]$ by $|B_t - x|$; this would give us a process $l(t, x)$, the local time at x, defined except on some null set that may depend on x. Now look again at Trotter's Theorem 5.9, which says that, almost surely, $l(\cdot, x)$ can be defined *simultaneously for all* x, and in such a way as to be jointly continuous—a far stronger result.

15. Kolmogorov's test. Suppose $h: \mathbb{R}^+ \to \mathbb{R}$ has the property that $t^{-1/2}h(t)\uparrow$ as $t\downarrow 0$, and let

$$\Lambda := \{B_t \leqslant h(t) \text{ near } 0\} = \bigcup_{\delta>0} \{B_s \leqslant h(s) \text{ for all } s \leqslant \delta\}.$$

Then *Kolmogorov's test* says that $\mathbf{P}(\Lambda) = 0$ or 1 according to whether the integral

$$\int_{0+} t^{-3/2} h(t) \exp\left[-\frac{h(t)^2}{2t}\right] dt \tag{15.1}$$

diverges or converges. The correct way to prove this result is by excursion theory, but it would be premature to discuss this here. The essence of the excursion theory ideas is contained in the fine proof due to Motoo [2]; see also Itô and McKean [1, p. 33]. We omit the proof.

(15.2) Remarks

(i) Since, for each $t > 0$,

$$\Lambda = \bigcup_{\delta \in \mathbb{Q} \cap (0,t]} \bigcap_{s \in \mathbb{Q} \cap (0,\delta]} \{B_s \leqslant h(s)\},$$

we see that $\Lambda \in \mathscr{F}_t^\circ$, and hence $\Lambda \in \mathscr{F}_{0+}^\circ$; by the Blumenthal 0–1 Law, therefore $\mathbf{P}(\Lambda) = 0$ or 1.

(ii) Since $\mathbf{P}[H_a \in dt]/dt = a(2\pi t^3)^{-1/2} \exp(-a^2/2t)$, the integral in (15.1) has a simple interpretation, making Kolmogorov's test easy to remember; the special case $h(t) = a > 0$ for all t will help you to remember that $\mathbf{P}(\Lambda) = 1$ if and only if the integral converges.

(iii) It is a simple consequence of Kolmogorov's test that

$$\mathbf{P}\left[\limsup_{t\downarrow 0} \frac{B_t}{\sqrt{2t \log\log(1/t)}} = 1\right] = 1, \tag{15.3}$$

the celebrated *Law of the Iterated Logarithm* (*LIL*) for Brownian motion. Prove (15.3) now as an exercise using Kolmogorov's test; we shall give a direct proof in the next section.

16. Brownian exponential martingales and the Law of the Iterated Logarithm. The main aim of this section is to give McKean's [1] neat proof of the classical

Law of the Iterated Logarithm (LIL) for Brownian motion. Recall the statement (10.3):

$$\text{(16.1)} \qquad \mathbf{P}\left[\limsup_{t\downarrow 0} \frac{B_t}{\sqrt{2t\log\log(1/t)}} = 1\right] = 1.$$

Because of the time-inversion property (3.5), this is equivalent to the statement

$$\text{(16.2)} \qquad \mathbf{P}\left[\limsup_{t\uparrow \infty} \frac{B_t}{\sqrt{2t\log\log t}} = 1\right] = 1.$$

Proof of (16.1). Write $h(t) = [2t\log\log(1/t)]^{1/2}$. The first part of the proof is to show that $\limsup_{t\downarrow 0} B_t/h(t)$ cannot be bigger than 1. Using the Doob submartingale maximal inequality (see Section II.70) on the exponential martingale $Z_t := \exp(\alpha B_t - \frac{1}{2}\alpha^2 t)$, we obtain

$$\text{(16.3)} \qquad \mathbf{P}\left[\sup_{s\leqslant t}(B_s - \tfrac{1}{2}\alpha s) > \beta\right] = \mathbf{P}\left[\sup_{s\leqslant t} Z_s > e^{\alpha\beta}\right] \leqslant e^{-\alpha\beta}\mathbf{E}Z_t = e^{-\alpha\beta}.$$

Now fix $\theta, \delta \in (0,1)$, and apply (16.3) with $t = \theta^n, \alpha = \theta^{-n}(1+\delta)h(\theta^n)$ and $\beta = \frac{1}{2}h(\theta^n)$. Since $\alpha\beta = \text{constant} + (1+\delta)\log n$,

$$\mathbf{P}\left[\sup_{s\leqslant \theta^n}(B_s - \tfrac{1}{2}\alpha s) > \beta\right] \leqslant \text{constant} \times n^{-(1+\delta)}$$

and so, by the Borel–Cantelli Lemma, it is almost surely true that, for all large enough n,

$$\sup_{s\leqslant \theta^n}[B_s - \tfrac{1}{2}s(1+\delta)\theta^{-n}h(\theta^n)] \leqslant \tfrac{1}{2}h(\theta^n).$$

Thus, for $\theta^{n+1} < t \leqslant \theta^n$, we have

$$B_t \leqslant \sup_{s\leqslant \theta^n} B_s \leqslant \tfrac{1}{2}(2+\delta)h(\theta^n) \leqslant \tfrac{1}{2}\theta^{-1/2}(2+\delta)h(t),$$

so that

$$\limsup_{t\downarrow 0} B_t/h(t) \leqslant \tfrac{1}{2}\theta^{-1/2}(2+\delta), \quad \text{a.s.}$$

Letting $\theta\uparrow 1, \delta\downarrow 0$ through countable sequences, we conclude that

$$\text{(16.4)} \qquad \limsup_{t\downarrow 0} B_t/h(t) \leqslant 1, \quad \text{a.s.}$$

We now turn to the second part of the proof, that $\mathbf{P}[\limsup B_t/h(t) \geqslant 1] = 1$. Once again, choosing $\theta \in (0,1)$, we let A_n be the event

$$A_n = \{B(\theta^n) - B(\theta^{n+1}) > (1-\theta)^{1/2}h(\theta_n)\}.$$

The events A_n are independent, and, since $B(\theta^n) - B(\theta^{n+1})$ has an $N(0, \theta^n(1-\theta))$

law, we have, with $a_n := \theta^{-n/2}h(\theta^n)$,

$$\begin{aligned}\mathbf{P}(A_n) &= \mathbf{P}(B_1 > a_n)\\ &= (2\pi)^{-1/2}\int_{a_n}^{\infty}\exp(-\tfrac{1}{2}y^2)\,dy\\ &\geqslant (2\pi)^{-1/2}\int_{a_n}^{\infty}(1-3y^{-4})\exp(-\tfrac{1}{2}y^2)\,dy\\ &= (2\pi)^{-1/2}\exp(-\tfrac{1}{2}a_n^2)a_n^{-1}(1-a_n^{-2})\\ &:= \gamma_n, \quad \text{say.}\end{aligned}$$

Since

$$\tfrac{1}{2}a_n^2 = \log n + \log\log(\theta^{-1})$$

we conclude that $\sum_n \gamma_n = +\infty$, by comparison with $\sum_n (n\log n)^{-1}$. Thus, almost surely, for infinitely many n,

$$B(\theta^n) - B(\theta^{n+1}) \geqslant (1-\theta)^{1/2}h(\theta^n).$$

But on applying (16.4) to $-B$, we conclude that, for all large enough n,

$$B(\theta^{n+1}) \geqslant -2h(\theta^{n+1}) > -4\theta^{1/2}h(\theta^n).$$

Thus, for all large enough n,

$$B(\theta^n) \geqslant [(1-\theta)^{1/2} - 4\theta^{1/2}]h(\theta^n),$$

and now the result follows by letting $\theta \downarrow 0$. □

(*16.5*) COROLLARY. *Let $Y_1, Y_2, \ldots$ be IID zero-mean random variables with variance* 1, *and let $S_n = \sum_{j=1}^n Y_j$. Then*

$$\mathbf{P}\left[\limsup_{n\to\infty}\frac{S_n}{\sqrt{2n\log\log n}} = 1\right] = 1.$$

Proof. Please review the Skorokhod embedding of Section 7. We are going to use the construction and notation of that section, and, in particular, we shall assume that $(S_n)_{n\geqslant 0} = (B(T_n))_{n\geqslant 0}$; that is, the random walk is embedded in B. By the Strong Law, $n^{-1}T_n \to \mathbf{E}T_1 = 1$, and so, with $h(t) := \sqrt{2t\log\log t}$,

$$\frac{h(T_n)}{h(n)} \to 1, \quad \text{a.s.} \quad (n\to\infty).$$

Trivially, then,

$$\limsup_{n\to\infty}\frac{S_n}{h(n)} = \limsup_{n\to\infty}\frac{B(T_n)}{h(T_n)} \leqslant \limsup_{t\to\infty}\frac{B(t)}{h(t)} = 1.$$

The other inequality needs a little more care, but is still not hard. Note that $\rho^2 := \limsup_{n\to\infty} S_n/h(n)$ is a tail measurable random variable, which is therefore constant a.s. (by the Kolmogorov 0–1 Law.) Suppose that $\rho < 1$; then, for all large enough n,

$$\frac{B(T_n)}{h(T_n)} < \rho. \tag{16.6}$$

This will contradict (16.2) provided we can prove that in the interval $[T_n, T_{n+1}]$ the Brownian motion cannot rise too far. To estimate this, using the explicit form of the embedding, we have, for $x > 0$,

$$\varphi(x) := \mathbf{P}\left[\sup_{t \leqslant T_1} B_t > x\right] = \gamma \int_{-\infty}^{0} F_-(da) \int_x^{\infty} F_+(db)(b-a)\frac{(-a)}{x-a},$$

since $(-a)/(x-a)$ is the probability that the Brownian motion reaches x before a. We aim to prove that, for any $\varepsilon > 0$,

$$\sum_{n \geqslant \varepsilon} \varphi(\varepsilon\sqrt{2n \log\log n}) < \infty, \tag{16.7}$$

which, since φ is clearly decreasing, will follow from

$$\sum_n \varphi(\varepsilon\sqrt{n}) < \infty. \tag{16.8}$$

The point of this is that if we let

$$A_n := \left\{ \sup_{T_n \leqslant u \leqslant T_{n+1}} (B_u - B_{T_n})/h(n) > \varepsilon \right\}$$

then

$$\mathbf{P}(A_n) = \varphi(\varepsilon h(n)),$$

and (16.7) will imply that almost surely only finitely many of the A_n occur. Thus, for all large enough n,

$$\sup_{T_n \leqslant u \leqslant T_{n+1}} \frac{B_u}{h(n)} \leqslant \frac{B(T_n)}{h(n)} + \varepsilon < \rho + \varepsilon,$$

whence for all large enough n

$$\sup_{T_n \leqslant u \leqslant T_{n+1}} B_u/h(u) < \rho + \varepsilon.$$

If we have chosen ε so small that $\rho + \varepsilon < 1$, this contradicts (16.2). The proof will therefore be complete once we have (16.8).

We write

$$\begin{aligned}\varphi(x) &= \gamma \int_{-\infty}^{0} F_-(da) \int_x^{\infty} F_+(db)(b-x)\frac{-a}{x-a} \\ &\quad + \gamma \int_{-\infty}^{0} F_-(da) \int_x^{\infty} F_+(db)(-a) \\ &:= \varphi_1(x) + \varphi_2(x), \quad \text{say.}\end{aligned}$$

Now $\varphi_2(x) = 1 - F(x)$, so

$$\sum_n \varphi_2(\varepsilon\sqrt{n}) < \infty \Leftrightarrow \int^\infty \varphi_2(\varepsilon\sqrt{x})dx < \infty \Leftrightarrow \int^\infty (1 - F(t))t\,dt < \infty;$$

and the last statement is true because F has a finite second moment.

As for φ_1, we have easily

$$\varphi_1(x) \sim x^{-1} \int_x^\infty F_+(db)(b - x)$$

and

$$\sum_n \varphi_1(\varepsilon\sqrt{n}) < \infty \Leftrightarrow \int dx \frac{1}{\varepsilon\sqrt{x}} \int_{\varepsilon\sqrt{x}}^\infty F_+(db)(b - \varepsilon\sqrt{x}) < \infty$$

$$\Leftrightarrow \int dt \int_t^\infty F_+(db)(b - t) < \infty$$

$$\Leftrightarrow \int^\infty b^2 F_+(db) < \infty;$$

and this last statement is true, again by the finiteness of the second moment of F. □

(*16.9*) *Remarks.* The Law of the Iterated Logarithm is just one aspect of a much bigger picture, and is a simple consequence of the following much deeper result.

(*16.10*) THEOREM (Strassen). *For* $n \geqslant 3$, *define*

$$X^{(n)}(t) := B(nt)(2n \log\log n)^{-1/2}, \qquad 0 \leqslant t \leqslant 1,$$

a random element of $C([0, 1])$. *With probability* 1, *the set of limit points of* $(X^{(n)})_{n \geqslant 3}$ *is the set*

$$K := \left\{ f \in C[0, 1] : f \text{ is absolutely continuous, } \int_0^1 f'(t)^2\,dt \leqslant 1 \right\}.$$

See Strassen [1] or Freedman [1] for a proof. You will not be surprised to learn that if (S_n) is a zero-mean unit-variance random walk, and

$$S_n(t) := (nt - j)S_{j+1} + (j + 1 - nt)S_j \quad (j/n \leqslant t \leqslant (j + 1)/n)$$

is the piecewise-linear interpolation of (S_n), then almost surely the set of limit points of $((2n \log\log n)^{-1/2} S_n(\cdot))_{n \geqslant 3}$ is K. This is the Strassen Invariance Principle. You will also not be surprised to learn that one deduces the Strassen invariance principle from Theorem 16.10 in the same way that we deduced the classical LIL, Corollary 16.5, from (16.2), namely by Skorokhod embedding of the random walk in Brownian motion.

The proof of Theorem 16.10 that you will find in either of the above references is specific to the Brownian situation, of course. But there is a sense in which this result can be seen as part of the much more general theory of *large deviations*, pioneered by Cramér, Schilder, Donsker and Varadhan, Ventcel and Freidlin, and developed further by Donsker, Varadhan, Stroock, among others. The excellent account by Deuschel and Stroock [1] contains a proof of Theorem 16.10 from a large-deviations point of view, and is delightfully clear.

3. BROWNIAN MOTION IN HIGHER DIMENSIONS

17. Some martingales for Brownian motion. By Brownian motion in $\mathbb{R}^d$ we mean a process $B_t := (B_t^1, \ldots, B_t^d)$ where each of the $(B_t^j)_{t \geq 0}$ $(j = 1, \ldots, d)$ is a Brownian motion, independent of all the others. To study Brownian motion in $\mathbb{R}^d$, we are going to need martingales, and the purpose of this section is to derive a result that gives us all the martingales we shall need. This result can be seen as a special case of general results in Markov process theory or in stochastic calculus, but we shall prove it here using the special structure of Brownian motion, since we do not yet have the general results.

(*17.1*) THEOREM. *Suppose that* $f: \mathbb{R}^+ \times \mathbb{R}^d \to \mathbb{R}$ is $C^{1,2}$, *and that there exists a constant K such that, for all* $t \geq 0$, $x \in \mathbb{R}^d$,

$$(17.2) \quad |f(t,x)| + \left|\frac{\partial f}{\partial t}(t,x)\right| + \sum_{j=1}^d \left|\frac{\partial f}{\partial x_j}(t,x)\right| + \sum_{i=1}^d \sum_{j=1}^d \left|\frac{\partial^2 f}{\partial x_i \partial x_j}(t,x)\right| \leq K e^{K(t+|x|)}$$

Then the process

$$(17.3) \qquad C_t^f := f(t,B_t) - f(0,B_0) - \int_0^t \mathscr{G} f(s,B_s)\,ds \quad \textit{is a martingale,}$$

where

$$(17.4) \qquad \mathscr{G} f(t,x) := \left(\frac{\partial f}{\partial t} + \frac{1}{2}\sum_{j=1}^d \frac{\partial^2 f}{\partial x_j^2}\right)(t,x).$$

Remarks. The class $C^{1,2}$ is, of course, the class of functions $f(t,x)$ with continuous partial derivatives of all orders up to 1 in t and up to 2 in x. The exponential growth condition (17.2) will be seen to be unnecessary provided we relax the statement (17.3) to say that C^f is a *local* martingale. We shall not digress to define this now. In dimension $d = 1$, the only functions of x for which $f(B_t)$ is a martingale are the linear functions, but in dimension $d \geq 2$ we shall see that there is a very rich family of f for which $f(B_t)$ is a martingale.

Proof. We must prove that, for $0 \leq s \leq t$,

$$\mathbf{E}[C_t^f - C_s^f \mid \mathscr{F}_s] = 0,$$

for which, by the independent-increments property of B, it will suffice to prove that, for any $x \in \mathbb{R}^d$ and $t \geqslant 0$

$$\mathbf{E}^x[C_t^f] = 0, \tag{17.5}$$

where $\mathbf{P}^x$ is the law of Brownian motion started at x. Without loss of generality, we can take $x = 0$ (and write $\mathbf{P}$ for $\mathbf{P}^0$), and we shall prove that, for $0 < \varepsilon < t$,

$$\mathbf{E}[C_t^f - C_\varepsilon^f] = 0. \tag{17.6}$$

Using the assumption (17.2), the fact that $\mathbf{P}[\sup_{u \leqslant t} |B_u| \geqslant a] \leqslant c\mathbf{P}[|B_1| \geqslant a/\sqrt{t}]$ (see (13.4)), and dominated convergence, (17.6) implies (17.5).

Letting $p_t(x) := (2\pi t)^{-d/2} \exp(-|x|^2/2t)$ denote the d-dimensional Brownian transition density, we observe that, for $t > 0$, $x \in \mathbb{R}^d$,

$$\frac{\partial p_t}{\partial t}(x) = \frac{1}{2} \sum_{j=1}^{d} \frac{\partial^2 p_t}{\partial x_j^2}(x). \tag{17.7}$$

Hence

$$\begin{aligned}\mathbf{E}[C_t^f - C_\varepsilon^f] &= E\left[f(t, B_t) - f(\varepsilon, B_\varepsilon) - \int_\varepsilon^t \mathcal{G} f(s, B_s)\, ds \right] \\ &= \int [p_t(x) f(t, x) - p_\varepsilon(x) f(\varepsilon, x)]\, dx \\ &\quad - \int_\varepsilon^t ds \int p_s(x) \left[\frac{\partial f}{\partial t}(s, x) + \tfrac{1}{2} \Delta f(s, x) \right] dx.\end{aligned}$$

But

$$\begin{aligned}\int p_s(x) \tfrac{1}{2} \Delta f(s, x)\, dx &= \int \tfrac{1}{2} \Delta p_s(x) f(s, x)\, dx, \\ &\qquad \text{(integrating twice by parts and using (17.2))} \\ &= \int \frac{\partial p_s}{\partial t}(x) f(s, x)\, dx,\end{aligned}$$

using (17.7). Thus

$$\begin{aligned}\mathbf{E}[C_t^f - C_\varepsilon^f] &= \int [p_t(x) f(t, x) - p_\varepsilon(x) f(\varepsilon, x)]\, dx \\ &\quad - \int_\varepsilon^t ds \int \left[p_s(x) \frac{\partial f}{\partial t}(s, x) + f(s, x) \frac{\partial p_s}{\partial t}(x) \right] dx \\ &= \int [p_t(x) f(t, x) - p_\varepsilon(x) f(\varepsilon, x)]\, dx - \int_\varepsilon^t ds \int \frac{\partial}{\partial t}(p_s(x) f(s, x))\, dx \\ &= \int [p_t(x) f(t, x) - p_\varepsilon(x) f(\varepsilon, x)]\, dx - \int \left\{ \int_\varepsilon^t ds \frac{\partial}{\partial t} [p_s(x) f(s, x)] \right\} dx \\ &= 0.\end{aligned}$$

□

Note that the interchange of the s-integral and the x-integral in the penultimate line B is only justified because we have ensured $s \geqslant \varepsilon > 0$ and so $p_s(x)$ and $\partial p_s(x)/\partial t$ are bounded; this is why we had to proceed to the natural (17.5) via the slightly clumsy (17.6).

18. Recurrence and transience in higher dimensions. As a first application of Theorem 17.1, we shall show in this section that Brownian motion is recurrent in dimension $d=2$, and transient in dimension $d \geqslant 3$. If B is Brownian motion in $\mathbb{R}^d$, let

$$H_a := \inf\{t > 0 : |B_t| = a\}.$$

(*18.1*) THEOREM. *For* $0 < a < |x| < b$,

$$\mathbf{P}^x(H_a < H_b) = \begin{cases} \dfrac{\log b - \log|x|}{\log b - \log a} & (d=2), \\ \dfrac{|x|^{2-d} - b^{2-d}}{a^{2-d} - b^{2-d}} & (d \geqslant 3). \end{cases} \tag{18.2}$$

Proof. Let $f: \mathbb{R}^d \to \mathbb{R}$ be a C_K^∞ function such that, for $a \leqslant |x| \leqslant b$,

$$f(x) = \begin{cases} \log|x| & (d=2), \\ |x|^{2-d} & (d \geqslant 3). \end{cases}$$

Then f satisfies the conditions of Theorem 17.1, and moreover,

$$\Delta f(x) = 0 \quad \text{for } a \leqslant |x| \leqslant b,$$

as is readily verified. So, using the Optional Sampling Theorem on the martingale C^f at the stopping time $\tau := H_a \wedge H_b$ yields, for $|x| \in (a,b)$,

$$\begin{aligned} 0 = \mathbf{E}^x[C_0^f] &= \mathbf{E}^x[C_\tau^f] \\ &= \mathbf{E}^x[f(B_{H_a}) : H_a < H_b] + \mathbf{E}^x[f(B_{H_b}) : H_b < H_a] - f(x), \end{aligned}$$

since $\tau \leqslant H_b < \infty$, a.s. (one dimensional Brownian motion certainly leaves $[-b, b]$ in finite time!). So, in the case $d=2$,

$$f(x) = \log|x| = \mathbf{P}^x[H_a < H_b] \log a + \mathbf{P}^x[H_b < H_a] \log b,$$

and rearrangement gives (18.2), with analogous reasoning for the case $d \geqslant 3$.

□

(*18.3*) COROLLARY. *Brownian motion in dimension* $d=2$ *is recurrent, and in dimension* $d \geqslant 3$ *it is transient; more precisely, for any* $0 < a < |x|$,

$$\mathbf{P}^x[H_a < \infty] = \begin{cases} 1 & (d=2) \\ (a/|x|)^{d-2} & (d \geqslant 3). \end{cases} \tag{18.4}$$

Proof. Since $\{H_a < \infty\} = \bigcup_n \{H_a < H_n\}$, (18.4) follows immediately from (18.2) by letting $b \uparrow \infty$.

(*18.5*) *Remarks.* It follows immediately that Brownian motion in the plane visits every non-empty open set U with probability 1, and, in fact, keeps returning to U: there is no last visit to U. In dimension greater than 2, any bounded set will ultimately be left for ever. Thus, in dimension $d \geqslant 3$, Brownian motion is transient. In dimension 2, Brownian motion is recurrent—or, more accurately, *neighbourhood*-recurrent, since Brownian motion in the plane does not hit points, as we now show.

(*18.6*) COROLLARY. *For Brownian motion in* $\mathbb{R}^2$, $|x| > 0$,

$$\mathbf{P}^x[H_0 < \infty] = 0.$$

Proof.

$$\begin{aligned}\{H_0 < \infty\} &= \bigcup_n \{H_0 \leqslant H_n\}\\ &= \bigcup_n \Big(\bigcap_{m > |x|^{-1}} \{H_{1/m} \leqslant H_n\} \Big)\end{aligned}$$

and

$$\begin{aligned}\mathbf{P}^x\Big[\bigcap_{m > |x|^{-1}} \{H_{1/m} \leqslant H_n\} \Big] &= \downarrow \lim_{m \to \infty} \mathbf{P}^x[H_{1/m} \leqslant H_n]\\ &= 0,\end{aligned}$$

using (18.2). □

(*18.7*) *Remarks.* We are seeing here the first signs of the important result that if $(B_t)_{t \geqslant 0}$ is Brownian motion in $\mathbb{R}^d$ $(d \geqslant 2)$ then $(|B_t|)_{t \geqslant 0}$ *is a diffusion process, called a d-dimensional Bessel process* (denoted BES(d)). We postpone discussion of this until we have a better idea of what a diffusion process is; see Section V.48 for a detailed analysis and Section VI.52 for the celebrated Ray–Knight Theorem, where Bessel processes enter in a wholly natural way to describe the diffusion property in the spatial parameter of Brownian local time $(l(\tau, x)_{x \geqslant 0})$ taken at some suitable stopping time τ.

The Bessel processes are the most important one-dimensional diffusions apart from Brownian motion; Pitman and Yor [1] provide a detailed study of many of their properties. See also Revuz and Yor [1].

The fact that the Bessel process is a diffusion in its own right is due to the fact that Brownian motion in $\mathbb{R}^d$ is *rotation-invariant*; if $R \in \mathrm{O}(d)$ is fixed then $(RB_t)_{t \geqslant 0}$ is a Brownian motion too. The proof is trivial. Lemma 14.1 now allows us to conclude that $|B_t|$ is a Markov process.

19. Some applications of Brownian motion to complex analysis. One of the richest areas of application of Brownian motion is complex analysis. The fundamental observation is that if f is an analytic function in some domain D then each of the functions $\mathrm{Re}\, f$ and $\mathrm{Im} f$ satisfies Laplace's equation $\Delta u = 0$ in D. Thus, in view of Theorem 17.1, $\mathrm{Re}\, f(Z_t)$ and $\mathrm{Im}\, f(Z_t)$ are local martingales, where $Z_t := X_t + iY_t$ is complex Brownian motion.

(19.1) *Remark.* Even if f were defined on the whole of $\mathbb{C}$, there would be no reason for $\mathrm{Re}\, f$ to satisfy the growth condition (17.2) (as an example, take $f(z) = \exp(z^3)$). This is not really a problem, because if we take an open $D_0 \subset D$, with $\bar{D}_0$ compact, and take $\tilde{f}$ to be C_K^∞, equal to f in D_0, then certainly $\mathrm{Re}\, \tilde{f}$ satisfies (17.2), and so $\mathrm{Re}\, f(Z_{t\wedge\tau})$ is a martingale, $\tau := \inf\{t : Z_t \notin D_0\}$. In all the applications we make to complex analysis, this sort of localisation will often be implicit, and we shall not dwell on the details.

(19.2) PROPOSITION (Maximum Modulus Theorem) *Suppose that $f: D \to \mathbb{C}$ is analytic, and that $D \supseteq \bar{D}_r := \{z \in \mathbb{C} : |z| \leqslant r\}$. Then*

$$\max\{|f(z)| : z \in \bar{D}_r\} = \max\{|f(z)| : |z| = r\}. \tag{19.3}$$

Proof. If f were constant, there would be nothing to prove; so suppose f is not constant, and take $z_0 \in \bar{D}_r$ such that

$$|f(z_0)| = \max\{|f(z)| : z \in \bar{D}_r\}.$$

Since f is bounded on $\bar{D}_r$, we may add $\lambda f(z_0)$ to f and assume (by taking $\lambda > 0$ large enough) that $f(\bar{D}_r)$ is contained in a half-space distant at least 1 from 0. Thus $\log f$ is a well-defined analytic function on D_r, continuous on $\bar{D}_r$. Hence $h := \mathrm{Re} \log f \equiv \log |f|$ is harmonic in D_r and continuous in $\bar{D}_r$. The Maximum Modulus Principle will follow once we show the stronger result:

(19.4) *if $U \subseteq \mathbb{C}$ is open, and $h: U \to \mathbb{R}$ is harmonic then h has no local maximum in U; if $z_0 \in D(z_0, \varepsilon) := \{z : |z - z_0| < \varepsilon\} \subset U$ and $h(z_0) \geqslant h(z)$ for all $z \in D(z_0, \varepsilon)$ then $h(z_0) = h(z)$ for all $z \in D(z_0, \varepsilon)$.*

The proof of this is a trivial application of the Optional Sampling Theorem. If $\tau := \inf\{t : |Z_t - z_0| = \delta\}$, where $Z_0 = z_0$ then, since $h(Z_t, \tau)$ is a martingale (Theorem 17.1), we have

$$\begin{aligned} h(z_0) &= \mathbf{E} h(Z_\tau) \\ &= \int_0^{2\pi} \frac{d\theta}{2\pi} h(z_0 + \delta e^{i\theta}), \end{aligned}$$

since, by symmetry, the exit distribution from $D(z_0, \delta)$ is uniform on the boundary. But, since h is continuous and $h(z_0 + \delta e^{i\theta}) \leqslant h(z_0)$ for $\delta < \varepsilon$, it must be that $h(z_0 + \delta e^{i\theta}) = h(z_0)$ for $\delta < \varepsilon$, establishing (19.4).

Thus if $|f|$ had a local maximum inside D_r, there would be a disc where

$\log|f|$ was constant, implying that $\log f$ (and therefore f) was constant in the disc, and hence throughout D. □

(*19.5*) PROPOSITION (Fundamental Theorem of Algebra). *Suppose that f: $\mathbb{C} \to \mathbb{C}$ is a non-constant polynomial. Then there exists $z_0 \in \mathbb{C}$ such that $f(z_0) = 0$.*

Proof. Suppose the contrary, that f is non-vanishing in $\mathbb{C}$. Then $g := 1/f$ is a well-defined analytic function, tending to zero at infinity (since f is a polynomial) and therefore bounded. Since f is non-constant, we may find disjoint discs D_1 and D_2 and $\alpha < \beta$ such that

$$\operatorname{Re} g(z_1) \leqslant \alpha < \beta \leqslant \operatorname{Re} g(z_2), \qquad z_i \in D_i.$$

Let Z be Brownian motion in $\mathbb{C}$, and consider the bounded martingale $\operatorname{Re} g(Z_t)$. The Martingale Convergence Theorem says that this is almost surely convergent, yet, by Corollary 18.3, the process Z keeps visiting D_1 and D_2; there is no last visit to D_1 or D_2. Hence $\liminf \operatorname{Re} g(Z_t) \leqslant \alpha < \beta \leqslant \limsup \operatorname{Re} g(Z_t)$, a contradiction. □

So far, we have not really been using the full strength of the Brownian-motion/complex-analysis combination; we have only applied *harmonic* functions to Brownian motion, and an *analytic* function comprises two very intimately related harmonic functions. See the theory of conformal martingales (Getoor and Sharpe [5]) and Chapter 4. The connection with Brownian motion goes much deeper, as the following result shows.

(*19.6*) THEOREM. *Let $f: D \to \tilde{D}$ be analytic, and let Z be Brownian motion in D. Then there is a Brownian motion $\tilde{Z}$ in $\tilde{D}$ such that*

$$f(Z_t) = \tilde{Z}\left(\int_0^t |f'(Z_u)|^2\, du\right).$$

Remarks. A Brownian motion in a domain $D \subseteq \mathbb{C}$ is only defined up until the first exit time from D. The importance of (19.7) is that *the image of Brownian motion under an analytic map is another Brownian motion* (to within a time change). See Section IV.34 for a proof. A common and powerful use of this is that the exit distribution for one domain gets mapped to the exit distribution of some other domain by an analytic map. Let us see an example of this.

(*19.7*) PROPOSITION (Poisson integral formula). *Let $\mathbf{H} = \{z \in \mathbb{C} : \operatorname{Im} z > 0\}$, $(Z_t)_{t \geq 0}$ be Brownian motion started at $\gamma = a + ib \in \mathbf{H}$, and $\tau := \inf\{u : Z_u \notin \mathbf{H}\}$. Then*

(19.8) $$\mathbf{P}[\operatorname{Re} Z_\tau \in dx] = \frac{b\, dx}{\pi[b^2 + (x-a)^2]} = \operatorname{Im}\left(\frac{1}{x-\gamma}\right)\frac{dx}{\pi}.$$

First proof. The stopping time τ is simply the first time that $Y := \operatorname{Im} Z$ hits zero. But this time has a density that we know (13.5). Meanwhile, $X = \operatorname{Re} Z$ is moving like an independent Brownian motion, so

$$\mathbf{P}[X_\tau \in dx]/dx = \int_0^\infty \frac{be^{-b^2/2t}}{\sqrt{2\pi t^3}} \frac{e^{-(x-a)^2/2t}}{\sqrt{2\pi t}} dt$$

$$= \frac{b}{\pi[b^2 + (x-a)^2]}. \qquad \square$$

Second proof. The map

$$z \mapsto w(z) := \frac{z - \gamma}{z - \bar{\gamma}}$$

maps $\mathbb{H}$ one–one onto $\mathbb{D} = \{w : |w| < 1\}$, and takes γ to 0. Thus $\{w(Z_t) : t < \tau\}$ is a time transformation of Brownian motion on $\mathbb{C}$ started at 0 and run until it exits the disc $\mathbb{D}$. Thus, for $x \in \mathbb{R}$,

$$\mathbf{P}(Z_\tau \in dx) = \frac{1}{2\pi} |w'(x)| dx,$$

which agrees with (19.8). $\square$

By using the Riemann Mapping Theorem, we can find the exit distribution from any connected simply connected domain by transforming the problem to one for Brownian motion in $\mathbb{D}$ started at 0. Suppose, for example, that we want the exit distribution from $\mathbb{D}$ for Brownian motion started at ζ in $\mathbb{D}$. The map

$$z \mapsto g(z) := \frac{f(z) - f(\gamma)}{f(z) - \overline{f(\gamma)}}, \quad \text{where } f(z) := i\frac{1+z}{1-z} \in \mathbb{H},$$

maps $\mathbb{D}$ one–one onto $\mathbb{D}$, taking γ to 0. Thus, if $z = e^{i\theta}$,

(19.9) $$\mathbf{P}^\zeta(\text{Brownian motion exists } \mathbb{D} \text{ in } d\theta) = \frac{1}{2\pi} |g'(z)| \, d\theta = \frac{1 - |\zeta|^2}{|z - \zeta|^n} \mu_n(d\theta)$$

where $n = 2$ and μ_2 is the normalized Lebesgue measure on $\partial\mathbb{D}$.

Of course, complex analysts are very familiar with the use of the Riemann Mapping Theorem to obtain exit distributions (or Poisson kernels or harmonic measure). Like them, we have to be sensible in a case such as that in which the domain is $\mathbb{D}\backslash(-1, 0]$ and has the wrong topology, the upper and lower parts of the cut needing to be separated.

Martin boundary theory will clarify the matter of correct topologies, show that (19.9) holds in all dimensions (check it when $n = 1$ now!), and prove that every positive harmonic function h on $\mathbb{D}_n := \{x \in \mathbb{R}^n : |x| < 1\}$ with $h(0) = 1$ has a unique representation

$$h(\zeta) = \int_{\partial \mathbf{D}_n} \frac{1-|\zeta|^2}{|z-\zeta|^n} \nu(dz),$$

where ν is a probability measure on $\partial\mathbf{D}_n$. When $h = 1$ on $\mathbf{D}_n$, γ is the normalized Lebesgue measure on $\partial\mathbf{D}_n$.

20. Windings of planar Brownian motion. Let $(Z_t)_{t\geq 0}$ be complex Brownian motion, $Z_0 \neq 0$. Then there is a continuous determination of $\theta_t := \arg(Z_t)$, and a unique one with $\theta_0 \in [0, 2\pi)$. The angle θ_t keeps tracks of the winding of Brownian motion about 0 up to time t. The earliest result on the windings of Brownian motion is the following remarkable theorem of Spitzer [2].

(*20.1*) THEOREM (Spitzer). *Suppose that* $Z_0 = 1$. *Then*

(20.2)
$$\frac{2\theta_t}{\log t} \xrightarrow{\mathscr{D}} C_1 \quad (t \to \infty),$$

where C_1 *is the standard Cauchy law with density* $[\pi(1+x^2)]^{-1}$.

Proof. Let $\mathbf{P}^z$ denote the law of complex Brownian motion started at $z \in \mathbb{C}$. For $x > 0$, and Brownian motion Z started at 1, $\tilde{Z}_t := xZ_{t/x^2}$ is Brownian motion started at x, and $\tilde{\theta}_t = \theta_{t/x^2}$. Thus an equivalent statement to (20.2) is

(20.3) *the* $\mathbf{P}^x$-*law of* $(\log 1/x)^{-1}\theta_1$ *converges to* C_1 *as* $x \downarrow 0$,

and it is this that we prove. The idea of the proof is to fix some very small disc of radius a about the origin, which (with high probability) Brownian motion will leave before time 1. As $x \downarrow 0$, the contribution to θ_1 that comes after hitting the circle of radius a is negligible, so we want the limiting behaviour of θ_T, where $T = \inf\{t : |Z_t| = a\}$. But, by Theorem 19.6, we could realize the Brownian motion Z started at $x > 0$ as a time change of $\exp(\zeta)$, where ζ is complex Brownian motion started at $\log x$. In particular, the argument of Z_T is equal to the imaginary part of ζ where ζ first hits the line $\operatorname{Re} z = \log a$, and the law of this is Cauchy with parameter $\log(a/x)$, from Proposition 19.7—this is where the Cauchy distribution comes from.

Now we implement this sketched proof. Suppose given $\varepsilon > 0$ and some bounded uniformly continuous test function $f : \mathbf{R} \to [0, 1]$. Choose $\delta > 0$ so small that $|x - y| \leq \delta \Rightarrow |f(x) - f(y)| \leq \frac{1}{4}\varepsilon$. Now fix $a \in (0, 1)$ so small that, for all $|z| \leq a$,

$$\mathbf{P}^z[|Z_t| \leq a \text{ for all } t \leq 1] \leq \tfrac{1}{4}\varepsilon.$$

As we explained above, for all x in $(0, a)$,

$$\text{the } \mathbf{P}^x\text{-law of } \frac{\theta_T}{\log(a/x)} \equiv \frac{\arg(Z_T)}{\log(a/x)} \text{ is } C_1.$$

Now pick K so large that

$$\mathbf{P}^a[|\theta_t| \geqslant K \text{ for some } t \leqslant 1] \leqslant \tfrac{1}{4}\varepsilon,$$

and $x_0 > 0$ so small that $K \leqslant \delta \log(1/x_0)$ and, for $x \leqslant x_0$,

$$\int_{-\infty}^{\infty} \left| f(v) - f\left(v \frac{\log(a/x)}{(\log(1/x))}\right)\right| C_1(dv) \leqslant \tfrac{1}{4}\varepsilon.$$

Then, for $x \leqslant x_0$,

$$\begin{aligned}
&\mathbf{E}^x \left| f\left(\frac{\theta_T}{\log(a/x)}\right) - f\left(\frac{\theta_1}{\log(1/x)}\right)\right| \\
&\leqslant \mathbf{E}^x \left| f\left(\frac{\theta_T}{\log(a/x)}\right) - f\left(\frac{\theta_T}{\log(1/x)}\right)\right| + \mathbf{E}^x \left| f\left(\frac{\theta_T}{\log(1/x)}\right) - f\left(\frac{\theta_1}{\log(1/x)}\right)\right| \\
&\leqslant \tfrac{1}{4}\varepsilon + \mathbf{P}^x[T \geqslant 1] + \mathbf{P}^x[|\theta_t - \theta_T| > K \text{ for some } T \leqslant t \leqslant 1, T \leqslant 1] \\
&\quad + \mathbf{E}^x\left[\left| f\left(\frac{\theta_T}{\log(1/x)}\right) - f\left(\frac{\theta_1}{\log(1/x)}\right)\right| : T \leqslant 1, |\theta_t - \theta_T| \leqslant K \quad \text{for } T \leqslant t \leqslant 1\right] \\
&\leqslant \varepsilon.
\end{aligned}$$

□

(20.4) Remarks

(i) The key to this simple proof of Spitzer's theorem is the Brownian Mapping Theorem 19.6. Spitzer's original proof was based on complicated calculations (see Itô and McKean [1, pp. 270–271]), but several proofs based on the Brownian Mapping Theorem have since been given; see Durrett [1] and Messulam and Yor [1].

(ii) If one takes Brownian motion in $\{z:|z| \geqslant 1\}$ reflected in the unit circle then the above argument goes through with the obvious changes to show that if Z starts at 1 then

(20.5) $$\frac{2\theta_t}{\log t} \xrightarrow{\mathscr{D}} S \quad (t \to \infty),$$

where S is the distribution with density $(2\pi \cosh y)^{-1}$ (remarkably, the Fourier transform of this distribution is sech θ: see Feller [1, Vol. 2, p. 503] who remarks that the distribution is 'of no importance'!) The result (20.5) was pointed out to us by Kalvis Jansons. The limit law S has all moments, in contrast to C_1, which is evidence that *windings that happen near zero make a large contribution.*

(iii) The study of Brownian windings has been carried a very long way in recent years by Pitman and Yor [3, 4] and Le Gall and Yor [1, 2],

As a sample of the kinds of results achieved, we give the asymptotic joint distribution of the windings about n points.

(20.6) THEOREM (Pitman and Yor [3]). *Let $z_1, \ldots, z_n$ be distinct points of $\mathbb{C}$,*

θ_t^j the winding of Z about z_j by time t. Then

$$\frac{2}{\log t}(\theta_t^1,\ldots,\theta_t^n) \xrightarrow{\mathscr{D}} (V_1,\ldots,V_n) \tag{20.7}$$

where $V_j = U + HY_j$, the variables Y_j are independent standard Cauchy independent of the pair (U,H), which have joint distribution characterised by

$$\mathbf{E}\exp(-\alpha H + ivU) = \left(\cosh v + \frac{\alpha \sinh v}{v}\right)^{-1} \tag{20.8}$$

for $\alpha \geqslant 0$, $v \in \mathbb{R}$.

(iv) For a proof using Brownian windings of Picard's Little Theorem (if $f:\mathbb{C}\to\mathbb{C}$ is analytic and non-constant then the range of f can omit at most one value), see Davis [1] and also Durrett [1]. For a proof of Picard's Great Theorem, see Davis [2].

21. Multiple points, cone points, cut points. In this section we give without proof a number of fascinating and beautiful results that illuminate the behaviour of the Brownian path, and provide a few references to an area in a bewildering state of development.

(*21.1*) THEOREM (Dvoretsky–Erdös–Kakutani [1–3]; Dvoretsky–Erdös–Kakutani–Taylor [1])

(*a*) *Brownian motion in two dimensions has points of all multiplicities $2, 3, \ldots, c$, where c denotes the multiplicity of the continuum.*
(*b*) *Brownian motion in three dimensions has double points but no triple points.*
(*c*) *Brownian motion in dimension greater than three has no double points.*

(*21.2*) *Remarks.* Numerous proofs of all or part of Theorem 21.1 have appeared since the first ones, many of them valid for more general Lévy processes; see Hawkes [1], Evans [1], Le Gall, Rosen and Shieh [1] and Rogers [5] for a sample. The existence of multiple points of Brownian motion has been but a small part of a much more profound study of the existence and properties of *intersection local time*, carried out by Dynkin, Le Gall, Rosen, Wolpert, Yor and others; we refer the interested reader to some of the papers cited in the bibliography for more information on these topics. Applications include questions related to quantum field theory (Dynkin [5, 7, 8, 9]), the asymptotics of the 'Wiener sausage' (Le Gall [4]) and the exact Hausdorff measure of the set of Brownian multiple points (Le Gall [5]).

For $\alpha \in (0, \pi)$ let C_α denote the wedge $\{re^{i\theta} : r \geqslant 0, |\theta| \leqslant \alpha\}$ of angle 2α. If Z is Brownian motion in $\mathbb{C}$ then a time t such that

$$Z_u \in Z_t - C_\alpha \quad \text{for all } u \leqslant t$$

is called a *cone time*, the set of all such being denoted H_α. A cone time t is a time at which the path-so-far lies in the shifted cone $-C_\alpha$ with vertex at the current position, Z_t. The position Z_t is then called a *cone point*. When do cone points exist? The following result of Burdzy [1] and Shimura [1] answers this.

(*21.3*) THEOREM (Burdzy [1]; Shimura [1]). *Cone points exist if and only if* $\alpha > \frac{1}{4}\pi$.

(*21.4*) *Remarks.* Le Gall [6] shows how this result follows easily from the criterion of Varadhan and Williams [1] for a reflecting Brownian motion in a wedge to hit the corner. Evans [2] and Le Gall [6] study the Hausdorff dimension of the set of Brownian cone points and the construction and properties of a local time on the set H_α of cone times.

And lastly in this rushed survey of interesting Brownian motion properties, we give the fine result of Burdzy [2] on cut points of the two-dimensional Brownian path.

(*21.5*) THEOREM (Burdzy [2]). *Let Z be Brownian motion in $\mathbb{C}$. Then almost surely there exists $t \in (0, 1)$ such that*

$$\{Z_s : 0 \leqslant s < t\} \cap \{Z_s : t < s \leqslant 1\} = \varnothing.$$

22. Potential theory of Brownian motion in $\mathbb{R}^d (d \geqslant 3)$. This brief section can do no more than provide some heuristic sketches of an immense topic; the books of Blumenthal and Getoor [1], Dellacherie and Meyer [1], Helms [1], Meyer [2] and Port and Stone [3] are just a few of the many written on this subject. We shall later provide proofs of some of the results listed here.

Basically, any transient Markov process has a potential theory, but we shall discuss here only Brownian motion in at least three dimensions. The key concept of potential theory is the *Green kernel* G defined by

$$Gf(x) = \int_0^\infty P_t f(x)\, dt, \quad f \in C_b(\mathbb{R}^d), \tag{22.1}$$

where $(P_t)_{t \geqslant 0}$ is the transition semigroup of our Markov process. For $BM(\mathbb{R}^d)$, we have

$$Gf(x) = \int g(x, y) f(y)\, dy, \tag{22.2}$$

where

$$g(x, y) = \int_0^\infty p_t(x, y)\, dt \tag{23.3}$$

$$= \int_0^\infty (2\pi t)^{-d/2} \exp\left(-\frac{|x-y|^2}{2t}\right) dt$$

$$= \tfrac{1}{2}\pi^{-d/2}|y-x|^{2-d}\int_0^\infty e^{-u}u^{-2+d/2}\,du$$

$$= \tfrac{1}{2}\pi^{-d/2}|y-x|^{2-d}\Gamma(\tfrac{1}{2}d-1).$$

(Note that our Green function is based on the probabilists' normalisation $\frac{1}{2}\Delta$, rather than the physicists' Δ.) The same calculation in dimension 1 or 2 gives an infinite answer, because Brownian motion in those dimensions is recurrent. The probabilistic interpretation

$$\text{(22.4)}\qquad GI_A(x) = \mathbf{E}^x\left[\int_0^\infty I_A(B_t)\,dt\right]$$

$$= \mathbf{E}^x[\text{time spent in } A \text{ by } B.]$$

is a useful one to bear in mind. A heuristic but suggestive calculation gives from the notion (4.4) of the generator $\mathscr{G} = \frac{1}{2}\Delta$ as the derivative of the semigroup that

$$\text{(22.5)}\qquad \dot{P}_t = \tfrac{1}{2}\Delta P_t \Rightarrow P_t = \exp(\tfrac{1}{2}t\Delta)$$

$$\Rightarrow \int_0^\infty P_t\,dt := G = (-\tfrac{1}{2}\Delta)^{-1},$$

and thinking of the Green kernel as the inverse of $-\mathscr{G}$ will never lead you astray, though usually a proof must be sought elsewhere.

(22.6) *Exercise.* For $f \in C_K^\infty(\mathbb{R}^d)$, prove that

$$\tfrac{1}{2}\Delta Gf = G(\tfrac{1}{2}\Delta f) = -f.$$

Hint. Use analysis to prove the first equality, then (22.4) and Theorem 17.1 for the second.

Fix now some pathwise-connected compact subset $K \subseteq \mathbb{R}^d$, which we think of in physical terms as a conducting body. A classical problem of electrostatics is to determine the *equilibrium charge distribution* for K, and the *equilibrium potential*; if a charge is placed on K, then the charge will flow in K very rapidly so as to equate the electrostatic potential everywhere within K. (If the potential were not constant, charge would flow between regions where it differed until everything was evened out.) This equilibrium charge distribution will minimise the *energy* of the charge, and the potential associated with it is called the *equilibrium potential.*

How are these physical concepts related to the probabilistic ones? The following theorems do not require connectedness of K.

(22.7) THEOREM (Hunt). *The function $P_K 1$ on $\mathbb{R}^d$ defined by*

$$\text{(22.8)}\qquad P_K 1(x) := \mathbf{P}^x(B_t \in K \text{ for some } t > 0)$$

is expressible as the potential

$$P_K 1 = G\mu_K \equiv \int g(\cdot, y)\mu_K(dy) \tag{22.9}$$

of a unique measure μ_K *on* K. *The measure* μ_K *is concentrated on* ∂K. *If* μ *is any other measure concentrated on* K, *and satisfying* $G\mu \leqslant 1$ *on* K, *then* $G\mu \leqslant G\mu_K$ on $\mathbb{R}^d$.

(The restriction in (22.8) to $t > 0$ is essential—consider the case $K = \{0\}$!)

The measure μ_K appearing in Theorem 22.7 is, of course, what we shall call the equilibrium charge distribution. The *capacity* $C(K)$ of K is defined to be $\mu_K(K)$, and is characterised by the extremal property

$$\begin{aligned} C(K) &:= \max\{\mu(K) : \mu \text{ is concentrated on } K, G\mu \leqslant 1 \text{ on } K\} \\ &= \max\{\mu(K) : \mu \text{ is concentrated on } K,\ G\mu \leqslant 1 \text{ on } \mathbb{R}^d\}. \end{aligned} \tag{22.10}$$

These concepts are neatly related to the concept of energy: the *energy* $\mathscr{E}(\mu)$ of a measure μ on $\mathbb{R}^d$ is defined by

$$\mathscr{E}(\mu) := \iint \mu(dx) g(x, y) \mu(dy). \tag{22.11}$$

Then

$$C(K)^{-1} = \min\{\mathscr{E}(\mu) : \mu \text{ concentrated on } K, \mu(K) = 1\}, \tag{22.12}$$

and the minimum is uniquely attained at $\mu = \mu_K / C(K)$.

There is a beautiful probabilistic interpretation of the equilibrium charge, due to Chung [2] and Getoor and Sharpe [1], which improves on Hunt's Theorem. Define

$$\sigma := \sup\{t > 0 : B_t \in K\},$$

with the convention that $\sup \varnothing = 0$.

(*22.13*) THEOREM. *For* $x \in \mathbb{R}^d$, $y \in K$, $t > 0$

$$\mathbf{P}^x(B_\sigma \in dy, \sigma \in dt) = p_t(x, y)\mu_K(dy)dt.$$

Since

$$\{\sigma > 0\} = \{B_t \in K \text{ for some } t > 0\},$$

(22.9) follows immediately by integrating with respect to t. We shall give a proof of Theorem 22.13, and most of Theorem 22.7, in Section VI.35, but before that in Section III.46 we give an argument based on time reversal that gives a clear intuitive picture of these results.

(*22.14*) *Remark*. It is true that

$$C(K) = \max\{\nu(K) : \nu \text{ concentrated on } K, G\nu \leqslant 1\},$$

and that

$$C(K) = \inf\{\mathscr{E}(\mu) : G\mu \geqslant 1 \text{ on } K\}.$$

Why is the second one inf, rather than min? Why not just take μ_K, whose energy we know is $C(K)$? Unfortunately, it is not in general true that $G\mu_K \geqslant 1$ on K, as the example where K is a ball together with a distant point illustrates. This is a typical feature of potential theory—it is essential to be very careful in order to make a statement that is totally correct, as you will see if you consult any of the references cited above.

We have seen transparent probabilistic interpretations of the equilibrium charge and equilibrium potential; now we provide an equally transparent interpretation of capacity.

(*22.15*) THEOREM (Spitzer–Kesten–Whitman). *Consider the 'Wiener sausage'* $(S_t)_{t\geqslant 0}$ *of* $(B_t)_{t\geqslant 0}$ *defined by*

$$S_t := \bigcup_{u\leqslant t} (K + B_u),$$

where K is some fixed compact set. Then

(22.16) $$t^{-1}|S_t| \to C(K), \quad \text{a.s.},$$

where $|A|$ denotes the Lebesgue measure of A. (Recall that the physicists' capacity of K will be *twice* ours.)

The proof of this is based on the Subadditive Ergodic Theorem of Hammersley and Kingman; see Durrett [3] for a well-motivated account and fascinating examples of this, as of many other topics. The verification of the conditions of the Subadditive Ergodic Theorem is a triviality, and the interest is in identifying the almost-sure limit, which is simply

(22.17) $$\gamma = \lim_{t\to\infty} t^{-1}\mathbf{E}|S_t|.$$

Now

(22.18)
$$\begin{aligned}
\mathbf{E}|S_t| &= \mathbf{E}\int dy\, I_{\{y\in S_t\}} \\
&\geqslant \int dy\, \mathbf{P}(y\in B_u + K \text{ for some } 0 < u \leqslant t) \\
&= \int dy\, \mathbf{P}(B_u \in y - K \text{ for some } 0 < u \leqslant t) \\
&= \int dy\, \mathbf{P}^y(B_u \in K \text{ for some } 0 < u \leqslant t) \\
&= \int dy\, \mathbf{P}^y(\tau \leqslant t), \\
&\qquad\qquad (\text{where } \tau := \inf\{t > 0 : B_t \in K\}) \\
&\geqslant \int dy\, \mathbf{P}^y(0 < \sigma \leqslant t) \\
&= tC(K),
\end{aligned}$$

from (22.13). Thus immediately $\gamma \geqslant C(K)$, and the proof will be complete if we can prove that

$$\int dy\, \mathbf{P}^y(\tau \leqslant t < \sigma) = o(t).$$

The key to proving this is to show by splitting the path at time t and using the reversibility of Brownian motion with respect to Lebesgue measure (compare Exercise II.E39.29) that

(22.19) $$\int dy\, \mathbf{P}^y(\tau \leqslant t < \sigma) = \int dz\, \mathbf{P}^z(\tau \leqslant t)\mathbf{P}^z(\sigma > 0).$$

From (22.3) and (22.9), we obtain that

$$P_K 1(z) := \mathbf{P}^z(\sigma > 0) \sim \tfrac{1}{2}\pi^{-1/2d}\Gamma(\tfrac{1}{2}d - 1)C(K)|z|^{2-d},$$

so it is not very surprising that (22.19) is of smaller order than (22.18), and this is indeed the case, though the verification involves us in techniques that lie ahead, so we shall leave these last steps.

Finally, no introduction to potential theory would be complete without a few words about the Dirichlet problem. The problem is this. Suppose given a bounded open connected set $D \subseteq \mathbb{R}^d$, and a bounded measurable $\varphi: \partial D \to \mathbb{R}$; can one find a harmonic function h on D such that $\lim_{x \to b} h(x) = \varphi(b)$ where $b \in \partial D$ and x converges in D to b? There is a simple solution to this problem using Brownian motion. For any Borel set A, and bounded measurable g, define (noting once again the '$t > 0$' condition)

(22.20) $$H_A := \inf\{t > 0 : B_t \in A\}, \quad P_A g(x) := \mathbf{E}[g(B(H_A)) : H_A < \infty].$$

Let $V = \mathbb{R}^d \backslash D$. Call a point b of ∂D a *regular boundary point* if D of b is regular for V for Brownian motion, that is, if

$$\mathbf{P}^b(H_V = 0) = 1.$$

It is not hard to show that if b is the tip of a cone lying entirely within V then b is regular. That the probabilistic definition of regular boundary point agrees with the classical definition in terms of Wiener's test is proved in Section 7.10 of Itô and McKean [1].

The optimal solution of the Dirichlet problem is the following.

(*22.21*) THEOREM (Wiener). *Let D be a bounded domain in $\mathbb{R}^n$. Let φ be a bounded measurable function on ∂D that is continuous at each regular boundary point. Then there exists a unique harmonic function h on D such that*

(22.22) $$\lim_{D \ni x \to b} h(x) = \varphi(b)$$

for every regular boundary point b.

Doob's idea is to prove Theorem 22.21 by establishing the explicit formula

$$h = P_V\varphi = P_{\partial D}\varphi \quad \text{in } D. \tag{22.23}$$

You will find a careful proof of all steps of this Theorem in Section 13.6 of Dynkin [2], among other places. The main points of the argument are first to prove (using the strong Markov property and rotational invariance of Brownian motion) that $h(x)$ is the average of h over any ball $B(x,r)$, centred at x with radius r, contained entirely in D; secondly to use this together with convolution with a smooth function to show that h is C^∞ inside D, and hence satisfies Laplace's equation; and thirdly to prove (22.22) by an ε-δ argument.

As for the uniqueness assertion, it is clear that if h is any solution, and $G \subset D$ is a subdomain, $\bar{G} \subset D$, then $P_{\partial G}h = h$ on G, since h is continuous on $\bar{G}$, and, by the Optional-Stopping Theorem,

$$h(x) = \mathbf{E}^x h(B(H_{\partial D})) \quad (x \in G).$$

Now we let $G \uparrow D$ and use the fact that

$$B(H_V) \text{ is regular for } V, \quad \mathbf{P}^x \text{ a.s.} \tag{22.24}$$

The proof of (22.24) needs Hunt's Theorem 22.7, though if every point of ∂V is regular, no proof is needed, of course. Itô and McKean [1] and Port and Store [3] give several complements to, and extensions of, Theorem 22.21.

Probabilistic potential theory is a big and important subject, essentially originating in Hunt's profound papers [1], which explain the basic principles. Blumenthal and Getoor [1] provide a complete account of Hunt's theory; other standard references include Dellacherie and Meyer [1], Fukushima [1], Kellogg [1], Port and Stone [3], Silverstein [1] and Helms [1]. Hunt emphasised the role of a *dual process*, which is a kind of time-reversal of the basic process being studied, though the self-duality of Brownian motion tends to obscure things here. See the papers of Mitro [1, 2] for a lucid account. In Chapter III, we shall use time reversal to present the Martin boundary for continuous-time chains; Martin boundary theory describes all possible positive harmonic functions or D, and is thus deeper than Theorem 22.21, which only characterises bounded harmonic functions with boundary regularity.

23. Brownian motion and physical diffusion. Brownian motion as we have being studying it is very closely related to what a physicist would understand by the term 'diffusion'; the connection is the celebrated *diffusion equation* of mathematical physics, which we shall now derive.

Consider the diffusion of some substance (for example, a dye) through a medium (which could be water, or a crystal). Let $\rho(t, x)$ be the concentration of dye at position x at time t, and let us suppose initially that the medium is isotropic (no preferred directions, uniform throughout space). Consider some plane in the medium, perpendicular to the x^1-direction, say; this plane is constantly traversed by molecules of dye, which pass from one side of the plane to the other. If the concentration to the left of the plane is higher than that to the right, there will be a net flux of particles of dye from left to right; and the greater the difference in concentration, the greater this flux from left to right will be. Fick's Law of diffusion says that the flux is equal to $-\frac{1}{2}a\,\partial\rho/\partial x_1$. More generally, the flux $F(t,x)$ is a vector quantity and obeys

$$F(t,x) = -\tfrac{1}{2}a\nabla\rho(t,x). \tag{23.1}$$

The vector field F specifies the direction and strength of the net flux of dye, and to find the flow of particles across a plane perpendicular to the unit vector u, we simply form the scalar product $u\cdot F$. Now if we consider a small volume V around the point x, the total amount of dye in V is $\int_V \rho(t,x)\,dx$, so

$$\begin{aligned}
&\text{rate of change of amount of dye in } V\\
&= \frac{\partial}{\partial t}\int_V \rho(t,x)\,dx\\
&= \text{integral of flux around } \partial V\\
&= -\int_{\partial V} F(t,x)\cdot dn\\
&= -\int_V \nabla\cdot F(t,x)\,dx,
\end{aligned}$$

by the Divergence Theorem. Since V is arbitrary, we deduce the diffusion equation

$$\frac{\partial\rho}{\partial t}(t,x) = \tfrac{1}{2}\nabla\cdot(a\nabla\rho)(t,x). \tag{23.2}$$

In the case $a \equiv 1$, we have the Kolmogorov (forward) equation for the evolution of the Brownian transition density:

$$\frac{\partial}{\partial t}p_t(x,y) = \tfrac{1}{2}\Delta_y p_t(x,y), \tag{23.5}$$

as we argued at (4.7). We have derived the diffusion equation (23.2) under the assumption that a is a constant, but it may equally well depend on position (if the diffusivity of the medium varies), and may even be a matrix-valued function of position. The latter could arise in a crystal, where the preferred directions of the crystal will tend to distort the concentration gradient and produce a flux that is not aligned exactly with the concentration gradient. The derivation of (23.2) remains unchanged.

(23.4) Remarks. The special case $a \equiv 1$ of the diffusion equation gives us the Kolmogorov (forward) equation for the Brownian transition density, so you

will be wondering whether the general statement of the diffusion equation (23.2) has a similar probabilistic interpretation. It does indeed, and the interpretation is the analogue of the interpretation given in Section 4 for Brownian motion. Without going into too much detail here, we are concerned with a diffusion whose infinitesimal generator $\mathscr{G}$ has adjoint

$$\mathscr{G}^* := \tfrac{1}{2}\nabla\cdot(a\nabla). \tag{23.5}$$

What this means is as explained is Section 4: there is a transition semigroup $(P_t)_{t\geq 0}$ such that $\mathscr{G}f = \lim_{t\downarrow 0} t^{-1}(P_t f - f)$, at least for some class of f. As a process, $(X_t)_{t\geq 0}$ satisfies

$$f(X_t) - f(X_0) - \int_0^t \mathscr{G}f(X_s)\,ds \quad \textit{is a martingale}$$

for all f in some class. This is a sample-path formulation of the statement (4.4):

$$\frac{d}{dt}P_t = \mathscr{G}P_t = P_t\mathscr{G}.$$

The Markov process X may be considered as describing the motion of a single particle of dye. (Note in passing that in general there is no closed-form expression for P_t, unlike the Brownian case, so an alternative prescription such as the generator $\mathscr{G}$ is very necessary!) Formally, the arguments establishing Kolmogorov's backward and forward equations (4.6) and (4.7) now run as before, but do note that the analogue of the forward equation (4.7) should read

$$\frac{\partial}{\partial t}p_t = \mathscr{G}_y^* p_t,$$

where $\mathscr{G}^*$ is the formal adjoint of $\mathscr{G}$:

$$\int f\mathscr{G}g = \int g\mathscr{G}^*f \quad \text{for } f, g \in C_K^\infty.$$

This distinction did not arise for Brownian motion, whose generator $\frac{1}{2}\Delta$ is self-adjoint, nor will it arise here if the matrix a is symmetric (since then $\mathscr{G}$ is again self-adjoint). It is the received wisdom of physics that a is symmetric and non-negative definite. Diffusions with 'divergence-form' generators (that is, of the form (23.5) with a symmetric non-negative definite) are particularly tractable because they are amenable to the theory of Dirichlet forms. The ideas of Dirichlet-form theory are drawn from physical notions about energy.

While we are discussing the physical aspects of diffusion, it is worth pointing out that no physicist would accept Brownian motion as a literal model for the movement of a particle, since the path has infinite variation. However, a more satisfactory model can be built by making the velocity v_t of the particle into Brownian motion—but an even more satisfactory model can be made by making

v solve

(23.6) $$dv_t = dB_t - \lambda v_t\,dt,$$

where $\lambda > 0$ is the *viscous drag coefficient*, and B is Brownian motion on $\mathbb{R}$, $B_0 = 0$. This permits the velocity to wriggle around, but makes it unlikely to get too big. The correct interpretation of (23.6) is of course

(23.7) $$v_t = v_0 + B_t - \lambda \int_0^t v_u\,du,$$

which is solved explicitly by

(23.8) $$v_t = v_0 e^{-\lambda t} + e^{-\lambda t} \int_0^t e^{\lambda s}\,dB_s$$

where the stochastic integral appearing on the right is here properly interpreted by integrating by parts:

(23.9) $$\int_0^t e^{\lambda s}\,dB_s := e^{\lambda t} B_t - \lambda \int_0^t e^{\lambda s} B_s\,ds.$$

The stochastic differential equation (23.6) is called the *Ornstein–Uhlenbeck* stochastic differential equation, and its solution is called the *Ornstein–Uhlenbeck* (*OU*) *process*. Assume for simplicity there we work in one dimension, and that v_0 is zero-mean Gaussian with variance $(2\lambda)^{-1}$, independent of B. Then v is a stationary zero-mean Gaussian process with covariance

(23.10) $$\operatorname{cov}(v_s, v_t) = (2\lambda)^{-1} \exp(-\lambda|t-s|)$$

The physicist would then take the process

$$X_t := \int_0^t v_s\,ds$$

as a model for the diffusion of a particle. This integrated *OU* process, being *non*-Markovian, is an altogether less obliging process than Brownian motion, and few of the functionals of X have closed-form distributions. However, it is not too hard to prove that

$$\left(\frac{\lambda}{\sqrt{n}} X(nt)\right)_{t \geq 0} \xrightarrow{\mathscr{D}} (B_t)_{t \geq 0} \quad (n \to \infty)$$

(where the sense of convergence in distribution is fully explained in Part 6 of Chapter II), so *for large time scales, Brownian motion may be accepted as a model of physical diffusion.*

(*23.11*) *Exercise.* Confirm that (23.8) solves (23.6), and that the covariance is (23.10).

4. GAUSSIAN PROCESSES AND LÉVY PROCESSES

Gaussian processes

24. Existence results for Gaussian processes. In general, a Gaussian process is a process $(X_t)_{t\in T}$ indexed by a general set, such that, for any $t_1,\ldots,t_n \in T$, $(X_{t_1},\ldots,X_{t_n})$ has a multivariate Gaussian distribution. Thus the distribution is specified by the mean and covariance:

(24.1) $$\mu(t) := \mathbf{E}X_t, \quad \rho(s,t) := \operatorname{cov}(X_s, X_t), \quad s,t\in T,$$

since we could write down the point density of $(X_{t_1},\ldots,X_{t_n})$ in terms of μ and ρ. It is customary to assume that $\mu\equiv 0$, since the general case can be reduced to this by taking the process $X'_t := X_t - \mu(t)$; we shall follow this custom and henceforth assume that $\mu \equiv 0$.

(*24.2*) PROPOSITION. *The function $\rho: T\times T\to\mathbb{R}$ is the covariance of a Gaussian process if and only if ρ is non-negative definite*:

(24.3) *for any $t_1,\ldots,t_n$, $(\rho(t_i,t_j))_{i,j=1}^n$ is a non-negative definite matrix.*

Proof. Necessity is immediate. Sufficiency uses the Daniell–Kolmogorov Theorem II.31.1. To check the conditions of that theorem, notice that the state space $\mathbb{R}$ is Polish, and that, for $J=\{t_1,\ldots,t_n\}\subset I=\{t_1,\ldots,t_N\}$ $(N>n)$, the law of $(X_{t_1},\ldots,X_{t_n})$ regarded as the projection of $(X_{t_1},\ldots,X_{t_N})$ down to $(X_{t_1},\ldots,X_{t_n})$ is just $N(0,V)$, where $V_{ij}=\rho(t_i,t_j)$ $(i,j=1,\ldots,n)$. But this *is* the law of $(X_{t_1},\ldots,X_{t_n})$ regarded as $\{X_t: t\in J\}$, and so the consistency condition of the Daniell–Kolmogorov Theorem holds. □

The limitation of Proposition 24.2 is that the condition (24.3) is not in general easy to check. One situation where this can be done is the following. The consequences are far-ranging.

(*24.4*) LEMMA. *Let E be a measurable space with a σ-finite measure m, and suppose that $(P_t)_{t\geqslant 0}$ is a sub-Markovian transition semigroup that has a density $p_t(\cdot,\cdot)$ with respect to m:*

$$(P_t f)(x) = \int p_t(x,y) f(y) m(dy), \quad f\in b\mathscr{B},\ t>0.$$

Suppose further that

(24.5) (i) $$p_t(x,y) = p_t(y,x) \quad \text{for } t>0,\ x,y\in E;$$

(24.5)(ii) $$g(x,y) = \int_0^\infty p_t(x,y)\,dt < \infty \quad \text{for all } x,y\in E.$$

Then g is the covariance of a Gaussian process on E.

Proof. We need only check the condition (24.3). But, for any $a \in \mathbf{R}^n$, $x_1, \ldots, x_n \in E$,

$$\begin{aligned}\sum_{i=1}^{n}\sum_{j=1}^{n} a_i g(x_i, x_j) a_j &= \int_0^\infty dt \sum_{i=1}^{n}\sum_{j=1}^{n} a_i p_t(x_i, x_j) a_j \\ &= \int_0^\infty dt \sum_{i=1}^{n}\sum_{j=1}^{n} \int m(dy) a_i p_{t/2}(x_i, y) p_{t/2}(y, x_j) a_j \\ &= \int_0^\infty dt \int m(dy) \left[\sum_{i=1}^{n} a_i p_{t/2}(x_i, y)\right]^2 \\ &\geqslant 0,\end{aligned}$$

using the symmetry of p_t at the last step. □

(We have looked ahead to Section III.3 for the definition of a sub-Markovian transition semigroup, but we have already seen the essentials in Section 4.)

As an example of a situation to which Lemma 24.4 would apply, consider a random walk on a finite connected graph G, when a jump from vertex i to a neighbouring vertex j takes place at unit rate, and where the process is killed at a constant rate δ. The measure m is simply the counting measure on the vertices of G.

As an example of a situation to which Lemma 24.4 would *not* apply, consider Brownian motion in $\mathbf{R}^d$ $(d \geqslant 3)$; the condition (24.5)(ii) fails for $x = y$. Broadly speaking, the finiteness of $g(x, y)$ for all $x \neq y$ is equivalent to the transience of the process, and $g(x, x) < \infty$ is equivalent to the process visiting x with positive probability (though, without enough conditions to ensure nice sample paths, we cannot yet begin to make rigorous sense of this in general.)

The argument of Lemma 24.4 is too pretty for us to abandon all hope of making it apply to Brownian motion. Some sense can be made of it, but we have to consider a *generalised random field*, indexed by the vector space

(24.6) $$\mathscr{D}(\mathscr{E}) := \{\sigma\text{-finite measures } \mu \text{ on } \mathbf{R}^d \text{ s.t. } \mathscr{E}(\mu, \mu) < \infty\},$$

where $\mathscr{E}$ is the energy functional encountered in Section 22:

(24.7) $$\mathscr{E}(\mu, \nu) := \iint \mu(dx) g(x, y) \nu(dy).$$

The inequality

(24.8) $$\mathscr{E}(\mu, \nu)^2 \leqslant \mathscr{E}(\mu, \mu)\mathscr{E}(\nu, \nu)$$

follows from the Cauchy–Schwarz inequality and a simple modification of the idea of Lemma 24.4, as you are invited to check, and provides a proof that $\mathscr{D}(\mathscr{E})$ is closed under addition.

We can now consider the Gaussian field $\{X_\mu : \mu \in \mathscr{D}(\mathscr{E})\}$, with covariance

$$\mathbf{E}(X_\mu X_\nu) = \mathscr{E}(\mu, \nu).$$

The intuitive interpretation

$$X_\mu = \int \mu(dx) X_x$$

guides us, but has no strict sense; there does not exist a process $(X_x)_{x \in \mathbb{R}^d}$. Nevertheless $(X_\mu)_{\mu \in \mathscr{D}(\mathscr{E})}$ is a perfectly good Gaussian field, whose existence is confirmed by checking the condition (24.3) and appealing to the Danniell–Kolmogorov Theorem. We shall say some more about these random fields whose covariance comes from a symmetric Green function in the next section. For more on the relevance to quantum field theory, see Symanzik [1], Brydges, Fröhlich and Spencer [1] and Dynkin [5].

One other existence result that we cannot do without is the celebrated theorem of Bochner. We consider now only the case where $T = \mathbb{R}^d$, and where the covariance structure is *stationary*: for all $x, y \in \mathbb{R}^d$,

$$\begin{aligned} \rho(x, y) &= \rho(0, y - x) \\ &:= \rho(y - x), \end{aligned}$$

for brevity. We give the full form of Bochner's Theorem.

(24.9) THEOREM (Bochner). *Let $\varphi: \mathbb{R}^d \to \mathbb{C}$ be bounded and continuous. Then the following are equivalent.*

(24.10) (i) *There exists a finite measure μ on $\mathbb{R}^d$ such that*

$$\varphi(\theta) = \int \mu(dx)\, e^{i\theta \cdot x}.$$

(24.10)(ii) *For any $a_1, \dots, a_n \in \mathbb{C}, x_1, \dots, x_n \in \mathbb{R}^d$,*

$$\sum_{l=1}^{n} \sum_{j=1}^{n} a_l \varphi(x_l - x_j) \bar{a}_j \geqslant 0.$$

(We say that φ is non-negative definite.*)*

Before proving this we apply it to the representation of a stationary Gaussian process on $\mathbb{R}^d$.

(24.11) COROLLARY. *Suppose that $\rho: \mathbb{R}^d \to \mathbb{R}$ is continuous. In order that ρ should be the covariance of a stationary Gaussian process on $\mathbb{R}^d$, it is necessary and sufficient that ρ may be represented in the form*

(24.12)
$$\rho(x) = \int F(d\theta) e^{i\theta \cdot x},$$

where F is a finite non-negative symmetric measure on $\mathbb{R}^d$. (The measure F is called the spectral measure *of the Gaussian process.)*

Proof. If the representation (24.12) holds, the criterion (24.3) is easy to prove. Conversely, if (24.3) holds then ρ is non-negative definite, continuous (by hypothesis) and bounded (since $\rho(x) \leqslant \rho(0) = \mathbf{E}X_0^2 < \infty$), so, by Bochner's Theorem, ρ is a Fourier transform of some non-negative measure, which is symmetric since ρ is. □

Proof of Theorem 24.9. The implication (24.10)(i)⇒(24.10)(ii) is trivial.

For the converse, the aim should be to get the inverse Fourier transform of φ. We approach this by taking some large integers $K, n > 0$, and (with $\delta := 1/n$) noticing that (24.10)(ii) implies, for any $\theta \in \mathbf{R}^d$,

$$0 \leqslant (2n+1)^{-2d} \sum\nolimits_{K,n} e^{-i\delta\theta\cdot l}\varphi(\delta l - \delta j)e^{i\delta\theta\cdot j}(2K)^{-d}$$

where $\sum_{K,n}$ denotes the sum over all pairs $(l, j) \in (\mathbf{Z}^d)^2$ such that $\|l\|_\infty := \sup\{|l_r| : r = 1, \ldots, d\} \leqslant Kn$, $\|j\|_\infty \leqslant Kn$. But, as $n \to \infty$ with K fixed, this expression converges to

$$\text{(24.13)} \qquad (2K)^{-d} \int_{\{\|x\|_\infty \leqslant K\}} dx \int_{\{\|y\|_\infty \leqslant K\}} dy\, e^{-i\theta\cdot(x-y)}\varphi(x-y)$$
$$= \int_{\{\|v\|_\infty \leqslant 2K\}} dv\, e^{-i\theta\cdot v}\varphi(v) \prod_{j=1}^d \left(1 - \frac{|v_j|}{2K}\right),$$

which is thus non-negative. (We use the continuity of φ to get the convergence.) But (24.13) is (to within powers of 2π) the inverse Fourier transform of $\varphi(v)\prod_{j=1}^d (1 - |v_j|/2K)$, and

$$\prod_{j=1}^d \left(1 - \frac{|v_j|}{2K}\right) = \mathbf{E}\exp\left(\frac{iv\cdot X}{2K}\right)$$

where the components of X are independent, with density

$$f(v) = (1 - \cos v)(\pi v^2)^{-1}.$$

Thus if $f_{2K}(v) := 2Kf(2Kv)$, we have that (24.13) is (a multiple of) $(\check{\varphi} * F_{2K})(\theta)$, where $F_{2K}(v) = \prod_{j=1}^d f_{2K}(v_j)$, and $\check{\varphi}$ is the inverse Fourier transform of φ. But the distributions with density F_{2K} converge weakly to the point mass at 0, and the density of $\check{\varphi} * F_{2K}$ is non-negative. Hence $\check{\varphi}$ is a non-negative measure, which is what we sought. □

(24.14) *Remark.* The function $\rho(x) = I_{\{x=0\}}$ $(x \in \mathbf{R}^d)$ is the covariance function of a Gaussian process $\{X_x : x \in \mathbf{R}^d\}$ for which $X_{x_1}, \ldots, X_{x_n}$ are independent $N(0,1)$ for any $x_1, \ldots, x_n$. However, ρ is not representable in the form (24.12). This may appear to be a limitation of the representation result Corollary 24.11, but it is not particularly grave; the process X cannot have a version with any sensible regularity properties, and so is essentially useless. It is a simple exercise to prove that if X is a stationary Gaussian process for which $x \mapsto X_x(\omega)$ is continuous for almost all ω then ρ must be continuous. Thus if we want a stationary Gaussian process with continuous paths, continuity of ρ is necessary, and, as we shall see next, we need only strengthen continuity to a mild form of Hölder continuity to obtain a sufficient condition.

(24.15) *Exercises*

(i) *Spectral measure of the one-dimensional Ornstein–Uhlenbeck process.* Confirm that by taking $F(d\theta) = (\lambda/\pi)(\lambda^2 + \theta^2)^{-1}\,d\theta$ in (24.12), we recover the (Ornstein–Uhlenbeck) covariance $\rho(x) = e^{-\lambda|x|}$.

(ii) *Lévy's Brownian motion.* With $F(d\theta) = (2\pi)^n \exp(-\frac{1}{2}t|\theta|^2)\,d\theta$ in (24.12), show that

$$\rho_t(x) = (2\pi t)^{-1/2} \exp\left(-\frac{|x|^2}{2t}\right)$$

defines a stationary covariance function on $\mathbf{R}^n$. Deduce that, for $\lambda = \frac{1}{2}\gamma^2 > 0$,

$$\gamma \int_0^\infty e^{-\lambda t}(2\pi t)^{-1/2} \exp\left(-\frac{|x|^2}{2t}\right) dt = e^{-\gamma|x|}$$

defines a stationary covariance function, for which

$$\mathbf{E}[(X_x - X_y)^2] = 2(1 - \exp\{-\gamma|x - y|\}).$$

Prove that there exists a Gaussian process $(Y_x : x \in \mathbf{R}^n)$, *Lévy's Brownian motion*, such that

$$Y_0 = 0, \quad \mathbf{E}[(Y_x - Y_y)]^2 = |x - y|.$$

A wonderful paper by McKean [4] discusses Markov properties of Lévy's Brownian motion, showing that it behaves very differently in even and odd dimensions.

25. Continuity results. Let $(X_t)_{t\in\mathbf{R}^n}$ be a stochastic process with values in a complete separable metric space (S, ρ). We say that X has a *continuous version* if there exists an S-valued stochastic process $(X'_t)_{t\in\mathbf{R}^n}$ such that

(25.1) (i) $t \mapsto X'_t(\omega)$ is continuous for almost all ω);
(25.1)(ii) $\rho(X'_t(\omega), X_t(\omega)) = 0$, a.s. for all $t \in \mathbf{R}^n$.

If a process has a continuous version, we generally discard the original process and work with the continuous version instead, because the original was too irregular to work with. We now give a simple but powerful result that is usually sufficient to decide when a process has a continuous version.

(25.2) THEOREM (Kolmogorov's Lemma). *If* $(X_t)_{t\in\mathbf{R}^n}$ *is a stochastic process* *with values in a complete separable metric space (S, ρ), and if there exist positive constants α, C, ε such that, for all $s, t \in \mathbb{R}^n$*

$$\mathbf{E}\rho(X_s, X_t)^\alpha \leqslant C|s-t|^{n+\varepsilon}, \tag{25.3}$$

then there exists a continuous version of X. This version is Hölder continuous of order θ for each $\theta < \varepsilon/\alpha$.

Proof. Let $\mathbb{D} := \bigcup_{k \geqslant 0} \mathbb{D}_k$ be the set of dyadic rational points in $\mathbb{R}^n$, where $\mathbb{D}_k := 2^{-k}\mathbb{Z}^n$. The idea of the proof is to show that the restriction of X to $\mathbb{D} \cap [0,1)^n$ is Hölder (θ) for any $\theta < \varepsilon/\alpha$; we then extend X by continuity to $[0,1)^n$, and then apply the same argument to a cube of arbitrary size. So we fix $\theta \in (0, \varepsilon/\alpha)$, and define

$$A_k := \{\text{for some } i, j \in \mathbb{Z}^n, |i-j| = 1, i2^{-k} \text{ and } j2^{-k} \in [0,1)^n, \text{ and } \rho(X(i2^{-k}), X(j2^{-k})) > 2^{-k\theta}\}.$$

Then

$$\begin{aligned}\mathbf{P}(A_k) &\leqslant \sum_{i \in [0, 2^k)^n} \sum_{\substack{j \in \mathbb{Z}^n \\ |j-i|=1}} \mathbf{P}(\rho(X(i2^{-k}), X(j2^{-k})) > 2^{-k\theta}) \\ &\leqslant 2^{nk} 2n 2^{k\theta\alpha} C 2^{-k(n+\varepsilon)} \\ &= 2nC2^{-k(\varepsilon-\theta\alpha)},\end{aligned}$$

and so by the Borel–Cantelli Lemma with probability 1 only finitely many of the A_k happen, so that, for some $K = K(\omega)$,

$$\rho(X(i2^{-k}), X(j2^{-k})) \leqslant K2^{-k\theta} \tag{25.4}$$

for all $k \in \mathbb{N}$, $i, j \in [0, 2^k]^n$, $|i-j| = 1$. All that remains is to extend (25.4) from neighbouring dyadic rationals to any. For this, let us assume for notational simplicity that $n = 1$; the general result is an immediate consequence of this. If we take $0 \leqslant x < y < 1$, with $x, y \in \mathbb{D}$, then, for some k, $2^{-k-1} < y - x \leqslant 2^{-k}$, and so there exists i such that

$$x \leqslant i2^{-k-1} < (i+1)2^{-k-1} \leqslant y,$$

and

$$\rho(X_y, X_x) \leqslant \rho(X_y, X((i+1)2^{-k-1})) + \rho(X(i2^{-k-1}), X_x) + 2K2^{-(k+1)\theta},$$

using (25.4). Now we similarly analyse the intervals $[x, i2^{-k-1}]$ and $[(i+1)2^{-k-1}, y]$ by chipping off the largest dyadic-rational intervals in each (of length at most 2^{-k-2} in each case). Continuing thus, we have

$$\begin{aligned}\rho(X_y, X_x) &\leqslant 2K2^{-(k+1)\theta} + 2K \sum_{r \geqslant k+2} 2^{-r\theta} \\ &\leqslant K'|y-x|^\theta.\end{aligned}$$

The Hölder continuity allows us to extend the process X now to the whole of $[0,1)^n$.

(25.5) *Remarks*

(i) A function that is Hölder continuous of order $\theta > 1$ is constant.
(ii) A more general and more powerful way of proving the existence of continuous versions has been discovered by Garsia, Rodemich and Rumsey [1], and is extremely useful. You will find nice accounts in Stroock and Varadhan [1] and Walsh [3].

We are going to use Kolmogorov's Lemma to derive sufficient conditions for the existence of a continuous version of a Gaussian process. These conditions are not necessary, but the gap is unimportant in most examples that arise. Necessary and sufficient conditions for the continuity of a Gaussian process are now known; work of Fernique, Dudley and others enabled Talagrand to reach the summit in [1]. The conditions are of a technical nature, so we refer the interested reader to Talagrand's paper, or to the books by Adler [1, 2], which are full of other interesting results on Gaussian processes.

(25.6) COROLLARY. *Let $(X_t)_{t\in\mathbb{R}^n}$ be a (zero-mean) Gaussian process with covariance function $\rho(s,t) := \mathbf{E}X_sX_t$. A sufficient condition for the existence of a continuous version is that ρ should be locally Hölder continuous: for each $N \in \mathbb{N}$ there exists $\theta = \theta(N) > 0$ and $C = C(N)$ such that, for $|t|, |s| \leqslant N$,*

$$|\rho(s,t) - \rho(t,t)| \leqslant C|s-t|^{\theta} \tag{25.7}$$

Proof. We have

$$\begin{aligned}\mathbf{E}(|X_t - X_s|^2) &= \rho(t,t) - 2\rho(t,s) + \rho(t,s)\\ &\leqslant 2C|t-s|^{\theta}.\end{aligned}$$

Since $X_t - X_s$ is Gaussian, there exist constants a_m such that $\mathbf{E}|X_t - X_s|^{2m} = a_m(\mathbf{E}|X_t - X_s|^2)^m$, and so

$$\mathbf{E}(|X_t - X_s|^{2m}) \leqslant (2C)^m a_m |s-t|^{m\theta}.$$

For large enough $m, m\theta > n$, and we can use Kolmogorov's Lemma. □

For a stationary Gaussian process, $\rho(s,t) = \rho(t-s)$, the condition (25.7) reduces to the Hölder continuity at 0 of ρ (which implies easily that ρ is Hölder continuous everywhere.) In the case of a stationary Gaussian process, it is often convenient to have a condition in terms of the measure F that represents ρ, (24.12). Here is one such.

(25.8) COROLLARY. *Suppose that, for some $\varepsilon \in (0,1)$,*

$$\int_{\mathbb{R}^n} |x|^{\varepsilon} F(dx) < \infty. \tag{25.9}$$

Then X has a continuous version.

Proof. Let $A := \int |x|^\varepsilon F(dx)$, $B_r = \{x \in \mathbf{R}^n : |x| \leqslant r\}$, and note that (25.9) implies

$$F(B_r^c) \leqslant Ar^{-\varepsilon}.$$

Let us now take the case $n = 1$, indicating later how the general case follows:

$$\begin{aligned} 0 \leqslant \rho(0) - \rho(x) &= \int (1 - \cos \theta x) F(d\theta) \\ &\leqslant \int (\tfrac{1}{2}(\theta x)^2 \wedge 2) F(d\theta) \\ &\leqslant \int_{-2/x}^{2/x} \tfrac{1}{2}\theta^2 x^2 F(d\theta) + 2F(B_{2/x}^c) \\ &\leqslant \tfrac{1}{2}x^2 \int_{-2/x}^{2/x} \theta^2 F(d\theta) + 2A(\tfrac{1}{2}x)^\varepsilon. \end{aligned}$$

But the estimation

$$\begin{aligned} \int_0^N \tfrac{1}{2}\theta^2 F(d\theta) &= \int_0^N yF((y, N])\,dy \\ &\leqslant A \int_0^N y^{1-\varepsilon}\,dy \\ &= A(2-\varepsilon)^{-1} N^{2-\varepsilon} \end{aligned}$$

gives us

$$0 \leqslant \rho(0) - \rho(x) \leqslant x^\varepsilon 2^{1-\varepsilon} A \frac{6-\varepsilon}{2-\varepsilon},$$

so that ρ is Hölder continuous at zero.

For general n, fix some unit vector v, and let F_v be the image of the measure F under the map $\theta \mapsto \theta \cdot v$. Thus the measure F_v satisfies the bound $F_v(\{x : |x| > r\}) \leqslant F(B_r^c) \leqslant Ar^{-\varepsilon}$, and so, as before, $0 \leqslant \rho(0) - \rho(x \cdot v) \leqslant x^\varepsilon 2^{1-\varepsilon} A(6-\varepsilon)/(2-\varepsilon)$. Since the constant does not depend on v, the Hölder continuity at 0 of ρ now follows. □

We therefore have quite useable criteria sufficient to ensure the existence of a continuous version of a stationary Gaussian process $\{X_t : t \in \mathbf{R}^n\}$. Can we obtain similarly simple criteria for the existence of a C^k version?

(25.10) THEOREM. *Let $\rho(x) = \int e^{i\theta \cdot x} F(d\theta)$ be the covariance function of a stationary Gaussian process on $\mathbf{R}^n$, let $\alpha = (\alpha_1, \ldots, \alpha_n)$ be a multi-index and let $\varepsilon \in (0, 1)$ be such that*

(25.11)
$$\int_{\mathbf{R}^n} \left(\prod_{j=1}^n \theta_j^{2\alpha_j} \right) |\theta|^\varepsilon F(d\theta) < \theta.$$

Then there is a version $\{X_t : t \in \mathbb{R}^n\}$ of the process for which $D^\alpha X(t)$ exists and is continuous. The process $\{D^\alpha X(t) : t \in \mathbb{R}^n\}$ is a stationary Gaussian process with spectral measure

(25.12)
$$F_\alpha(d\theta) := \left(\prod_{j=1}^{n} \theta_j^{2\alpha_j}\right) F(d\theta).$$

Proof. It is clearly sufficient prove only the case $\alpha = (0, \ldots, 0, 1)$. For notational convenience, we write a point of $\mathbb{R}^n$ as (τ, t), where $\tau \in \mathbb{R}^{n-1}$ and $t \in \mathbb{R}$. We build a stationary Gaussian process $\{(\xi_\tau, Y_{\tau,t}) : (\tau, t) \in \mathbb{R}^n\}$ with zero mean and covariance structure

(25.13) (i)
$$\mathbf{E}(\xi_0 \xi_\tau) = \mathbf{E}(X_0 X_{\tau,0}) = \rho(\tau, 0),$$

(25.13) (ii)
$$\mathbf{E}(\xi_0 Y_{\tau,t}) = \frac{\partial \rho}{\partial t}(\tau, t),$$

(25.13)(iii)
$$\mathbf{E}(Y_0 Y_{\tau,t}) = -\frac{\partial^2 \rho}{\partial t^2}(\tau, t).$$

(The fact that $\int \theta_n^2 F(d\theta) < \infty$ implies that the first two partial derivatives of ρ with respect to t exist and are continuous.) In order to see that (25.13) really does give the covariance of a Gaussian process, we see at the same time *why* (25.13) was chosen. Indeed, if we fix some $h > 0$ and consider the process

$$(\xi_\tau, Y_{\tau,t}^h) = (X_{\tau,0}, h^{-1}(X_{\tau,t+h} - X_{\tau,t}))$$

then the process clearly exists, so its covariance is non-negative definite and satisfies

(25.14) (i)
$$\mathbf{E}(\xi_0 \xi_\tau) = \rho(\tau, 0),$$

(25.14) (ii)
$$\mathbf{E}(\xi_0 Y_{\tau,t}^h) = h^{-1}[\rho(\tau, t+h) - \rho(\tau, t)],$$

(25.14)(iii)
$$\mathbf{E}[Y_0^h Y_{\tau,t}^h] = h^{-2}[2\rho(\tau, t) - \rho(\tau, t+h) - \rho(\tau, t-h)].$$

Thus the limiting form of the covariance (25.13) is *also* non-negative definite, and therefore *is* the covariance structure of some Gaussian process. In view of the integrability assumption (25.11) and Corollary 25.8, there is a continuous version of $(Y_{\tau,t})$, and the spectral measure of $(Y_{\tau,t})$ is just $\theta_n^2 F(d\theta)$, since the covariance function of Y is $-\partial^2\rho/\partial t^2$. There is a continuous version of ξ because there is a continuous version of X. Now we simply define

(25.15)
$$\tilde{X}_{\tau,t} := \xi_\tau + \int_0^t Y_{\tau,s}\, ds.$$

It is immediate that $\tilde{X}$ is a continuous Gaussian process, with a continuous derivative with respect to t, and it is a simple exercise to confirm that $\tilde{X}$ has the same covariance as X. □

(25.16) Remarks

(i) John Kent [3] has obtained attractive sufficient conditions in terms of the covariance structure of an *arbitrary* (non-Gaussian) stationary process for the existence of a continuous version. His result is as follows. If ρ is C^n, and $p_n(h)$ is the polynomial of degree n given by the Taylor expansion of ρ about O, and if there exists $\gamma > 0$ such that

(25.17) $$|\rho(h) - p_n(h)| = O(r^n/|\log r|^{3+\gamma}) \quad \text{as } r = |h| \to 0,$$

then there exists a continuous version of the random field $\{X_t : t \in \mathbb{R}^n\}$.

(ii) Everything we have done in this section and the previous section goes through with minor modification for vector-valued Gaussian processes. Thus if $\{X_t : t \in \mathbb{R}^d\}$ is a k-vector stationary Gaussian process, we have that, for each $j = 1, \ldots, d$, $\{X_t^j : t \in \mathbb{R}^d\}$ is a stationary real Gaussian process, and so its covariance can be represented as in (24.12):

$$\rho_{jj}(t) := \mathbf{E}X^j(0)X^j(t) = \int F_{jj}(d\theta)e^{i\theta \cdot t}.$$

By considering more generally the real stationary Gaussian process $a \cdot X_t$, where $a \in \mathbb{R}^k$ is fixed, we deduce the representation

(25.18) $$\rho_{jl}(t) := \mathbf{E}X^j(0)X^l(t) = \int F_{jl}(d\theta)e^{i\theta \cdot t},$$

where, for each Borel $B \subseteq \mathbb{R}^d$, $(F_{jl}(B))$ is a non-negative definite matrix. Likewise, the condition.

$$\sum_j \int F_{jj}(d\theta)|\theta|^{2r+\varepsilon} < \infty$$

will ensure the existence of a C^r version of the vector Gaussian random field.

(25.19) Example: Brownian bridge. A Brownian bridge is an $\mathbb{R}^n$-valued Gaussian process $(X_t)_{0 \leqslant t \leqslant T}$ such that, for $s, t \in [0, T]$,

(25.20) $$\mathbf{E}X_t = at, \qquad \mathbf{E}(X_s X_t^*) - staa^* = \left(s \wedge t - \frac{st}{T}\right)I,$$

where $a \in \mathbb{R}^n$ is fixed, and $T > 0$ is fixed. *Does such a process exist, and does it have a continuous version*? To answer this, we may assume without loss of generality that $a = 0$ (because we could always add the function $t \mapsto at$ to the zero-mean process), and that $n = 1$ (because we could construct each component of the motion separately). There are many ways of proving that such a process exists and has a continuous version (look at Theorem IV. 40.3 for four different representations!) but for now we use the methods developed for general Gaussian processes to prove this.

First, if η is any bounded signed measure on $[0, T]$, we see that

$$\int_{[0,T]} \eta(dx) \int_{[0,T]} \eta(dy)\rho(x, y) := \int_{[0,T]} \eta(dx) \int_{[0,T]} \eta(dy)\left(x \wedge y - \frac{xy}{T}\right)$$
$$= \int_0^T dv \int_v^T \eta(dx) \int_v^T \eta(dy) - \frac{1}{T}\left[\int_0^T dv \int_v^T \eta(dx)\right]^2$$
$$\geqslant 0,$$

by the Cauchy–Schwarz inequality. Hence, by Proposition 24.2, the function $\rho(s, t) := s \wedge t - st/T$ is the covariance of a stationary Gaussian process. (We could also use Lemma 24.4, since ρ is (a multiple of) the Green function of Brownian motion in $[0, T]$, killed when it exits $(0, T)$; this approach is less elementary, though.)

As to the existence of a continuous version, the condition (25.7) of Corollary 25.6 is trivial to verify, and delivers the result immediately.

(*25.21*) *Exercise.* If B is Brownian motion, verify directly that

$$X(t) := at + \frac{T - t}{T} B\left(\frac{tT}{T - t}\right)$$

satisfies (25.20), and conclude that there exists a continuous version of the Brownian bridge.

(*25.22*) *Example: Brownian sheet.* The Brownian sheet is a real-valued two-parameter zero-mean Gaussian process $\{B(s, t) : s, t \geqslant 0\}$ such that

$$\rho((s, t), (u, v)) := \mathbf{E}[B(s, t)B(u, v)] = (s \wedge u)(t \wedge v).$$

The existence of such a Gaussian process follows because

$$\int_{(\mathbb{R}^+)^2} \eta(ds, dt) \int_{(\mathbb{R}^+)^2} \eta(du, dv)(s \wedge u)(t \wedge v) = \int_{(\mathbb{R}^+)^2} dx\, dy \left[\int_{s=x}^{\infty} \int_{t=y}^{\infty} \eta(ds, dt)\right]^2 \geqslant 0$$

proves that the covariance function ρ is non-negative definite (Proposition 24.2). the continuity follows again easily from Corollary 25.6, since ρ is Lipschitz continuous.

It is worth remarking that the process

$$X_\tau(t) := X(\tau, t) := e^{-\tau/2} B(e^\tau, t)$$

is a continuous Gaussian process such that, for each $\tau \in \mathbb{R}$, $(X_\tau(t))_{t \geqslant 0}$ is a standard Brownian motion. Moreover, X_τ is a stationary process, as is easily verified. This 'Brownian motion of Brownian motions' arises in many contexts; see the expository papers in Williams [13], and Walsh [3].

(*25.23*) *Exercise.* Satisfy yourself that an n-parameter Brownian 'sheet' $\{B(t_1,\ldots, t_n): t_n\in\mathbb{R}^+\}$ can be defined just as easily.

26. Isotropic random flows. The study of turbulent fluid flow using stochastic methods has a long history, and has involved many great names in probability and fluid dynamics, including Kolmogorov [1, 2] Taylor [1], Batchelor [1], Itô [8] and Yaglom [1]. The first objective of this work is to construct and classify Gaussian random fields $U:\mathbb{R}^d\to\mathbb{R}^d$ that are not only stationary (with respect to all shifts of the parameter $x\in\mathbb{R}^d$), but also *isotropic*, which means that, for each $G\in O(d)$,

(26.1) $$(GU(x))_{x\in\mathbb{R}^d} \overset{\mathcal{D}}{=} (U(Gx))_{x\in\mathbb{R}^d}.$$

Physically, this means that the random field U 'looks the same' in all coordinate systems. If we assume as usual that U is zero-mean, and define the covariance function

(26.2) $$\rho^{jk}(x) := \mathbf{E}[U^j(0)U^k(x)]$$

as before, then the condition (26.1) is equivalent to

(26.3) $$\rho(Gx) = G\rho(x)G^T, \quad \forall G\in O(d),\ \forall x\in\mathbb{R}^d.$$

In terms of the spectral measure representation (25.18), if we could assume that F had a smooth density $F_{jk}(d\theta) = f_{jk}(\theta)\,d\theta$, then the isotropy condition (26.1) would be equivalent to

(26.4) $$f(G\theta) = Gf(\theta)G^T, \quad \forall G\in O(d),\ \forall\theta\in\mathbb{R}^d.$$

(The assumption that F has a smooth density is harmless; if ρ is an isotropic covariance then $\rho^\varepsilon(x) := e^{-\varepsilon|x|^2}\rho(x)$ is another isotropic covariance with a spectral measure that *does* have a smooth density.) For concreteness, we shall from now on assume that, for some $\varepsilon>0$,

(26.5) $$\sum_j \int F_{jj}(d\theta)|\theta|^{2+\varepsilon} < \infty,$$

so that the random field U has a C^1 version. What does an isotropic random field look like? The next result partly answers this.

(*26.6*) PROPOSITION. *If ρ is isotropic then it may be represented in the form*

(26.7) $$\rho^{jk}(x) = \rho_L(r)\frac{x^jx^k}{r^2} + \rho_N(r)\left(\delta^{jk} - \frac{x^jx^k}{r^2}\right),$$

where $r := |x|$, and ρ_L, ρ_N are two continuous functions such that $\rho_L(0) = \rho_N(0)$.

Proof. Take $e_1 = (1, 0, \ldots, 0)^T\in\mathbb{R}^d$ and note that by (26.3),

(26.8) $$\rho(re_1) = G\rho(re_1)G^T$$

for all $G \in O(d)$ for which $Ge_1 = e_1$. But the G that fix e_1 are exactly those expressible as

$$G = \left(\begin{array}{c|c} 1 & 0 \\ \hline 0 & R \end{array} \right), \tag{26.9}$$

where $R \in O(d-1)$. If (26.8) holds for all G of the form (26.9), it is easy to deduce that $\rho(re_1)$ must be of the form

$$\rho(re_1) = \left(\begin{array}{c|c} \rho_L(r) & 0 \\ \hline 0 & \rho_N(r) I_{d-1} \end{array} \right) = \rho_L(r) e_1 e_1^T + \rho_N(r)(I - e_1 e_1^T),$$

where I_{d-1} is the $(d-1) \times (d-1)$ identity matrix, and ρ_L, ρ_N are some continuous functions. The result now follows by rotating the generic $x \in \mathbf{R}^d$ to be a multiple of e_1. □

This result is far from a complete characterisation of an isotropic covariance, since we know little about the functions ρ_L and ρ_N. The following result leads us to a complete description of ρ_L and ρ_N.

(*26.10*) COROLLARY. *Let σ be the surface Lebesgue measure on S^{d-1}. Then the spectral measure $(F_{jk}(\cdot))_{j,k=1,\dots,d}$ is the spectral measure of an isotropic covariance if and only if there exist measures μ_P and μ_S on $(0, \infty)$ and a constant $\gamma \geqslant 0$ such that, for any $h \in C_K^\infty(\mathbf{R}^d)$,*

$$\int h(\theta) F_{jk}(d\theta) = \int_{S^{d-1}} \sigma(du) \int_{(0,\infty)} h(ru)[u^j u^k \mu_P(dr) + (\delta^{jk} - u^j u^k)\mu_S(dr)] + \gamma h(0) \delta^{jk}. \tag{26.11}$$

Proof. Suppose first that F is isotropic. If F has a smooth density f then f satisfies (26.4). But then Proposition 26.6 implies that

$$f_{jk}(\theta) = \varphi_P(|\theta|) \theta^j \theta^k |\theta|^{-2} + \varphi_S(|\theta|)(\delta^{jk} - \theta^j \theta^k |\theta|^{-2}) \tag{26.12}$$

for some smooth φ_P and φ_S. Thus, for $h \in C_K^\infty(\mathbf{R}^d)$,

$$\int h(\theta) F_{jk}(d\theta) = \int_{S^{d-1}} \sigma(du) \int_0^\infty r^{d-1}\, dr\, h(ru) f_{jk}(ru),$$

which is of the form (26.11) with $\gamma = 0$, $\mu_P(dr) := r^{d-1} \varphi_P(r)\, dr$, and $\mu_S(dr) := r^{d-1} \theta_S(r)\, dr$. Moreover, for radially symmetric h,

$$\sum_j \int h(\theta) F_{jj}(d\theta) = c_d \int_0^\infty h(r)[\mu_P(dr) + (d-1)\mu_S(dr)], \tag{26.13}$$

where $c_d := \sigma(S^{d-1})$. To dispense with the assumption of a smooth density, let F^ε be the spectral measure corresponding to covariance $e^{-\varepsilon|x|^2}\rho(x)$ and observe that (26.13) with $h \equiv 1$ shows that the measures μ_P^ε and μ_S^ε have bounded total mass, and taking $h(\theta) = |\theta|^2$ in (26.13) shows that μ_P^ε and μ_S^ε have bounded second moment and so are tight. Taking a weakly convergent subsequence, we derive the limiting form (26.11), the constant γ appearing because in the limit μ_P and μ_S could put mass on 0.

For the converse statement, if F has the form (26.11) then

$$(26.14)\quad \rho^{jk}(x) = \int_{S^{d-1}} \sigma(du) \int_0^\infty \cos(ru\cdot x)[u^j u^k \mu_P(dr) + (\delta^{jk} - u^j u^k)\mu_S(dr)] + \gamma\delta^{jk}$$

$$:= \delta^{jk} \int_0^\infty g(rx)\mu_S(dr) + \int_0^\infty g_{jk}(rx)(\mu_P - \mu_S)(dr) + \gamma\delta^{jk},$$

where

$$g(v) := \int_{S^{d-1}} \cos(v\cdot u)\sigma(du),$$

$$g_{jk}(v) := \int_{S^{d-1}} u^j u^k \cos(v\cdot u)\sigma(du).$$

Note that we can express

$$g(v) = (2\pi)^{(d-1)/2}|v|^{-(d-2)/2} J_{(d-2)/2}(|v|) := \psi(|v|),$$

$$g_{jk}(v) = -\frac{\partial^2}{\partial v^j \partial v^k} g(v)$$

$$= -\frac{\psi'(|v|)}{|v|}\left(\delta^{jk} - \frac{v^j v^k}{|v|^2}\right) - \psi''(|v|)\frac{v^j v^k}{|v|^2}$$

$$= (2\pi)^{(d-1)/2}\left[\delta^{jk}|v|^{-d/2} J_{d/2}(|v|) - |v|^{-(d-2)/2} J_{(d+2)/2}(|v|)\frac{v^j v^k}{|v|^2}\right],$$

where $J_\nu(\cdot)$ is the standard Bessel function (see Watson [1]), using the well-known identities

$$J_{\nu-1}(z) + J_{\nu+1}(z) = \frac{2\nu}{z} J_\nu(z),$$

$$J_{\nu-1}(z) - J_{\nu+1}(z) = 2J'_\nu(z).$$

Abbreviating $(2\pi)^{(d-1)/2}$ to α_d, $\frac{1}{2}(d-2)$ to ν and setting

$$H_\nu(x) = \alpha_d J_\nu(|x|)|x|^{-\nu}, \qquad x \in \mathbf{R}^d,$$

we obtain from (26.14), after some calculations,

$$(26.15)\quad \rho^{jk}(x) = \gamma\delta^{jk} + \left(\delta^{jk} - \frac{x^j x^k}{|x|^2}\right)\left\{\int_0^\infty \mu_S(dr)[H_\nu(rx) - H_{\nu+1}(rx)] + \int_0^\infty \mu_P(dr) H_{\nu+1}(rx)\right\} + \frac{x^j x^k}{|x|^2}\left\{\int_0^\infty \mu_S(dr)(d-1)H_{\nu+1}(rx) + \int_0^\infty \mu_P(dr)[H_{\nu+1}(rx) - |rx|^2 H_{\nu+2}(rx)]\right\}.$$

Thus we have expressed ρ^{jk} in the form (26.7) with

$$(26.16)\quad \rho_L(|x|) := \rho_{PL}(|x|) + \rho_{SL}(|x|), \qquad \rho_N(|x|) := \rho_{PN}(|x|) + \rho_{SN}(|x|),$$

where

$$(26.17)\ \text{(i)}\quad \rho_{PL}(|x|) = \int_0^\infty [H_{\nu+1}(rx) - |rx|^2 H_{\nu+2}(rx)]\mu_P(dr),$$

$$(26.17)\ \text{(ii)}\quad \rho_{SL}(|x|) = \int_0^\infty (d-1)H_{\nu+1}(rx)\mu_S(dr),$$

$$(26.17)\text{(iii)}\quad \rho_{PN}(|x|) = \int_0^\infty H_{\nu+1}(rx)\mu_P(dr),$$

$$(26.17)\text{(iv)}\quad \rho_{SN}(|x|) = \int_0^\infty [H_\nu(rx) - H_{\nu+1}(rx)]\mu_S(dr)$$

□

Thus Corollary 26.10 not only characterises completely the spectral measures of an isotropic covariance, but also gives a representation (26.16), (26.17) for the possible isotropic covariances themselves. Let us now explain the choice of the subscripts P and S for the measures; *P stands for 'potential', S stands for 'solenoidal'*. Indeed, the general isotropic covariance can be expressed, according to (26.15), as

$$(26.18)\quad \rho(x) = \gamma\delta^{jk} + \rho_P(x) + \rho_S(x),$$

where

$$\rho_P^{jk}(x) := \left(\delta^{jk} - \frac{x^j x^k}{|x|^2}\right)\rho_{PN}(|x|) + \frac{x^j x^k}{|x|^2}\rho_{PL}(|x|),$$

$$\rho_S^{jk}(x) := \left(\delta^{jk} - \frac{x^j x^k}{|x|^2}\right)\rho_{SN}(|x|) + \frac{x^j x^k}{|x|^2}\rho_{SL}(|x|),$$

are the covariance of two isotropic Gaussian random fields, as is $\gamma\,\delta^{jk}$ (the latter being the covariance of the trivial field $U(x) = Y$, $\forall x$, where $Y \sim N(0, \gamma I)$). *Thus we have decomposed the general isotropic Gaussian random field into the sum of three independent isotropic Gaussian random fields, with covariances ρ_P, ρ_S and γI respectively*. The random field with covariance ρ_S actually is solenoidal (that is, divergence-free), as we confirm by computing ($\partial_j := \partial/\partial x^j$)

$$\begin{aligned} \mathbf{E}[\operatorname{div} U(0)]^2 &= \mathbf{E}\Big(\sum_j \partial_j U^j(0)\Big)^2 \\ &= \sum_j \sum_k \mathbf{E}[\partial_j U^j(0) \partial_k U^k(0)] \\ &= -\sum_j \sum_k \partial_j \partial_k \rho^{jk}(0) \\ &= \sum_j \sum_k \int F_{jk} d(\theta) \theta^j \theta^k \\ &= 0 \end{aligned}$$

if F is given by (26.11) with $\mu_P = 0$ and $\gamma = 0$.

Next, we check that the random field with covariance ρ_P is actually a potential (that is, curl-free: $\partial_j U^k = \partial_k U^j$, $\forall j, k$.) For this, we just need to compute

$$\begin{aligned} \mathbf{E}[\partial_j U^k(0) - \partial_k U^j(0)]^2 &= -\partial_j \partial_j \rho^{kk}(0) + 2\partial_j \partial_k \rho^{jk}(0) - \partial_k \partial_k \rho^{jj}(0) \\ &= \int [\theta_j^2 F_{kk}(d\theta) - 2\theta_j \theta_k F_{jk}(d\theta) + \theta_k^2 F_{kk}(d\theta)] \\ &= 0 \end{aligned}$$

if F is given by (26.11) with $\mu_S = 0$ and $\gamma = 0$. If a C^1 vector field U is curl-free, then there is some real C^2 function $V := \mathbf{R}^d \to \mathbf{R}$ such that $U = \operatorname{grad} V$. In this context, it is natural to ask whether, for a curl-free isotropic Gaussian random field U, we could find a stationary Gaussian process $(V(x))_{x \in \mathbf{R}^d}$ such that $U = \operatorname{grad} V$. This cannot always be done; as an exercise, check that the condition that permits such a representation is

$$\int \sum_j F_{jj}(d\theta)/|\theta|^2 := c_d \int_0^\infty r^{-2} \mu_P(dr) < \infty.$$

The interest of this decomposition is that the solenoidal flows preserve Lebesgue measure and so correspond to an incompressible flow of fluid.

Let us recall that the reason that isotropic Gaussian fields are of interest is as a model for turbulent fluid flow; and any turbulent fluid flow must be expected to be changing in time, so should be modelled by some random field $(U(t, x): t \in \mathbf{R}, x \in \mathbf{R}^d)$ with values in $\mathbf{R}^d$. In view of what we have already done, we can set up a simple model for such a time-varying Gaussian random field by taking

$$\mathbf{E}[U^j(t, x) U^k(s, y)] = \gamma(s, t) \rho^{jk}(x - y), \tag{26.19}$$

where ρ is an isotropic covariance, and γ is the covariance of a real Gaussian process indexed by $\mathbb{R}$. The reason we do not assume that γ is the covariance of a stationary Gaussian process is that we may wish to use $\gamma(s,t) := s \wedge t$, the Brownian-motion covariance. Indeed, if we do so then, for each x, $(U(t,x))_{t\geq 0}$ is a Brownian motion in $\mathbb{R}^d$, and the correlation between the different Brownian motions $U(\cdot, x)$ is given by ρ. This suggests the intriguing notion of studying the motion of a particle dropped into this turbulent flow; if the particle starts at x, and is at position $X_t(x)$ at time t, then, in some sense, we should have infinitesimally

$$d(X_t(x)) = dU(t, X_t(x)). \tag{26.20}$$

Good sense can be made of this; see Baxendale and Harris [1] and Kunita [3]. It will not come as any surprise that the process $(X_t(x))_{t\geq 0}$ is a Brownian motion. But much more interesting is to study the *flow* $(X_t(x))_{t\geq 0, x\in\mathbb{R}^d}$, which tells us not only how individual particles move, but also how they move relative to each other. There are many fascinating and beautiful results here; see Baxendale [1], Baxendale and Harris [1]. Carverhill [1, 2] Harris [1], Kunita [3], Le Jan [1–4], Le Jan and Watanabe [1] and the many references therein for a view of what is known. We have already looked ahead way beyond the scope of this volume, and must now end our discussion of isotropic random fields; we hope that what we have said will help you get started on papers such as Baxendale and Harris [1], Le Jan [4] and Yaglom [1].

(*26.21*) *Exercise.* Confirm that (26.19) is the covariance of a Gaussian process.

27. Dynkin's Isomorphism Theorem. Since our entire discussion of Gaussian processes is by way of a digression motivated by interest, we make no apologies for describing briefly here (a caricature of) Dynkin's work on Gaussian processes and local-time fields, even though we shall develop it no further. This is appropriate for the mysterious but powerful result that we are about to discuss, since it is clear that its full potential is yet to be explored; we point to the paper of Marcus and Rosen [1], where continuity results for local-time fields are deduced from continuity results for Gaussian processes using Dynkin's ideas, and to the paper of Sheppard [1], which deduces the classical Ray–Knight Theorem on Brownian local time using the Dynkin's result.

We shall consider a sub-Markovian process $(X_t)_{t\geq 0}$ with values in a finite set I, and a transition semigroup $(P_t)_{t\geq 0}$ that is symmetric and assumed to be integrable and irreducible:

$$0 < g(x,y) := \int_0^\infty p_t(x,y)\,dt < \infty, \quad \forall x, y \in I.$$

Let Q denote the Q-matrix of the chain X: $Q = \dot{P}(0)$.

We shall also consider a zero-mean Gaussian process $(\varphi_x)_{x\in E}$ with covariance $\mathbf{E}(\varphi_x\varphi_y) = g_{xy}$, independent of X, and introduce the notation $\mathbf{P}^{x\to y}$ for the law of X started at x and conditioned to die at y; precisely, this is the process with the sub-Markov transition function

$$p_t^y(a,b) = p_t(a,b)g(b,y)/g(a,y). \tag{27.1}$$

(27.2) THEOREM (Dynkin). *If $F: \mathbf{R}^I \to \mathbf{R}^+$ is any bounded measurable function then, for any $x, y \in I$,*

$$\mathbf{E}\varphi_x\varphi_y F(\tfrac{1}{2}\varphi^2) = \mathbf{E}^{x\to y}F(\tfrac{1}{2}\varphi^2 + T)g_{xy}, \tag{27.3}$$

where $T_a := \int_0^\infty I_{\{X_t = a\}}\,dt$ is the occupation field of the chain X.

(27.4) *Clarification.* On the left-hand side of (27.3), the expectation is only over the Gaussian field φ. On the right-hand side, X has law $\mathbf{P}^{x\to y}$, and φ is independent of X.

Proof. It clearly suffices to prove the theorem only for F of the form

$$F(\xi) = \exp\left(-\sum_{a\in E}\lambda_a\xi_a\right),$$

which we now assume. Let Λ denote the diagonal matrix $(\mathrm{diag}(\lambda_a))$.

The inclusion of the weighting $F(\frac{1}{2}\varphi^2)$ changes the law of the Gaussian field from $N(0, G)$ to $N(0, (\Lambda - Q)^{-1})$; in particular,

$$\frac{\mathbf{E}\varphi_x\varphi_y F(\frac{1}{2}\varphi^2)}{\mathbf{E}F(\frac{1}{2}\varphi^2)} = (\Lambda - Q)^{-1}(x,y). \tag{27.5}$$

This leaves us just to confirm that

$$g_{xy}\mathbf{E}^{x\to y}F(T) = (\Lambda - Q)^{-1}(x,y), \tag{27.6}$$

since then (27.3) follows immediately from (27.5) and (27.6). Under $\mathbf{P}^{x\to y}$, the process X is a Markov chain with Q-matrix

$$\tilde{Q} := D^{-1}QD,$$

where $D = \mathrm{diag}(g(a,y))$, as we deduce immediately from (27.1). The problem is thus to evaluate

$$\mathbf{E}^x\exp\left[-\int_0^\zeta \lambda(\tilde{X}_s)\,ds\right], \tag{27.7}$$

where $\tilde{X}$ is a chain with Q-matrix $\tilde{Q}$, dying at ζ, whereupon it is sent to a graveyard ∂. The interpretation of (27.7) is that the process $\tilde{X}$ is also being killed at rate $\lambda(\cdot)$ and being sent to a graveyard ∂', and (27.7) is the probability that the process ends in ∂ rather than ∂'. The Q-matrix of the process on

$I \cup \{\partial, \partial'\}$ is thus

$$\begin{array}{c} \\ \\ \end{array}\begin{pmatrix} \tilde{Q}-\Lambda & -\tilde{Q}1 & \Lambda 1 \\ 0 & 0 & 0 \\ 0 & 0 & 0 \end{pmatrix}, \quad \text{columns } I,\ \partial,\ \partial'$$

and from this we compute immediately

$$\begin{aligned} \mathbf{P}^x(\tilde{X} \text{ ends in } \partial) &= (\Lambda - \tilde{Q})^{-1}(-\tilde{Q}1)_x \\ &= D^{-1}(\Lambda - Q)^{-1}D(-\tilde{Q}1)_x. \end{aligned}$$

But

$$\begin{aligned} (\tilde{Q}1)_x &= \sum_a \tilde{q}_{xa} = \sum_a q_{xa}g_{ay}/g_{xy} \\ &= -\delta_{xy}/g_{xy}, \end{aligned}$$

so that

$$\mathbf{P}^x(\tilde{X} \text{ ends in } \partial) = \frac{1}{g_{xy}}(\Lambda - Q)^{-1}(x, y),$$

completing the proof. □

Lévy processes

28. Lévy processes. The aim of this section is to give the briefest of introductions to the theory of Lévy processes. Inclusion of this material is justified because Brownian motion is a Lévy process; Lévy processes also provide one of the most important examples of the Markov processes studied in Chapter III and the semimartingales of Chapter VI.

(*28.1*) DEFINITION. *A process* $(X_t)_{t\geq 0}$ *with values in* $\mathbf{R}^d$ *is called a* Lévy process *(or* process with stationary independent increments*) if it has the properties*

(28.2) (i) *for almost all* ω, $t \to X_t(\omega)$ *is right continuous on* $[0, \infty)$, *with left limits on* $(0, \infty)$;

(28.2) (ii) *for* $0 \leq t_0 \leq t_1 < \cdots < t_n$, *the random variables* $Y_j := X_{t_j} - X_{t_{j-1}}$ $(j = 1, \ldots, n)$ *are independent*;

(28.2)(iii) *the law of* $X_{t+h} - X_t$ *depends on* h, *but not on* t.

The analytic theory of the semigroups associated with Lévy processes is the same as the theory of infinitely divisible distributions.

(*28.3*) DEFINITION. *A probability measure* μ *on* $\mathbf{R}^d$ *is infinitely divisible if, for* *each n, there is a probability μ_n on $\mathbb{R}^d$ such that if $V_1, \ldots, V_n$ are independent with law μ_n then*

$$V_1 + \cdots + V_n \stackrel{\mathcal{D}}{=} \mu.$$

It is clear that if X is a Lévy process then the law of X_1 is infinitely divisible. The converse, that any infinitely divisible law is the law of X_1 for some Lévy process, X, will follow from the central result of the theory of Lévy processes.

(*28.4*) THEOREM (Lévy–Khinchin representation) *For each $b \in \mathbb{R}^d$, each non-negative definite symmetric $d \times d$ matrix Σ and each measure ν on $\mathbb{R}^d \backslash \{0\}$ satisfying the integrability condition*

$$\int (|x|^2 \wedge 1)\nu(dx) < \infty, \tag{28.5}$$

the function

$$\varphi(\theta) = \exp[\psi(\theta)], \quad \theta \in \mathbb{R}^d, \tag{28.6}$$

is the characteristic function of an infinitely divisible law (here we write

$$\begin{aligned} \psi(\theta) := {} & ib \cdot \theta - \tfrac{1}{2}\theta^T \Sigma \theta \\ & + \int (e^{i\theta \cdot x} - 1 - i\theta \cdot x I_{(|x| \leq 1)})\nu(dx), \end{aligned} \tag{28.7}$$

the characteristic exponent *of the law.) Moreover, the characteristic function of any infinitely divisible law on $\mathbb{R}^d$ may be represented in this way, with the representing triple (b, Σ, ν) being uniquely determined.*

Proof. We refer the reader to Section 9.5 of Breiman [1] or Section XVII.2 of Feller [1] for the one-dimensional case. For the general situation, see Fristedt [1]. □

The measure ν in (28.7) is called the *Lévy measure* of the infinitely divisible law. An important extension of the definition of infinite divisibility is the following.

(*28.8*) THEOREM. Let *X be a random variable with the property that, for each n*

$$X \stackrel{\mathcal{D}}{=} \sum_{j=1}^{n} X_{nj}, \tag{28.9}$$

where the X_{nj} are independent, and, for each $\varepsilon > 0$,

$$\lim_{n \to \infty} \sup_{j \leq n} \mathbf{P}(|X_{nj}| > \varepsilon) = 0. \tag{28.10}$$

Then X is infinitely divisible.

Proof. See Breiman [1].

The *uniform asymptotic negligibility condition* (28.10) is commonly encountered in limit theory; it is the simplest condition one could impose to prevent any one of the summands X_{nj} from contributing noticeably to the sum. As an example, the first passage times of one-dimensional diffusions are clearly infinitely divisible using Theorem 28.8, but not obviously so using Definition 28.3. As an exercise, check that if, for each n, the (X_{nj}) have a common distribution then (28.10) holds.

We now show how any infinitely divisible law (with characteristic exponent ψ of the form (28.7)) may be realised as the law of X_1 for some Lévy process X. The law of X_t must be given by

$$\mathbf{E}\exp(i\theta\cdot X_t)=\exp[t\psi(\theta)],$$

the right-hand side being the characteristic function of an infinitely divisible law by Theorem 28.4. Thus if μ_t is the law of X_t, for any $t,s\geqslant 0$

$$\mu_{t+s}=\mu_t*\mu_s,$$

so we may define a Markovian semigroup $(P_t)_{t\geqslant 0}$ by

$$P_tf(x):=\int f(x+y)\mu_t(dy),\quad \forall f\in C_0(\mathbf{R}^d).$$

It is easy to see that, for each $t\geqslant 0$, P_t: $C_0(\mathbf{R}^d):=\{$continuous functions on $\mathbf{R}^d$ which vanish at $\infty\}\to C_0(\mathbf{R}^d)$. Hence $(P_t)_{t\geqslant 0}$ is a Feller–Dynkin semigroup (see Section III.6), and by Theorem III.7.17, there exists a process X with paths that are right-continuous on $[0,\infty)$ with left limits on $(0,\infty)$ and with transition semigroup $(P_t)_{t\geqslant 0}$, and such that $X_0=0$. It is now clear that X is a Lévy process, and $X_1\sim\mu_1$ as claimed.

The result to which we have just appealed is quite technical, but once we have got it, it allows us to pass from an analytical description of the process (in terms of the convolution semigroup of infinitely divisible laws) to a sample-path description, which is far more powerful. To understand what we have gained by doing this, we are going to prove Lévy's Theorem that the only continuous Lévy processes are drifting Brownian motions (indeed, without sample paths, what does this theorem mean?!) First, though, we discuss the 'building-block' Lévy process—the *compound Poisson process.*

Consider a process that is constructed from a standard Poisson counting process $(N_t)_{t\geqslant 0}$ of rate $\lambda>0$, and an independent sequence $Y_1, Y_2,\ldots$ of IID random variables with distribution function F as follows: we define

$$X_t=\sum_{j=1}^{N_t}Y_j. \tag{28.11}$$

It is clear that the paths of X are right-continuous with left limits. Moreover, the increments of X over disjoint intervals are independent, and $X_{t+s}-X_t$ has the law of a sum of a Poisson number of copies of Y_1:

$$\mathbf{E}\exp[i\theta\cdot(X_{t+s}-X_t)] = \sum_{n\geq 0}\frac{(\lambda s)^n e^{-\lambda s}}{n!}\int e^{i\theta\cdot x}F(dx)$$

$$= \exp\left[s\int(e^{i\theta\cdot x}-1)\lambda F(dx)\right].$$

Thus X is a Lévy process, and if we compare with (28.7), we see that in the representation of X, $\Sigma \equiv 0$, and ν has finite mass. We can now prove Lévy's Theorem.

(*28.12*) THEOREM (Lévy). *If X is a continuous Lévy process in $\mathbf{R}^d$ then X is expressible in the form*

(28.13) $$X_t = \sigma B_t + bt$$

for some $b\in\mathbf{R}^d$, and σ a $d\times d$ matrix.

Proof. Let us suppose that the characteristic exponent of X is given by (28.7); our task is then to prove that $\nu\equiv 0$. Fixing $\varepsilon\in(0,1)$, we construct on some suitable probability space two independent Lévy processes X^ε and Y^ε with characteristic exponents

$$\psi_\varepsilon(\theta) := ib\cdot\theta - \tfrac{1}{2}\theta^T\Sigma\theta + \int_{|x|\leq\varepsilon}(e^{i\theta\cdot x}-1-i\theta\cdot x)\nu(dx),$$

$$\psi^\varepsilon(\theta) := \psi(\theta) - \psi_\varepsilon(\theta)$$

respectively. Then

$$\psi^\varepsilon(\theta) = \int_{|x|>\varepsilon}\{e^{i\theta\cdot x}-1-i\theta I_{(|x|\leq 1)}\}\,\nu(dx)$$

is the characteristic exponent of a *compound Poisson process*, since $\nu(\{x: |x|>\varepsilon\})<\infty$. Thus Y^ε has *only finitely many jumps in any bounded time interval*. Since X^ε and Y^ε are independent, their sum is a Lévy process with characteristic exponent $\psi_\varepsilon + \psi^\varepsilon = \psi$, so that $X^\varepsilon + Y^\varepsilon$ is (a copy of) X. Now, since X^ε is a Lévy process, has right-continuous paths with left limits, there can be only countably many discontinuities of X^ε in $[0,1]$. Let $\mathscr{D}_\varepsilon$ denote this countable random set. Since the jumps of Y^ε come at the times of a Poisson process independent of X^ε, we have immediately that, almost surely, *no jump time of Y^ε falls in $\mathscr{D}_\varepsilon$*. Thus any jump time of Y^ε will actually be a jump time of $X^\varepsilon + Y^\varepsilon = X$; but X is supposed to be continuous. The only possibility then is that Y^ε *has no jumps*, which is to say

$$\nu(\{x:|x|>\varepsilon\}) = 0.$$

Since $\varepsilon>0$ was arbitrary, the expression (28.7) for the characteristic exponent of X collapses to

$$\psi(\theta) = ib\cdot\theta - \tfrac{1}{2}\theta^T \Sigma \theta,$$

and taking $\sigma = \Sigma^{1/2}$ yields the form (28.13), as required. □

We conclude this section with a few examples of common Lévy processes (or, equivalently, infinitely divisible distributions). The Gaussian distribution we already know about.

(*28.14*) *Stable processes.* A real-valued Lévy process X is said to be *stable* of index $\alpha \in (0,2]$ if, for any $c > 0$,

(28.15) $$(X_{ct})_{t\geqslant 0} \overset{\mathscr{D}}{=} (c^{1/\alpha} X_t)_{t \geqslant 0}.$$

The family of all (real-valued) stable (α) processes is given by the characteristic exponents

(28.16) $$\psi(\theta) = -c|\theta|^\alpha [1 - i\beta \operatorname{sgn}(\theta) \tan \tfrac{1}{2}\pi\alpha]$$

where $1 < \alpha < 2$, $-1 \leqslant \beta \leqslant 1$, or $0 < \alpha < 1$, $-1 \leqslant \beta \leqslant 1$. For $\alpha = 1$, the exponent is of the form $\psi(\theta) = -c|\theta| + i\mu\theta$.

The representation (b, Σ, ν) in terms of c, β and α is, for $0 < \alpha < 1$,

(28.17) $$\frac{\nu(dx)}{dx} = |x|^{-1-\alpha} [\tfrac{1}{2}(1+\beta) I_{(x>0)} + \tfrac{1}{2}(1-\beta) I_{(x<0)}] \frac{\alpha c}{\Gamma(1-\alpha)\cos\frac{1}{2}\alpha\pi},$$

$$\Sigma = 0, \qquad b = \int_{-1}^{1} x\,\nu(dx) = \frac{\beta}{1-\alpha} \frac{\alpha c}{\Gamma(1-\alpha)\cos\frac{1}{2}\alpha\pi},$$

and, for $1 < \alpha < 2$, exactly the same formula is valid (now, note, $\cos\frac{1}{2}\alpha\pi < 0$).

The representation of $\psi(\theta) = -c|\theta| + i\mu\theta$ in the case $\alpha = 1$ is achieved by

(28.18) $$\frac{\nu(dx)}{dx} = \frac{c}{\pi x^2}, \qquad \Sigma = 0, \qquad b = \mu.$$

The case $\alpha = 2$ is just the Brownian case.

One special case is that of the *symmetric Cauchy process*:

(28.19) $$\psi(\theta) = -|\theta|.$$

The Cauchy process arises naturally in two-dimensional Brownian motion; if X and Y are independent BM_0 processes, $\tau_t := \inf\{u : X_u = t\}$, then $Y(\tau_t)$ is a Cauchy process, as you can easily verify from the fact that, for $\lambda > 0$,

$$\mathbf{E}\exp(-\lambda\tau_t) = \exp(-t\sqrt{2\lambda}).$$

(See (9.1).) The *asymmetric Cauchy process* has

(28.20) $$\psi(\theta) = -\tfrac{1}{2}\pi|\theta| - i\theta\log|\theta|$$

and

(28.21) $$\nu(dx) = x^{-2}\,dx\, I_{(x>0)}, \qquad b = \Sigma = 0.$$

Note that the asymmetric Cauchy process is *not* stable.

In view of the interpretation of ν as the 'jump measure', it is natural (and correct) to think that the asymmetric Cauchy process Z can only jump upward; and it is natural (and incorrect) to think that, for some large enough $c > 0$, $Z_t + ct$ is increasing. The reason that this is incorrect is somewhat mysterious, and is to do with the fact that, for any $t > 0$, $\sum_{s \leqslant t} |\Delta Z_s| = +\infty$, almost surely. We discuss this further in Section VI.2, but raise the issue here to emphasise that Lévy processes for which

$$\int (|x|^2 \wedge 1)\nu(dx) < \infty, \qquad \int (|x| \wedge 1)\nu(dx) = +\infty$$

are the most mysterious by all. We shall prove straight away that we cannot make $Z_t + ct$ increasing, however large c is, by characterising all increasing Lévy processes.

(*28.22*) *Subordinators.* A subordinator is simply an increasing Lévy process; examples include $X_t = t$ and compound Poisson processes with positive jumps. We shall prove the following characterisation.

(*28.23*) THEOREM. *The distribution F on $\mathbb{R}^+$ is infinitely divisible if and only if there is a representation*

$$\int_{[0,\infty)} e^{-\lambda x} F(dx) = \exp\left[-c\lambda - \int_0^\infty (1 - e^{-\lambda x})\mu(dx) \right] \tag{28.24}$$

for some $c \geqslant 0$, and measure μ on $(0, \infty)$ satisfying the integrability condition

$$\int_0^\infty (x \wedge 1)\mu(dx) < \infty. \tag{28.25}$$

(*28.26*) *Remarks*

(i) The function

$$\gamma(\lambda) := c\lambda + \int_0^\infty (1 - e^{\lambda x})\mu(dx) \tag{28.27}$$

is called the *Laplace exponent* of the infinitely divisible law. (Our notation γ is not standard.)

(ii) The method of proof of Theorem 28.23 contains most of the essential steps of the proof of Theorem 28.4, but simplified by the fact that we work with Laplace transforms.

(iii) The measure (28.21) does not satisfy the integrability condition (28.25), so the law of the asymmetric Cauchy process with positive drift c cannot be an infinitely divisible law on $\mathbb{R}^+$.

Proof of Theorem 28.23. Suppose first that F is infinitely divisible, $F(0) = 0$.

Define

$$\tilde{F}(\lambda) := \int_{[0,\infty)} e^{-\lambda x} F(dx) \in (0, 1],$$

and observe that, for each $n \in \mathbb{N}$, there is an nth root F_n of F with

$$\tilde{F}_n(\lambda) := \int_{[0,\infty)} e^{-\lambda x} F_n(dx) = [\tilde{F}(\lambda)]^{1/n}.$$

Thus as $n \to \infty$, $\tilde{F}_n(\lambda) \to 1$ uniformly on compact sets, and

(28.28)
$$\begin{aligned} \log \tilde{F}(\lambda) &= n \log \tilde{F}_n(\lambda) \\ &= n \log \{1 - [1 - \tilde{F}_n(\lambda)]\} \\ &\leqslant -n[1 - \tilde{F}_n(\lambda)]. \end{aligned}$$

But since $\tilde{F}_n(\lambda) \to 1$ uniformly on compacts, we can assert that, for any $\varepsilon > 0$, and $K > 0$, there is some n_0 such that, for $n \geqslant n_0$, $\lambda \leqslant K$,

$$1 - \tilde{F}_n(\lambda) \leqslant \delta,$$

where $\delta > 0$ is such that, for all $0 \leqslant x \leqslant \delta$,

$$\log(1 - x) \geqslant -(1 + \varepsilon)x.$$

Hence, for $n \geqslant n_0$, $\lambda \leqslant K$, we have

(28.29)
$$\log \tilde{F}(\lambda) = n \log \tilde{F}_n(\lambda) \geqslant -n(1 + \varepsilon)[1 - \tilde{F}_n(\lambda)].$$

The conclusion from (28.28) and (28.29) is that

$$n[1 - \tilde{F}_n(\lambda)] \to -\log \tilde{F}(\lambda) \quad (n \to \infty).$$

But

$$\begin{aligned} n[1 - \tilde{F}_n(\lambda)] &= n \int_{(0,\infty)} (1 - e^{-\lambda x}) F_n(dx) \\ &= \int_{(0,\infty)} \frac{1 - e^{-\lambda x}}{1 - e^{-x}} n(1 - e^{-x}) F_n(dx). \end{aligned}$$

Thus the measures $m_n(dx) := n(1 - e^{-x}) F_n(dx)$ on $(0, \infty)$ have bounded total mass (indeed, $m_n(0, \infty) \to -\log \tilde{F}(1)$), so there is a subsequence down which $m_n \Rightarrow m$, a measure on $[0, \infty]$, and we conclude that

$$n[1 - \tilde{F}_n(\lambda)] \to m(\{0\})\lambda + \int_{(0,\infty)} \frac{1 - e^{-\lambda x}}{1 - e^{-x}} m(dx) + m(\{\infty\}).$$

Writing $c := m(\{0\})$ and $\mu(dx) := (1 - e^{-x})^{-1} m(dx)$ gives the form (28.24) for $-\log \tilde{F}(\lambda) := \gamma(\lambda)$, except for the presence of $m(\{\infty\})$. But letting $\lambda \downarrow 0$ shows that, in fact, $m(\{\infty\}) = 0$, and the proof is complete. □

(28.30) *The gamma process.* Many of the common families of distributions of statistics are actually infinitely divisible, the gamma distribution included. Since the gamma law is concentrated on $\mathbb{R}^+$, we see from Theorem 28.23 that if X_t is a gamma random variable with scale parameter α and shape parameter t (that is, X_t has density $x^{t-1}e^{-\alpha x}\alpha^t/\Gamma(t)$) then

(28.31) $$\mathbf{E}\, e^{-\lambda X_t} = \alpha^t(\lambda+\alpha)^{-t} = \exp\left[-c\lambda t - t\int_0^\infty (1-e^{-\lambda x})\mu(dx)\right]$$

for some $c \geqslant 0$, and μ satisfying (28.25). The reader should have no difficulty in confirming that

(28.32) $$\mu(dx) = e^{-\alpha x}\frac{dx}{x}, \qquad c = 0,$$

makes (28.31) true. The fact that the mean and variance of a gamma law are proportional to t is now obvious from the interpretation of the gamma law as the law of a Lévy process.

(28.33) Among other common distributions, the t-distribution, the lognormal, and reciprocals of gammas are all infinitely divisible; see respectively Grosswald [1], Thorin [1] and Bondesson [1].

Although the Lévy–Khinchin representation looks like all that there is to say about infinite divisibility, it is unfortunately rare that one can exhibit the characteristic function in a sufficiently explicit form to be able to decide infinite divisibility. For that reason, attention has focused on various subclasses of the infinitely divisible laws; see Bondesson [1] for a survey. Pitman and Yor [1] succeeded in explaining the analytical results of Ismail and Kelker [1], for example that, for $\nu > -1$,

$$\lambda \mapsto \lambda^{\nu/2}2^\nu I_\nu(\sqrt{\nu})/\Gamma(\nu+1)$$

is the Laplace transform of an infinitely divisible law, by finding a diffusion additive functional with this law, which was thus obviously infinitely divisible. Pitman and Yor also discovered a whole host of other infinitely divisible laws. This opened a whole new vein in the study of infinite divisibility, and it is fair to say that it is still far from exhausted.

29. Fluctuation theory and Wiener–Hopf factorisation. At the very beginning of this chapter, one of the reasons we gave for studying Brownian motion was that it was sufficiently concrete that many calculations can be done explicitly; in particular, the law of the maximum by time t, or the law of the first passage time to a level, were quite easy to derive using the reflection principle. Generalising to Lévy processes, these problems become very much harder to answer (despite the explicit Lévy–Khinchin representation), and lead us into the realm of Wiener–Hopf factorisation. The whole area is notorious for the lack of good closed-form answers, although various general formulae are known. We shall here present without proof a selection of largely classical results that illustrate what is known. We have drawn extensively on the excellent surveys by Bingham [1] and Fristedt [1].

Let X be a (real-valued) Lévy process, $\bar{X}_t := \sup_{s\leqslant t} X_s$, $\underline{X}_t := \inf_{s\leqslant t} X_s$ and let $\tau_a := \inf\{t>0: X_t > a\}$. Take T to be an exponential random variable of mean η^{-1}, independent of X.

(*29.1*) THEOREM (Spitzer, Rogozin, Pecherskii)

(29.2) (i) $$\mathbf{E}\, e^{i\theta X(T)} = \eta[\eta - \psi(\theta)]^{-1}$$

(29.2) (ii) $$= \mathbf{E}\, e^{i\theta \bar{X}(T)} \mathbf{E}\, e^{i\theta\{X(T)-\bar{X}(T)\}}$$

(29.2)(iii) $$= \mathbf{E}\, e^{i\theta \bar{X}(T)} \mathbf{E}\, e^{i\theta \underline{X}(T)}.$$

Moreover, we have the Spitzer–Rogozin identity

(29.3) $$\mathbf{E}\, e^{i\theta \bar{X}(T)} = \exp\left[\int_0^\infty \frac{e^{-\eta t}\, dt}{t} \int_0^\infty (e^{i\theta x} - 1)\mathbf{P}(X_t \in dx)\right],$$

with the analogous expression for the characteristic function of $\underline{X}(T)$.

(*29.4*) *Remarks.* The equality (29.2)(i) is immediate from the definitions. The equality (29.2)(ii) follows because $\underline{X}(T) \overset{\mathcal{D}}{=} X(T) - \bar{X}(T)$; draw a picture of the sample path and turn it upside down! The equality (29.2)(ii) is the profound statement; although evidently $X_T = \bar{X}_T + (X_T - \bar{X}_T)$, the factorisation (29.2)(ii) follows from the less obvious fact that

(29.5) $$\bar{X}_T \text{ and } \bar{X}_T - X_T \text{ are independent.}$$

This fundamental fact is best understood by excursion theory, and the account given by Greenwood and Pitman [2] is the definitive reference. The excursion-theoretic standpoint is the key to most of distributional identities concerning Lévy processes. We omit the very important identity of Fristedt [1] because we have not the necessary notions yet to state it, but record here a simple but useful identity; for $\lambda \geqslant 0$,

(29.6)
$$\mathbf{E}\, e^{-\lambda \bar{X}(T)} = \eta\left[\eta + \kappa(\eta)\lambda + \int_{(-\infty,0]} \mathbf{P}(\underline{X}(T)\in dy)\int_{-y}^{\infty} \nu(dx)(1 - e^{-\lambda(x+y)})\right]^{-1},$$

where κ is a non-negative increasing function. See Rogers [5] for a derivation and the explanation of the significance of κ, which is related to Fristedt's identity. Spitzer's book [1] gives the random-walk version of (29.3), which was extended to Lévy processes by Rogozin [1].

Although the Wiener–Hopf factor (29.3) can rarely be computed in closed form, the identity (29.3) does yield useful information. By extending from $\theta \in \mathbb{R}$

to $\theta \in \bar{\mathbf{H}} = \{z \in \mathbf{C}: \operatorname{Im} z \geqslant 0\}$ and letting $\theta = i\lambda \to i\infty$, $\lambda > 0$, we deduce that

(29.7) $\mathbf{P}(\tau_0 > 0) = 0$ *or* 1 *according as* $\int_{0+} (dt/t)\mathbf{P}(X_t > 0) = +\infty$ *or* $< +\infty$.
Rogozin [1] deduces the test

(29.8) $\mathbf{P}(\bar{X}_\infty = +\infty) = 0$ *or* 1 *according as* $\int^\infty (dt/t)\mathbf{P}(X_t > 0) < \infty$ *or* $= +\infty$.

If one restricts attention to the case of spectrally one-sided Lévy processes (those for which the measure ν in the Lévy–Khinchin representation (28.7), puts no mass on one or other of the half-lines) then more complete results may be obtained, simply because, if $\nu(\mathbf{R}^+) = 0$, say, the law of $\bar{X}(T)$ is exponential. This is probabilistically obvious (why?—recall that X has no upward jumps), but can be seen immediately from (29.6). It may not be easy to work out the rate of the exponential, but at least one has in principle both of the Wiener–Hopf factors, and some expressions for them. For example, if X were spectrally positive, and $-\underline{X}(T) \sim \exp[\beta(\eta)]$, then, from (29.6),

$$\mathbf{E}\exp[-\lambda \bar{X}(T)] = \eta\left[\eta + \lambda\kappa(\eta) + \int_0^\infty \nu(dx)\int_0^x \beta e^{-\beta y}(1 - e^{\lambda(x-y)})\,dy\right]^{-1},$$

and it can be shown that $\kappa(\eta) > 0$ if and only if $\sigma > 0$ (that is, there is a Brownian component; see Rogers [5, Theorem 3]).

Distributional results for Lévy processes continue to be discovered; see Doney [1,2] for particularly interesting recent ones.

30. Local times of Lévy processes. Of the many sample-path properties of Lévy processes, some of the most interesting are to do with existence and properties of a local-time process. In this section, we give a brief introduction to the main results in this area, which are due to Kesten, Bretagnolle, Blumenthal and Getoor, Barlow and Hawkes. Throughout, we assume that the (real-valued) Lévy process X is *not a compound Poisson process*, this case being a trivial complication.

Define $C \subseteq \mathbf{R}$ by

$$C := \{x \in \mathbf{R} : \mathbf{P}(X_t = x \text{ for some } t > 0) > 0\}.$$

(30.1) THEOREM (Kesten, Bretagnolle). *Either* $C = \varnothing$, *or else* $\operatorname{Leb}(C) > 0$. *If the latter alternative obtains then C is one of* $\mathbf{R}$, $(0, \infty)$ *or* $(-\infty, 0)$. *A necessary and sufficient condition for* $\operatorname{Leb}(C) > 0$ *is*

(30.2) $$\int_{\mathbf{R}} \operatorname{Re}\left[\frac{1}{1 - \psi(\theta)}\right] d\theta < \infty.$$

The original proof of this result is to be found to Kesten [1]; Bretagnolle [1] was able to simplify Kesten's approach by working directly with the potentials of singletons, rather than little intervals.

Let m_t (respectively m) denote the occupation measure by time t (respectively the 1-discounted occupation measure):

$$m_t(A) := \int_0^t I_A(X_s)\,ds, \qquad m(A) := \int_0^\infty e^{-t} I_A(X_t)\,dt.$$

Let L_t (respectively L) denote the density of m_t, (respectively m), if such densities exist. The following attractive result decides when an occupation density exists.

(30.3) THEOREM (Hawkes [3]). *A local time exists if and only if*

(30.4)
$$\int_{\mathbf{R}} \operatorname{Re}\left[\frac{1}{1-\psi(\theta)}\right] d\theta < \infty.$$

Moreover, $L(\cdot)$ is almost surely square-integrable.

Proof.

$$\begin{aligned}\mathbf{E}|\hat{m}(\theta)|^2 &= \mathbf{E}\int_0^\infty\int_0^\infty e^{-s-t} e^{i\theta X_t - i\theta X_s}\,ds\,dt.\\ &= \operatorname{Re}\left[\frac{1}{1-\psi(\theta)}\right].\end{aligned}$$

Thus the condition (30.4) (which is the same as (30.2)) implies that almost surely $\hat{m} \in L^2(\mathbf{R})$, and so m has an $L^2(\mathbf{R})$ density.

Conversely, if a local time exists, the range of X has positive Lebesgue measure almost surely, implying $\operatorname{Leb}(C) > 0$, and hence (30.2), by Theorem 30.1. □

Millar and Tran [1] show that the local time of the asymmetric Cauchy process (which exists by applying the test (30.4) to the characteristic exponent (28.20)) has the property that $x \mapsto L_t(x)$ is unbounded on every interval; thus the local time is far from continuous in general.

Under the condition

(30.5)
$$0 \text{ is regular for } \{0\},$$

Blumenthal and Getoor [2] proved the existence of a jointly measurable process $\{L(t,x): t \geq 0,\ x \in \mathbf{R}\}$ that is an occupation density and such that, for each x, $t \mapsto L(t,x)$ is continuous increasing and, indeed, an additive functional; see Section III.16). Thus the condition (30.5) ensures the continuity of L in the time variable, leaving just the continuity in the space variable. (The condition (30.5) means $\mathbf{P}^0(H_0 = 0) = 1$, where, as usual, $H_x := \inf\{t > 0 : X_t = x\}$. It can be shown (see Rogozin [1]) that (30.5) is equivalent to

(30.6)
$$\int_{-\infty}^{\infty} \operatorname{Re}\left[\frac{1}{1-\psi(\theta)}\right] d\theta < \infty$$

and either

$$\sigma^2 > 0 \quad or \quad \int (|x| \wedge 1)\gamma(dx) = +\infty.$$

Let us assume the (30.4) holds, and that the function φ is defined by

$$\varphi(x) := \left\{ \frac{1}{\pi} \int (1 - \cos\theta x)\,\mathrm{Re}\left[\frac{1}{1-\psi(\theta)}\right] d\theta \right\}^{1/2}.$$

Let $\bar{\varphi}$ be the monotone rearrangement of φ, that is,

$$\bar{\varphi}(t) = \inf\{s : \eta(s) > t\},$$

where

$$\eta(s) := \mathrm{Leb}\{u : \varphi(u) \leqslant t\}.$$

The sufficiency of the following condition is due to Barlow and Hawkes [1], the necessity to Barlow [5].

(30.7) THEOREM (Barlow, Barlow and Hawkes). *Assume (30.4) and (30.5). There exists a jointly continuous version of the local-time process* $\{L(t,x) : t \geqslant 0, x \in \mathbb{R}\}$ *if and only if*

$$\int_{0+} \frac{\bar{\varphi}(u)\,du}{u\sqrt{\log(1/u)}} < \infty. \tag{30.8}$$

The condition (30.8) was suggested by results in the continuity of Gaussian processes; see Hawkes [3] for a well-motivated survey.

Barlow [5] also establishes the following modulus-of-continuity result.

(30.9) THEOREM. *If* $\varphi(x) = |x|^\alpha f(x)$, *where* f *is slowly varying at* 0, *then, for a proper subinterval* I *of* $\mathbb{R}$,

$$\lim_{\delta \downarrow 0} \sup_{\substack{a,b \in I \\ |a-b| < \delta \\ 0 \leqslant s \leqslant t}} \frac{|L(s,a) - L(s,b)|}{\varphi(b-a)(-\log|b-a|)^{1/2}} = 2\left[\sup_{x \in I} L(t,x)\right]^{1/2}.$$

We have already mentioned important work of Marcus and Rosen [1] which utilises the Dynkin isomorphism theorem in deep studies of local-time.

CHAPTER II

Some Classical Theory

This chapter is a reminder of what every probabilist should know, with the emphasis on things that tend to be neglected. The considerable length of the chapter—and it is now much more extensive than in the first edition—should be sufficient guarantee that 'reminder' is used in the usual 'courtesy' sense! Because things are now developed in strictly logical order, you may sometimes have to wait a little time for applications. (We very occasionally cheat just a little in Exercises by using things that you may feel are not yet proved with full rigour, but we always clear these up later.) *Exercises are very much part of the text—please do them!*

So many reminders of standard definitions are included that we break with our usual definition format except when we wish to give special emphasis to particularly important material that may not be so familiar.

1. BASIC MEASURE THEORY

The basic results of measure theory are summarised here, with commentary, but mostly without proofs. A full account, with all results proved, may be found, for example, in Williams [15], referred to as [W] throughout this chapter; that account has the advantage that its notation and terminology are the same as those used here. Neveu [1] is a marvellous account of measure theory for probabilists; and, for the definitive account of the full theory, including Choquet capacitability theory (which is needed for the Début and Section Theorems), see Volume 1 of Dellacherie and Meyer [1]. If you have studied *Measure for Measure* from such classics as Halmos [1] or Dunford and Schwartz [1] then this part of the chapter can serve to remind you of probabilists' language.

Measurability and measure

1. Measurable spaces; σ-algebras; π-systems; d-systems. Measurability will, in a sense, be much more important to us than measure. The emphasis in probability is therefore very different from that of courses that aim straight for the Dominated-Convergence Theorem.

(*1.1*) *Algebra; σ-algebra; σ($\mathcal{C}$); measurable space.* Let S be a set. A collection Σ_0 of subsets of S is called an *algebra on S* (or *algebra of subsets of S*) if the following three conditions hold:

$$S \in \Sigma_0,$$
$$F \in \Sigma_0 \Rightarrow F^c := S \setminus F \in \Sigma_0,$$
$$F, G \in \Sigma_0 \Rightarrow F \cup G \in \Sigma_0.$$

Note that $\emptyset = S^c \in \Sigma_0$ and that $F, G \in \Sigma_0 \Rightarrow F \cap G = (F^c \cup G^c)^c \in \Sigma_0$. Thus an algebra on S is a family of subsets of S stable under finitely many set operations. (*Note*: Some authors use 'field' for 'algebra' and 'σ-field' for 'σ-algebra'.)

A collection Σ of subsets of S is called a *σ-algebra on S* (or *σ-algebra of subsets of S*) if Σ is an algebra on S such that *whenever* $F_n \in \Sigma (n \in \mathbf{N})$,

$$\bigcup_n F_n \in \Sigma.$$

Note that if Σ is a σ-algebra on S and $F_n \in \Sigma$ for $n \in \mathbf{N}$, then $\bigcap_n F_n = (\bigcup_n F_n^c)^c \in \Sigma$. Thus a σ-algebra on S is a family of subsets of S 'stable under any countable collection of set operations'. A pair (S, Σ), where S is a set and Σ is a σ-algebra on S, is called a *measurable space.* An element of Σ is called a *Σ-measurable* subset of S.

Let $\mathcal{C}$ be a class of subsets of S. Then $\sigma(\mathcal{C})$, the *σ-algebra generated by* $\mathcal{C}$, is the smallest σ-algebra Σ on S such that $\mathcal{C} \subseteq \Sigma$. It is the intersection of all σ-algebras on S that have $\mathcal{C}$ as a subclass. (Obviously, the class $\mathcal{P}(S)$ of *all* subsets of S is a σ-algebra that extends $\mathcal{C}$.)

(*1.2*) *Borel σ-algebras; $\mathcal{B}(S)$; $\mathcal{B} = \mathcal{B}(\mathbf{R})$.* Let S be a topological space. Then $\mathcal{B}(S)$, the Borel σ-algebra on S, is the σ-algebra generated by the family of open subsets of S. With slight abuse of notation,

$$\mathcal{B}(S) := \sigma(\text{open sets}). \tag{1.3}$$

It is standard shorthand that $\mathcal{B} := \mathcal{B}(\mathbf{R})$. The σ-algebra $\mathcal{B}$ is the most important of all σ-algebras. Every subset of $\mathbf{R}$ that you meet in everyday use is an element of $\mathcal{B}$; and indeed it is difficult (but possible!) to find a subset of $\mathbf{R}$ constructed explicitly (without the Axiom of Choice) that is not in $\mathcal{B}$. To construct a subset of $\mathbf{R}$ that is not Lebesgue-measurable (as we do in Exercise E20.6b), one must use the Axiom of Choice. See Durrett [3, p. 411].

Elements of $\mathcal{B}$ can be quite complicated, and it is not possible to write down the 'generic' element of $\mathcal{B}$ in practicable fashion. However, the collection

$$\pi(\mathbf{R}) := \{(-\infty, x] : x \in \mathbf{R}\} \tag{1.4}$$

(not a standard notation, but a key example of a 'π-system') is very easy to understand, and it is often the case that all we need to know about $\mathcal{B}$ is the almost obvious result that

$$\mathcal{B} = \sigma(\pi(\mathbf{R})). \tag{1.5}$$

(*1.6*) *π-systems and d-systems.* Here we develop the point just made into a very useful technique.

Let S be a set. A collection $\mathcal{I}$ of subsets of S is called a *π-system* if $\mathcal{I}$ is stable under finite intersections:

$$\text{whenever } A, B \in \mathcal{I}, \text{ we have } A \cap B \in \mathcal{I}.$$

A collection $\mathcal{D}$ of subsets of S is called a *d-system* (*on S*) if

$$S \in \mathcal{D},$$
$$\text{if } A, B \in \mathcal{D} \text{ and } A \subseteq B \text{ then } B \backslash A \in \mathcal{D},$$
$$\text{if } A_n \in \mathcal{D} \text{ and } A_n \uparrow A \text{ then } A \in \mathcal{D}.$$

Recall that $A_n \uparrow A$ means $A_n \subseteq A_{n+1} (\forall n)$ and $\bigcup A_n = A$.

(*1.7*) PROPOSITION. *A collection Σ of subsets of S is a σ-algebra if and only if Σ is both a π-system and a d-system.*

Proof. The 'only if' part is trivial, so we prove only the 'if' part.

Suppose that Σ is both a π-system and a d-system, and that E, F and $E_n (n \in \mathbb{N})$ are in Σ. Then $E^c := S \backslash E \in \Sigma$, and

$$E \cup F = S \backslash (E^c \cap F^c) \in \Sigma.$$

Hence $G_n := E_1 \cup \cdots \cup E_n \in \Sigma$, and, since $G_n \uparrow \bigcup E_k$, we see that $\bigcup E_k \in \Sigma$.

If $\mathscr{C}$ is a class of subsets of S, we define $d(\mathscr{C})$ to be the intersection of all d-systems that contain $\mathscr{C}$. Obviously, $d(\mathscr{C})$ is a d-system, the smallest d-system containing $\mathscr{C}$. It is also obvious that

$$d(\mathscr{C}) \subseteq \sigma(\mathscr{C}).$$

(*1.8*) LEMMA (Dynkin). *If $\mathcal{I}$ is a π-system, then*

$$d(\mathcal{I}) = \sigma(\mathcal{I}).$$

Thus any d-system that contains a π-system contains the σ-algebra generated by that π-system.

Proof. Because of Proposition 1.7, we need only prove that $d(\mathcal{I})$ is a π-system.

Step 1: Let $\mathcal{D}_1 := \{B \in d(\mathcal{I}) : B \cap C \in d(\mathcal{I}), \forall C \in \mathcal{I}\}$. Because $\mathcal{I}$ is a π-system, $\mathcal{D}_1 \supseteq \mathcal{I}$. It is easily checked that $\mathcal{D}_1$ inherits the d-system structure from $d(\mathcal{I})$. [For, clearly, $S \in \mathcal{D}_1$. Next, if $B_1, B_2 \in \mathcal{D}_1$ and $B_1 \subseteq B_2$, then, for C in $\mathcal{I}$,

$$(B_2 \backslash B_1) \cap C = (B_2 \cap C) \backslash (B_1 \cap C);$$

and, since $B_2 \cap C \in d(\mathcal{I}), B_1 \cap C \in d(\mathcal{I})$ and $d(\mathcal{I})$ is a d-system, we see that $(B_2 \backslash B_1) \cap C \in d(\mathcal{I})$, so that $B_2 \backslash B_1 \in \mathcal{D}_1$. Finally, if $B_n \in \mathcal{D}_1$ $(n \in \mathbb{N})$ and $B_n \uparrow B$ then, for $C \in \mathcal{I}$,

$$(B_n \cap C) \uparrow (B \cap C)$$

so that $B \cap C \in d(\mathcal{I})$ and $B \in \mathcal{D}_1$.] We have shown that $\mathcal{D}_1$ is a d-system containing $\mathcal{I}$, so that (since $\mathcal{D}_1 \subseteq d(\mathcal{I})$ by definition) $\mathcal{D}_1 = d(\mathcal{I})$.

Step 2: Let $\mathcal{D}_2 := \{A \in d(\mathcal{I}) : A \cap B \in d(\mathcal{I}),\ \forall B \in d(\mathcal{I})\}$. Step 1 showed that $\mathcal{D}_2$ contains $\mathcal{I}$. But, just as in Step 1, we can prove that $\mathcal{D}_2$ inherits the d-system structure from $d(\mathcal{I})$, and that therefore $\mathcal{D}_2 = d(\mathcal{I})$. But the fact that $\mathcal{D}_2 = d(\mathcal{I})$ says that $d(\mathcal{I})$ is a π-system. □

2. Measurable functions. This section is largely a matter of acquainting you with our notation.

(*2.1*) DEFINITION (Measurable functions, $\mathrm{m}(\Sigma_1/\Sigma_2)$). *Suppose that (S_1, Σ_1) and (S_2, Σ_2) are measurable spaces, and that h is a map*

$$h : S_1 \to S_2.$$

Then h is called Σ_1/Σ_2-measurable *(or just* measurable *when Σ_1 and Σ_2 are understood), and we write $h \in \mathrm{m}(\Sigma_1/\Sigma_2)$, if*

$$h^{-1} : \Sigma_2 \to \Sigma_1;$$

that is, if the inverse image

$$h^{-1}(A) := \{s \in S : h(s) \in A\}$$

of every set $A \in \Sigma_2$ is in Σ_1.

This definition is exactly analogous to the definition of continuity.

(*2.2*) PROPOSITION. *The map h^{-1} preserves all set operations:*

$$h^{-1}(\bigcup_\alpha A_\alpha) = \bigcup_\alpha h^{-1}(A_\alpha), \quad h^{-1}(A^c) = (h^{-1}(A))^c, \quad \textit{etc.}$$

Proof. This is just definition chasing. □

(*2.3*) PROPOSITION. *If $\mathcal{C} \subseteq \Sigma_2, \sigma(\mathcal{C}) = \Sigma_2$ and $h^{-1} : \mathcal{C} \to \Sigma_1$, then $h \in \mathrm{m}(\Sigma_1/\Sigma_2)$.*

Proof. Let $\mathcal{E}$ be the class of elements F in Σ_2 such that $h^{-1}(F) \in \Sigma_1$. By (2.2), $\mathcal{E}$ is a σ-algebra, and, by hypothesis, $\mathcal{E} \supseteq \mathcal{C}$. □

(*2.4*) PROPOSITION (Composition Lemma). *If (S_1, Σ_1), (S_2, Σ_2) and (S_3, Σ_3) are measurable spaces, and if h_1 is measurable from (S_1, Σ_1) to (S_2, Σ_2) and h_2 is measurable from (S_2, Σ_2) to (S_3, Σ_3), then $h_2 \circ h_1$ is measurable from (S_1, Σ_1) to (S_3, Σ_3).*

Proof. This is obvious. □

(2.5) $\mathbf{R}$*-valued functions;* $\mathrm{m}\Sigma$; $(\mathrm{m}\Sigma)^+$; $\mathrm{b}\Sigma$. Let (S, Σ) be a measurable space. A function $h: S \to \mathbf{R}$ is called Σ*-measurable*, and we write $h \in \mathrm{m}\Sigma$, if $h^{-1}: \mathscr{B} \to \Sigma$, that is, if $h \in \mathrm{m}(\Sigma_1/\mathscr{B})$. We write $(\mathrm{m}\Sigma)^+$ for the class of non-negative elements in $\mathrm{m}\Sigma$, and $\mathrm{b}\Sigma$ for the class of bounded Σ-measurable functions on S.

Note. Because lim sups of sequences even of finite-valued functions may be infinite, and for other reasons, it is convenient to extend these definitions to functions h taking values in $[-\infty, \infty]$ in the obvious way: h is called Σ-*measurable* if $h^{-1}: \mathscr{B}[-\infty, \infty] \to \Sigma$. Which of the various results stated for real-valued functions extend to functions with values in $[-\infty, \infty]$, and what these extensions are, should be obvious.

(2.6) PROPOSITION. *Our function* $h: S \to \mathbf{R}$ *is* Σ*-measurable if and only if*

$$\{h \leqslant c\} := \{s \in S : h(s) \leqslant c\} \in \Sigma \quad (\forall c \in \mathbf{R}).$$

Proof. Take $\mathscr{C}$ to be the class $\pi(\mathbf{R})$ of intervals of the form $(-\infty, c], c \in \mathbf{R}$, and apply 2.3.

Note. Obviously, similar results apply in which $\{h \leqslant c\}$ is replaced by $\{h > c\}$, $\{h \geqslant c\}$ etc.

(2.7) LEMMA. *Sums and products of measurable* $\mathbf{R}$*-valued functions are measurable: in other words,* $\mathrm{m}\Sigma$ *is an algebra over* $\mathbf{R}$. *Thus if* $\lambda \in \mathbf{R}$ *and* h, h_1, $h_2 \in \mathrm{m}\Sigma$, *then*

$$h_1 + h_2 \in \mathrm{m}\Sigma, \quad h_1 h_2 \in \mathrm{m}\Sigma, \quad \lambda h \in \mathrm{m}\Sigma.$$

Example of proof. Let $c \in \mathbf{R}$. Then for $s \in S$ it is clear that $h_1(s) + h_2(s) > c$ if and only if for some rational q we have

$$h_1(s) > q > c - h_2(s).$$

In other words,

$$\{h_1 + h_2 > c\} = \bigcup_{q \in \mathbf{Q}} (\{h_1 > q\} \cap \{h_2 > c - q\}),$$

a countable union of elements of Σ. □

(2.8) LEMMA (measurability of infs, lim infs of functions). *Let* $(h_n : n \in \mathbf{N})$ *be a sequence of elements of* $\mathrm{m}\Sigma$. *Then*

$$\text{(i) } \inf h_n, \quad \text{(ii) } \liminf h_n, \quad \text{(iii) } \limsup h_n$$

are Σ*-measurable (into* $([-\infty, \infty], \mathscr{B}[-\infty, \infty])$*), but we shall still write* $\inf h_n \in \mathrm{m}\Sigma$

(for example)). Further,

$$\text{(iv)}\ \{s:\lim h_n(s) \text{ exists in } \mathbb{R}\}\in\Sigma.$$

Proof. (i) $\{\inf h_n \geqslant c\} = \bigcap_n \{h_n \geqslant c\}$.

(ii) Let $L_n(s) := \inf\{h_r(s): r \geqslant n\}$. Then $L_n \in \mathrm{m}\Sigma$, by part (i). But

$$L(s) := \liminf h_n(s) = \uparrow\lim L_n(s) = \sup L_n(s),$$

and $\{L \leqslant c\} = \bigcap_n \{L_n \leqslant c\} \in \Sigma$.

(iii) This part is now obvious.

(iv) This is also clear because the set on which $\lim h_n$ exists in $\mathbb{R}$ is

$$\{\limsup h_n < \infty\} \cap \{\liminf h_n > -\infty\} \cap g^{-1}(\{0\}),$$

where

$$g := \limsup h_n - \liminf h_n. \qquad \square$$

(2.9) *σ-algebra generated by a collection of functions on S.* This important idea is analogous to the weakest topology that makes every function in a given family continuous, etc.

Generally, if we have a collection $(Y_\gamma: \gamma \in C)$ of maps $Y_\gamma: \Omega \to \mathbb{R}$, then

$$\mathcal{Y} := \sigma(Y_\gamma: \gamma \in C)$$

is defined to be the smallest σ-algebra $\mathcal{Y}$ on Ω such that each map Y_γ $(\gamma \in C)$ is $\mathcal{Y}$-measurable. Clearly,

$$\sigma(Y_\gamma: \gamma \in C) = \sigma(\{\omega \in \Omega: Y_\gamma(\omega) \in B\}: \gamma \in C, B \in \mathcal{B}).$$

(2.10) *Borel functions.* A function h from a topological space S to $\mathbb{R}$ is called *Borel* if h is $\mathcal{B}(S)$-measurable. The most important case is when S itself is $\mathbb{R}$.

(2.11) PROPOSITION. *If S is topological and $h: S \to \mathbb{R}$ is continuous, then h is Borel.*

Proof. Take $\mathcal{C}$ to be the class of open subsets of $\mathbb{R}$, and apply Proposition 2.3. □

3. Monotone-Class Theorems. The following elementary Monotone-Class Theorem allows us to deduce results about general measurable functions from results about indicators of elements of π-systems.

(3.1) THEOREM. *Let $\mathcal{H}$ be a class of bounded functions from a set S into $\mathbb{R}$ satisfying the following conditions:*

(i) $\mathcal{H}$ is a vector space over $\mathbb{R}$;

(ii) the constant function 1 is an element of $\mathcal{H}$;

(iii) if (f_n) is a sequence of non-negative functions in $\mathscr{H}$ such that $f_n \uparrow f$, where f is a bounded function on S, then $f \in \mathscr{H}$.

Suppose further that $\mathscr{H}$ contains the indicator function of every set in some π-system $\mathscr{I}$. Then $\mathscr{H}$ contains every bounded $\sigma(\mathscr{I})$-measurable function on S.

Sketch of proof. Let $\mathscr{D}$ be the class of subsets D of S such that $I_D \in \mathscr{H}$. Then the listed properties of $\mathscr{H}$ guarantee that $\mathscr{D}$ is a d-system. Since $\mathscr{D}$ contains $\mathscr{I}$ by hypothesis, Dynkin's Lemma shows that $\mathscr{D}$ contains $\sigma(\mathscr{I})$.

Suppose that f is a $\sigma(\mathscr{I})$-measurable function such that for some K in $\mathbb{N}$,

$$0 \leqslant f(s) \leqslant K, \quad \forall s \in S.$$

For $n \in \mathbb{N}$, define

$$f_n(s) := \sum_{i=0}^{K2^n} i2^{-n} I_{D(n,i)}(s),$$

where

$$D(n,i) := \{s : i2^{-n} \leqslant f(s) < (i+1)2^{-n}\}.$$

Since f is $\sigma(\mathscr{I})$-measurable, every $D(n,i) \in \sigma(\mathscr{I})$, so that $I_{D(n,i)} \in \mathscr{H}$. Since $\mathscr{H}$ is a vector space, every $f_n \in \mathscr{H}$. But $0 \leqslant f_n \uparrow f$, so that $f \in \mathscr{H}$.

If $f \in b\sigma(\mathscr{I})$, we may write $f = f^+ - f^-$, where $f = \max(f, 0)$ and $f^- = \max(-f, 0)$. Then $f^+, f^- \in b\sigma(\mathscr{I})$ and $f^+, f^- \geqslant 0$, so that $f^+, f^- \in \mathscr{H}$ by what we established above. □

For certain applications, it is very useful to have more sophisticated forms of monotone-class theorem. Here is one.

(3.2) THEOREM. *Let $\mathscr{H}$ be a vector space of bounded real-valued functions on a set S. Suppose that $\mathscr{H}$ contains constant functions, is closed under uniform convergence, and has the following property: for a uniformly bounded sequence (f_n) of non-negative functions in $\mathscr{H}$ such that $f_n(s) \uparrow f(s)$ $(\forall s)$, we must have $f \in \mathscr{H}$.*

If $\mathscr{H}$ contains a subset $\mathscr{C}$ that is closed under multiplication, then $\mathscr{H}$ contains every bounded $\sigma(\mathscr{C})$-measurable function from S to $\mathbb{R}$.

You are invited to prove this in Section 13. The hypothesis that $\mathscr{H}$ is closed under uniform convergence (usually one that is easily verified) may be dropped. See Dellacherie and Meyer [1].

4. Measures; the uniqueness lemma; almost everywhere; a.e. (μ, Σ). This is fairly familiar material, but watch the use of π-systems in Lemma 4.6.

(4.1) *Set functions: additivity; σ-additivity; monotone convergence.* Let S be a set, let Σ_0 be an *algebra* on S, and let

$$\mu_0 : \Sigma_0 \to [0, \infty]$$

be a 'non-negative set function'. Then μ_0 is called *additive* if $\mu_0(\emptyset) = 0$ and, for $F, G \in \Sigma_0$,

$$F \cap G = \emptyset \Rightarrow \mu_0(F \cup G) = \mu_0(F) + \mu_0(G).$$

The map μ_0 is called *countably additive* (or *σ-additive*) if $\mu(\emptyset) = 0$ and if, whenever $(F_n : n \in \mathbb{N})$ is a sequence of disjoint sets in Σ_0 with union $F = \bigcup F_n$ in Σ_0 (note that this is an assumption since Σ_0 need not be a σ-algebra), then

$$\mu_0(F) = \sum_n \mu_0(F_n).$$

(4.2) LEMMA. *Suppose that μ_0 is additive on (S, Σ_0), where Σ_0 is an algebra on S. Then μ_0 is σ-additive on Σ_0 if and only if whenever $F_n \in \Sigma_0$ $(n \in \mathbb{N})$ and $F_n \uparrow F$, where $F \in \Sigma_0$, we have $\mu_0(F_n) \uparrow \mu_0(F)$.*

Recall that $F_n \uparrow F$ means $F_n \subseteq F_{n+1}$ $(\forall n \in \mathbb{N})$, $\bigcup F_n = F$. Lemma 4.2 is *the* fundamental property of measure.

Proof of 'only if' part. Write $G_1 := F_1, G_n := F_n \setminus F_{n-1}$ $(n \geqslant 2)$. Then the sets G_n $(n \in \mathbb{N})$ are *disjoint*, and

$$\mu_0(F_n) = \mu_0(G_1 \cup G_2 \cup \cdots \cup G_n) = \sum_{k \leqslant n} \mu_0(G_k) \uparrow \sum_{k < \infty} \mu_0(G_k) = \mu_0(F). \qquad \square$$

It is now obvious how to prove the 'if' part.

(4.3) LEMMA. *If $\mu_0(S) < \infty$ then μ_0 is σ-additive on Σ_0 if and only if whenever $G_n \in \Sigma_0 (n \in \mathbb{N})$ and $G_n \downarrow \emptyset$, we have $\mu(G_n) \downarrow 0$.*

You prove this!

(4.4) *Measure space; finite and σ-finite measures.* Let (S, Σ) be a measurable space, so that Σ is a σ-algebra on S. A map

$$\mu : \Sigma \to [0, \infty]$$

is called a ***measure*** on (S, Σ) if μ is countably additive. The triple (S, Σ, μ) is then called a ***measure space***.

Now let (S, Σ, μ) be a measure space. Then μ (or indeed the measure space (S, Σ, μ)) is called

finite if $\mu(S) < \infty$,

σ-finite if there is a sequence $(S_n : n \in \mathbb{N})$ of elements of Σ such that

$$\mu(S_n) < \infty \ (\forall n \in \mathbb{N}) \quad \text{and} \quad \bigcup S_n = S.$$

Intuition is usually good for finite measures, and adapts well for σ-finite measures. However, measures which are not σ-finite can be rather crazy.

An element F of Σ is called *μ-null* if $\mu(F)=0$. It is easily shown by the use of Lemma 4.2 that a countable union of μ-null sets is μ-null.

A statement $\mathscr{S}$ about points s of S is said to hold *μ-almost everywhere* (a.e. (μ)) if

$$F:=\{s:\mathscr{S}(s) \text{ is false}\}\in\Sigma \quad \text{and} \quad \mu(F)=0;$$

If we wish to emphasise that $F\in\Sigma$, we say that S holds a.e. (μ,Σ).

(4.5) *A fundamental uniqueness lemma.* The point here is that σ-algebras are 'difficult', but π-systems are 'easy': one can often write down in closed form the general element of a π-system, while the general element even of $\mathscr{B}$ is impossibly complicated.

(4.6) LEMMA. *Let $\mathscr{I}$ be a π-system on a set S, and let $\Sigma:=\sigma(\mathscr{I})$. Suppose that μ_1 and μ_2 are measures on (S,Σ) such that $\mu_1(S)=\mu_2(S)<\infty$ and $\mu_1=\mu_2$ on $\mathscr{I}$. Then*

$$\mu_1=\mu_2 \quad \textit{on } \Sigma.$$

Proof. The class of elements F of Σ for which $\mu_1(F)=\mu_2(F)$ is a d-system containing $\mathscr{I}$. Dynkin's Lemma 1.8 now gives the result. □

(4.7) COROLLARY (The Uniqueness Lemma). *If two probability measures agree on a π-system, then they agree on the σ-algebra generated by that π-system.*

This result will play an extremely important role.

5. Carathéodory's Extension Theorem. The following result underpins the existence of every non-trivial probabilistic model. We shall see its use in the celebrated Daniell–Kolmogorov Theorem on the existence of stochastic processes.

(5.1) THEOREM (Carathéodory). *Let S be a set, let Σ_0 be an algebra on S, and let*

$$\Sigma:=\sigma(\Sigma_0).$$

If μ_0 is a countably additive map $\mu_0:\Sigma_0\to[0,\infty]$, then there exists a measure μ on (S,Σ) such that

$$\mu=\mu_0 \quad \textit{on } \Sigma_0.$$

If $\mu_0(S)<\infty$ then, by Lemma 4.6, this extension is unique—an algebra is a π-system!

For a proof see for example [W; A1.5–1.8].

(5.2) *Lebesgue measure* Leb *on* $((0,1], \mathcal{B}(0,1])$. Let $S = (0,1]$. For $F \subseteq S$, say that $F \in \Sigma_0$ if F may be written as a finite union

$$F = (a_1, b_1] \cup \cdots \cup (a_r, b_r], \tag{5.3}$$

where $r \in \mathbb{N}$, $0 \leqslant a_1 \leqslant b_1 \leqslant \cdots \leqslant a_r \leqslant b_r \leqslant 1$. Then Σ_0 is an algebra on $(0,1]$ and

$$\Sigma := \sigma(\Sigma_0) = \mathcal{B}(0,1].$$

(We write $\mathcal{B}(0,1]$ instead of $\mathcal{B}((0,1])$.) For F as in (5.3), let

$$\mu_0(F) = \sum_{k \leqslant r} (b_k - a_k).$$

Then μ_0 is well-defined and additive on Σ_0 (this is easy). Moreover (see [W, A1.9], μ_0 is *countably* additive on Σ_0. (To prove this is not trivial. Our proof of the Daniell–Kolmogorov Theorem will remind you how it is done.) Hence, by Theorem 5.1, there exists a unique measure μ on $((0,1], \mathcal{B}(0,1])$ extending μ_0 on Σ_0. This measure μ is called the *Lebesgue measure* on $((0,1], \mathcal{B}(0,1])$ or (loosely) the Lebesgue measure on $(0,1]$. We shall often denote μ by Leb. The Lebesgue measure (still denoted by Leb) on $([0,1], \mathcal{B}[0,1])$ is of course obtained by a trivial modification, the set $\{0\}$ having Lebesgue measure 0.

In a similar way, we can construct the (σ-finite) Lebesgue measure (which we also denote by Leb) on $\mathbb{R}$ (more strictly, on $(\mathbb{R}, \mathcal{B}(\mathbb{R}))$).

6. Inner and outer μ-measures; completion. Results in this section should be proved as an (easy) exercise. They are, however, important.

Let (S, Σ, μ) be a measure space. For $G \subseteq S$, define the *inner μ-measure* $\mu_*(G)$ *of* G via

$$\mu_*(G) := \sup\{\mu(F) : F \in \Sigma; F \subseteq G\},$$

and the *outer μ-measure* $\mu^*(G)$ of G via

$$\mu^*(G) := \inf\{\mu(H) : H \in \Sigma; H \supseteq G\}.$$

The function μ^* on $\mathcal{P}(S)$ is *sub-σ-additive* (or *countably sub-additive*) in that

$$\mu^*\left(\bigcup G_n\right) \leqslant \sum \mu^*(G_n)$$

for any sequence $(G_n : n \in \mathbb{N})$.

If $\mu_*(G) = \mu^*(G)$, we say that G is *μ-measurable*, and write $G \in \Sigma^\mu$. Then Σ^μ is a σ-algebra, and we can extend μ to a *measure*, still denoted by μ, on Σ^μ by writing

$$\mu(G) := \mu_*(G) = \mu^*(G) \quad \text{for } G \text{ in } \Sigma^\mu.$$

The triple (S, Σ^μ, μ) is called the *completion* of (S, Σ, μ). The σ-algebra Σ^μ is the smallest σ-algebra that extends Σ and contains every set of outer μ-measure 0.

We end this section with a lemma that is very significant for probability theory. Its proof is an easy exercise.

(6.1) LEMMA. *Suppose that $\mu(S)=1$ and that G is a subset of S with $\mu^*(G)=1$. Then for $F\in\Sigma$, $\mu^*(G\cap F)=\mu(F)$. Moreover, $(G,\mathcal{G},\mu^*)$ is a measure space, where $\mathcal{G}$ is the class of subsets of G of the form $G\cap F$, where $F\in\Sigma$.*

Integration

7. Definition of the integral $\int f\,d\mu$. Let (S,Σ,μ) be a measure space.

(7.1) *Notation etc:* $\mu(f):=:\int f\,d\mu$; $\mu(f;A)$. We are interested in defining for suitable elements f in $m\Sigma$ the integral of f with respect to μ, for which we shall use the *alternative notations*

$$\mu(f):=:\int_S f(s)\mu(ds):=:\int_S f\,d\mu.$$

It is worth mentioning now that we shall also use the following equivalent notations for $A\in\Sigma$:

$$\int_A f(s)\mu(ds):=:\int_A f\,d\mu:=:\mu(f;A):=\mu(f\mathrm{I}_A)$$

(with a true definition on the extreme right!) It should be clear that, for example,

$$\mu(f;f\geqslant x):=\mu(f;A),\quad \text{where } A=\{s\in S:f(s)\geqslant x\}.$$

(7.2) *Integrals of non-negative simple functions;* SF^+. If A is an element of Σ, we define

$$\mu_0(\mathrm{I}_A):=\mu(A)\leqslant\infty.$$

The use of μ_0 rather than μ signifies that we currently have only a naive integral defined for simple functions.

An element f of $(m\Sigma)^+$ is called *simple*, and we shall then write $f\in\mathrm{SF}^+$, if f may be written as a finite sum

$$f=\sum_{k=1}^{m}a_k\mathrm{I}_{A_k}, \tag{7.3}$$

where $a_k\in[0,\infty]$ and $A_k\in\Sigma$. We then define

$$\mu_0(f)=\sum a_k\mu(A_k)\leqslant\infty\quad(\text{with } 0.\infty:=0=:\infty.0). \tag{7.4}$$

Of course, it needs to be checked that $\mu_0(f)$ is well defined, since f will have many different representations of the form (7.3), and we must ensure that they yield the same value of $\mu_0(f)$ in (7.4).

(7.5) *Integrals of non-negative functions.* For $f\in(m\Sigma)^+$, we define

$$\mu(f):=\sup\{\mu_0(h):h\in\mathrm{SF}^+,h\leqslant f\}\leqslant\infty.$$

Clearly, for $f \in \mathrm{SF}^+$, we have $\mu(f) = \mu_0(f)$.

(7.6) DEFINITION (μ-integrable functions; $\mathcal{L}^1(S, \Sigma, \mu)$). *For* $f \in \mathrm{m}\Sigma$, *we write* $f = f^+ - f^-$, *where*

$$f^+(s) := \max(f(s), 0), \quad f^-(s) := \max(-f(s), 0).$$

Then $f^+, f^- \in (\mathrm{m}\Sigma)^+$ *and* $|f| = f^+ + f^-$.

For $f \in \mathrm{m}\Sigma$, *we say that* f *is* μ*-integrable, and write*

$$f \in \mathcal{L}^1(S, \Sigma, \mu),$$

if

$$\mu(|f|) = \mu(f^+) + \mu(f^-) < \infty,$$

and then we define

$$\int f\, d\mu := \mu(f) := \mu(f^+) - \mu(f^-).$$

Note that, for $f \in \mathcal{L}^1(S, \Sigma, \mu)$,

$$|\mu(f)| \leqslant \mu(|f|),$$

the familiar rule that *the modulus of the integral is less than or equal to the integral of the modulus.*

(7.7) LEMMA (Linearity). *For* $\alpha, \beta \in \mathbb{R}$ *and* $f, g \in \mathcal{L}^1(S, \Sigma, \mu)$,

$$\alpha f + \beta g \in \mathcal{L}^1(S, \Sigma, \mu)$$

and

$$\mu(\alpha f + \beta g) = \alpha\mu(f) + \beta\mu(g).$$

(7.8) *Note.* There is a slight problem here. For some s, the expression $(\alpha f + \beta g)(s)$ may lead to the undefined $\infty - \infty$. Please defer worrying about this until Section 10.

8. Convergence theorems. We recall the standard results.

(8.1) THEOREM (The Monotone-Convergence Theorem). *If* (f_n) *is a sequence of elements of* $(\mathrm{m}\Sigma)^+$ *such that* $f_n \uparrow f$, *then*

$$\mu(f_n) \uparrow \mu(f) \leqslant \infty,$$

or, in other notation,

$$\int_S f_n(s)\mu(ds) \uparrow \int_S f(s)\mu(ds).$$

[W, A5] contains a proof. This theorem is really all there is to integration theory. We shall see that other key results such as the Fatou Lemma and the Dominated-Convergence Theorem follow trivially form it.

(8.2) LEMMA (The Fatou Lemma for functions). *For a sequence* (f_n) *in* $(\mathrm{m}\Sigma)^+$,

$$\mu(\liminf f_n) \leqslant \liminf \mu(f_n).$$

Proof. We have

(8.3) $$\liminf_n f_n = \uparrow\lim g_k, \quad \text{where } g_k := \inf_{n \geqslant k} f_n.$$

For $n \geqslant k$, we have $f_n \geqslant g_k$, so that $\mu(f_n) \geqslant \mu(g_k)$, whence

$$\mu(g_k) \leqslant \inf_{n \geqslant k} \mu(f_n);$$

and, on combining this with an application of Theorem 8.1 to (8.3), we obtain

$$\mu\left(\liminf_n f_n\right) = \uparrow\lim_k \mu(g_k) \leqslant \uparrow\lim_k \inf_{n \geqslant k} \mu(f_n)$$
$$=: \liminf_n \mu(f_n).$$ □

(8.4) Lemma ('Reverse Fatou' Lemma). *If* (f_n) *is a sequence in* $(\mathrm{m}\Sigma)^+$ *such that for some* g *in* $(\mathrm{m}\Sigma)^+$, *we have* $f_n \leqslant g, \forall n$, *and* $\mu(g) < \infty$, *then*

$$\mu(\limsup f_n) \geqslant \limsup \mu(f_n).$$

Proof. Apply Lemma 8.2 to the sequence $(g - f_n)$. □

(8.5) THEOREM (The Dominated-Convergence Theorem). *Suppose that* $f_n, f \in \mathrm{m}\Sigma$, *that* $f_n(s) \to f(s)$ *for* μ*-almost every* s *in* S, *and that the sequence* (f_n) *is dominated by an element* g *of* $\mathcal{L}^1(S, \Sigma, \mu)^+$:

$$|f_n(s)| \leqslant g(s), \quad \forall s \in S, \ \forall n \in \mathbf{N},$$

where $\mu(g) < \infty$. *Then*

$$f_n \to f \text{ in } \mathcal{L}^1(S, \Sigma, \mu); \quad \textit{that is, } \mu(|f_n - f|) \to 0;$$

whence

$$\mu(f_n) \to \mu(f).$$

Note. This theorem is central to many applications of measure theory. For us, it will be superseded by the uniform-integrability result: Theorem 21.2.

Proof. We have $|f_n - f| \leqslant 2g$, where $\mu(2g) < \infty$, so, by the reverse Fatou Lemma 8.4,

$$\limsup \mu(|f_n - f|) \leqslant \mu(\limsup |f_n - f|) = \mu(0) = 0.$$

Since

$$|\mu(f_n)-\mu(f)|=|\mu(f_n-f)|\leqslant\mu(|f_n-f|),$$

the theorem is proved. □

Here is a useful result.

(8.6) LEMMA (Scheffé's Lemma). *Suppose that* $f_n, f\in\mathcal{L}^1(S,\Sigma,\mu)$ *and that* $f_n\to f$ (a.e.(μ)). *Then*

$$\mu(|f_n-f|)\to 0 \quad \textit{if and only if} \quad \mu(|f_n|)\to\mu(|f|).$$

Exercise. Prove Scheffé's Lemma. Consider first the case in which f_n and f are non-negative. Note that then $(f_n-f)^-\leqslant f$.

(8.7) *The standard machine.* What we call the standard machine is a much cruder alternative to the Monotone-Class Theorem.

The idea is that to prove that a 'linear' result is true for all functions h in a space such as $\mathcal{L}^1(S,\Sigma,\mu)$,

(i) we first show the result is true for the case when h is an indicator function—which it normally is by definition;
(ii) we then use linearity to obtain the result for h in SF^+;
(iii) we then use the Monotone-Convergence Theorem to obtain the result for $h\in(\mathrm{m}\Sigma)^+$, integrability conditions on h usually being superfluous at this stage;
(iv) finally, we show, by writing $h=h^+-h^-$ and using linearity, that the claimed result is true.

When it works, it is easier to 'watch the standard machine work' than to appeal to the monotone-class result, though there are times when the greater subtlety of the Monotone-Class Theorem is essential.

9. The Radon–Nikodým Theorem; absolute continuity; $\lambda\ll\mu$ notation; equivalent measures. Let (S,Σ,μ) be a measure space. If $f\in(\mathrm{m}\Sigma)^+$, then, by linearity and the Monotone-Convergence Theorem,

(9.1) $$(f\mu)(F):=\mu(f;F):=\mu(f\mathrm{I}_F)\quad(F\in\Sigma)$$

defines a measure $f\mu$ on (S,Σ). Note that

(9.2) $$\mu(F)=0\quad\text{implies that}\quad(f\mu)(F)=0.$$

The Radon–Nikodým Theorem is a very important converse for *σ-finite* measures.

(9.3) THEOREM (The Radon–Nikodým Theorem) and DEFINITION

(absolute continuity, $\ll$, $d\lambda/d\mu$). *Let* (S,Σ) *be a measurable space, and let* μ *and* λ *be* σ*-finite measures on* (S,Σ). *Then the following statements are equivalent:*

(i) *for* $F\in\Sigma$, $\mu(F)=0$ *implies that* $\lambda(F)=0$;
(ii) $\lambda = f\mu$ *for some* f *in* $(\mathrm{m}\Sigma)^+$.

Suppose that statements (i) and (ii) hold. We then say that λ *is* absolutely continuous *relative to* μ. *The function* f *is defined uniquely modulo* μ*-null sets: we say that* f is a version of the (Radon–Nikodým) density of λ relative to μ, *and write*

$$f = \frac{d\lambda}{d\mu} \quad \text{a.e.}(\mu).$$

You can see the relevance of σ-finiteness if you consider μ to be a measure that counts the number of elements in a set.

For the classical proof see Halmos [1]. Meyer [2] and [W, Section 14.13] are among the books that give a martingale proof.

(9.4) LEMMA. *Suppose that* λ *and* μ *are finite measures on a measurable space* (S,Σ). *Then* $\lambda \ll \mu$ *if and only if for* $\varepsilon>0$ *we can find a* $\delta>0$ *such that*

$$F\in\Sigma \text{ and } \mu(F)<\delta \text{ imply that } \lambda(F)<\varepsilon.$$

(9.5) LEMMA and DEFINITION (equivalent measures). *Again suppose that* λ *and* μ *are finite measures on a measurable space* (S,Σ). *Suppose further that* $\lambda\ll\mu$ *and* $\mu\ll\lambda$. *We then say that* λ and μ are equivalent. *Note that* a.e.(μ) *and* a.e.(λ) *now mean the same thing: we write* a.e. *If* f *is a version of* $d\lambda/d\mu$ *and* g *is a version of* $d\mu/d\lambda$ *then* $0<f<\infty$ a.e. *and* $g=1/f$ a.e..

You are invited to prove these lemmas in Section 13.

10. Inequalities; $\mathcal{L}^p$ and L^p spaces ($p\geqslant 1$). Continue to let (S,Σ,μ) be a measure space, and let $p\in[1,\infty)$. For $f\in\mathrm{m}\Sigma$, write $f\in\mathcal{L}^p := \mathcal{L}^p(S,\Sigma,\mu)$ if

$$\|f\|_p := \{\mu(|f|^p)\}^{1/p} < \infty.$$

(10.1) LEMMA (Minkowski's Inequality). *We have*

$$\|f+g\|_p \leqslant \|f\|_p + \|g\|_p.$$

(10.2) LEMMA (Hölder's Inequality). *If* $p>1$ *and* $q>1$ *satisfy* $p^{-1}+q^{-1}=1$ *then, for* $f,g\in\mathrm{m}\Sigma$,

$$|\mu(fg)| \leqslant \mu(|fg|) \leqslant \|f\|_p\|g\|_q.$$

The *Schwarz inequality* is the case when $p=q=2$.

The best way to view these classical inequalities is as consequences of Jensen's inequality (18.3). See [W, 6.13]. An immediate consequence of Jensen's inequality ([W, 6.7]) is the following result.

(10.3) *If μ is a finite measure and $1 \leqslant p \leqslant r$ then, for $f \in m\Sigma$,*

$$\|f\|_p \leqslant \mu(S)^c \|f\|_r, \quad \textit{where } c := p^{-1} - r^{-1}.$$

We now need some standard results from functional analysis. See, for example, Dunford and Schwartz [1] or Halmos [1].

Define an equivalence relation on $\mathscr{L}^p$ as follows:

$$f \equiv g \quad \text{if and only if} \quad \|f - g\|_p = 0,$$

equivalently,

$$f \equiv g \quad \text{if and only if} \quad f = g, \quad \text{a.e.}(\mu).$$

Let $[f]$ be the equivalence class in $\mathscr{L}^p$ containing f, and let L^p be the set of equivalence classes. Since, for $f \in \mathscr{L}^p, \mu(|f| = \infty) = 0$, every equivalence class $[f]$ contains a representative f^* taking only finite values. With obvious notation, we can define

$$\alpha[f] + \beta[g] = [\alpha f^* + \beta g^*], \qquad \|[f]\|_p := \|f\|_p.$$

The use of equivalence classes avoids both the $\infty - \infty$ problem mentioned in (7.8) and the associated lack of associativity. The set L^p now becomes a normed vector space. But more is true:

(10.4) *L^p is a Banach space; in particular, L^p is a complete metric space under the distance*

$$d([f],[g]) := \|[f - g]\|_p.$$

The following characterisation of the dual of L^p is important.

(10.5) LEMMA. *For $p > 1$, the dual space $(L^p)^*$ of L^p is the space L^q where $p^{-1} + q^{-1} = 1$; if Λ is a bounded linear functional on L^p, then there exists $g \in L^q$ such that*

$$\Lambda(f) = \mu(fg), \quad \forall f \in L^p.$$

What happens when $p = 1$ and $q = \infty$? We say that $f \in \mathscr{L}^\infty$ if the *μ-essential supremum norm of f* is finite:

$$\begin{aligned}\|f\|_\infty &:= \mu\text{-ess sup}(f) := \sup\{x \geqslant 0 : \mu(|f| \geqslant x) > 0\}\\ &:= \inf\{x \geqslant 0 : \mu(|f| \geqslant x) = 0\} < \infty.\end{aligned}$$

Build L^∞ in the obvious way. Then

(10.6) *$(L^1)^* = L^\infty$; but, except in trivial cases, $(L^\infty)^*$ will be much bigger than L^1.*

A key application is to combine these results with the Hahn–Banach Theorem as follows.

(*10.7*) LEMMA. *Let* (S, Σ, μ) *be a measure space, let* $p \in [1, \infty)$, *and let* V *be a vector subspace of* L^p. *Define* $q \in (1, \infty]$ *by* $p^{-1} + q^{-1} = 1$. *Suppose that, for* $g \in L^q$,

$$(\mu(fg) = 0, \forall f \in V) \Rightarrow (g = 0).$$

Then V *is dense in* L^p.

Here is a way of putting this into practice.

(*10.8*) LEMMA. *Let* (S, Σ, μ) *be a finite measure space, and let* $\mathscr{I}$ *be a* π*-system on* S *such that* $\sigma(\mathscr{I}) = \Sigma$. *Let* $p \in [1, \infty)$. *Let* V *be the vector subspace of* L^p *spanned by the indicator functions of elements of* $\mathscr{I}$. *Then* V *is dense in* L^p.

You are invited to prove this in Section 13.

(*10.9*) **Important discussion;** $\mathscr{L}^p$ **versus** L^p. If we ask the problem 'Does a certain real-valued stochastic process $\{X_t : t \geqslant 0\}$ have continuous paths?', we are asking about the map $t \mapsto X(t, \omega)$. The question forces us to regard random variables as true functions, not as equivalence classes. All interesting problems in continuous time are invisible to the 'elegant' equivalence-class approach of functional analysis. So, we must use $\mathscr{L}^p$ rather than L^p. The $\infty - \infty$ problem will not worry us, because whenever we need to subtract random variables, they will have all values finite.

Product structures

11. Product σ-algebras. Product structures are especially important in probability theory because of their close connection with the concept of independence.

(*11.1*) *Finite-product σ-algebras.* This is an area in which we really need the Monotone-Class Theorems; the standard machine is not good enough.

Let (S_1, Σ_1) and (S_2, Σ_2) be measurable spaces. Let S denote the Cartesian product $S := S_1 \times S_2$. For $i = 1, 2$, let ρ_i denote the ith coordinate map, so that

$$\rho_1(s_1, s_2) := s_1, \quad \rho_2(s_1, s_2) := s_2.$$

The fundamental definition of $\Sigma = \Sigma_1 \times \Sigma_2$ is as the σ-algebra

$$\Sigma = \sigma(\rho_1, \rho_2). \tag{11.2}$$

Thus Σ is generated by sets of the form

$$\rho_1^{-1}(B_1) = B_1 \times S_2 \quad (B_1 \in \Sigma_1)$$

together with sets of the form

$$\rho_2^{-1}(B_2) = S_1 \times B_2 \quad (B_2 \in \Sigma_2).$$

Generally, a *product σ-algebra is generated by Cartesian products in which one factor is allowed to vary over the σ-algebra corresponding to that factor, and all other factors are whole spaces.* In the case of our product of *two* factors, we have

$$(11.3) \qquad (B_1 \times S_2) \cap (S_1 \times B_2) = B_1 \times B_2,$$

and you can easily check that

$$\mathscr{I} = \{B_1 \times B_2 : B_i \in \Sigma_i\}$$

is a π-system generating $\Sigma = \Sigma_1 \times \Sigma_2$. A similar remark would apply for a *countable* product $\prod \Sigma_n$, but you can see that, since we may only take *countable* intersections in analogues of (11.3), products of uncountable families of σ-algebras cause problems. The fundamental definition analogous to (11.2) still works.

(11.4) LEMMA. *Let $\mathscr{H}$ denote the class of functions $f: S \to \mathbf{R}$ that are in* bΣ *and that are such that*

for each s_1 in S_1, the map $s_2 \mapsto f(s_1, s_2)$ is Σ_2-measurable on S_2,

for each s_2 in S_2, the map $s_1 \mapsto f(s_1, s_2)$ is Σ_1-measurable on S_1.

Then $\mathscr{H} = \text{b}\Sigma$.

Proof. It is clear that if $A \in \mathscr{I}$ then $I_A \in \mathscr{H}$. Verification that $\mathscr{H}$ satisfies the hypotheses of the Monotone-Class Theorem 3.1 is straightforward. Since $\Sigma = \sigma(\mathscr{I})$, the result follows. □

Extension of the above concepts to general finite products is obvious, as are canonical identifications:

$$\begin{aligned}(S_1 \times S_2, \Sigma_1 \times \Sigma_2) \times (S_3, \Sigma_3) &= (S_1, \Sigma_1) \times (S_2 \times S_3, \Sigma_2 \times \Sigma_3) \\ &= (S_1 \times S_2 \times S_3, \Sigma_1 \times \Sigma_2 \times \Sigma_3).\end{aligned}$$

12. Product measure; Fubini's Theorem. We continue with the notation of the preceding section. We suppose that for $i = 1, 2$, μ_i is a *finite* measure on (S_i, Σ_i). We know from the preceding section that, for $f \in \Sigma$, we may define the integrals

$$\mathrm{I}_1^f(s_1) := \int_{S_2} f(s_1, s_2) \mu_2(ds_2), \qquad \mathrm{I}_2^f(s_2) := \int_{S_1} f(s_1, s_2) \mu_1(ds_1).$$

(12.1) LEMMA. *Let $\mathscr{H}$ be the class of elements in* bΣ *such that the following*

property holds:

$$\mathrm{I}_1^f(\cdot)\in \mathrm{b}\Sigma_1 \quad\text{and}\quad \mathrm{I}_2^f(\cdot)\in \mathrm{b}\Sigma_2 \quad\text{and}\quad \int_{S_1}\mathrm{I}_1^f(s_1)\mu_1(ds_1)=\int_{S_2}\mathrm{I}_2^f(s_2)\mu_2(ds_2).$$

Then $\mathscr{H}=\mathrm{b}\Sigma$.

Proof. If $A\in\mathscr{I}$ then, trivially, $I_A\in\mathscr{H}$. Verification of the conditions of the Monotone-Class Theorem 3.1 is straightforward. □

For $F\in\Sigma$ with indicator function $f:=\mathrm{I}_F$, we now define

$$\mu(F):=\int_{S_1}\mathrm{I}_1^f(s_1)\mu_1(ds_1)=\int_{S_2}\mathrm{I}_2^f(s_2)\mu_2(ds_2).$$

(*12.2*) THEOREM (Fubini's Theorem; Product measure). *Recall that, for* $i=1,2$, μ_i *is a finite measure on* (S_i,Σ_i). *The set function* μ *is a measure on* (S,Σ) *called the* product measure *of* μ_1 *and* μ_2, *and we write* $\mu=\mu_1\times\mu_2$ *and*

$$(S,\Sigma,\mu)=(S_1,\Sigma_1,\mu_1)\times(S_2,\Sigma_2,\mu_2).$$

Moreover, μ *is the unique measure on* (S,Σ) *for which*

(12.3) $$\mu(A_1\times A_2)=\mu_1(A_1)\mu_2(A_2),\qquad A_i\in\Sigma_i.$$

If $f\in(\mathrm{m}\Sigma)^+$, *then, with the obvious definitions of* I_1^f *and* I_2^f, *we have*

(12.4) $$\mu(f)=\int_{S_1}\mathrm{I}_1^f(s_1)\mu_1(ds_1)=\int_{S_2}\mathrm{I}_2^f(s_2)\mu_2(ds_2),$$

in $[0,\infty]$. *If* $f\in\mathrm{m}\Sigma$ *and* $\mu(|f|)<\infty$, *then* (12.4) *is valid (with all terms in* $\mathbf{R}$*).*

Proof. The fact that μ is a measure is a consequence of linearity and the Monotone-Convergence Theorem. The fact that μ is then uniquely specified by (12.3) is obvious from the Uniqueness Lemma 4.7 and the fact that $\sigma(\mathscr{I})=\Sigma$.

The result (12.4) is automatic for $f=\mathrm{I}_A$, where $A\in\mathscr{I}$. The Monotone-Class Theorem 3.1 shows that it is therefore valid for $f\in\mathrm{b}\Sigma$, and in particular for f in the SF^+ space for (S,Σ,μ). The Monotone-Convergence Theorem then shows that it is valid for $f\in(\mathrm{m}\Sigma)^+$; and linearity shows that (12.4) is valid if $\mu(|f|)<\infty$.

(*12.5*) (Extension). *All of Fubini's Theorem will work if the* (S_i,Σ_i,μ_i) *are* σ*-finite measure spaces.*

We can prove this by breaking up σ-finite spaces into countable unions of disjoint finite blocks.

(*12.6*) *Lebesgue measure on* $\mathbf{R}^n$. We have $\mathscr{B}(\mathbf{R}^n)=\mathscr{B}^n:=\mathscr{B}(\mathbf{R})^n$. For further study of such matters, see the exercises in the next section. We define the Lebesgue measure on $\mathbf{R}^n$ as Leb^n, but often denote this by Leb when n is understood.

13. Exercises. All of these exercises play a part later, Hints are given at the end of this section. The *number* of an exercise indicates when (that is, at the end of which section in the main text) you can attempt the question.

(*E13.2a*) *R-functions on* $\mathbb{R}$. By an R-function F on $\mathbb{R}$, we mean a right-continuous function on $\mathbb{R}$ such that the left limit $F(x-)$ exists for every $x \in \mathbb{R}$. Prove that an R-function is Borel-measurable.

(*E13.2b*) Prove that if S is a metric space, and if $C = C_b(S)$, the space of bounded continuous functions on S, then $\sigma(C) = \mathcal{B}(S)$.

(*E13.3a*) Prove the Monotone-Class Theorem 3.2.

(*E13.3b*) Let $X: S \to \mathbb{R}$. Prove that $\sigma(X) = \{X^{-1}(B): B \in \mathcal{B}\}$. Prove that if $Y: S \to \mathbb{R}$, then Y is $\sigma(X)$-measurable if and only if $Y = f(X)$ for some Borel-measurable f on $\mathbb{R}$.

(*E13.5a*) *From Lebesgue to Lebesgue–Stieltjes measure.* Let F be a right-continuous non-decreasing function on $\mathbb{R}$. Let $a := \inf_x F(x)$ and $b := \sup_x F(x)$. For $y \in I := [a, b) \cap \mathbb{R}$, define

$$\phi(y) := \inf\{x : F(x) \geqslant y\}.$$

Prove that ϕ is a left-continuous (therefore Borel) function on I. Assume that Lebesgue measure μ exists on $(I, \mathcal{B}(I))$, and define

$$\mu_F(B) := (\mu \circ \phi^{-1})(B) := \mu\{y : \phi(y) \in B\} \quad (B \in \mathcal{B}).$$

Prove that μ_F is a measure on $(\mathbb{R}, \mathcal{B})$, and that it is the unique such measure with

$$\mu_F(u, v] = F(v) - F(u) \quad (-\infty < u < v < \infty).$$

(*E13.5b*) *Functions of finite variation.* For a sub-interval $(a, b]$ of $\mathbb{R}$, we define the *total variation* $V_F(a, b]$ *of* F *over* $(a, b]$ to be

$$V_F(a, b] := \sup\left\{ \sum_{i=1}^{n} |F(t_i) - F(t_{i-1})| : n \in \mathbb{N}, a < t_1 < t_2 < \cdots < t_n \leqslant b \right\}.$$

A function F is said to be of finite variation (or an FV function) if it is an R-function such that $V_F(a, b] < \infty$ for every finite subinterval $(a, b]$ of $\mathbb{R}$. Let F be an FV function. Prove that, for $a \in \mathbb{R}$,

(i) $b \mapsto V_F(a, b]$ is an R-function on $[a, \infty)$,
(ii) the functions $b \mapsto V_F(a, b] + F(b)$ and $b \mapsto V_F(a, b] - F(b)$ are non-decreasing on $[a, \infty)$.

It is now trivial that F is an FV function if and only if it is the difference of two non-decreasing R-functions. This makes clear how an FV function F induces a signed measure μ_F, the difference of two measures.

(*E13.5c*) *Continuous FV functions.* Let F be a continuous FV function. Prove that $V_F(a,b]$ is continuous in b for $b \geqslant a$. Define the *quadratic variation* $Q_F(a,b]$ *over the interval* $(a,b]$ by

$$Q_F(a,b] := \sup\left\{\sum_{i=1}^{n} |F(t_i) - F(t_{i-1})|^2 : n\in\mathbf{N}, a \leqslant t_1 < t_2 < \cdots < t_n \leqslant b\right\}.$$

Prove that $Q_F(a,b] = 0$ for all intervals $(a,b]$. (Whence, Brownian paths are not of finite variation.)

(*E13.6a*) Prove Lemma 6.1.

(*E13.6b*) *Bad subsets of* $[0,1]$. Many of our counterexamples start off: 'Take a subset of $[0,1]$ of outer (Lebesgue) measure 1 and inner measure 0.' Here is a way of constructing such a set. At the moment, this exercise is rather hard. We shall see later (Exercise 60.50) that martingales make it easy.

Let ρ be an irrational number. Define an equivalence relation on $[0,1]$ by saying that $x_1 \equiv x_2$ if and only if $x_1 - x_2 = m + n\rho$ *for some* $m, n\in\mathbf{Z}$, in other words, if $x_1 - x_2 = n\rho \bmod 1$ for some $n\in\mathbf{Z}$. Use the Axiom of Choice to create a set A with one element from each equivalence class. Define

$$B := \{(2n\rho + \alpha) \bmod 1 : n\in\mathbf{Z}, \alpha\in A\},$$
$$C := \{(\rho + \beta) \bmod 1 : \beta\in B\}.$$

Note that $B\cap C = \varnothing, B\cup C = [0,1], B = \{(\rho + \gamma) \bmod 1 : \gamma\in C\}$, so that each of B and C is 'half' of $[0,1]$. Prove that B has outer Lebesgue measure 1 and inner Lebesgue measure 0.

(*E13.8*) Prove Scheffé's Lemma 8.6.

(*E13.9a*) Prove Lemma 9.4.

(*E13.9b*) Prove Lemma 9.5.

(*E13.9c*) *Integrals under change of measure.* Let (S,Σ,μ) be a measure space, let $f\in(\mathrm{m}\Sigma)^+$, and let $\lambda = (f\mu)$, the measure in (9.1). Show that, for $g\in\mathrm{m}\Sigma$, we have $g\in\mathcal{L}^1(S,\Sigma,\lambda)$ if and only if $fg\in\mathcal{L}^1(S,\Sigma,\mu)$, and then $\lambda(g) = \mu(fg)$.

(*E13.9d*) *Absolutely continuous functions on* $\mathbf{R}$. A function F on $\mathbf{R}$ is called *absolutely continuous* if there exists a function f in $\mathcal{L}^1_{\mathrm{loc}}(\mathbf{R},\mathcal{B},\mathrm{Leb})$, in the sense that $fI_{[a,b]}\in\mathcal{L}^1(\mathbf{R},\mathcal{B},\mathrm{Leb})$ for all finite subintervals $[a,b]$ of $\mathbf{R}$, such that

$$F(b) - F(a) = \int_a^b f(x)\,dx \quad (\infty < a \leqslant b < \infty).$$

We then call f a *derivative* of F (obviously, in an extension of the Newton–Leibniz sense) and write $F' = f$, a.e. Prove that an absolutely continuous function

f is continuous. Prove that if F is an absolutely continuous non-decreasing function, then the measure μ_F of Exercise 5a is absolutely continuous with respect to Lebesgue measure with $d\mu_F/d\,\text{Leb} = F'$, a.e.

(*E13.9e*) Show that if F is an absolutely continuous function then it is a continuous FV function with

$$V_F(a,b] = \int_a^b |F'(x)|\,dx.$$

The last part is tricky now, but easy using martingale theory.

(*E13.10a*) Prove Lemma 10.8.

Topology and measure

The ideas behind the following exercises are important throughout the book.

There is always a possibility of 'conflict' between topology, which allows certain uncountable operations (the union of an uncountable number of open sets is still open), and measure theory, which allows only countably many operations. For measures on *separable* metric spaces, things are as one might hope. Recall that a metric space (S,ρ) is called *separable* if it has a countable dense subset.

(*E13.11a*) Let (S,ρ) be a *separable* metric space. Prove that $\rho: S\times S\to\mathbf{R}$ is $\mathcal{B}(S)\times\mathcal{B}(S)$ measurable. Show that any subset A of S has a countable dense subset.

(*E13.11b*) Let (S_1,Σ_1) and (S_2,Σ_2) be measurable spaces. Define

$$(S,\Sigma) := (S_1,\Sigma_1)\times(S_2,\Sigma_2).$$

Prove that the map

$$(x,y) = ((x_1,x_2),(y_1,y_2)) \mapsto (x_1,y_1)$$

of $S\times S$ into $S_1\times S_1$ is $(\Sigma\times\Sigma)/(\Sigma_1\times\Sigma_1)$-measurable.

(*E13.11c*) Let (S_1,ρ_1) and (S_2,ρ_2) be *separable* metric spaces. Then the product topology on $S := S_1\times S_2$ arises from the metric

$$\rho(x,y) = \rho((x_1,x_2),(y_1,y_2)) := \rho_1(x_1,y_1) + \rho_2(x_2,y_2).$$

Define $\Sigma_1 := \mathcal{B}(S_1), \Sigma_2 := \mathcal{B}(S_2)$ and $\Sigma := \Sigma_1\times\Sigma_2$. Prove that ρ is $(\Sigma\times\Sigma)$-measurable. Deduce that

$$\mathcal{B}(S_1\times S_2) = \mathcal{B}(S_1)\times\mathcal{B}(S_2).$$

(*Generalisation*). Suppose that, for $n\in\mathbf{N}$, (S_n, ρ_n) in a separable metric space. The product topology on $\prod_{n\in\mathbf{N}} S_n$ arises from the metric ρ, where

$$\rho(x, y) := \sum_{n\in\mathbf{N}} 2^{-n} \frac{\rho(x_n, y_n)}{1+\rho(x_n, y_n)}.$$

Show that S is separable. Check (at least in principle!) that $\mathscr{B}(S) = \prod \mathscr{B}(S_n)$.

(*E13.11d*) Let S be the *non-separable* metric space with cardinality greater than that of the continuum, and with the discrete metric

$$\rho(x, y) := \begin{cases} 1 & \text{if } x \neq y, \\ 0 & \text{if } x = y. \end{cases}$$

Let $\Delta := \{(x, y)\in S \times S : x = y\}$. Then Δ is closed and therefore $\Delta\in\mathscr{B}(S \times S)$. Convince yourself that it is plausible (it is true!) that $\Delta\notin\mathscr{B}(S) \times \mathscr{B}(S)$ and that ρ is (therefore) not $(\mathscr{B}(S) \times \mathscr{B}(S))$-measurable.

Hints for selected exercises

(*H13.2a*) F is the limit of step functions $F(2^{-n}([2^n t] + 1))$, $[x]$ denoting $\sup\{n\in\mathbf{Z} : n \leqslant x\}$.

(*H13.2b*) If F is a closed set and ρ is the distance function, then

$$\max(0, 1 - n\rho(x, F)) \downarrow I_F.$$

(*H13.3a*) Consider the π-system of sets of the form

$$(*) \qquad \bigcap_{i=1}^{n} \{s : c_i(s)\in(a_i, b_i)\},$$

where $n\in\mathbf{N}$ and, for $1 \leqslant i \leqslant n$, $c_i\in\mathscr{C}$ and $-\infty < a_i < b_i < \infty$. If we can prove that the indicator function of every set of the form $(*)$ is in $\mathscr{H}$ then the desired result will follow from Theorem 3.1.

For any open subinterval of $\mathbf{R}$, we can find continuous functions g_m on $\mathbf{R}$ with $g_m \uparrow I_{(a,b)}$. Let $c\in\mathscr{C}$. By the Weierstrass theorem, we can find polynomials $p_{m,k}$ such that $p_{m,k} \to g_m$ uniformly on $[-\|c\|, \|c\|]$. Then $p_{m,k}\circ c \to g_m\circ c$ uniformly on S, whence $g_m\circ c\in\mathscr{H}$; and now it follows that $I_{(a,b)}\circ c\in\mathscr{H}$. The rest is easy.

(*H13.3b*) Since, $\{X^{-1}(B) : B\in\mathscr{B}\}$ is a σ-algebra, the first part is easy. Let $\mathscr{H}$ consist of functions of the form $f\circ X$, where f is bounded Borel-measurable. Then $\mathscr{H}$ satisfies conditions (i)–(iii) of Theorem 3.1.

(*H13.5a*) We have $u < \phi(y) \leqslant v$ if and only if $F(u) < y \leqslant F(v)$.

(*H13.5b*) $V_F(a, b]$ (note the 'open at a') is decreasing as $b \downarrow\downarrow a$. Suppose that, for some a,

$$\lim_{b \downarrow\downarrow a} V_F(a, b] = \varepsilon_0 > 0.$$

Find a partition $a < t_0 < t_1 < \cdots < t_n = b$ on which

$$\sum_{i=1}^{n} |F(t_i) - F(t_{i-1})| \geqslant V_F(a, b] - \tfrac{1}{2}\varepsilon_0.$$

Now choose a partition of $(a, t_0]$ on which the analogous sum is at least $\frac{3}{4}\varepsilon_0$ to arrive at a contradiction.

(*H13.5c*) We have

$$\sum |F(t_i) - F(t_{i-1})|^2 \leqslant \sup[|F(t_j) - F(t_{j-1})|] \sum |F(t_i) - F(t_{i-1})|.$$

(*H13.6b*) Wait until the exercises on discrete-parameter martingales.

(*H13.8*) The hint was given after Lemma 8.6.

(*H13.9a*) Peep ahead to the proof of Lemma 20.1 to get the *idea*.

(*H13.9c*) If $g = I_F$ $(F \in \Sigma)$, then $\lambda(g) = \mu(fg)$ by definition. Now use the standard machine.

(*H13.9e*) It is obvious that $V_F \leqslant \int, \ldots$, but why do we have equality? Again wait for martingales to come to the rescue.

(*H13.11a*) Let (s_n) be a sequence dense in S. Then

$$\rho(x, y) = \inf_n [\rho(x, s_n) + \rho(s_n, y)].$$

For each n, $x \mapsto \rho(x, s_n)$ is continuous, and $(x, y) \mapsto \rho(x, s_n)$ is $\mathscr{B}(S) \times \mathscr{B}(S)$-measurable.

(*H13.11c*) See (E13.11b).

(*H13.11d*) See Billingsley [2].

2. BASIC PROBABILITY THEORY

Probability and expectation

14. Probability triple; almost surely (a.s.), a.s.(P), a.s.(P, $\mathscr{F}$). By a *probability triple*, we mean a measure space $(\Omega, \mathscr{F}, \mathbf{P})$ of total mass $\mathbf{P}(\Omega) = 1$.

From now on, $(\Omega, \mathcal{F}, \mathbf{P})$ will always denote a probability triple. Unless otherwise stated,

$$\mathcal{L}^p := \mathcal{L}^p(\Omega, \mathcal{F}, \mathbf{P}), \qquad L^p := L^p(\Omega, \mathcal{F}, \mathbf{P}),$$

and $\|\cdot\|_p$ will refer to these spaces.

An element E of $\mathcal{F}$ will be called an *event*, and $\mathbf{P}(E)$ will be called the *probability of the event* E. A statement S about outcomes ω in Ω is said to be true *almost surely* (a.s.) if

$$F := \{\omega : S(\omega) \text{ is true}\} \in \mathcal{F} \quad \text{and} \quad \mathbf{P}(F) = 1.$$

If we wish to emphasise which probability measure we are talking about, we write 'a.s.(P)', and if we further wish to emphasise that the truth set of S is in $\mathcal{F}$, we write 'a.s.(P, $\mathcal{F}$)'. It is easy shown that if $F_n \in \mathcal{F}$ $(n \in \mathbb{N})$ and $\mathbf{P}(F_n) = 1, \forall n$, then $\mathbf{P}(\bigcap_n F_n) = 1$.

The intuitive meaning. We assume that the intuitive meaning is familiar to you from more elementary books. Chance is regarded as having chosen a particular point ω (the actual realisation) of Ω 'according to the law P' before the experiment modelled by $(\Omega, \mathcal{F}, \mathbf{P})$ is performed. For an event F, F occurs in reality if and only if the chosen ω is in F.

15. lim sup E_n; First Borel–Cantelli Lemma. Suppose now that $(E_n : n \in \mathbb{N})$ is a sequence of events. We define

$$\begin{aligned}\limsup E_n &:= \bigcap_m \bigcup_{n \geqslant m} E_n \\ &= \{\omega : \text{for every } m, \exists n(\omega) \geqslant m \text{ such that } \omega \in E_{n(\omega)}\} \\ &= \{\omega : \omega \in E_n \text{ for infinitely many } n\}.\end{aligned}$$

Two important results relate to $\limsup E_n$.

(*15.1*) LEMMA (Reverse Fatou Lemma for sets):

$$\mathbf{P}(\limsup E_n) \geqslant \limsup \mathbf{P}(E_n).$$

Proof. Let $G_m := \bigcup_{n \geqslant m} E_n$. Then $G_m \downarrow G$, where $G := \limsup E_n$. By result (4.3), $\mathbf{P}(G_m) \downarrow \mathbf{P}(G)$. But, clearly,

$$\mathbf{P}(G_m) \geqslant \sup_{n \geqslant m} \mathbf{P}(E_n).$$

Hence

$$\mathbf{P}(G) \geqslant \downarrow \lim_m \left\{ \sup_{n \geqslant m} \mathbf{P}(E_n) \right\} =: \limsup \mathbf{P}(E_n). \qquad \square$$

(*15.2*) LEMMA (First Borel–Cantelli Lemma). *Let $(E_n : n \in \mathbb{N})$ be a sequence of* *events such that* $\sum_n \mathbf{P}(E_n) < \infty$. *Then*

$$\mathbf{P}(\limsup E_n) = \mathbf{P}(E_n, \text{i.o.}) = 0.$$

Proof. We have, for each m,

$$\mathbf{P}(G) \leqslant \mathbf{P}(G_m) \leqslant \sum_{n \geqslant m} \mathbf{P}(E_n).$$

(Convince yourself of the rigour.) Now let $m \uparrow \infty$. □

16. Law of random variable; distribution function; joint law

(*16.1*) DEFINITION (Random variable; law). *Let* $(E, \mathcal{E})$ *be a measurable space. By an* $(E, \mathcal{E})$-valued random variable X *(or* E-valued random variable X, *when* $\mathcal{E}$ *is understood)* carried by our probability triple $(\Omega, \mathcal{F}, \mathbf{P})$, *we mean an* $(\mathcal{F}/\mathcal{E})$*-measurable map* X *from* Ω *to* E, *so that* $X^{-1}: \mathcal{E} \to \mathcal{F}$.

By the law Λ_X of X, *we mean the probability measure* $\Lambda_X := \mathbf{P} \circ X^{-1}$ *on* $(E, \mathcal{E})$, *so that*

$$\Lambda_X(A) = \mathbf{P}(X \in A) := \mathbf{P}\{\omega : X(\omega) \in A\} \quad (A \in \mathcal{E}).$$

Suppose that for $i = 1, 2, (E_i, \mathcal{E}_i)$ is a measurable space and that X_i is an $(E_i, \mathcal{E}_i)$-valued random variable. Let $E := E_1 \times E_2$, $\mathcal{E} := \mathcal{E}_1 \times \mathcal{E}_2$ and

$$X(\omega) := (X_1(\omega), X_2(\omega)) \in E.$$

Check that X is an $(E, \mathcal{E})$-valued random variable. The joint law Λ_{X_1, X_2} of X_1 and X_2 is then defined to be the law of X. There are obvious extensions, some of which we study in great detail later.

If our variable X is $\mathbb{R}$-valued (that is, $(\mathbb{R}, \mathcal{B})$-valued) then it follows from the Uniqueness Lemma 4.7 and the fact that $\mathcal{B} = \sigma(\pi(\mathbb{R}))$, where $\pi(\mathbb{R})$ is at (1.4), that the law of X is determined by the *distribution function* F_X of X:

$$F_X(x) := \mathbf{P}(X \leqslant x) \quad (x \in \mathbb{R}).$$

17. Expectation; $\mathbf{E}(X; F)$. We introduce some notation used throughout the book.

(*17.1*) DEFINITION (Expectation; $\mathbf{E}(X)$) *For a random variable*

$$X \in \mathcal{L}^1 = \mathcal{L}^1(\Omega, \mathcal{F}, \mathbf{P}),$$

we define the expectation $\mathbf{E}(X)$ of X *by*

$$\mathbf{E}(X) := \int_\Omega X \, dP = \int_\Omega X(\omega) \mathbf{P}(d\omega).$$

We also define $\mathbf{E}(X)$ $(\leqslant \infty)$ *for* $X \in (\mathrm{m}\mathcal{F})^+$. *In earlier notation,* $\mathbf{E}(X) = \mathbf{P}(X)$.

(*17.2*) DEFINITION (Notation $\mathbf{E}(X;F)$). *For $X\in\mathcal{L}^1$ (or $(m\mathcal{F})^+$) and $F\in\mathcal{F}$, we define*

$$\mathbf{E}(X;F):=\int_F X(\omega)\mathbf{P}(d\omega):=\mathbf{E}(XI_F),$$

where, as ever,

$$I_F(\omega):=\begin{cases}1 & \text{if } \omega\in F,\\ 0 & \text{if } \omega\notin F.\end{cases}$$

(*17.3*) LEMMA. *If $h\in(\mathrm{m}\mathcal{E})^+$, then*

$$\mathbf{E}h(X)=\int_E h(x)\Lambda_X(dx)\leqslant\infty. \tag{17.4}$$

For $h\in\mathrm{m}\mathcal{E}$, $h(X)\in\mathcal{L}^1(\Omega,\mathcal{F},\mathbf{P})$ if and only if $h\in\mathcal{L}^1(E,\mathcal{E},\Lambda_X)$, and then

$$\mathbf{E}h(X)=\int_E h(x)\Lambda_X(dx).$$

Proof. Use the standard machine at (8.7). If $h=I_A$ for some $A\in\mathcal{E}$ then (17.4) is true by definition of Λ_X; etc. □

An important case of this lemma is when $(E,\mathcal{E})=(\mathbf{R},\mathcal{B})$ and h is the identity function on $\mathbf{R}$.

18. Inequalities: Markov, Jensen, Schwarz, Tchebychev. These inequalities will be used repeatedly in estimates.

(*18.1*) LEMMA (Markov's inequality). *Suppose that $Z\in\mathrm{m}\mathcal{F}$ and that $g:\mathbf{R}\to[0,\infty]$ is $\mathcal{B}$-measurable and non-decreasing. (We know that $g(Z)=g\circ Z\in(\mathrm{m}\mathcal{F})^+$.) Then*

$$\mathbf{E}g(Z)\geqslant\mathbf{E}(g(Z);Z\geqslant c)\geqslant g(c)\mathbf{P}(Z\geqslant c).$$

Examples

$$\text{for } Z\in(\mathrm{m}\mathcal{F})^+,\quad c\mathbf{P}(Z\geqslant c)\leqslant\mathbf{E}(Z)\quad(c>0),$$
$$\text{for } X\in\mathcal{L}^1,\quad c\mathbf{P}(|X|\geqslant c)\leqslant\mathbf{E}(|X|)\quad(c>0).$$

Considerable strength can often be obtained by choosing the optimum θ for c in

$$\mathbf{P}(Y>c)\leqslant e^{-\theta c}\mathbf{E}(e^{\theta Y})\quad(\theta>0,c\in\mathbf{R}).$$

(*18.2*) *Jensen's inequality for convex functions.* A function $c:G\to\mathbf{R}$, where G is an open subinterval of $\mathbf{R}$, is called *convex* on G if its graph lies below any of its chords: for $x, y \in G$ and $0 \leqslant p = 1 - q \leqslant 1$,

$$c(px + qy) \leqslant pc(x) + qc(y).$$

Then c is automatically continuous on G. If c is twice-differentiable on G then c is convex if and only if $c'' \geqslant 0$. Important examples of convex functions are $|x|, x^2$ and $e^{\theta x} (\theta \in \mathbb{R})$.

(*18.3*) THEOREM (Jensen's inequality). *Suppose that $c: G \to \mathbb{R}$ is a convex function on an open subinterval G of $\mathbb{R}$ and that X is a random variable such that*

$$\mathbf{E}(|X|) < \infty, \quad \mathbf{P}(X \in G) = 1, \quad \mathbf{E}|c(X)| < \infty.$$

Then

$$\mathbf{E}c(X) \geqslant c(\mathbf{E}(X)).$$

See [W; Section 6.6] for a full proof. The point is that, since there is a supporting hyperplane for c at $(\mu, c(\mu))$, where $\mu = \mathbf{E}(X)$, there exists an m in $\mathbb{R}$ such that

$$c(X) \geqslant m(X - \mu) + c(\mu);$$

and Jensen's inequality follows on taking expectations.

(*18.4*) LEMMA (Monotonicity of norms). *If $1 \leqslant p \leqslant r$ and $Y \in \mathrm{m}\mathcal{F}$, then*

$$\| Y \|_p \leqslant \| Y \|_r.$$

Proof. Apply Jensen's inequality with $X = |Y|^p$ and $c(x) = x^{r/p}$ $(x > 0)$. □

(*18.5*) *Familiar facts.* We recall three results that are consequences of Jensen's inequality (see Section 24): for $p > 1$ and $p^{-1} + q^{-1} = 1$,

$$\begin{aligned} \text{Hölder:} \quad & |\mathbf{E}(XY)| \leqslant \mathbf{E}|XY| \leqslant \|X\|_p \|Y\|_q, \\ \text{Schwarz:} \quad & |\mathbf{E}(XY)| \leqslant \mathbf{E}|XY| \leqslant \|X\|_2 \|Y\|_2, \\ \text{Minkowski:} \quad & \|X + Y\|_p \leqslant \|X\|_p + \|Y\|_p. \end{aligned}$$

(*18.6*) *Variance; covariance; Tchebychev's inequality.* If $X, Y \in \mathcal{L}^2$ then, by the monotonicity of norms, $X, Y \in \mathcal{L}^1$, so that we may define

$$\mu_X := \mathbf{E}(X), \quad \mu_Y := \mathbf{E}(Y).$$

Since the constant functions with values μ_X and μ_Y are in $\mathcal{L}^2$, we see that

$$\tilde{X} := X - \mu_X, \quad \tilde{Y} := Y - \mu_Y$$

are in $\mathcal{L}^2$. By the Schwarz inequality, $\tilde{X}\tilde{Y} \in \mathcal{L}^1$, and so we may define

$$\mathrm{Cov}(X, Y) := \mathbf{E}(\tilde{X}\tilde{Y}) = \mathbf{E}[(X - \mu_X)(Y - \mu_Y)].$$

The Schwarz inequality further justifies expanding out the product in the final

[ ] bracket to yield the alternative formula

$$\mathrm{Cov}(X,Y) = \mathbf{E}(XY) - \mu_X\mu_Y.$$

As you know, the *variance* of X is defined by

$$\mathrm{Var}(X) := \mathbf{E}[(X-\mu_X)^2] = \mathbf{E}(X^2) - \mu_X^2 = \mathrm{Cov}(X,X).$$

You also know *Tchebychev's inequality*:

(18.7) $$c^2\mathbf{P}(|X-\mu_X| \geqslant c) \leqslant \mathrm{Var}(X) \quad (c>0).$$

19. Modes of convergence of random variables. Let $(X_n : n \in \mathbb{N})$ be a sequence of random variables and let X be a random variable, all carried by our triple $(\Omega, \mathcal{F}, \mathbf{P})$ and all $\mathbb{R}$-valued.

Recall that we say that $X_n \to X$ *almost surely* if

(19.1) $$\mathbf{P}(X_n \to X) = 1.$$

We say that $X_n \to X$ *in probability* if, for every $\varepsilon > 0$,

(19.2) $$\mathbf{P}(|X_n - X| > \varepsilon) \to 0 \quad \text{as } n \to \infty.$$

We say that $X_n \to X$ *in* $\mathcal{L}^p$ if each X_n is in $\mathcal{L}^p$ and $X \in \mathcal{L}^p$ and

$$\| X_n - X \|_p \to 0 \quad \text{as} \quad n \to \infty,$$

or, equivalently,

(19.3) $$\mathbf{E}(|X_n - X|^p) \to 0 \quad \text{as } n \to \infty.$$

Some relationships between these modes of convergence will now be stated. Regard the proofs as exercises. See [W; EA13.1]. Convergence in probability is the weakest of the above forms of convergence. Thus

(19.4) $$(X_n \to X, \text{a.s.}) \Rightarrow (X_n \to X \text{ in prob})$$

(19.5) $$(X_n \to X \text{ in } \mathcal{L}^p) \Rightarrow (X_n \to X \text{ in prob}).$$

No other implication between any two of our three forms of convergence is valid. But, of course, for $r \geqslant p \geqslant 1$, monotonicity of norms shows that

(19.6) $$(X_n \to X \text{ in } \mathcal{L}^r) \Rightarrow (X_n \to X \text{ in } \mathcal{L}^p).$$

'Fast convergence in probability' does imply almost sure convergence:

(19.7) $$\left(\sum_n \mathbf{P}(|X_n - X| > \varepsilon) < \infty, \forall \varepsilon > 0 \right) \Rightarrow (X_n \to X, \text{ a.s.}).$$

Property (19.7) is used in proving the following result:

(19.8) $X_n \to X$ in probability if and only if every subsequence of (X_n) contains a further subsequence along which we have almost sure convergence to X.

Uniform integrability and $\mathcal{L}^1$ convergence

20. Uniform integrability. We begin with a lemma.

(20.1) LEMMA. *Suppose that $X \in \mathcal{L}^1 = \mathcal{L}^1(\Omega, \mathcal{F}, \mathbf{P})$. Then, given $\varepsilon > 0$, there exists a $\delta > 0$ such that for $F \in \mathcal{F}$, $\mathbf{P}(F) < \delta$ implies that $\mathbf{E}(|X|; F) < \varepsilon$.*

Proof. If the conclusion is false, then, for some $\varepsilon_0 > 0$, we can find a sequence (F_n) of elements of $\mathcal{F}$ such that

$$\mathbf{P}(F_n) < 2^{-n} \quad \text{and} \quad \mathbf{E}(|X|; F_n) \geqslant \varepsilon_0.$$

Let $H := \limsup F_n$. Then the First Borel–Cantelli Lemma shows that $\mathbf{P}(H) = 0$, but the 'Reverse Fatou' Lemma 8.4 shows that

$$\mathbf{E}(|X|; H) \geqslant \varepsilon_0;$$

and we have arrived at the required contradiction. □

(20.2) COROLLARY. *Suppose that $X \in \mathcal{L}^1$ and that $\varepsilon > 0$. Then there exists K in $[0, \infty)$ such that*

$$\mathbf{E}(|X|; |X| > K) < \varepsilon.$$

Proof. Let δ be as in Lemma (34.1). Since $K\mathbf{P}(|X| > K) \leqslant \mathbf{E}(|X|)$, we can choose K such that $\mathbf{P}(|X| > K) < \delta$. □

(20.3) DEFINITION (UI family). *A class $\mathcal{C}$ of $\mathbb{R}$-valued random variables is called* uniformly integrable (UI) *if, given $\varepsilon > 0$, there exists K in $[0, \infty)$ such that*

$$\mathbf{E}(|X|; |X| > K) < \varepsilon, \quad \forall X \in \mathcal{C}.$$

We note that for such a class $\mathcal{C}$, we have (with K_1 relating to $\varepsilon = 1$), for every $X \in \mathcal{C}$,

$$\mathbf{E}(|X|) = \mathbf{E}(|X|; |X| > K_1) + \mathbf{E}(|X|; |X| \leqslant K_1) \leqslant 1 + K_1.$$

Thus *a UI family is bounded in $\mathcal{L}^1$.*

It is not true that a family bounded in $\mathcal{L}^1$ is UI.

(20.4) *Example.* Take $(\Omega, \mathcal{F}, \mathbf{P}) = ([0, 1], \mathcal{B}[0, 1], \text{Leb})$. Let

$$E_n = (0, n^{-1}), \qquad X_n = n\mathbf{I}_{E_n}.$$

Then $\mathbf{E}(|X_n|) = 1$, $\forall n$, so that (X_n) is bounded in $\mathcal{L}^1$. However, for any $K > 0$, we have, for $n > K$,

$$\mathbf{E}(|X_n|; |X_n| > K) = n\mathbf{P}(E_n) = 1,$$

so that (X_n) is not UI. Here $X_n \to 0$ a.s., but $\mathbf{E}(X_n) \not\to 0$. □

We now give two simple sufficient conditions for the UI property.

(20.5) LEMMA. *Suppose that $\mathscr{C}$ is a class of random variables that is bounded in $\mathscr{L}^p$ for some $p > 1$; thus, for some $A \in [0, \infty)$,*

$$\mathbf{E}(|X|^p) < A, \quad \forall X \in \mathscr{C}.$$

Then $\mathscr{C}$ is UI.

Proof. If $v \geqslant K > 0$ then $v \leqslant K^{1-p} v^p$ (obviously!). Hence, for $K > 0$ and $X \in \mathscr{C}$, we have

$$\mathbf{E}(|X|; |X| > K) \leqslant K^{1-p} \mathbf{E}(|X|^p; |X| > K) \leqslant K^{1-p} A.$$

The result follows. □

(20.6) LEMMA. *Suppose that $\mathscr{C}$ is a class of random variables that is dominated by an integrable non-negative variable Y: $|X(\omega)| \leqslant Y(\omega), \forall X \in \mathscr{C}$ and $\mathbf{E}(Y) < \infty$. Then $\mathscr{C}$ is UI.*

Proof. It is obvious that, for $K > 0$ and $X \in \mathscr{C}$, $\mathbf{E}(|X|; |X| > K) \leqslant \mathbf{E}(Y; Y > K)$, and now it is only necessary to apply (20.2) to Y. □

(20.7) LEMMA. *A class $\mathscr{C}$ of random variables is UI if and only if the following two conditions hold:*

(i) $\mathscr{C}$ is bounded in $\mathscr{L}^1$;
(ii) given $\varepsilon > 0$, there exists $\delta > 0$ such that, whenever $X \in \mathscr{C}$, and $F \in \mathscr{F}$ is such that $\mathbf{P}(F) < \delta$, we have $\mathbf{E}(|X|; F) < \varepsilon$.

(20.8) LEMMA. *If $\mathscr{C}$ and $\mathscr{D}$ are UI families of random variables, then*

$$\mathscr{C} + \mathscr{D} := \{X + Y : X \in \mathscr{C}, Y \in \mathscr{D}\}$$

is UI.

Proofs of Lemmas 20.7 and 20.8 are left as easy exercises.

21. $\mathscr{L}^1$ convergence. We begin with what is (in view of (19.8)) a consequence of the Dominated-Convergence Theorem.

(21.1) THEOREM (Bounded-Convergence Theorem). *Let (X_n) be a sequence of random variables, and let X be a random variable. Suppose that $X_n \to X$ in probability and that, for some K in $[0, \infty)$, we have for every n and ω,*

$$|X_n(\omega)| \leqslant K.$$

Then

$$\mathbf{E}(|X_n - X|) \to 0.$$

Proof. You check that $\mathbf{P}(|X| \leqslant K) = 1$. Let $\varepsilon > 0$ be given. Choose n_0 such that

$$\mathbf{P}(|X_n - X| > \tfrac{1}{3}\varepsilon) < \varepsilon/3K \quad \text{when } n \geqslant n_0.$$

Then, for $n \geqslant n_0$,

$$\begin{aligned}\mathbf{E}(|X_n - X|) &= \mathbf{E}(|X_n - X|; |X_n - X| > \tfrac{1}{3}\varepsilon) + \mathbf{E}(|X_n - X|; |X_n - X| \leqslant \tfrac{1}{3}\varepsilon) \\ &\leqslant 2K\mathbf{P}(|X_n - X| > \tfrac{1}{3}\varepsilon) + \tfrac{1}{3}\varepsilon \leqslant \varepsilon.\end{aligned}$$

The proof is finished. □

(*21.2*) THEOREM (A necessary and sufficient condition for $\mathscr{L}^1$ convergence). *Let (X_n) be a sequence in $\mathscr{L}^1$, and let $X \in \mathscr{L}^1$. Then $X_n \to X$ in $\mathscr{L}^1$, or, equivalently $\mathbf{E}(|X_n - X|) \to 0$, if and only if the following two conditions are satisfied:*

(i) $X_n \to X$ in probability;
(ii) the sequence (X_n) is UI.

It is of course the 'if' part of the theorem that is useful. Since the result is 'best possible', it must improve on the Dominated-Convergence Theorem for our $(\Omega, \mathscr{F}, \mathbf{P})$ triple; and, of course, the result (20.6) makes this explicit.

Proof of 'if' part. Suppose that conditions (i) and (ii) are satisfied. For $K \in [0, \infty)$, define a function $\varphi_K : \mathbf{R} \to [-K, K]$ as follows:

$$\varphi_K(x) := \begin{cases} K & \text{if } x > K, \\ x & \text{if } |x| \leqslant K, \\ -K & \text{if } x < -K. \end{cases}$$

Let $\varepsilon > 0$ be given. By the UI property of the (X_n) sequence and (20.2), we can choose K so that

$$\mathbf{E}\{|\varphi_K(X_n) - X_n|\} < \tfrac{1}{3}\varepsilon, \quad \forall n; \qquad \mathbf{E}\{|\varphi_K(X) - X|\} < \tfrac{1}{3}\varepsilon.$$

But, since $|\varphi_K(x) - \varphi_K(y)| \leqslant |x - y|$, we see that $\varphi_K(X_n) \to \varphi_K(X)$ in probability; and, by Theorem 21.1, we can choose n_0 such that, for $n \geqslant n_0$,

$$\mathbf{E}\{|\varphi_K(X_n) - \varphi_K(X)|\} < \tfrac{1}{3}\varepsilon.$$

The triangle inequality therefore implies that, for $n \geqslant n_0$, $\mathbf{E}(|X_n - X|) < \varepsilon$, and the proof is complete. □

Independence

22. Independence of σ-algebras and of random variables. Here are the key definitions of independence.

Sub-σ-algebras $\mathscr{G}_1, \mathscr{G}_2, \ldots$ of $\mathscr{F}$ are called *independent* if, whenever $G_i \in \mathscr{G}_i$ $(i \in \mathbf{N})$ and $i_1, \ldots, i_n$ are distinct,

$$\mathbf{P}(G_{i_1} \cap \cdots \cap G_{i_n}) = \prod_{k=1}^{n} \mathbf{P}(G_{i_k}).$$

Random variables $X_1, X_2, \ldots$ are called *independent* if the σ-algebras

$$\sigma(X_1), \sigma(X_2), \ldots$$

are independent.

Events $E_1, E_2, \ldots$ are called *independent* if the σ-algebras $\mathscr{E}_1, \mathscr{E}_2, \ldots$ are independent, where

$$\mathscr{E}_n \text{ is the } \sigma\text{-algebra } \{\emptyset, E_n, \Omega\backslash E_n, \Omega\}.$$

Since $\mathscr{E}_n = \sigma(I_{E_n})$, it follows that events $E_1, E_2, \ldots$ are independent if and only if the random variables $I_{E_1}, I_{E_2}, \ldots$ are independent.

(22.1) *The π-system Lemma.* We know from elementary theory that events $E_1, E_2, \ldots$ are independent if and only if whenever $n \in \mathbf{N}$ and $i_1, \ldots, i_n$ are distinct,

$$\mathbf{P}(E_{i_1} \cap \cdots \cap E_{i_n}) = \prod_{k=1}^{n} \mathbf{P}(E_{i_k}),$$

corresponding results involving complements of the E_i etc., being consequences of this.

We now use the Uniqueness Lemma 4.7 to obtain a significant generalisation of this idea, allowing us to study independence via (manageable) π-systems rather than (awkward) σ-algebras.

Let us concentrate on the case of two σ-algebras.

(22.2) LEMMA. *Suppose that $\mathscr{G}$ and $\mathscr{H}$ are sub-σ algebras of $\mathscr{F}$, and that $\mathscr{I}$ and $\mathscr{J}$ are π-systems with*

$$\sigma(\mathscr{I}) = \mathscr{G}, \quad \sigma(\mathscr{J}) = \mathscr{H}.$$

Then $\mathscr{G}$ and $\mathscr{H}$ are independent if and only if $\mathscr{I}$ and $\mathscr{J}$ are independent in that

$$\mathbf{P}(I \cap J) = \mathbf{P}(I)\mathbf{P}(J), \quad I \in \mathscr{I}, \; J \in \mathscr{J}.$$

Proof. Suppose that $\mathscr{I}$ and $\mathscr{J}$ are independent. For fixed I in $\mathscr{I}$, the *measures* (check that they *are* measures!)

$$H \mapsto \mathbf{P}(I \cap H) \quad \text{and} \quad H \mapsto \mathbf{P}(I)\mathbf{P}(H)$$

on $(\Omega, \mathscr{H})$ have the same total mass $\mathbf{P}(I)$, and agree on $\mathscr{J}$. Therefore, by the Uniqueness Lemma 4.7, they agree on $\sigma(\mathscr{J}) = \mathscr{H}$. Hence

$$\mathbf{P}(I \cap H) = \mathbf{P}(I)\mathbf{P}(H), \quad I \in \mathscr{I}, \; H \in \mathscr{H}.$$

Thus, for fixed H in $\mathscr{H}$, the measures

$$G \mapsto \mathbf{P}(G \cap H) \quad \text{and} \quad G \mapsto \mathbf{P}(G)\mathbf{P}(H)$$

on $(\Omega, \mathscr{G})$ have the same total mass $\mathbf{P}(H)$, and agree on $\mathscr{I}$. They therefore agree on $\sigma(\mathscr{I}) = \mathscr{G}$; and this is what we set out to prove. □

Suppose now that X and Y are two real-valued random variables on $(\Omega, \mathscr{F}, \mathbf{P})$ such that, whenever $x, y \in \mathbf{R}$,

$$\mathbf{P}(X \leqslant x; Y \leqslant y) = \mathbf{P}(X \leqslant x)\mathbf{P}(Y \leqslant y). \tag{22.3}$$

Now, (22.3) says that the π-systems $\pi(X) := \{X^{-1}((-\infty, x]) : x \in \mathbf{R}\}$ and $\pi(Y)$ are independent. Hence $\sigma(X)$ and $\sigma(Y)$ are independent: that is, X and Y are independent in our new 'abstract' sense.

In the same way, we can prove that random variables $X_1, X_2, \ldots, X_n$ are independent if and only if

$$\mathbf{P}(X_k \leqslant x_k : 1 \leqslant k \leqslant n) = \prod_{k=1}^{n} \mathbf{P}(X_k \leqslant x_k),$$

and all the familiar things from elementary theory.

(22.4) *Independence and product measure.* This is the ultimate form of the 'independence means multiply' idea. Suppose that, for $i = 1, 2$, $(E_i, \mathscr{E}_i)$ is a measurable space and that X_i is an $(E_i, \mathscr{E}_i)$-valued random variable. Recall from Section 16 the definitions of the laws Λ_{X_1} and Λ_{X_2} of X_1 and X_2, and of the joint law Λ_{X_1, X_2} on $(E, \mathscr{E}) := (E_1 \times E_2, \mathscr{E}_1 \times \mathscr{E}_2)$.

(22.5) THEOREM. *The variables X_1 and X_2 are independent if and only if*

$$\Lambda_{X_1, X_2} = \Lambda_{X_1} \times \Lambda_{X_2}.$$

If X_1 and X_2 are independent and if, for $i = 1, 2$, $h_i \in (\mathrm{m}\mathscr{E})^+$ then

$$\mathbf{E}h_1(X_1)h_2(X_2) = \mathbf{E}h_1(X_1).\mathbf{E}h_2(X_2) \leqslant \infty. \tag{22.6}$$

You prove the first statement. Result (22.6) follows from Fubini's Theorem together with the ideas in (17.3): if we define h on E via $h(x) := h_1(x_1)h_2(x_2)$, where $x = (x_1, x_2)$, then

$$\begin{aligned} \mathbf{E}h_1(X_1)h_2(X_2) &= \int_E h(x)\Lambda_{X_1, X_2}(dx) \\ &= \int_{E_1}\int_{E_2} h_1(x_1)h_2(x_2)\Lambda_1(dx_1)\Lambda_2(dx_2) \\ &= \mathbf{E}h_1(X_1).\mathbf{E}h_2(X_2). \end{aligned}$$

There are obvious generalisations of the theorem.

If X and Y are independent elements of $\mathscr{L}^1(\Omega, \mathscr{F}, \mathbf{P})$, then $XY \in \mathscr{L}^1$ and $\mathbf{E}(XY) = \mathbf{E}(X)\mathbf{E}(Y)$. If, further, $X, Y \in \mathscr{L}^2$, then

$$\mathrm{Var}(X + Y) = \mathrm{Var}(X) + \mathrm{Var}(Y);$$

and so on, and so forth...

23. Existence of families of independent variables. From Section 1.6 on Ciesielski's construction of Brownian motion onwards, we have required models that support families of independent variables with prescribed laws. Theorem 26.1 gives the elegant and proper way of doing this; and the strength of that theorem is needed in, for example, the direct construction of Poisson measures in Section 37. However, it is often the case that all we require is a model that supports the existence of a sequence of *real-valued* random variables with prescribed distribution functions. We now recall briefly the well-known trick for achieving this; [W] gives some more details.

Let

$$(\Omega, \mathcal{F}, \mathbf{P}) = ([0,1], \mathcal{B}[0,1), \text{Leb}).$$

Expand ω in Ω in binary, and write $X(\omega) := \omega$:

$$X(\omega) := \omega = \cdot\omega_1\omega_2\omega_3\cdots = \sum 2^{-k}\omega_k.$$

(Conventions made about dyadic rationals are irrelevant.) Then the variables $(\pi_n : n \in \mathbf{N})$, where $\pi_n(\omega) := \omega_n$ are independent coin-tossing variables, each taking the values 0 and 1 with probability $\frac{1}{2}$ each. Thus

$$X_1(\omega) := \cdot\omega_1\omega_3\omega_6\ldots,$$

$$X_2(\omega) := \cdot\omega_2\omega_5\omega_9\ldots,$$

$$X_3(\omega) := \cdot\omega_4\omega_8\omega_{13}\ldots$$

etc. defines an independent sequence of variables each with the same distribution as X, that is, each with the uniform distribution on $[0,1]$.

If $(F_n : n \in \mathbf{N})$ is a sequence of distribution functions on $\mathbf{R}$ then the definitions

$$Y_n := \sup\{y : F_n(y) \leqslant X_n\}$$

produce a sequence of independent variables (Y_n), Y_n having distribution function F_n.

All that is needed to prove these statements in this section is the Uniqueness Lemma 4.7.

Note that we now have all the theoretical equipment for the proof of the existence of Wiener measure in Section I.6.

24. Exercises. Do all the exercises preceding Exercise E9.1 in [W].

Of course, the exercises in our Section 13 have important consequences for probability. For example, E13.11a shows that if X and Y are random variables taking values in $(S, \mathcal{B}(S))$, where (S, ρ) is a *separable* metric space, then $\rho(X, Y)$ is a real-valued random variable.

3. STOCHASTIC PROCESSES

The Daniell–Kolmogorov Theorem

25. $(E^T, \mathcal{E}^T)$; σ-algebras on function space; cylinders and σ-cylinders. Let $(E, \mathcal{E})$ be a measurable space, and let T be a set. Recall that E^T is the set of all functions

f from T to E. For $t \in T$, define $\pi_t : E^T \to E$ to be the evaluation map

$$\pi_t(f) := f(t). \tag{25.1}$$

(25.2) DEFINITION (the σ-algebra $\mathscr{E}^T$). *Define the σ-algebra*

$$\mathscr{E}^T := \sigma\{\pi_t : t \in T\}$$

on E^T. Thus $\mathscr{E}^T$ is the smallest σ-algebra on E^T such that each π_t is $(\mathscr{E}^T/\mathscr{E})$-measurable.

For $\varnothing \neq S \subseteq T$, define $\pi_S : E^T \to E^S$ to be the restriction map

$$\pi_S(f) := f|_S. \tag{25.3}$$

Then (check!) π_S is $(\mathscr{E}^T/\mathscr{E}^S)$-measurable.

(25.4) DEFINITION (cylinder, special cylinder). *We say that a subset F of E^T is a* cylinder *if F has the form*

$$F = \pi_S^{-1} A_S = A_S \times E^{T \setminus S}, \tag{25.5}$$

for some non-empty finite *subset S of T and some A_S in $\mathscr{E}^S$. We say that F is a* special cylinder *if it has the form*

$$\bigcap_{t \in S} \pi_t^{-1} H_t = \left(\prod_{t \in S} H_t \right) \times E^{T \setminus S}, \tag{25.6}$$

where S is a finite subset of T and $H_t \in \mathscr{E}$ for $t \in S$.

(25.7) LEMMA. *The cylinder sets form an algebra that generates $\mathscr{E}^T$. The special cylinder sets form a π-system that generates $\mathscr{E}^T$.*

The proof is left as a simple exercise.

(25.8) DEFINITION (σ-cylinder). *We say that F is a* σ-cylinder *if it has the form*

$$F = \pi_S^{-1} A_S$$

for some non-empty countable *subset S of T and some A_S in $\mathscr{E}^S$.*

(25.9) LEMMA. *$\mathscr{E}^T$ is precisely the collection of σ-cylinders: thus membership of an element of $\mathscr{E}^T$ imposes restriction on the values of f only at countably many t-values.*

This result is very important. Its proof is easy. One need only show that the σ-cylinders form a σ-algebra. The key point is that if $((S(n))$ is a sequence of countable subsets of T and for each n, $A(n) \in \mathscr{E}^{S(n)}$, then

$$\bigcap_n \pi_{S(n)}^{-1} A(n) = \pi_S^{-1} A,$$

where $S = \bigcup S(n)$ and

$$A = \bigcap_n [A(n) \times E^{S \setminus S(n)}] \in \mathscr{E}^S.$$

The notation introduced in this section will be carried forward for some time.

26. Infinite products of probability triples. Section 25 *had* to be the first section of this discussion of stochastic processes. The present section requires the definition of $\mathscr{E}^T$, but otherwise would belong more properly at the end of part 2 of this chapter.

(*26.1*) THEOREM. *For each t in T, let μ_t be a probability measure on $(E, \mathscr{E})$. Then there exists a unique probability measure μ on $(E^T, \mathscr{E}^T)$ such that, whenever S is a finite subset of T and $H_t \in \mathscr{E}$ for $t \in S$,*

(26.2)
$$\mu\left(\bigcap_{t \in S} \pi_t^{-1} H_t\right) = \prod_{t \in S} \mu_t(H_t).$$

The uniqueness of μ is an immediate consequence of Lemma 4.7 and the fact that the special cylinders form a π-system that generates $\mathscr{E}^T$.

The existence of μ is quite a deep matter. Fubini's Theorem implies the existence of a finitely additive measure μ_0 on the algebra of cylinder sets such that the analogue of (26.2) holds for μ_0. Carathéodory's Theorem 5.1 shows that we need only (!) prove that μ_0 is σ-additive on the collection of cylinder sets. After a little thought about Lemma 25.9, we realise that we need only prove the result for the case when T is countable. Compare Observation 30.2 below. A leisurely proof is given in [W; Chapter A9] (you can regard the $(\mathbb{R}, \mathscr{B})$ there as a notation for $(E, \mathscr{E})$ for this purpose). You should compare and contrast that proof with the proof given below for the Daniell–Kolmogorov Theorem, for which topological assumptions are necessary. *When $(E, \mathscr{E})$ does have suitable topological properties, as is always the case in practice, Theorem 26.1 follows from the DK Theorem.*

Note that

(*26.3*) *the $(E, \mathscr{E})$-valued random variables $(\pi_t : t \in T)$ on the probability triple $(E^T, \mathscr{E}^T, \mu)$ are independent, π_t having law μ_t.*

The problem of the existence of 'completely independent' stochastic processes (and, in particular, of independent, identically distributed (IID) sequences) is therefore settled. We now turn to the study of more general processes.

27. Stochastic process; sample function; law. There are many different ways, all important, of regarding a stochastic process.

(*27.1*) CLASSICAL DEFINITION (stochastic process; state-space; parameter set; carrier triple). *Let T be a set, $(E, \mathcal{E})$ a measurable space and $(\Omega, \mathcal{F}, \mathbf{P})$ a probability triple. The traditional definition of a* stochastic process with time-parameter set T, state-space $(E, \mathcal{E})$ and carrier triple $(\Omega, \mathcal{F}, \mathbf{P})$ *is as a collection $\{X_t : t \in T\}$ of $(E, \mathcal{E})$-valued random variables carried by our triple $(\Omega, \mathcal{F}, \mathbf{P})$.*

Thus, we have, for each t, the picture

$$X_t : \Omega \to E, \quad X_t^{-1} : \mathcal{E} \to \mathcal{F}.$$

(27.2) DEFINITION (sample function; sample path; realisation). *Let X be as in (27.1). For $\omega \in \Omega$, the map (element of E^T)*

$$X(\omega): T \to E$$
$$t \mapsto X_t(\omega)$$

is called the sample function of X *(or, especially when T is a time-parameter set, the* sample path of X*) or* realisation of X corresponding to ω.

This leads to an alternative view of X, namely as the map

(27.3)
$$X : \Omega \to E^T, \quad X^{-1} : \mathcal{E}^T \to \mathcal{F},$$
$$\omega \mapsto X(\omega);$$

in other words, as an $(E^T, \mathcal{E}^T)$-valued random variable. You can easily check that X is $(\mathcal{F}/\mathcal{E}^T)$-measurable as a map from Ω to E^T if and only if each X_t is $(\mathcal{F}/\mathcal{E})$-measurable as a map from Ω to E.

(27.4) DEFINITION (law of stochastic process). *The* law *of the stochastic process X in Definition 27.1 is the probability measure*

(27.5)
$$\mu := \mathbf{P} \circ X^{-1} \quad \text{on } (E^T, \mathcal{E}^T);$$

in other words, it is the law of the $(E^T, \mathcal{E}^T)$-valued random variable X.

If we wish to emphasise the role of $(\Omega, \mathcal{F}, \mathbf{P})$ and/or T, we shall use such notations as

$$(\Omega, \mathcal{F}, \mathbf{P}; X) \quad \text{or} \quad (\Omega, \mathcal{F}, \mathbf{P}; \{X_t : t \in T\})$$

to signify our process X.

28. Canonical process. Let X be the process of Section 27, and let μ be its law.

The process

(28.1)
$$(E^T, \mathcal{E}^T, \mu; \pi_t : t \in T)$$

trivially has the same law μ as X; it is called the *canonical process* with law μ.

A canonical process is completely determined by its law; and, for a canonical process, the idea that a sample point 'is' the outcome of the experiment is restored. Canonical processes are certainly nice. However, probability theory gets most of its depth from being able to construct (certainly non-canonical!) processes from other processes by time transformation, or as solutions of SDEs, etc.

Important note on terminology. It is important that we are currently working with the space E^T of *all* functions from a set T into a space E that carries a measurable structure $\mathscr{E}$. When we speak of canonical Brownian motion, we usually mean the set-up

$$(C, \mathscr{A}, \mathbf{W}; \pi_t : t \in \mathbb{R}^+),$$

where $C = C([0, \infty); \mathbb{R})$ is the space of *continuous* paths $w: [0, \infty) \to \mathbb{R}$, $\pi_t: C \to \mathbb{R}$ is the evaluation map $\pi_t(\omega) = \omega(t)$, $\mathscr{A} = \sigma\{\pi_t : t \in \mathbb{R}\}$ and $\mathbf{W}$ is Wiener measure.

You must keep in mind that *at the moment, all paths are allowed.*

29. Finite-dimensional distributions, sufficiency; compatibility. We continue with the notation of the last few sections.

For a non-empty subset S of T, define $\pi_S X : \Omega \to E^S$ via

(29.1) $$(\pi_S X)(\omega) := \pi_S(X(\omega)) = X(\omega)|_S$$

and

(29.2) $$\mu_S := \mathbf{P} \circ (\pi_S X)^{-1} \quad \text{on } (E^S, \mathscr{E}^S).$$

(29.3) DEFINITION (Fin(T), finite-dimensional distributions). *Let* Fin(T) *denote the set of non-empty finite subsets of T. The probability measures*

$$\{\mu_S : S \in \text{Fin}(T)\}$$

are called the finite-dimensional distributions of X.

For $S \in \text{Fin}(T)$,

(29.4) $$\mu_S = \mu \circ \pi_S^{-1} \text{ on } (E^S, \mathbf{E}^S).$$

If we know the finite-dimensional distributions of X then we know the value of μ on all cylinders, and hence, by Lemmas 4.7 and 25.7, we know μ on the whole of $(E^T, \mathscr{E}^T)$:

(29.5) *the finite-dimensional distributions determine the law.*

(This is what is meant by 'sufficiency' in the title of this section: it has nothing to do with Fisher's brilliant concept in statistics). Do Exercise E38.29.

Note that if $U, V \in \text{Fin}(T)$ and $U \subseteq V$, *and if* π_U^V denotes the restriction map from E^V to E^U, then we have the *compatibility condition* or *projective property*:

(29.6) $$\mu_U = \mu_V \circ (\pi_U^V)^{-1}.$$

The fundamental Daniell–Kolmogorov Theorem considers the following problem. Suppose that we have a family of probability measures as in (29.3) that satisfies the compatibility condition (29.6): does there exist a measure μ on $(E^T, \mathscr{E}^T)$ such that (29.4) holds? In the language of category theory, we are asking whether a projective system has a projective limit. Rather surprisingly, the answer is 'Not in general'. In order to obtain a positive result, we have to make a topological assumption about the measurable space $(E, \mathscr{E})$: we need the following inner regularity with respect to compact sets.

(29.7) LEMMA. *Let J be a compact metric space, and let B be a Borel subset of J. If m is a finite measure on J and $\varepsilon > 0$, then there exists a compact subset K of B such that*

$$m(K) > m(B) - \varepsilon.$$

This standard result is proved in Section 81.

30. The Daniell–Kolmogorov (DK) Theorem; 'compact metrisable' case. The DK Theorem is the essential first step in constructing stochastic processes. The general case of the theorem is given in the next section. It is well worth presenting the present 'simple' case on its own.

Recall that Fin(T) is the family of non-empty finite subsets of T.

(30.1) THEOREM (Daniell, Kolmogorov). *Let E be a compact metrisable space, and let $\mathscr{E} = \mathscr{B}(E)$. Let T be a set. Suppose that for each S in* Fin(T), *there exists a probability measure μ_S on $(E^S, \mathscr{E}^S)$, and that the measures $\{\mu_S : S \in \text{Fin}(T)\}$ are* compatible *or* projective *in that*

$$\mu_U = \mu_V \circ (\pi_U^V)^{-1} \tag{30.2}$$

holds whenever $U, V \in \text{Fin}(T)$ and $U \subseteq V$. Here π_U^V is the restriction map from E^V into E^U. Then there exists a unique measure μ on $(E^T, \mathscr{E}^T)$ such that

$$\mu_S = \mu \circ \pi_S^{-1} \quad on \quad (E^S, \mathscr{E}^S), \tag{30.3}$$

where π_S is the restriction map from E^T to E^S.

Start of Proof. For any cylinder set F, we have for some $S \in \text{Fin}(T)$ and some $A_S \in \mathbf{E}^S$,

$$F = \pi_S^{-1} A_S = A_S \times E^{T \setminus S}. \tag{30.4}$$

For such an F, set

$$\mu_0(F) := \mu_S(A_S). \tag{30.5}$$

The compatibility condition (30.3) guarantees that this definition is independent of the particular representation of F used in (30.4). Moreover, it is obvious that

μ_0 is finitely additive on the algebra $\mathscr{C}$ of cylinder sets. We need only show that

(30.6) *μ_0 is countably additive on $\mathscr{C}$*

since then Carathéodory's Theorem does the rest.

The result (4.3) makes it clear that *Theorem 30.1 is implied by the following lemma.*

(30.7) LEMMA. *Suppose that*

(i) $F_n \in \mathscr{C} \quad (n \in \mathbb{N}); \quad F_n \supseteq F_{n+1} \quad (\forall n);$
(ii) for some $\varepsilon > 0, \quad \mu_0(F_n) > 2\varepsilon \quad (\forall n).$

Then $\bigcap_n F_n \neq \varnothing$.

Proof of Lemma. Let (F_n) satisfy the hypotheses of Lemma 30.7. We have

(30.8) $$F_n = \pi_{S(n)}^{-1} A_n = A_n \times E^{T \setminus S(n)}$$

for some $S(n) \in \mathrm{Fin}(T)$ and some A_n in $\mathscr{E}^{S(n)}$. Now, $\mu_{S(n)}$ is a probability measure on the compact metrisable space $E^{S(n)}$, and so, by Lemma 29.7, there is a compact subset K_n of A_n such that

$$\mu_{S(n)}(K_n) > \mu_{S(n)}(A_n) - 2^{-n}\varepsilon.$$

In other words,

(30.9) $$\mu_0(H_n) > \mu_0(F_n) - 2^{-n}\varepsilon,$$

where

(30.10) $$H_n := K_n \times E^{T \setminus S(n)}.$$

Note that H_n is compact, by Tychonov's Theorem.

You can easily combine the hypotheses of Lemma 30.7 with (30.9) to prove that

$$\mu_0(H_1 \cap \cdots \cap H_n) > \varepsilon, \quad \forall n.$$

Thus

(30.11) $$H_1 \cap \cdots \cap H_n \neq \varnothing, \quad \forall n.$$

If $\bigcap_k H_k = \varnothing$ then $\bigcup_k H_k^c = E^T$, whence the fact that E is compact forces

$$\bigcup_{k \leq n} H_k^c = E^T \quad \text{for some} \quad n,$$

contradicting (30.11). Hence $\bigcap_k H_k \neq \varnothing$, whence, *a fortiori*, $\bigcap_k F_k \neq \varnothing$. Thus Lemma 30.7 and Theorem 30.1 are true. □

Now do Exercise E38.30.

31. The Daniell–Kolmogorov Theorem: general case. The case now to be presented is not the most general known, but it is good enough for us.

(*31.1*) THEOREM (Daniell, Kolmogorov). *Theorem 30.1 remains true if the assumption that E is compact metrisable is replaced by the assumption that E is a Lusin space, that is, E is homeomorphic to a Borel subset of a compact metrisable space.*

Remark. Of course, $\mathbb{R}^n$ is homeomorphic to a Borel (indeed, open) subset of a compact metric space. For example, use stereographic projection of the S^n sphere in $\mathbb{R}^{n+1}$.

The following observation will prove useful.

(*31.2*) OBSERVATION. *In proving the theorem, we may assume that T is countable.*

Justification of Observation. All of the remarks made up to and including the statement of Lemma 30.7 transfer to the present case. Proving Lemma 30.7 for a fixed sequence (F_n) as in (30.8) is identical to proving the same result when $T = \bigcup S(n)$. □

Proof of Theorem 31.1. We suppose that $E \in \mathscr{J}$, where $\mathscr{J} := \mathscr{B}(J)$, J being a compact metrisable space.

We do (as we may) assume that T is countable.

We derive the theorem directly from Theorem 30.1, making no further use of Lemma 30.7.

For each S in $\mathrm{Fin}(T)$, we extend μ_S on $(E^S, \mathscr{E}^S)$ to $\hat{\mu}_S$ on $(J^S, \mathscr{J}^S)$ in the obvious way:

$$\hat{\mu}_S(\hat{A}_S) := \mu_S(\hat{A}_S \cap E^S) \quad (\hat{A}_S \in \mathscr{J}^S).$$

Define $\hat{\mu}_0$ on the algebra $\hat{\mathscr{C}}$ of cylinder sets associated with $(J, \mathscr{J})$ via the obvious analogue of (30.5). Since J is compact metrisable, we know from Theorem 30.1 that $\hat{\mu}_0$ has a unique countably additive extension $\hat{\mu}$ to $(J^T, \mathscr{J}^T)$.

Now, T is countable. We may therefore find a sequence $(T(k))$ of finite sets with $T(k) \uparrow T$. But then

$$\begin{aligned}\hat{\mu}(E^T) &= \downarrow\lim \hat{\mu}(E^{T(k)} \times J^{T \setminus T(k)}) = \downarrow\lim \hat{\mu}_{T(k)}(E^{T(k)}) \\ &= \downarrow\lim 1 = 1,\end{aligned}$$

and

$$\mu(L) := \hat{\mu}(L \cap E^T) \quad (L \in \mathscr{E}^T)$$

obviously defines the required probability measure on $(E^T, \mathscr{E}^T)$ asserted by the theorem.

The proof of the DK Theorem is finished. □

Discussion. Suppose that E is not compact, and that T is uncountable. Then E^T is not an element of $\mathscr{J}^T$, and we cannot say that $\hat{\mu}(E^T)=1$. If however, $\hat{F}$ is any element of $\mathscr{J}^T$ such that $E^T \subset \hat{F}$ then, for some countable subset S of T, $\hat{F} \supseteq E^S \times J^{T\setminus S}$, and $\hat{\mu}(\hat{F})=1$. Thus the outer $\hat{\mu}$-measure $\hat{\mu}^*(\hat{F})$ of $\hat{F}$ is equal to 1, and what is happening is that

$$\mu(F)=\hat{\mu}^*(F) \quad (F\in\mathscr{E}^T).$$

This kind of thing will keep on happening!

32. Gaussian processes; pre-Brownian motion. In this section and the next we look at some applications of the DK Theorem. We shall soon see in Section 34 that these applications are, as yet, extremely unsatisfactory.

(32.1) Gaussian processes. Let T be a parameter set. Let $m: T\to\mathbb{R}$, and let V be a symmetric non-negative-definite function from $T\times T$ to $\mathbb{R}$, so that, for any finite subset S of T and any function f on S,

$$\sum_{r\in S}\sum_{s\in S} V(r,s)f(r)f(s)\geqslant 0.$$

We know from elementary theory that, for $S\in\operatorname{Fin}(T)$, there exists a unique measure μ_S on $(\mathbb{R}^S,\mathscr{B}^S)$ such that, for $\theta\in\mathbb{R}^S$,

(32.2)
$$\int_{\mathbb{R}^S}\exp\left[i\sum_{s\in S}\theta(s)f(s)\right]\mu_S(df)$$
$$=\exp\left[i\sum_{s\in S}\theta(r)m(r)-\frac{1}{2}\sum_{r\in S}\sum_{s\in S}\theta(r)V(r,s)\theta(s)\right].$$

Indeed, if the restriction V_S of V to $S\times S$ is strictly positive-definite then μ_S has density

(32.3)
$$(2\pi)^{-|S|/2}(\det V_S)^{-1/2}\exp\left\{-\frac{1}{2}\sum_{r\in S}\sum_{s\in S}[f(r)-m(r)](V_S)^{-1}(r,s)[f(s)-m(s)]\right\}$$

relative to the Lebesgue measure on $\mathbb{R}^S$.

We also know from elementary theory that the measures $\{\mu_S:S\in\operatorname{Fin}(T)\}$ are compatible (projective) in the sense of the DK Theorem. Hence we can construct the *Gaussian process* $(\mathbb{R}^T,\mathscr{B}^T,\mu;\pi_t:t\in T)$ *with mean function* m *and covariance function* V; this has the projective limit μ as its law. That m is the mean function and V is the covariance function is confirmed by

$$\mu(\pi_t)=m(t),\qquad \mu(\pi_s\pi_t)-\mu(\pi_s)\mu(\pi_t)=V(s,t),\quad \forall(s,t).$$

(32.4) Orthogonality and independence. If T_1 and T_2 are disjoint subsets of T

and $V(t_1, t_2) = 0$ whenever $t_i \in T_i$, $i = (1, 2)$, then $\{\pi_{t_1} : t_1 \in T_1\}$ and $\{\pi_{t_2} : t_2 \in T_2\}$ are independent processes.

(*32.5*) *Pre-Brownian motion.* If we take

$$T = [0, \infty), \qquad m(t) \equiv 0, \qquad V(s, t) \equiv \min(s, t)$$

then, as we already know, V is positive-definite. We call the associated canonical process $(\mathbb{R}^T, \mathscr{B}^T, \mu; \pi_t : t \in T)$ *pre-Brownian motion.* We adjoint the suffix 'pre-' because this process has all possible functions from $(0, \infty)$ to $\mathbb{R}$ as paths, not just continuous functions.

33. Pre-Poisson set functions. Let $(W, \mathscr{W}, \lambda)$ be a σ-finite measure space such that every singleton set $\{x\}$ $(x \in W)$ is in $\mathscr{W}$. By a *pre-Poisson set function* on $(W, \mathscr{W})$ with intensity measure λ, we mean a $(\mathbb{Z}^+ \cup \{\infty\})$-valued process (if one exists) $\{\Lambda(B) : B \in \mathscr{W}\}$ with the following properties:

(33.1) (i) for every B in $\mathscr{W}$, $\Lambda(B)$ is a $\mathbb{Z}^+$-valued random variable with the Poisson distribution of parameter $\lambda(B)$:

$$\operatorname{Prob}(\Lambda(B) = k) = \frac{e^{-\lambda(B)}\lambda(B)^k}{k!} \quad \text{if } \lambda(B) < \infty,$$

$$\operatorname{Prob}(\Lambda(B) = \infty) = 1 \quad \text{if } \lambda(B) = \infty;$$

(33.1) (ii) if $B_1, \ldots, B_n$ are disjoint elements of $\mathscr{W}$ then $\Lambda(B_1), \ldots, \Lambda(B_n)$ are independent random variables:

(33.1)(iii) whenever B_1 and B_2 are disjoint elements of $\mathscr{W}$,

$$\mathbf{P}[\Lambda(B_1 \cup B_2) = \Lambda(B_1) + \Lambda(B_2)] = 1.$$

If S is a finite subset of $\mathscr{W}$ then we can easily specify the desired law μ_S of $\{\Lambda(B) : B \in S\}$. We know from elementary theory that the sum of independent Poisson variables is again Poisson, from which it follows that the family $\{\mu_S : S \in \operatorname{Fin}(\mathscr{W})\}$ is projective. Hence, we can construct a canonical pre-Poisson set function

$$((Z^+ \cup \{\infty\})^{\mathscr{W}}, \mathscr{P}(Z^+ \cup \{\infty\})^{\mathscr{W}}, \mu; \pi_B : B \in \mathscr{W}) \tag{33.2}$$

with intensity measure λ.

Beyond the DK Theorem

34. Limitations of the DK Theorem. Under its hypothesis, the DK Theorem 47.1 provides us with a canonical process

$$(E^T, \mathscr{E}^T, \mu; \pi_t : t \in T)$$

that has all possible functions in E^T as sample functions. Moreover, we know from Lemma 25.9 that $F \in \mathscr{E}^T$ if and only if F is a σ-cylinder, that is, if and only if $F = \pi_S^{-1} A_S$ for some *countable* set S and some A_S in $\mathscr{E}^S$.

(34.1) *Difficulties with path continuity.* Consider the canonical pre-Brownian process

$$(\mathbb{R}^{[0,\infty)}, \mathscr{B}^{[0,\infty)}, \mu; \pi_t : t \in [0, \infty))$$

in (32.5). Let C be the set of continuous functions from $[0, \infty)$ to $\mathbb{R}$. Life would be simple if it were the case that $C \in \mathscr{B}^{[0,\infty)}$ and $\mu(C) = 1$. However, $C \notin \mathscr{B}^{[0,\infty)}$ because C is not a σ-cylinder.

Suppose that $F \in \mathscr{B}^{[0,\infty)}$ and $F \subseteq C$. Then

(i) $F = \pi_S^{-1} A_S$ for some countable set S and some A in $\mathscr{B}^S$;
(ii) every element of F is continuous on $[0, \infty)$.

Since property (i) tells us nothing about the behaviour of elements of F off the set S, we conclude that $F = \emptyset$. we have proved that

(34.2) $$F \in \mathscr{B}^{[0,\infty)} \text{ and } F \subseteq C \text{ imply that } F = \emptyset.$$

Thus C has inner μ-measure 0, and completion certainly will not help us.

(*34.3*) *Difficulties with Poisson measures.* Recall the canonical pre-Poisson set function in Section 33. What we really want is that $B \mapsto \pi_B(\omega)$ is a measure for each ω. Let $\mathscr{M} \subseteq (\mathbb{Z}^+ \cup \{\infty\})^{\mathscr{W}}$ be the set of $(\mathbb{Z}^+ \cup \{\infty\})$-valued measures on $(W, \mathscr{W})$. Life would be simple if it were the case that $\mathscr{M} \in (\mathbb{Z}^+ \cup \{\infty\})^{\mathscr{W}}$ and $\mu(\mathscr{M}) = 1$. However, you can easily prove, in analogy with (34.2), that if $\mathscr{W}$ is uncountable, as it will be in every case of interest, then

(34.4) $$F \in (\mathbb{Z}^+ \cup \{\infty\})^{\mathscr{W}} \text{ and } F \subseteq \mathscr{M} \text{ imply that } F = \emptyset.$$

Note that if $\mathscr{M}_0$ denotes the set of finitely additive measures on $(W, \mathscr{W})$ then, if $\mathscr{W}$ is uncountable, the analogue of (34.4) will hold for $\mathscr{M}_0$.

35. The role of outer measures. Again let $(E, \mathscr{E})$ be a measurable space and T a set. Let μ be a probability measure on $(E^T, \mathscr{E}^T)$.

Let $G \subseteq E^T$. Think of G as a class of good sample functions. Thus G might be C in the context of (34.1), or $\mathscr{M}$ in the context of (34.3). In many contexts, G will be the set of *right*-continuous paths on $(0, \infty)$, which is a more natural class for probability theory than the set of continuous paths.

Before reading the next lemma, please reread Lemma 6.1.

(*35.1*) LEMMA and DEFINITION. *A process with law* μ *exists with all its sample functions in* G *if and only if the outer* μ*-measure* $\mu^*(G)$ *of* G *is* 1. *Then*

$$(G, \mathscr{G}, \mu^*; \pi_t : t \in T)$$

is the canonical process with path-space G and law μ.

(35.2) COROLLARY. *Because of Wiener's Theorem I.6.1, we have* $\mu^*(C)=1$ *if* μ *is the pre-Brownian law.*

It is never at all easy in practice to decide whether or not $\mu^*(G)=1$. Lemma 35.1 is more a matter of clarifying structure than a useful tool.

36. Modification; indistinguishability. Let Y be a stochastic process with parameter set T and state-space $(E,\mathscr{E})$ carried by the triple $(\Omega,\mathscr{F},\mathbf{P})$.

(36.1) *Important discussion.* To indicate the way in which things develop, suppose that Y is an $(\mathbb{R},\mathscr{B})$-valued process with time-parameter set $[0,\infty)$ and law μ, carried by a triple $(\Omega,\mathscr{F},\mathbf{P})$. It is often possible, by making heavy use of the structure of μ, to show that there exists a set Ω_G in $\mathscr{F}$ with $\mathbf{P}(\Omega_G)=1$ such that, for every $\omega\in\Omega_G$, the map

$$q\mapsto Y_q(\omega) \text{ from } \mathbb{Q}\cap[0,\infty) \text{ to } \mathbb{R}$$

has a right-continuous extension $t\mapsto X_t(\omega)$ from $[0,\infty)$ to $\mathbb{R}$. If $\omega\notin\Omega_G$, set $X_t(\omega)=0$ for all t. Then all paths of X are right-continuous. It is often further possible to show, again by using the structure of the particular μ, that $\mathbf{P}(X_t=Y_t)=1$, $\forall t\in T$. Then X will be a process with law μ, all paths of which are right-continuous. (The set Ω_C of ω for which $t\mapsto X_t(\omega)$ is continuous will then be an element of $\mathscr{F}$, so that $\mathbf{P}(\Omega_C)$ is meaningful.)

The 'regularisation' method just described, and due to Kolmogorov and Doob, is one of the most powerful and widely used ways of obtaining processes with right-continuous paths. Often, however, we use *direct methods of construction* such as Ciesielski's proof of the existence of path-continuous Brownian motion in Section I.6. Having obtained this Brownian motion, we can then construct path-continuous diffusions by solving SDEs; and so on.

The good sense of the following definition is now evident.

(36.2) DEFINITION (modification). *A process X is called a* modification of *Y if X has the same state-space, parameter set and carrier triple, and also*

$$\mathbf{P}(X_t=Y_t)=1 \quad \textit{for every } t\in T.$$

Clearly, two processes that are modifications of each other have the same law.

(36.3) *Note.* It is not necessarily the case (even if T is a singleton set) that if Y has law μ and $G\subseteq E^T$ satisfies $\mu^*(G)=1$ then Y has a modification with all paths in G. See Exercise E38.36.

The most-stringent form of 'near-equality' of processes will now be introduced.

(36.4) DEFINITION (indistinguishable processes). *Let X and Y be two processes with the same state-space, parameter set T and carrier triple. We say that X and*

Y are indistinguishable *(*or *are* equal modulo indistinguishability*) if*

$$\mathbf{P}(X_t = Y_t \quad \textit{for all} \quad t \in T) = 1.$$

(36.5) PROPOSITION. *The following statements hold.*

(i) Two indistinguishable processes are modifications of each other.
(ii) If the parameter set is countable then two processes that are modifications of each other are indistinguishable.
(iii) If X and Y are right-continuous processes with values in some Hausdorff space $(E, \mathscr{B}(E))$ then X and Y are modifications of each other if and only if they are indistinguishable.

37. Direct construction of Poisson measures and subordinators, and of local time from the zero set; heuristics; Azéma's martingale. Because Poisson measures are the foundation for excursion theory, this topic is very important for us.

As in Section 33, let $(W, \mathscr{W}, \lambda)$ be a σ-finite measure space in which all singleton sets belong to $\mathscr{W}$. We want to construct a pre-Poisson process Λ on $\mathscr{W}$ with intensity measure λ such that $B \mapsto \Lambda(B)$ is a measure. We follow Kingman [2]; see also Kingman [4].

First suppose that $\lambda(W) < \infty$. Use Theorem 26.1 to construct a sequence

$$(N, Z_1, Z_2, \ldots)$$

of independent variables on some triple $(\Omega, \mathscr{F}, \mathbf{P})$ where

(i) N has the Poisson distribution with parameter $\lambda(W)$;
(ii) each Z_k takes values in $(W, \mathscr{W})$ and has law $\lambda/\lambda(W)$.

Thus

$$\mathbf{P}(N = m) = e^{-\lambda(W)} \lambda(W)^m / m! \quad (m = 0, 1, 2, \ldots),$$
$$\mathbf{P}(Z_k \in B) = \lambda(B)/\lambda(W) \quad (B \in \mathscr{W}, k \in \mathbf{N}).$$

We now define Λ to be the *measure* on $(W, \mathscr{W})$ with

$$\Lambda(B, \omega) := \sum_{k=1}^{N(\omega)} I_B(Z_k(\omega)) \quad (B \in \mathscr{W}).$$

Then, for $B \in \mathscr{W}$ and $r, m \in \mathbf{Z}^+$,

$$\begin{aligned} \mathbf{P}[\Lambda(B) = r; \Lambda(W \backslash B) = m] &= \mathbf{P}[N = r + m] \mathbf{P}[\Lambda(B) = r | N = r + m] \\ &= \frac{e^{-\lambda(W)} \lambda(W)^{r+m}}{(r+m)!} \frac{(r+m)!}{r! m!} \left[\frac{\lambda(B)}{\lambda(W)} \right]^r \left[1 - \frac{\lambda(B)}{\lambda(W)} \right]^m \\ &= \frac{e^{-\lambda(B)} \lambda(B)^r}{r!} \frac{e^{-\lambda(W \backslash B)} \lambda(W \backslash B)^m}{m!}, \end{aligned}$$

so that $\Lambda(B)$ and $\Lambda(W\backslash B)$ are independent and have Poisson distributions with parameters $\lambda(B)$ and $\lambda(W\backslash B)$ respectively. It is easy to give a full proof that Λ has the required 'pre-Poisson'—and, indeed, now proper *Poisson-measure*—structure.

Now consider the case when λ is assumed only σ-finite. Write $\lambda = \sum_{n\in\mathbb{N}} \lambda_n$, where each λ_n is a finite measure on $(W, \mathscr{W})$. Use Theorem 26.1 and the construction just described for the case 'λ finite' to construct independent Poisson measures Λ_n on $(W, \mathscr{W})$, Λ_n having intensity measure λ_n. Then it is easily verified that $\Lambda := \sum \Lambda_n$ is a Poisson measure with intensity measure λ.

Of course, it now follows (but it hardly matters) that $\mu^*(\mathscr{M}) = 1$ in the context of (34.3).

Subordinators. Recall from Section I.28 that a *Lévy process* is a right-continuous process with stationary independent increments and that a *subordinator* is a Lévy process with non-decreasing paths. [*Note.* There are never enough symbols to go round in mathematics. When we combine different ideas, we often find conflict of commonly used notations. In the following discussion, we adjust the notation of Section I.28 so that it does not conflict with that which we have recently been using.]

Let X be a subordinator with $X(0) = 0$. The distribution of $X(1)$ is infinitely divisible: for each n, it is the sum of n independent random variables each with the distribution of $X(1/n)$. For $\theta > 0$, we have

$$\mathbf{E}\exp[-\theta X(t)] = \exp[-t\Psi(\theta)], \tag{37.1}$$

where we can regard the *Laplace exponent* Ψ as defined by (37.1) with $t = 1$. From Theorem I.28.3, we have, for $\theta > 0$,

$$\Psi(\theta) = c\theta + \int_{(0,\infty)} (1 - e^{-\theta x})\nu(dx), \tag{37.2}$$

for some $c \geqslant 0$ and some measure ν on $(0, \infty)$, the *Lévy measure* of X, with

$$\int \min(x, 1)\nu(dx) < \infty, \tag{37.3}$$

the condition guaranteeing that, for some (then all) $\theta > 0$, the integral appearing on the right-hand side of (37.2) is finite.

(37.4) THEOREM (Lévy, Itô). *Let ν satisfy the condition (37.3). Let Λ be a Poisson measure on* $(0, \infty) \times (0, \infty)$ *with intensity measure* Leb $\times \nu$. *Let* $c \geqslant 0$. *Define*

$$X(t) := ct + \int_{s\in(0,t]} \int_{x\in(0,\infty)} x\Lambda(ds \times dx). \tag{37.5}$$

Then X is a subordinator with Laplace exponent Ψ as at (37.2).

Heuristic proof. It is (truly!) obvious from the independence properties of the Poisson measure that X is a subordinator. Formally, the number $J(t, dx)$ of jumps of size between x and $x + dx$ made by X during time-interval $[0, t]$ is Poisson with parameter $\beta := t\nu(dx)$. Now $X(t) = ct + \int xJ(t, dx)$, the 'sum' of independent bits. Moreover,

$$\mathbf{E}\exp[-\theta xJ(t,dx)] = \frac{\sum_{n=0}^{\infty} e^{-\theta xn} e^{-\beta}\beta^n}{n!} = \exp[-\beta(1-e^{-\theta x})]$$

$$= \exp[-t(1-e^{-\theta x})\nu(dx)],$$

and the result follows.

Exercise. Make this heuristic proof rigorous. First consider the compound Poisson process (see Section I.28) obtained by removing all jumps of size less than ε from X.

The process $H^+ = \{H_a^+ : a \geqslant 0\}$ for Brownian motion. Let B be a *path-continuous* Brownian motion on $\mathbb{R}$, starting at 0. For $a \geqslant 0$, define

(37.6) $$H^+(a) := \inf\{t > 0 : B_t > a\}.$$

Then H^+ is a right-continuous non-decreasing process; and it is clear from the strong Markov theorem and the spatial homogeneity of Brownian motion that H^+ is a subordinator. From (I.9.1), with the c there equal to 0, we have

(37.7) $$\mathbf{E}\exp(-\theta H_a^+) = \exp[-a\Psi(\theta)],$$

where

(37.8) $$\Psi(\theta) = (2\theta)^{1/2} = \int_0^\infty (1 - e^{-\theta x})(2\pi x^3)^{-1/2}\,dx,$$

so that our c equals 0 and $\nu(dx) = (2\pi x^3)^{-1/2}$.

Define the continuous non-decreasing process

$$S_t := \sup_{s \leqslant t} B_s$$

as usual. Recall from (I.14) Lévy result that $Y := S - B$ defines a reflecting Brownian motion Y. The jumps of H^+ correspond to intervals of constancy of S and to the intervals between visits to 0 by Y. Let $\mathscr{Z} := \{t : Y_t = 0\}$, the zero-set for the reflecting Brownian motion Y. For $t \geqslant 0$ and $\varepsilon > 0$, let $N(t, \varepsilon)$ denote the number of component intervals of $[0, t]\backslash\mathscr{Z}$ with length greater than ε. We know that $N(H_a^+, \varepsilon)$ has a Poisson distribution of mean $a\nu(\varepsilon, \infty) = a(\frac{1}{2}\pi\varepsilon)^{-1/2}$. It is easy to prove by using martingale techniques and exploiting monotonicity (see Exercise (79.71c)) that, almost surely,

(37.9) $$(\tfrac{1}{2}\pi\varepsilon)^{1/2} N(H_a^+, \varepsilon) \to a = S(H_a^+) \quad (\varepsilon \to 0)$$

uniformly on compact a-intervals, and it follows that, almost surely,

$$(\tfrac{1}{2}\pi\varepsilon)^{1/2}N(t,\varepsilon)\to S(t) \quad (\varepsilon\to 0)$$

uniformly on compact a-intervals. *We have therefore constructed the local time $\ell = S$ for the reflecting Brownian motion Y directly from the set $\mathscr{Z}$ of times at which $Y = 0$.*

A striking and difficult result due to Lévy, Wendel, Taylor and Hawkes (see Hawkes [2]) tells us that $\ell(t)$ is the Hausdorff h-measure of $\mathscr{Z}\cap[0,t]$ associated with the function

$$h(\delta):=[2\delta\log\log(1/\delta)]^{1/2}.$$

Heuristics. Chapter II is, as you will agree, following a Definition–Lemma–Theorem approach. For the remainder of this section, however, we cast off the shackles of rigour—for interest's sake. We shall return later to many of the points considered here.

The intervals comprising the set $[0,\infty)\backslash\mathscr{Z}$ are the excursion intervals of Y (away from 0). The lengths of these intervals are determined by the measure ν. We might therefore conclude as a heuristic principle that, given that an excursion interval is of length at least α, the probability that it, is of length at least γ is

(37.10)
$$\frac{\nu(\gamma,\infty)}{\nu(\alpha,\infty)}=\left(\frac{\alpha}{\gamma}\right)^{1/2}.$$

We are now going to work with our BM_0 process B rather than Y. The zero-set for B is the same as that for the reflecting Brownian motion $|B|$, and so has the same structure as $\mathscr{Z}$.

Let $t>0$, and define

(37.11)
$$\alpha_t:=t-\sup\{s\leqslant t: B_s=0\},$$
$$\beta_t:=\inf\{u\geqslant t: B_u=0\}-t.$$

You might guess on the basis of (37.10) that

(37.12)
$$\mathbf{P}(\beta_t>\beta\,|\,\alpha_t=\alpha)=\left(\frac{\alpha}{\alpha+\beta}\right)^{1/2},$$

and Exercise (37b) in Section 38 shows that you would be right.

Azéma's martingale. Define Azéma's process J and its natural filtration $\{\mathscr{J}_t\}$ by

(37.13)
$$J_t:=\operatorname{sgn}(B_t)(2\alpha_t)^{1/2},\qquad \mathscr{J}_t:=\sigma\{J_s: s\leqslant t\}.$$

(37.14) THEOREM (Azéma). $\{J_t\}$ and $\{J_t^2-t\}$ *are martingales relative to* $\{\mathscr{J}_t\}$.

These processes are *not* martingales relative to the natural filtration of B.

Let us see how Azéma's Theorem ties in with (37.12).

(37.15) *Suppose that* $\alpha_t = \alpha$ *for some fixed* t *and* α *with* $0 < \alpha < t$. Let $u > \alpha$, and let T be the first time after time t that α_T is either u or 0. Then, *either*, and with probability $(\alpha/u)^{1/2}$

$$\alpha_T = u \quad \text{and} \quad T - t = u - \alpha,$$

or, and with probability $(\alpha/4v^3)^{1/2}\,dv$, for some v with $\alpha < v < u$,

$$\alpha_T = 0 \quad \text{and} \quad T - t \in (v, v + dv) - \alpha.$$

Elementary calculations now show that, conditionally on (37.15),

$$\mathbf{E}J_T = J_t, \quad \mathbf{E}(J_T^2 - T) = J_t^2 - t,$$

and these results start to make Azéma's Theorem plausible.

A much better explanation of Azéma's result is provided by the following facts: for $x \in \mathbb{R}$ and $s \leqslant t$,

$$(37.16) \qquad \mathbf{P}(B_t \in dx; \alpha_t \in d\alpha) = \left[\frac{1}{2\pi(t-\alpha)}\right]^{1/2} \frac{|x|}{(2\pi\alpha^3)^{1/2}} \exp\left(-\frac{x^2}{2\alpha}\right) dx\, d\alpha,$$
$$= \frac{1}{\pi}\left[\frac{1}{(t-\alpha)\alpha}\right]^{1/2} \frac{|x|}{2\alpha} \exp\left(-\frac{x^2}{2\alpha}\right) dx\, d\alpha,$$

whence

$$(37.17) \qquad \mathbf{P}(B_t \in dx | \alpha_t \in d\alpha,\ \operatorname{sgn}(B_t) > 0) = \frac{x}{\alpha} \exp\left(-\frac{x^2}{2\alpha}\right) dx \quad (x > 0).$$

See Exercise (37a) in the next section.

If ξ is a positive random variable with the probability density function on the right-hand side of (37.17) then $\mathbf{E}(\xi) = (\tfrac{1}{2}\pi\alpha)^{1/2}$ and $\mathbf{E}(\xi^2) = 2\alpha$. This strongly suggests that

$$(37.18) \qquad \mathbf{E}(B_t | \mathcal{J}_t) = \tfrac{1}{2}\pi^{1/2} J_t, \quad \mathbf{E}(B_t^2 - t | \mathcal{J}_t) = J_t^2 - t;$$

and hence $\{J_t\}$ and $\{J_t^2 - t\}$ inherit their martingale properties relative to $\{\mathcal{J}_t\}$ from those of $\{B_t\}$ and $\{B_t^2 - t\}$ relative to the natural filtration of B. See Exercise E79.71b.

If ℓ now denotes local time at 0 for B, and $\tau_t := \inf\{u : \ell(u) > t\}$, then the processes R, where R_t is the sum of the moduli of the jumps of J by time τ_t, is clearly a subordinator. However, the number of jumps of modulus greater than ε made by J by time τ_t equals the number of jumps of modulus greater than $\tfrac{1}{2}\varepsilon^2$ made by α by time τ_t; and this, like $N(H_t^+, \varepsilon^2)$, has a Poisson distribution of mean $t(\tfrac{1}{4}\pi)^{-1/2}\varepsilon^{-1}$. Thus the Lévy measure ν_R of R satisfies $\nu_R(\varepsilon, \infty) = t(\tfrac{1}{4}\pi)^{-1/2}\varepsilon^{-1}$; but this fails the integrability condition (37.3). You see the problem:

(37.19) *Azéma's martingale J is not of finite variation.*

Azéma's martingale has been the source of much interest recently, particularly to workers in quantum probability. See Azéma [1], Azéma and Yor [3], Emery [2], Meyer [12] and Revuz and Yor [1].

38. Exercises. (*E38.25*) Extend exercise E13.3b as follows. Show that if $(E, \mathscr{E})$ is a measurable space and T is a parameter set, then a function $\xi: E^T \to \mathbf{R}$ is $\mathscr{E}^T$-measurable if and only if $\xi = f \circ \pi_S$ for some countable subset S of T and some $\mathscr{E}^S$-measurable function $f: E^S \to \mathbf{R}$.

(*E38.29*) *Time-reversal for Brownian motion.* Fix $t \geqslant 0$. We consider Brownian motion with time-parameter set $[0, t]$. Let $\Omega := C([0, t]; \mathbf{R})$, and, for $\omega \in \Omega$ and $s \in [0, t]$, write $B_s(\omega) := \omega(s)$, and define $\mathscr{A} := \sigma(B_s : s \leqslant t)$. Let $\mathbf{P}^x$ be the law of Brownian motion starting at x, so that $\mathbf{P}^x$ is the unique measure on $(\Omega, \mathscr{A})$ such that for $n \in \mathbf{N}$, for $0 = s_0 < s_1 < \cdots < s_n$ and $x_0, x_1, \ldots, x_n \in \mathbf{R}$ with $x_0 = 0$, we have

$$\mathbf{P}\left(\bigcap_{i=1}^{n} \{B(s_i) \in dx_i\}\right) = \prod_{i=1}^{n} p(s_i - s_{i-1}, x_{i-1}, x_i)\, dx_i,$$

this making rigorous sense when integrated over $x_1, x_2, \ldots, x_n$ in a Borel subset of $\mathbf{R}^n$.

Let $\hat{}$ be the time-reversal map on Ω, so that

$$\hat{\omega}(s) := \omega(t - s) \quad (0 \leqslant s \leqslant t).$$

For $\xi \in \mathrm{m}\mathscr{A}$, define $\hat{\xi}(\omega) := \xi(\hat{\omega})$. Prove that, for $\xi \in (\mathrm{m}\mathscr{A})^+$,

$$\int_{x \in \mathbf{R}} \mathbf{E}^x(\xi)\, dx = \int_{y \in \mathbf{R}} \mathbf{E}^y(\hat{\xi})\, dy \quad (\leqslant \infty).$$

There are some measurability questions involved, which we shall study in detail later—do not fret over these.

Hint. First take $\xi = I_A(B_0) I_H(B_s) I_C(B_t)$, where $0 \leqslant s \leqslant t$, and A, H and C are Borel subsets of $\mathbf{R}$. Because $\int \mathbf{P}^x\, dx$ is not a finite measure, you will have to do some truncation. But the idea is the same as that used to show that the finite-dimensional distributions determine the law of a process.

(*E38.30*) *Lebesgue measure from coin tossing.* Show that one can reverse the argument in Section 23 as follows. Use the DK theorem to construct a sequence of independent variables $(\xi_k : k \in \mathbf{N})$ each taking the values 0 and 1 with probability $\frac{1}{2}$ each. Define $X := \sum 2^{-n} \xi_n$. Then Lebesgue measure on $([0, 1], \mathscr{B}[0, 1])$ is the law of X.

(*E38.36*) The object of this exercise is to confirm the point made in Note 36.3.

Let $\Omega := [0, 1]$, $\mathscr{H} := \mathscr{B}[0, 1]$, $\mu :=$ Leb on $(\Omega, \mathscr{H})$, and let μ^* be the outer measure associated with $(\mu, \mathscr{H})$. Let G be a subset of Ω with $\mu^*(G) = 1$ and $\mu^*(G^c) = 1$, where, of course, $G^c := \Omega \setminus G$. Define $\mathscr{F} := \sigma(\mathscr{H}, G)$ and

$$\mathbf{P}(F) := \mu^*(G^c \cap F) \quad (F \in \mathscr{F}).$$

Prove that $(\Omega, \mathscr{F}, \mathbf{P})$ is a probability triple. See Lemma 6.1. Let $Y(\omega) = \omega$ for

$\omega \in \Omega$. Prove that Y has law μ, but that, even though $\mu^*(G) = 1$, there is no modification of Y taking all its values in G.

(*E38.37a*) Use Exercise E38.29 and the result (I.9.2) (with c there equal to 0) to prove (37.15). Now deduce (37.16).

Hint. Consider $\xi := f(B_0)g(\alpha_t)h(B_t)\eta$, where

$$\eta := \begin{cases} 1 & \text{if } B_s = 0 \text{ for some } s \in [0, t], \\ 0 & \text{otherwise.} \end{cases}$$

(*E38.37b*) Prove (37.11).

(*E38.37c*) Modify the last two exercises to cope with the case when the Brownian motion B has drift c.

4. DISCRETE-PARAMETER MARTINGALE THEORY

Again, we follow [W] very closely. There, you will find the same notation, all proofs not given here, and many illustrative examples. Neveu [5] gives a fine broader picture of the scope of discrete-parameter martingale theory. Of course, Doob [1] is the classic account. In that account, Doob emphasises the debt we owe to Sparre Andersen and Jessen for their work on uniformly integrable martingales.

After revising the theory of conditional expectation (due, of course, to Kolmogorov), we concentrate on the Upcrossing Lemma, the Submartingale Inequality, results on uniform integrability, the Optional-Stopping Theorem, the Optional-Sampling Theorem (all, of course, due to Doob), and the 'Downward' Convergence Theorem for supermartingales due to Lévy and Doob. These results are central to the extension to the continuous-parameter theory, which occupies Part 5 of this chapter and dominates the remainder of both volumes.

CONVENTION: Until further notice, all random variables are $(\mathbb{R}, \mathcal{B})$-valued.

Conditional expectation

39. Fundamental theorem and definition. The following theorem and definition constitute the greatest of Kolmogorov's many contributions to the subject.

(*39.1*) THEOREM and DEFINITION (a version of the conditional expectation $\mathbf{E}(X \mid \mathcal{G})$). *Let* $(\Omega, \mathcal{F}, \mathbf{P})$ *be a triple, and* X *a random variable with* $\mathbf{E}(|X|) < \infty$.

Let $\mathcal{G}$ be a sub-σ-algebra of $\mathcal{F}$. Then there exists a random variable Y such that

(i) *Y is $\mathcal{G}$ measurable;*
(ii) $\mathbf{E}(|Y|) < \infty$;
(iii) *for every set $\mathcal{G}$ in $\mathcal{G}$ (equivalently, for every set G in some π-system that contains Ω and generates $\mathcal{G}$), we have* $\mathbf{E}(Y;G) = \mathbf{E}(X;G)$.

Moreover, if $\tilde{Y}$ is another random variable with these properties then $\tilde{Y} = Y$, a.s., that is, $\mathbf{P}[\tilde{Y} = Y] = 1$. A random variable Y with properties (i)–(iii) is called a version of the conditional expectation $\mathbf{E}(X|\mathcal{G})$ of X given $\mathcal{G}$, *and we write* $Y = \mathbf{E}(X|\mathcal{G})$, a.s.

The Radon–Nikodým proof will be given shortly.

(39.2) *The intuitive meaning.* An experiment has been performed. The only information available to you regarding which sample point ω has been chosen is the set of values $Z(\omega)$ for every $\mathcal{G}$-measurable random variable Z, or, equivalently, the values $I_G(\omega)$ for every $G \in \mathcal{G}$. Then $Y(\omega) = \mathbf{E}(X|\mathcal{G})(\omega)$ is regarded as (almost surely equal to) the 'expected value of $X(\omega)$ given this information'.

Note that if $\mathcal{G}$ is the trivial σ-algebra $\{\emptyset, \Omega\}$ (which contains no information) then $E(X|\mathcal{G})(\omega) = \mathbf{E}(X)$ for all ω.

Proof of Theorem 39.1. Existence. Suppose that $X \in \mathcal{L}^1(\Omega, \mathcal{F}, \mathbf{P})$. Consider first the case when $X \geqslant 0$. Then, as we saw in Section 9, the map $G \mapsto \mathbf{E}(X;G)$ is a *finite measure* on $(\Omega, \mathcal{G})$ that is absolutely continuous with respect to $\mathbf{P}$. Hence, by the Radon–Nikodým Theorem 9.3, there exists a Y in $\mathcal{L}^1(\Omega, \mathcal{G}, \mathbf{P})$ such that

$$\mathbf{E}(Y;G) = \mathbf{E}(X;G) \tag{39.3}$$

for all $G \in \mathcal{G}$. The existence of Y is therefore established; and the general case when $X \in \mathcal{L}^1(\Omega, \mathcal{F}, \mathbf{P})$ follows by linearity.

Uniqueness. If Y and $\tilde{Y}$ are in $\mathcal{L}^1(\Omega, \mathcal{G}, \mathbf{P})$ and $\mathbf{E}(Y - \tilde{Y}; G) \leqslant 0$ for every G in $\mathcal{G}$ then $Y \leqslant \tilde{Y}$, a.s. For consider $\mathbf{E}(Y - \tilde{Y}; G_n)$, where $G_n := \{\omega : (Y - \tilde{Y})(\omega) > n^{-1}\}$, etc.

The π-system formulation. Suppose that $X \in \mathcal{L}^1(\Omega, \mathcal{F}, \mathbf{P})$, $Y \in \mathcal{L}^1(\Omega, \mathcal{G}, \mathbf{P})$, and that $E(Y;G) = \mathbf{E}(X;G)$ for every G in some π-system containing Ω and generating $\mathcal{G}$. By the Dominated-Convergence Theorem 8.5, the class of sets G in $\mathcal{G}$ for which (39.3) holds is a d-system in the sense at (1.6). By Dynkin's Lemma 1.8, this class must coincide with $\mathcal{G}$. □

40. Notation; agreement with elementary usage. We often write $\mathbf{E}(X|Z)$ for $\mathbf{E}(X|\sigma(Z))$, $\mathbf{E}(X|Z_1, Z_2, \ldots)$ for $\mathbf{E}(X|\sigma(Z_1, Z_2, \ldots))$, etc.

The case of two RVs will suffice to illustrate the connection between the abstract definition 39.1 and elementary conditional expectation. So suppose that X and Z are RVs that have a joint probability density function (pdf) $f_{X,Z}(x,z)$. Then $f_Z(z)=\int_{\mathbf{R}} f_{X,Z}(x,z)\,dx$ acts as a probability density function for Z. Define the *elementary conditional pdf* $f_{X|Z}$ of X given Z via

$$f_{X|Z}(x|z):=\begin{cases}\dfrac{f_{X,Z}(x,z)}{f_Z(z)} & \text{if } f_Z(z)\neq 0,\\ 0 & \text{otherwise.}\end{cases}$$

Let h be a Borel function on $\mathbf{R}$ such that

$$\mathbf{E}|h(X)|=\int_{\mathbf{R}}|h(x)|f_X(x)\,dx<\infty,$$

where of course $f_X(x)=\int_{\mathbf{R}} f_{X,Z}(x,z)\,dz$ gives a pdf for X. Set

$$g(z):=\int_{\mathbf{R}} h(x)f_{X|Z}(x|z)\,dx.$$

Then $Y:=g(Z)$ *is a version of the conditional expectation of* $h(X)$ *given* $\sigma(Z)$.

Proof. The typical element of $\sigma(Z)$ has the form $\{\omega:Z(\omega)\in B\}$, where $B\in\mathcal{B}$. Hence we must show that $\mathbf{E}[h(X)I_B(Z)]=\mathbf{E}[g(Z)I_B(Z)]$. But this follows from Fubini's Theorem. □

41. Properties of conditional expectation: a list. This is the same list of properties as in Section 9.7 (and on the back cover!) of [W]. All Xs satisfy $\mathbf{E}(|X|)<\infty$ in this list of properties. Of course, $\mathcal{G}$ and $\mathcal{H}$ denote sub-σ-algebras of $\mathcal{F}$. (The use of 'c' to denote 'conditional' in (*cMon*) etc. is obvious.)

(41)(a) If Y is any version of $\mathbf{E}(X|\mathcal{G})$ then $\mathbf{E}(Y)=\mathbf{E}(X)$.
(41)(b) If X is $\mathcal{G}$ measurable then $\mathbf{E}(X|\mathcal{G})=X$, a.s.
(41)(c) (*Linearity*) $\mathbf{E}(a_1X_1+a_2X_2|\mathcal{G})=a_1\mathbf{E}(X_1|\mathcal{G})+a_2\mathbf{E}(X_2|\mathcal{G})$, a.s.

Clarification. If Y_1 is a version of $\mathbf{E}(X_1|\mathcal{G})$ and Y_2 is a version of $\mathbf{E}(X_2|\mathcal{G})$, then $a_1Y_1+a_2Y_2$ is a version of $\mathbf{E}(a_1X_1+a_2X_2|\mathcal{G})$.

(41)(d) (*Positivity*) If $X\geqslant 0$ then $\mathbf{E}(X|\mathcal{G})\geqslant 0$, a.s.
(41)(e) (*cMon*) If $0\leqslant X_n\uparrow X$ then $\mathbf{E}(X_n|\mathcal{G})\uparrow\mathbf{E}(X|\mathcal{G})$, a.s.
(41)(f) (*cFatou*) If $X_n\geqslant 0$ then $\mathbf{E}[\liminf X_n|\mathcal{G}]\leqslant\liminf\mathbf{E}[X_n|\mathcal{G}]$, a.s.
(41)(g) (*cDom*) If $|X_n(\omega)|\leqslant V(\omega)$, $\forall n$, $\mathbf{E}V<\infty$, and $X_n\to X$, a.s., then

$$\mathbf{E}(X_n|\mathcal{G})\to\mathbf{E}(X|\mathcal{G}),\quad \text{a.s.}$$

(41)(h) (*cJensen*) If $c:\mathbf{R}\to\mathbf{R}$ is convex and $\mathbf{E}|c(X)|<\infty$ then

$$\mathbf{E}[c(X)|\mathcal{G}]\geqslant c(\mathbf{E}[X|\mathcal{G}]),\quad \text{a.s.}$$

Important corollary. $\|\mathbf{E}(X|\mathcal{G})\|_p\leqslant\|X\|_p$ for $p\geqslant 1$.

(41)(i) (*Tower Property*) If $\mathscr{H} \subseteq \mathscr{G} \subseteq \mathscr{F}$ then

$$\mathbf{E}[\mathbf{E}(X|\mathscr{G})|\mathscr{H}] = \mathbf{E}[X|\mathscr{H}], \quad \text{a.s.}$$

Note. We shorten the left-hand side to $\mathbf{E}[X|\mathscr{G}|\mathscr{H}]$ for tidiness.

(41)(j) (*'Taking out what is known'*) If Z is $\mathscr{G}$-measurable and bounded then

$$(*) \qquad \mathbf{E}[ZX|\mathscr{G}] = Z\mathbf{E}[X|\mathscr{G}], \quad \text{a.s.}$$

If $p > 1$, $p^{-1} + q^{-1} = 1$, $X \in \mathscr{L}^p(\Omega, \mathscr{F}, \mathbf{P})$ and $Z \in \mathscr{L}^q(\Omega, \mathscr{G}, \mathbf{P})$ then $(*)$ again holds. If $X \in (m\mathscr{F})^+$, $Z \in (m\mathscr{G})^+$, $\mathbf{E}(X) < \infty$ and $\mathbf{E}(ZX) < \infty$ then $(*)$ holds.
(41)(k) (*Role of independence*) if $\mathscr{H}$ is independent of $\sigma(\sigma(X), \mathscr{G})$ then

$$\mathbf{E}[X|\sigma(\mathscr{G}, \mathscr{H})] = \mathbf{E}(X|\mathscr{G}), \quad \text{a.s.}$$

In particular, if X is independent of $\mathscr{H}$ then $\mathbf{E}(X|\mathscr{H}) = \mathbf{E}(X)$, a.s.

For proofs of all the above properties, see Section 9.8 of [W]. Do Exercise E60.41 now.

42. The role of versions; regular conditional probabilities and pdfs. If we consider conditional expectation as a map from $L^1(\Omega, \mathscr{F}, \mathbf{P})$ to $L^1(\Omega, \mathscr{G}, \mathbf{P})$, these spaces being the proper Banach spaces of equivalence classes of functions, then this map is truly uniquely defined with no untidy 'almost sure' qualifications and no need for 'versions'. So why not do this in this 'elegant' way?

The answer is that the ability to choose 'good' versions—of the *functions*, not the equivalence classes—is absolutely crucial to the whole theory. We shall repeatedly see cases where we get the good results by modifying random variables on null sets. Thus, for example, we want modifications of martingales that have right-continuous paths; and concepts of path regularity are meaningless if we work with equivalence classes.

In the remainder of this section, and in the next, we consider a rather important case that in some, though not all, respects parallels our earlier discussion of Poisson measures.

(*42.1*) DEFINITION (a version of conditional probability, $\mathbf{P}(F|\mathscr{G})$). *Let* $F \in \mathscr{F}$ *and let* $\mathscr{G} \subseteq \mathscr{F}$. *We call any version of* $\mathbf{E}(I_F|\mathscr{G})$ *a* version of the conditional probability of F given $\mathscr{G}$, *and write* $\mathbf{P}(F|\mathscr{G}) = \mathbf{E}(I_F|\mathscr{G})$, *a.s.*

By (41) (c) and the 'cMon' result (41)(e), we can show that *for a fixed sequence* (F_n) of disjoint elements of $\mathscr{F}$, we have

$$(42.2) \qquad \mathbf{P}(\bigcup F_n|\mathscr{G}) = \sum \mathbf{P}(F_n|\mathscr{G}), \quad \text{a.s.}$$

Except in trivial cases, there are *uncountably* many sequences of disjoint sets; and it is therefore not at all clear that we can choose a good modification $\{(\mathbf{P}|\mathscr{G})(F): F \in \mathscr{F}\}$ of the process $\{\mathbf{P}(F|\mathscr{G}): F \in \mathscr{F}\}$. Let us formulate what we mean by a good modification.

(*42.3*) **DEFINITION** (regular conditional probability given $\mathscr{G}$). *Let* $(\Omega, \mathscr{F}, \mathbf{P})$ *be a triple and let* $\mathscr{G}$ be a sub-σ-algebra of $\mathscr{F}$. *By a* regular conditional probability $(\mathbf{P}|\mathscr{G})(\cdot,\cdot)$ given $\mathscr{G}$, *we mean a map*

$$(42.4)(a) \qquad (\mathbf{P}|\mathscr{G}): \mathscr{F} \times \Omega \to [0,1]$$

such that

(42.4)(b) *for* $F \in \mathscr{F}$, *the function* $\omega \mapsto (\mathbf{P}|\mathscr{G})(F,\omega)$ *is a version of* $\mathbf{P}(F|\mathscr{G})$; *for almost every* ω, *the map*

$$(42.4)(c) \qquad F \mapsto (\mathbf{P}|\mathscr{G})(F,\omega)$$

is a probability measure on $\mathscr{F}$.

It is known, and is proved in Section 89, that *regular conditional probabilities exist under most conditions encountered in practice*, but, as we shall see in the next section, they do not always exist.

Note. The elementary conditional pdf $f_{X|Z}(x|z)$ of Section 40 *is* a regular conditional pdf for X given Z in that for every A in $\mathscr{B}$,

$$\omega \mapsto \int_A f_{X|Z}(x|Z(\omega))\,dx \quad \text{is a version of} \quad \mathbf{P}(X \in A|Z).$$

Proof. Take $h = I_A$ in Section 40.

43. A counterexample. This counterexample, for which Halmos, Dieudonné, Andersen and Jessen share the credit, exhibits a situation in which no regular conditional probability given $\mathscr{G}$ exists. It helps emphasise why we need some extra 'topological' hypothesis such as that used for the positive result in Section 89.

Take

$$(\Omega, \mathscr{G}) := ([0,1], \mathscr{B}[0,1]).$$

Let μ denote Lebesgue measure on $(\Omega, \mathscr{G})$. Let Z be a subset of Ω of inner μ-measure 0 and outer μ-measure 1. (We are assuming the Axiom of Choice!) Let $\mathscr{F}$ be the smallest σ-algebra on Ω extending $\mathscr{G}$ and containing Z, so that a typical element Λ of $\mathscr{F}$ may be written (with Z^c denoting $[0,1]\backslash Z$)

$$\Lambda = (Z \cap A) \cap (Z^c \cap B), \quad \text{where} \quad A, B \in \mathscr{G}.$$

The fact that Z and Z^c have outer measure 1 implies (see Lemma 6.1) that

$$\mu^*(Z \cap \Lambda) = \mu^*(Z \cap A) = \mu(A),$$
$$\mu^*(Z^c \cap \Lambda) = \mu^*(Z^c \cap B) = \mu(B).$$

Hence we can define a probability measure $\mathbf{P}$ on $(\Omega, \mathscr{F})$ by

$$P(\Lambda) := \tfrac{1}{2}\mu^*(Z \cap \Lambda) + \tfrac{1}{2}\mu^*(Z^c \cap \Lambda) = \tfrac{1}{2}\mu(A) + \tfrac{1}{2}\mu(B).$$

Assume that $(\mathbf{P}|\mathcal{G}): \mathcal{F} \times \Omega \to [0,1]$ is a regular conditional probability given $\mathcal{G}$. We shall show that this assumption leads to a contradiction.

Let $\Gamma \in \mathcal{G}$. Then, for $G \in \mathcal{G}$,

$$\mathbf{E}((\mathbf{P}|\mathcal{G})(Z \cap \Gamma); G) = \mathbf{P}(Z \cap \Gamma \cap G) = \tfrac{1}{2}\mu^*(Z \cap \Gamma \cap G) = \tfrac{1}{2}\mu(\Gamma \cap G) = \mathbf{E}(\tfrac{1}{2}I_\Gamma; G),$$

so that

$$(\mathbf{P}|\mathcal{G})(Z \cap \Gamma, \omega) = \tfrac{1}{2}I_\Gamma(\omega), \quad \text{a.s.}$$

Since $\mathcal{G}$ is generated by a countable π-system $\mathcal{I}$, and since

$$\Gamma \mapsto (\mathbf{P}|\mathcal{G})(Z \cap \Gamma, \omega), \qquad \Gamma \mapsto \tfrac{1}{2}I_\Gamma(\omega)$$

are *measures* for every ω (the first because of our assumption), the set

$$\begin{aligned} J := &\{\omega : (\mathbf{P}|\mathcal{G})(Z \cap \Gamma)(\omega) = \tfrac{1}{2}I_\Gamma(\omega), \forall \Gamma \in \mathcal{G}\} \\ = &\{\omega : (\mathbf{P}|\mathcal{G})(Z \cap \Gamma)(\omega) = \tfrac{1}{2}I_\Gamma(\omega), \forall \Gamma \in \mathcal{I}\} \end{aligned}$$

is in $\mathcal{G}$, and $\mathbf{P}(J) = \mu(J) = 1$. (The argument now takes on a 'Russell's paradox' appearance. The set J is itself an element of $\mathcal{G}$....) If $\omega \in J$ then

$$(\mathbf{P}|\mathcal{G})(Z \cap J, \omega) = \tfrac{1}{2}I_J(\omega) \neq \tfrac{1}{2}I_{J\setminus\{\omega\}}(\omega) = (\mathbf{P}|\mathcal{G})(Z \cap [J\setminus\{\omega\}], \omega),$$

so that $Z \cap J \neq Z \cap [J \setminus \{\omega\}]$; in other words, $\omega \in Z$. Hence J, which is an element of $\mathcal{G}$ of measure 1, is a subset of Z, contradicting the fact that Z has inner measure 0.

44. A uniform-integrability property of conditional expectations. The reason that the martingale and UI properties tie in so well is the following.

(*44.1*) THEOREM (Doob). *Let* $X \in \mathcal{L}^1(\Omega, \mathcal{F}, \mathbf{P})$. *Then the family*

$$\{\mathbf{E}(X|\mathcal{G}) : \mathcal{G} \text{ is a sub-}\sigma\text{-algebra of } \mathcal{F}\}$$

is uniformly integrable.

Clarification. Because of the business of versions, a formal description of the family in question would be the set of all random variables Y with the property that for some σ-algebra $\mathcal{G} \subseteq \mathcal{F}$, $Y = \mathbf{E}(X|\mathcal{G})$, a.s.

Proof. Let $\varepsilon > 0$ be given. Use Lemma 20.1 to choose $\delta > 0$ such that, for $F \in \mathcal{F}$,

$$\mathbf{P}(F) < \delta \quad \text{implies that} \quad \mathbf{E}(|X|; F) < \varepsilon.$$

Choose K so that $K^{-1}\mathbf{E}(|X|) < \delta$.

Now let $\mathcal{G} \subseteq \mathcal{F}$ and let Y be a version of $\mathbf{E}(X|\mathcal{G})$. By the 'cJensen' property in Section 41,

$$|Y| \leqslant \mathbf{E}(|X|\,|\mathcal{G}), \quad \text{a.s.} \tag{44.2}$$

Hence, as in fact we already know, $\mathbf{E}(|Y|) \leqslant \mathbf{E}(|X|)$, and

$$K\mathbf{P}(|Y| > K) \leqslant \mathbf{E}(|Y|) \leqslant \mathbf{E}(|X|),$$

so that $\mathbf{P}(|Y| > K) < \delta$. But $\{\omega : |Y(\omega)| > K\} \in \mathcal{G}$, and, from (44.2) and the definition of conditional expectation,

$$\mathbf{E}(|Y|; |Y| > K) \leqslant \mathbf{E}(|X|; |Y| > K) < \varepsilon,$$

and this is the desired uniform-integrability property. □

(Discrete-parameter) martingales and supermartingales

45. Filtration, filtered space; adapted process; natural filtration. Let $(\Omega, \mathcal{F}, \mathbf{P})$ be given.

(*45.1*) DEFINITIONS (filtration, filtered space). *By a* filtration *on* $(\Omega, \mathcal{F}, \mathbf{P})$, *we mean an increasing family* $\{\mathcal{F}_n : n \in \mathbf{Z}^+\}$ *of sub-σ-algebras of* $\mathcal{F}$:

$$\mathcal{F}_0 \subseteq \mathcal{F}_1 \subseteq \cdots \subseteq \mathcal{F}_\infty := \sigma\left(\bigcup_n \mathcal{F}_n\right) \subseteq \mathcal{F}.$$

The set-up $(\Omega, \mathcal{F}, \mathbf{P}, \{\mathcal{F}_n : n \in \mathbf{Z}^+\})$ *is then called a* filtered space.

(*45.2*) CONVENTION. *Until further notice, we assume given a filtered space*

$$(\Omega, \mathcal{F}, \mathbf{P}, \{\mathcal{F}_n : n \in \mathbf{Z}^+\}).$$

All of our martingales, supermartingales etc. will be defined relative to this set-up.

(*45.3*) DEFINITION (adapted process). *A process* $X = \{X_n : n \in \mathbf{Z}^+\}$ *carried by* $(\Omega, \mathcal{F}, \mathbf{P})$ *is said to be* adapted *(to our given filtration) if, for every* $n \in \mathbf{Z}^+$, X_n *is* $\mathcal{F}_n$*-measurable.*

(*45.4*) DEFINITION (natural filtration). *Let* $W = \{W_n : n \in \mathbf{Z}\}$ *be a stochastic process carried by our triple* $(\Omega, \mathcal{F}, \mathbf{P})$. *The* natural filtration $\{\mathcal{W}_n : n \in \mathbf{Z}^+\}$ *of* W *is defined to be the smallest filtration relative to which* W *is adapted, so that*

$$\mathcal{W}_n = \sigma(W_0, W_1, \ldots, W_n).$$

(*45.5*) *The intuitive meanings.* The information about the chosen ω that is available to us at time n consists of the values $Z_n(\omega)$ for every $\mathcal{F}_n$-measurable random variable Z_n. A process X is adapted if the value $X_n(\omega)$ is known to us at time n. Usually, $\{\mathcal{F}_n\}$ is the natural filtration $\{\mathcal{W}_n : n \in \mathbf{Z}^+\}$ of some process W, and then the information about ω which we have at time n consists of the values $W_0(\omega), W_1(\omega), \ldots, W_n(\omega)$. A process X is then adapted if and only if, for each n, $X_n = f_n(W_0, \ldots, W_n)$ for some $f \in \mathrm{m}\mathcal{B}^{n+1}$. See E38.25.

46. Martingale; supermartingale; submartingale. As already explained, these concepts are defined relative to our given filtered space in (45.2).

(*46.1*) THE KEY DEFINITIONS. *A process X is called a* martingale *if*

(i) X is adapted;
(ii) $\mathbf{E}(|X_n|)<\infty, \quad \forall n;$
(iii) $\mathbf{E}[X_n|\mathcal{F}_{n-1}]=X_{n-1}, \quad$ *a.s.* $\quad (n\geqslant 1).$

A supermartingale *is defined similarly, except that (iii) is replaced by*

$$\mathbf{E}[X_n|\mathcal{F}_{n-1}]\leqslant X_{n-1}, \quad \textit{a.s.} \quad (n\geqslant 1),$$

and a submartingale *is defined with (iii) replaced by*

$$\mathbf{E}[X_n|\mathcal{F}_{n-1}]\geqslant X_{n-1}, \quad \textit{a.s} \quad (n\geqslant 1).$$

A *super*martingale '*de*creases on average'; a *sub*martingale '*in*creases on average'. The 'p' points down, the 'b' up! Of course, Chapter I has explained how 'superharmonic' corresponds to 'local supermartingale', which was the reason for this choice of terminology.

Note that X is a supermartingale if and only if $-X$ is a submartingale, and that X is a martingale if and only if it is both a supermartingale and a submartingale. It is important to note that a process X for which $X_0\in\mathcal{L}^1(\Omega,\mathcal{F}_0,\mathbf{P})$ is a martingale (respectively, supermartingale, submartingale) if and only if the process $X-X_0=(X_n-X_0:n\in\mathbf{Z}^+)$ has the same property. So we can focus attention on processes that are null at 0.

If X is, for example, a supermartingale, then the Tower Property of conditional expectations, (41)(i), shows that, for $m<n$,

$$\mathbf{E}[X_n|\mathcal{F}_m]=\mathbf{E}[X_n|\mathcal{F}_{n-1}|\mathcal{F}_m]\leqslant\mathbf{E}[X_{n-1}|\mathcal{F}_m]\leqslant\cdots\leqslant X_m, \quad \text{a.s.}$$

(*46.2*) *Gambling interpretation.* Think of X_n-X_{n-1} as your winnings per unit stake on a gambling game. The game is unfavourable to you if X is a supermartingale, favourable to you if X is a submartingale, and fair if X is a martingale.

47. Previsible process; gambling strategy; a fundamental principle. We now study the discrete-parameter analogue of *stochastic integrals.*

(*47.1*) DEFINITION (previsible process). *We call a process* $(C_n:n\in\mathbf{N})$ previsible *if, for each* $n\in\mathbf{N}$, C_n *is* $\mathcal{F}_{n-1}$*-measurable.*

Note that C_0 is not defined.

Think of C_n as your stake on game n. You have to decide on the value of C_n based on the history up to (and including) time $n-1$. This is the intuitive significance of the 'previsible' character of C. Your winnings on game n are

$C_n(X_n - X_{n-1})$ and *your total winnings up to time n* are

(47.2)
$$Y_n = \sum_{1 \leqslant k \leqslant n} C_k(X_k - X_{k-1}) =: (C \bullet X)_n.$$

Note that $(C \bullet X)_0 = 0$, and that

$$Y_n - Y_{n-1} = C_n(X_n - X_{n-1}).$$

(*47.3*) DEFINITION (martingale transform, stochastic integral). *The process* $C \bullet X$ *is called the* martingale transform of X by C, *or the* (discrete) stochastic integral of C with respect to X.

(*47.4*) THEOREM. You can't beat the system!

(i) Let C be a bounded non-negative previsible process, so that, for some K in $[0, \infty)$, $|C_n(\omega)| \leqslant K$ *for every n and every* ω. *Let X be a supermartingale (respectively martingale). Then* $C \bullet X$ *is a supermartingale (martingale) null at 0.*

(ii) If C is a bounded previsible process and X is a martingale, then $(C \bullet X)$ *is a martingale null at 0.*

(iii) In (i) and (ii) the boundedness condition on C may be replaced by the condition $C_n \in \mathscr{L}^2, \forall n$, *provided we also insist that* $X_n \in \mathscr{L}^2, \forall n$.

Proof of (i). Write Y for $C \bullet X$. Since C_n is bounded non-negative and $\mathscr{F}_{n-1}$ measurable, we have, from (41)(j),

$$\mathbf{E}[Y_n - Y_{n-1} | \mathscr{F}_{n-1}] = C_n \mathbf{E}[X_n - X_{n-1} | \mathscr{F}_{n-1}] \leqslant 0 \quad (\text{resp. } = 0).$$

Proofs of (ii) and (iii) are now obvious. (Look again at (41)(j).) □

48. Doob's Upcrossing Lemma. Doob's use of upcrossings is one of the most sparkling things in the theory.

(*48.1*) DEFINITION (number of upcrossings). *Let X be a supermartingale. The* number $U_N(X; [a, b])(\omega)$ of upcrossings of $[a, b]$ made by $n \mapsto X_n(\omega)$ by time N *is defined to be the largest k in* $\mathbf{Z}^+$ *such that we can find*

$$0 \leqslant s_1 < t_1 < s_2 < t_2 < \cdots < s_k < t_k \leqslant N$$

with

$$X_{s_i}(\omega) < a, \qquad X_{t_i}(\omega) > b \quad (1 \leqslant i \leqslant k).$$

Regard $X_n - X_{n-1}$ as representing your winnings per unit stake on game n. Consider your total-winnings process $Y := C \bullet X$ under the previsible strategy C described as follows:

Pick two numbers a and b with $a < b$.
Repeat

Wait until X gets below a
Play unit stakes until X gets above b and stop playing
Until False (that is, forever!).

To be more formal (and to prove inductively that C is previsible), define

$$C_1 := I_{\{X_0 < a\}},$$

and, for $n \geqslant 2$,

$$C_n := I_{\{C_{n-1}=1\}} I_{\{X_{n-1} \leqslant b\}} + I_{\{C_{n-1}=0\}} I_{\{X_{n-1} < a\}}.$$

The fundamental inequality (recall that $Y_0(\omega) := 0$)

$$(48.2) \qquad Y_N(\omega) \geqslant (b-a)U_N(X;[a,b])(\omega) - [X_N(\omega) - a]^-$$

is now obvious: every upcrossing of $[a,b]$ increases the Y-value by at least $b-a$, while the $[X_N(\omega)-a]^-$ overemphasises the loss during the last 'interval of play'. (Draw a picture, or see [W].)

(*48.3*) THEOREM (Doob's Upcrossing Lemma). *Let X be a supermartingale. Let $U_N(X;(a,b])$ be the number of upcrossings of $[a,b]$ by time N. Then*

$$(b-a)\mathbf{E}U_N(X;[a,b]) \leqslant \mathbf{E}[(X_N-a)^-].$$

Very Important Note. The number of steps does not feature directly on the right-hand side; only the final variable X_N appears. It is the fact that we get a bound independent of the number of steps that makes this result so powerful.

Proof. The process C is previsible, bounded and non-negative, and $Y = C \bullet X$. Hence Y is a supermartingale, and $\mathbf{E}(Y_N) \leqslant 0$. The result now follows from (48.2). □

(*48.4*) COROLLARY. *Let X be a supermartingale that is bounded in $\mathcal{L}^1$ in that $\sup_n \mathbf{E}(|X_n|) < \infty$. Let $a, b \in \mathbf{Q}$ with $a < b$. Then, with $U_\infty(X;[a,b]) := \uparrow\lim_N U_N(X; [a,b])$,*

$$(b-a)\mathbf{E}U_\infty(X;[a,b]) \leqslant |a| + \sup_n \mathbf{E}(|X_n|) < \infty,$$

so that $\mathbf{P}(U_\infty(X;[a,b]) = \infty) = 0$.

Proof. By Lemma 48.3, we have, for $N \in \mathbf{N}$,

$$(b-a)\mathbf{E}U_N(X;[a,b]) \leqslant |a| + \mathbf{E}(|X_N|) \leqslant |a| + \sup_n \mathbf{E}(|X_n|).$$

Now let $N \uparrow \infty$, using the Monotone-Convergence Theorem. □

49. Doob's Supermartingale-Convergence Theorem. Doob's proof is worthy of the result.

(*49.1*) THEOREM (Doob's Supermartingale-Convergence Theorem). *Let X be a supermartingale bounded in $\mathscr{L}^1$: $\sup_n \mathbf{E}(|X_n|) < \infty$. Then, almost surely, $X_\infty := \lim X_n$ exists and is finite. For definiteness, we define $X_\infty(\omega) := \limsup X_n(\omega), \forall \omega$, so that X_∞ is $\mathscr{F}_\infty$-measurable and $X_\infty = \lim X_n$, a.s.*

Proof (Doob). Write (noting the use of $[-\infty, \infty]$):

$$\begin{aligned}\Lambda &:= \{\omega : X_n(\omega) \text{ does not converge to a limit in } [-\infty, \infty]\}\\ &= \{\omega : \liminf X_n(\omega) < \limsup X_n(\omega)\}\\ &= \bigcup_{\{a,b\in\mathbb{Q}:a<b\}} \{\omega : \liminf X_n(\omega) < a < b < \limsup X_n(\omega)\}\\ &=: \bigcup \Lambda_{a,b} \quad \text{(say).}\end{aligned}$$

But

$$\Lambda_{a,b} \subseteq \{\omega : U_\infty(X; [a,b])(\omega) = \infty\},$$

so that, by (11.4), $\mathbf{P}(\Lambda_{a,b}) = 0$. Since Λ is a countable union of sets $\Lambda_{a,b}$, we see that $\mathbf{P}(\Lambda) = 0$, whence

$$X_\infty := \lim X_n \text{ exists a.s. in } [-\infty, \infty].$$

But Fatou's Lemma shows that

$$\mathbf{E}(|X_\infty|) = \mathbf{E}(\liminf |X_n|) \leqslant \liminf \mathbf{E}(|X_n|) \leqslant \sup \mathbf{E}(|X_n|) < \infty,$$

so that $\mathbf{P}(X_\infty \text{ is finite}) = 1$. □

Note. There are other proofs for the discrete-parameter case. None of these is as probabilistic, and none shares the central importance of this one for the continuous-parameter case.

(*49.2*) COROLLARY. *If X is a non-negative supermartingale, then $X_\infty := \lim X_n$ exists almost surely.*

Proof. X is obviously bounded in $\mathscr{L}^1$, since $\mathbf{E}(|X_n|) = \mathbf{E}(X_n) \leqslant \mathbf{E}(X_0)$. □

50. $\mathscr{L}^1$ convergence and the UI property. It is important to know when supermartingales converge in $\mathscr{L}^1$.

(*50.1*) THEOREM. *Let X be a supermartingale bounded in $\mathscr{L}^1$, so that $X_\infty := \lim X_n$ exists a.s. Then $X_n \to X_\infty$ in $\mathscr{L}^1$ if and only if $X = \{X_n : n \in \mathbb{Z}^+\}$ is uniformly integrable, and then, for $n \in \mathbb{Z}^+$,*

$$\mathbf{E}(X_\infty | \mathscr{F}_n) \leqslant X_n, \quad \text{a.s.} \tag{50.2}$$

with a.s. equality if X is a (UI) martingale.

Proof. Because of Theorem 21.2 on the equivalence of $\mathcal{L}^1$ convergence and the UI property, all that remains is to prove that (50.2) holds if $X_n \to X_\infty$ *in* $\mathcal{L}^1$. But then, for $F \in \mathcal{F}_n$, and $r \geqslant n$,

$$\mathbf{E}(X_r; F) \leqslant \mathbf{E}(X_n; F),$$

and (50.2) follows on letting $r \to \infty$.

(*50.3*) THEOREM (Lévy's 'Upward' Theorem). *Let* $\xi \in \mathcal{L}^1(\Omega, \mathcal{F}, \mathbf{P})$, *and, for* $n \geqslant 0$, *define* $M_n := \mathbf{E}(\xi \mid \mathcal{F}_n)$, a.s. *Then M is a UI martingale and*

$$M_n \to \eta := \mathbf{E}(\xi \mid \mathcal{F}_\infty),$$

almost surely and in $\mathcal{L}^1$.

Proof. We know that M is a martingale because of the Tower Property 41(i). We know from Theorem 44.1 that M is UI. Hence $M_\infty := \lim M_n$ exists a.s. and in $\mathcal{L}^1$, and it remains only to prove that $M_\infty = \eta$, a.s., where $\eta := \mathbf{E}(\xi \mid \mathcal{F}_\infty)$. However, for $F \in \mathcal{F}_n$,

$$\mathbf{E}(\eta; F) = \mathbf{E}(M_n; F) = \mathbf{E}(M_\infty; F),$$

so that $\mathbf{E}(\eta; F) = \mathbf{E}(\xi; F)$ for all F in the π-system $\bigcup \mathcal{F}_n$ that generates $\mathcal{F}_\infty$. By property 39.1(iii) in the definition of conditional expectation, the result follows.

(*50.4*) THEOREM (Kolmogorov's 0–1 Law). *Let* $X_1, X_2, \ldots$ *be a sequence of independent RVs. Define*

$$\mathcal{T}_n := \sigma(X_{n+1}, X_{n+2}, \ldots), \qquad \mathcal{T} := \bigcap_n \mathcal{T}_n.$$

Then if $F \in \mathcal{T}$, $\mathbf{P}(F) = 0$ *or* 1.

Proof. Define $\mathcal{F}_n := \sigma(X_1, X_2, \ldots, X_n)$. Let $F \in \mathcal{T}$, and let $\eta := \mathrm{I}_F$. Since $\eta \in \mathrm{b}\mathcal{F}_\infty$, Lévy's Upward Theorem shows that

$$\eta = \mathbf{E}(\eta \mid \mathcal{F}_\infty) = \lim \mathbf{E}(\eta \mid \mathcal{F}_n), \quad \text{a.s.}$$

However, for each n, η is $\mathcal{T}_n$-measurable, and hence is independent of $\mathcal{F}_n$. Hence, by 41(k),

$$\mathbf{E}(\eta \mid \mathcal{F}_n) = \mathbf{E}(\eta) = \mathbf{P}(F), \quad \text{a.s.}$$

Hence $\eta = \mathbf{P}(F)$, a.s.; and since η only takes the values 0 and 1, the result follows. □

For another nice application of the Upward Theorem, see Exercise 60.50.

51. The Lévy–Doob Downward Theorem. This theorem is crucial for the continuous-parameter theory. We follow the account at T.V.21 in Meyer [2].

(*51.1*) THEOREM (Lévy–Doob Downward Theorem). *Suppose that* $(\Omega, \mathcal{F}, \mathbf{P})$ *is a probability triple, and that* $\{\mathcal{G}_n : n \in -\mathbf{N}\}$ *is a collection of sub-σ-algebras of*

$\mathcal{F}$ such that (for $k, n \in \mathbf{N}$)

$$\mathcal{G}_{-\infty} := \bigcap_k \mathcal{G}_{-k} \subseteq \cdots \subseteq \mathcal{G}_{-(n+1)} \subseteq \mathcal{G}_{-n} \subseteq \cdots \subseteq \mathcal{G}_{-1}.$$

Let $X = \{X_n : n \in -\mathbf{N}\}$ be a supermartingale relative to $\{\mathcal{G}_n : n \in -\mathbf{N}\}$, so that

$$\mathbf{E}(X_n | \mathcal{G}_m) \leqslant X_m, \quad \text{a.s.} \quad (m \leqslant n \leqslant -1).$$

Assume that $\sup_{n \leqslant -1} \mathbf{E}(X_n) < \infty$. *Then the process X is UI, and the limit*

$$X_{-\infty} := \lim_{n \to -\infty} X_n$$

exists a.s. and in $\mathcal{L}^1$. Further, for $n \leqslant -1$,

$$\mathbf{E}(X_n | \mathcal{G}_{-\infty}) \leqslant X_{-\infty}, \quad \text{a.s.,}$$

with a.s. equality if X is a martingale.

Proof. We prove the UI property; the existence (a.s. and $\mathcal{L}^1$) of $X_{-\infty}$ then follows from the Upcrossing Lemma just as in the case of the Supermartingale-Convergence Theorem 49.1.

Let $\varepsilon > 0$ be given. Since

$$\uparrow \lim_{n \downarrow -\infty} \mathbf{E}X_n < \infty,$$

there exists k such that

(51.2) $$0 \leqslant \mathbf{E}(X_n) - \mathbf{E}(X_k) \leqslant \tfrac{1}{2}\varepsilon \text{ for all } n \leqslant k.$$

Now, for $n \leqslant k$ and $\lambda > 0$,

$$\begin{aligned} \mathbf{E}(|X_n|; |X_n| > \lambda) &= -\mathbf{E}(X_n; X_n < -\lambda) + \mathbf{E}(X_n) - \mathbf{E}(X_n; X_n \leqslant \lambda) \\ &\leqslant -\mathbf{E}(X_k; X_n < -\lambda) + \mathbf{E}(X_n) - \mathbf{E}(X_k; X_n \leqslant \lambda), \end{aligned}$$

by the supermartingale property. Hence, by (51.2),

$$\mathbf{E}(|X_n|; |X_n| > \lambda) \leqslant \mathbf{E}(|X_k|; |X_n| > \lambda) + \tfrac{1}{2}\varepsilon.$$

Since $X_k \in \mathcal{L}^1$, Lemma 20.1 shows that we can find $\delta > 0$ such that

$$\mathbf{P}(F) < \delta \quad \text{implies that} \quad \mathbf{E}(|X_k|; F) < \tfrac{1}{2}\varepsilon.$$

But $\mathbf{P}(|X_n| > \lambda) \leqslant \lambda^{-1}\mathbf{E}(|X_n|)$, and, since $X^- := \max\{(-X), 0\}$ is a submartingale by 'cJensen',

$$\mathbf{E}(|X_n|) = \mathbf{E}(X_n) + 2\mathbf{E}(X_n^-) \leqslant \sup_n \mathbf{E}(X_n) + 2\mathbf{E}(X_{-1}^-).$$

We may therefore choose K such that

$$\begin{aligned} \mathbf{P}(|X_n| > K) < \delta \quad &\text{whenever } n \leqslant k, \\ \mathbf{E}(|X_j|; |X_j| > K) < \varepsilon \quad &\text{whenever } j > k. \end{aligned}$$

Then $\mathbf{E}(|X_n|; |X_n| > K) < \varepsilon$ for every $n \leqslant -1$, so that X is UI. □

Kolmogorov's Strong Law of Large Numbers is a consequence.

(*51.3*) THEOREM (SLLN). *Let $X_1, X_2, \ldots$ be independent, identically distributed random variables, with $\mathbf{E}(|X_k|) < \infty$ for some (then every) k. Let μ be the common value of $\mathbf{E}(X_n)$. Write $S_n := X_1 + X_2 + \cdots + X_n$. Then*

$$n^{-1}S_n \to \mu \quad \text{a.s. and in } \mathscr{L}^1.$$

Proof. Define

$$\mathscr{G}_{-n} := \sigma(S_n, S_{n+1}, S_{n+2}, \ldots), \quad \mathscr{G}_{-\infty} := \bigcap_n \mathscr{G}_{-n}.$$

Then, for $n \geqslant 1$,

$$\mathbf{E}(X_1|\mathscr{G}_{-n}) = \mathbf{E}(X_2|\mathscr{G}_{-n}) = \cdots = \mathbf{E}(X_n|\mathscr{G}_{-n}) = n^{-1}\mathbf{E}(S_n|\mathscr{G}_{-n}) = n^{-1}S_n, \quad \text{a.s.}$$

Hence $L := \lim n^{-1}S_n$ exists a.s. and in $\mathscr{L}^1$. For definiteness, define $L := \limsup n^{-1}S_n$ for every ω. Then, for each k,

$$L = \limsup \frac{X_{k+1} + \cdots + X_{k+n}}{n}$$

so that $L \in m\mathscr{T}_k$, where $\mathscr{T}_k = \sigma(X_{k+1}, X_{k+2}, \ldots)$. By Kolmogorov's 0–1 law, $\mathbf{P}(L = c) = 1$ for some c in $\mathbf{R}$. But

$$c = \mathbf{E}(L) = \lim \mathbf{E}(n^{-1}S_n) = \mu. \qquad \square$$

Remarks. See Meyer [2] for important extensions and applications of the results given so far in this chapter. These extensions include the *Hewitt–Savage 0–1 Law*, *de Finetti's Theorem* on exchangeable random variables, and the *Choquet–Deny Theorem* on bounded harmonic functions for random walks on groups.

52. Doob's Submartingale and $\mathscr{L}^p$ Inequalities. Many uses are made of the inequalities in this section. We return to the standard situation in which our time-parameter set is $\mathbf{Z}^+$.

(*52.1*) THEOREM (Doob's Submartingale Inequality). *Let Z be a non-negative submartingale. Then, for $c > 0$ and $n \in \mathbf{Z}^+$,*

$$c\mathbf{P}\left(\sup_{k \leqslant n} Z_k \geqslant c\right) \leqslant \mathbf{E}\left(Z_n; \sup_{k \leqslant n} Z_k \geqslant c\right) \leqslant \mathbf{E}(Z_n).$$

Important Notes. The number n of steps does not feature directly in the last two expressions. This is what gives the result its power. The proof will show that the assumption that Z is non-negative is not needed for the first inequality.

Proof. Let $F := \{\sup_{k \leqslant n} Z_k \geqslant c\}$. Then F is a *disjoint* union

$$F = F_0 \cup F_1 \cup \cdots \cup F_n,$$

where

$$F_0 := \{Z_0 \geqslant c\},$$
$$F_k := \{Z_0 < c\} \cap \{Z_1 < c\} \cap \cdots \cap \{Z_{k-1} < c\} \cap \{Z_k \geqslant c\}.$$

Now, $F_k \in \mathscr{F}_k$, and $Z_k \geqslant c$ on F_k. Hence

$$\mathbf{E}(Z_n; F_k) \geqslant \mathbf{E}(Z_k; F_k) \geqslant c\mathbf{P}(F_k).$$

Summing over k now yields the result. □

The main reason for the usefulness of the above theorem is the following.

(52.2) LEMMA. *If M is a martingale, c is a convex function and $\mathbf{E}|c(M_n)| < \infty, \forall n$, then $c(M)$ is a submartingale.*

Proof. Apply the conditional form of Jensen's inequality in Table 41. □

In preparation for Doob's $\mathscr{L}^p$ inequality, we now establish a consequence of Hölder's inequality.

(52.3) LEMMA. *Suppose that X and Y are non-negative random variables such that*

$$c\mathbf{P}(X \geqslant c) \leqslant \mathbf{E}(Y; X \geqslant c) \quad \textit{for every} \quad c > 0.$$

Then, for $p > 1$ and $p^{-1} + q^{-1} = 1$, we have

$$\| X \|_p \leqslant q \| Y \|_p.$$

Proof. We obviously have

$$L := \int_{c=0}^{\infty} pc^{p-1}\mathbf{P}(X \geqslant c)\,dc \leqslant \int_{c=0}^{\infty} pc^{p-2}\mathbf{E}(Y; X \geqslant c)\,dc =: R. \tag{52.4}$$

Using Fubini's Theorem with non-negative integrands, we obtain

$$L = \int_{c=0}^{\infty} \left(\int_{\Omega} \mathrm{I}_{\{X \geqslant c\}}(\omega)\mathbf{P}(d\omega) \right) pc^{p-1}\,dc$$
$$= \int_{\Omega} \left(\int_{c=0}^{X(\omega)} pc^{p-1}\,dc \right) P(d\omega) = \mathbf{E}(X^p).$$

Exactly similarly, we find that

$$R = \mathbf{E}(qX^{p-1}Y).$$

We apply Hölder's inequality to conclude that

$$\mathbf{E}(X^p) \leqslant \mathbf{E}(qX^{p-1}Y) \leqslant q \| Y \|_p \| X^{p-1} \|_q. \tag{52.3}$$

Suppose that $\| Y \|_p < \infty$, and suppose for now that $\| X \|_p < \infty$ also. Then, since $(p-1)q = p$, we have

$$\| X^{p-1} \|_q = \mathbf{E}(X^p)^{1/q},$$

so (52.5) implies that $\| X \|_p \leqslant q \| Y \|_p$. For general X, note that the hypothesis remains true for $X \wedge n$. Hence $\| X \wedge n \|_p \leqslant q \| Y \|_p$ for all n, and the result follows using the Monotone-Convergence Theorem. □

(52.6) THEOREM (Doob's $\mathscr{L}^p$ inequality). *Let $p > 1$ and define q so that $p^{-1} + q^{-1} = 1$. Let Z be a non-negative submartingale bounded in $\mathscr{L}^p$, and define (this is standard notation)*

$$Z^* := \sup_{k \in \mathbf{Z}^+} Z_k.$$

Then $Z^* \in \mathscr{L}^p$, and indeed

(52.7)
$$\| Z^* \|_p \leqslant q \sup_r \| Z_r \|_p.$$

The submartingale Z is therefore dominated by the element Z^* of $\mathscr{L}^p$. Also, $Z_\infty := \lim Z_n$ exists a.s. and in $\mathscr{L}^p$, and

$$\| Z_\infty \|_p = \sup_r \| Z_r \|_p = \uparrow \lim_r \| Z_r \|_p.$$

(b) *If Z is of the form $|M|$, where M is a martingale bounded in $\mathscr{L}^p$, then $M_\infty := \lim M_n$ exists a.s. and in $\mathscr{L}^p$, and of course $Z_\infty = |M_\infty|$, a.s.*

Proof. For $n \in \mathbf{Z}^+$, define $Z_n^* := \sup_{k \leqslant n} Z_k$. From Doob's Submartingale Inequality and the above Lemma, we see that

$$\| Z_n^* \|_p \leqslant q \| Z_n \|_p \leqslant q \sup_r \| Z_r \|_p.$$

Property (52.7) now follows from the Monotone-Convergence Theorem. Since $(-Z)$ is a supermartingale bounded in $\mathscr{L}^p$, and therefore in $\mathscr{L}^1$, we know that $Z_\infty := \lim Z_n$ exists a.s. However,

$$|Z_n - Z|^p \leqslant (2Z^*)^p \in \mathscr{L}^1,$$

so that the Dominated-Convergence Theorem shows that $Z_n \to Z$ in $\mathscr{L}^p$. Jensen's inequality shows that $\| Z_r \|_p$ is non-decreasing in r, and all the rest is straightforward. □

53. Martingales in $\mathscr{L}^2$; orthogonality of increments. Let $M = (M_n : n \geqslant 0)$ be a *martingale in $\mathscr{L}^2$* in that each M_n is in $\mathscr{L}^2$ so that $\mathbf{E}(M_n^2) < \infty, \forall n$. Then for $s, t, u, v \in \mathbf{Z}^+$, with $s \leqslant t \leqslant u \leqslant v$, we know from properties (a) and (j) of Table 41 that for $Z \in \mathscr{L}^2(\mathscr{F}_u)$,

$$\mathbf{E}(ZM_v) = \mathbf{E}\mathbf{E}(ZM_v | F_u) = \mathbf{E}[Z\mathbf{E}(M_v | \mathscr{F}_u)] = \mathbf{E}(ZM_u),$$

so that $M_v - M_u$ is orthogonal to $\mathcal{L}^2(\mathcal{F}_u)$ and, in particular,

$$\mathbf{E}[(M_t - M_s)(M_v - M_u)] = 0. \tag{53.1}$$

Hence the formula

$$M_n = M_0 + \sum_{k=1}^{n} (M_k - M_{k-1})$$

expresses M_n as the sum of orthogonal terms, and Pythagoras's theorem yields

$$\mathbf{E}(M_n^2) = \mathbf{E}(M_0^2) + \sum_{k=1}^{n} \mathbf{E}[(M_k - M_{k-1})^2]. \tag{53.2}$$

The following theorem is therefore obvious.

(*53.3*) THEOREM. *Let M be a martingale for which $M_n \in \mathcal{L}^2, \forall n$. Then M is bounded in $\mathcal{L}^2$ if and only if*

$$\sum \mathbf{E}[(M_k - M_{k-1})^2] < \infty;$$

and when this obtains, Theorem 52.6 implies that

$$M_n \to M_\infty \text{ almost surely and in } \mathcal{L}^2.$$

54. Doob decomposition. In the following theorem, the statement that 'A is a previsible process null at 0' means of course that $A_0 = 0$ and $A_n \in \mathrm{m}\mathcal{F}_{n-1}$ $(n \in \mathbb{N})$.

(*54.1*) THEOREM (Doob decomposition). *(i) Let $(X_n : n \in \mathbb{Z}^+)$ be an adapted process with $X_n \in \mathcal{L}^1, \forall n$. Then X has a Doob decomposition*

$$X = X_0 + M + A, \tag{54.2}$$

where M is a martingale null at 0, and A is a previsible process null at 0. Moreover, this decomposition is unique modulo indistinguishability in the sense that if $X = X_0 + \tilde{M} + \tilde{A}$ is another such decomposition then

$$\mathbf{P}(M_n = \tilde{M}_n, A_n = \tilde{A}_n, \forall n) = 1.$$

(ii) The process X is a submartingale if and only if A is an increasing process in the sense that

$$\mathbf{P}(A_n \leqslant A_{n+1}, \forall n) = 1.$$

Proof. If X has a Doob decomposition as at (54.2) then, since M is a martingale and A is previsible, we have, almost surely,

$$\begin{aligned}\mathbf{E}(X_n - X_{n-1} | \mathcal{F}_{n-1}) &= \mathbf{E}(M_n - M_{n-1} | \mathcal{F}_{n-1}) + \mathbf{E}(A_n - A_{n-1} | \mathcal{F}_{n-1}) \\ &= 0 + (A_n - A_{n-1}).\end{aligned}$$

Hence

$$A_n = \sum_{k=1}^{n} \mathbf{E}(X_k - X_{k-1} | \mathcal{F}_{k-1}), \text{ a.s.,} \tag{54.3}$$

and if we use (54.3) to *define* A, we obtain the required decomposition of X. The 'submartingale' result in Part (ii) of the theorem is now obvious. □

(54.4) *Remark.* The *Doob–Meyer decomposition*, which expresses a submartingale in continuous time as the sum of a local martingale and a previsible increasing process, is a deep result that is the foundation stone for stochastic-integral theory. It is proved in full generality in Chapter VI.

A useful estimate. The following estimate, which makes no positivity hypothesis, is often useful.

(54.5) LEMMA. *If X is a submartingale or supermartingale then, for $N \in \mathbf{Z}^+$ and $c > 0$,*

$$c\mathbf{P}\left(\sup_{k \leqslant N} |X_k| \geqslant 3c\right) \leqslant 4E(|X_0|) + 3E(|X_N|).$$

Proof. Let X be a submartingale with Doob decomposition

$$X = X_0 + M + A,$$

where A is increasing. Then

$$\sup_{k \leqslant N} |X_k| \leqslant |X_0| + \sup_{k \leqslant N} |M_k| + \sup_{k \leqslant N} |A_k| \leqslant |X_0| + \sup_{k \leqslant N} |M_k| + A_N.$$

Thus, using the fact that $|M|$ is a non-negative submartingale and the Submartingale Inequality, we have, for $c > 0$,

$$\begin{aligned} c\mathbf{P}\left(\sup_{k \leqslant N} |X_k| \geqslant 3c\right) &\leqslant c\mathbf{P}(|X_0| \geqslant c) + c\mathbf{P}\left(\sup_{k \leqslant N} |M_k| \geqslant c\right) + c\mathbf{P}(A_N \geqslant c) \\ &\leqslant \mathbf{E}(|X_0|) + \mathbf{E}(|M_N|) + \mathbf{E}(A_N) \\ &\leqslant \mathbf{E}(|X_0|) + \mathbf{E}(|X_N - X_0 - A_N|) + \mathbf{E}(A_N) \\ &\leqslant \mathbf{E}(|X_0|) + \mathbf{E}(|X_N|) + \mathbf{E}(|X_0|) + 2E(A_N) \\ &\leqslant 2\mathbf{E}(|X_0|) + \mathbf{E}(|X_N|) + 2\mathbf{E}(X_N - X_0) \\ &\leqslant 4\mathbf{E}(|X_0|) + 3\mathbf{E}(|X_N|). \end{aligned}$$

If X is a supermartingale, apply the result just obtained to the submartingale $(-X)$. □

55. The $\langle M \rangle$ and $[M]$ processes. The continuous-parameter analogues of these processes allow the stochastic integral to be defined. In both discrete and continuous time, a host of celebrated inequalities (Burkholder–Davis–Gundy, John–Nirenberg etc.) are associated with them. See, for example, Garsia [1] and Neveu [5] for the discrete case.

(55.1) DEFINITION (the angle-brackets process $\langle M\rangle$). *Let M be a martingale in $\mathcal{L}^2$ and null at 0. Then M^2 is a submartingale with (essentially unique) Doob decomposition*

$$M^2 = N + A,$$

where N is a martingale and A is a previsible increasing process, both N and A being null at 0. The process A is written $\langle M\rangle$, and called the angle-brackets process *of M.*

Define $A_\infty := \uparrow\lim A_n$, a.s.. Since $\mathbf{E}(M_n^2) = \mathbf{E}(A_n)$, we see that

(55.2) $$M \text{ is bounded in } \mathcal{L}^2 \text{ if and only if } \mathbf{E}(A_\infty) < \infty.$$

It is important to note that

(55.3) $$A_n - A_{n-1} = \mathbf{E}(M_n^2 - M_{n-1}^2 \mid \mathcal{F}_{n-1}) = \mathbf{E}[(M_n - M_{n-1})^2 \mid \mathcal{F}_{n-1}].$$

This is reflected in the following result.

(55.4) THEOREM and DEFINITION (the $[M]$ process). *Again, let M be a martingale in $\mathcal{L}^2$ and null at 0. Define*

$$[M]_n := \sum_{k=1}^{n} (M_k - M_{k-1})^2.$$

Then

(55.5) $$M^2 - [M] = C \cdot M, \quad \text{where} \quad C_n := 2M_{n-1},$$

and

$$V := M^2 - [M]$$

is a martingale. If M is bounded in $\mathcal{L}^2$ then the martingale V is uniformly integrable.

Proof. The result (55.5) is elementary, and implies that V is a martingale because of Theorem 47.4(iii). If M is bounded in $\mathcal{L}^2$ then, by Doob's $\mathcal{L}^2$ Inequality, the process M^2 is dominated by M^{*2}, which is in $\mathcal{L}^1$, and the process $[M]$ is dominated by $[M]_\infty$, which is also in $\mathcal{L}^1$. Hence V is dominated in $\mathcal{L}^1$, and is therefore UI. □

Stopping times, optional stopping and optional sampling

56. Stopping time. The discrete-parameter theory is easy. The continuous-parameter theory is more challenging.

(56.1) DEFINITION (stopping time). *A map $T:\Omega \to \{0, 1, 2, \ldots; \infty\}$ is called a* stopping time *if*

(56.2) $$\{T \leqslant n\} = \{\omega : T(\omega) \leqslant n\} \in \mathcal{F}_n, \quad \forall n \leqslant \infty,$$

equivalently, if

$$\text{(56.3)} \qquad \{T=n\}=\{\omega: T(\omega)=n\}\in\mathcal{F}_n, \quad \forall n\leqslant\infty.$$

Note that T can be ∞. The equivalence of (56.2) and (56.3) is trivial.

Example. Suppose that (A_n) is an adapted process, and that $B\in\mathcal{B}$. Let

$$T=\inf\{n\geqslant 0: A_n\in B\}=\text{time of first entry of } A \text{ into set } B.$$

By convention, $\inf(\varnothing)=\infty$, so that $T=\infty$ if A never enters set B. Obviously,

$$\{T\leqslant n\}=\bigcup_{k\leqslant n}\{A_k\in B\}\in\mathcal{F}_n,$$

so that T is a stopping time.

Conversely, if T is a stopping time, and we define the *process* $I_{[T,\infty)}$ by writing, for $0\leqslant n<\infty$ and $\omega\in\Omega$,

$$I_{[T,\infty)}(n,\omega):=\begin{cases}1 & \text{if } n\geqslant T(\omega),\\ 0 & \text{otherwise,}\end{cases}$$

then $I_{[T,\infty)}$ is adapted (check!), and T is the first entry time of this process into the set $\{1\}$.

57. Optional-stopping theorems. Let X be a supermartingale, and let T be a stopping time. For $n\geqslant 1$, regard X_n-X_{n-1} as your fortune per unit stake on game n. Suppose that you always bet 1 unit and quit playing at (immediately after) time T. Then your 'stake process' is $C^{(T)}$, where, for $n\in\mathbb{N}$,

$$C_n^{(T)}=I_{\{n\leqslant T\}}, \qquad \text{so that} \quad C_n^{(T)}(\omega)=\begin{cases}1 & \text{if } n\leqslant T(\omega),\\ 0 & \text{otherwise.}\end{cases}$$

Your 'winnings process' is the process with value at time n equal to

$$(C^{(T)}\bullet X)_n=X_{T\wedge n}-X_0.$$

If X^T denotes the process X *stopped at* T,

$$X_n^T(\omega):=X_{T(\omega)\wedge n}(\omega),$$

then

$$C^{(T)}\bullet X=X^T-X_0.$$

Now $C^{(T)}$ is clearly bounded (by 1) and non-negative. Moreover, $C^{(T)}$ is previsible because $C_n^{(T)}$ can only be 0 or 1 and, for $n\in\mathbb{N}$,

$$\{C_n^{(T)}=0\}=\{T\leqslant n-1\}\in\mathcal{F}_{n-1}.$$

Theorem 47.4 now yields the following result.

(57.1) THEOREM (stopped supermartingales are supermartingales). *The following results hold.*

(i) If X is a supermartingale and T is a stopping time, then the stopped process $X^T = (X_{T \wedge n} : n \in \mathbf{Z}^+)$ is a supermartingale, so that, in particular,

(57.2) $$\mathbf{E}(X_{T \wedge n}) \leqslant \mathbf{E}(X_0), \quad \forall n.$$

(ii) If X is a martingale and T is a stopping time, then X^T is a martingale, so that, in particular,

(57.3) $$\mathbf{E}(X_{T \wedge n}) = \mathbf{E}(X_0), \quad \forall n.$$

It is important to notice that this theorem imposes no extra integrability conditions whatsoever (except of course for those implicit in the definition of supermartingale and martingale).

But we have to be careful. Let X be a simple random walk on $\mathbf{Z}$ (with probabilities $\frac{1}{2}$ of jumping to a nearest neighbour), starting at 0. Then X is a martingale relative to its natural filtration. Let T be the stopping time:

$$T := \inf\{n : X_n = 1\}.$$

It is well known that $\mathbf{P}(T < \infty) = 1$. However, even though (57.3) holds for every n, we have $1 = \mathbf{E}(X_T) \neq \mathbf{E}(X_0) = 0$. It is important to know when we *can* say that

$$\mathbf{E}(X_T) \leqslant \mathbf{E}(X_0)$$

for a martingale X. The following theorem gives some sufficient conditions.

(57.4) THEOREM (Doob's Optional-Stopping Theorem). *The following results hold. (a) Let T be a stopping time. Let X be a supermartingale. Then X_T is integrable and*

$$\mathbf{E}(X_T) \leqslant \mathbf{E}(X_0)$$

in each of the following situations:

(i) T is bounded (for some N in $\mathbf{N}$, $T(\omega) \leqslant N, \forall \omega$);

(ii) X is bounded (for some K in $\mathbf{R}^+$, $|X_n(\omega)| \leqslant K$ for every n and every ω) and T is a.s. finite;

(iii) $\mathbf{E}(T) < \infty$, and, for some K in $\mathbf{R}^+$,

$$|X_n(\omega) - X_{n-1}(\omega)| \leqslant K, \quad \forall (n, \omega).$$

(b) If any of the conditions (i)–(iii) holds and X is a martingale then $\mathbf{E}(X_T) = \mathbf{E}(X_0)$.

Proof of (a). We know that $X_{T \wedge n}$ is integrable, and

(57.5) $$\mathbf{E}(X_{T \wedge n} - X_0) \leqslant 0.$$

For (i), we can take $n = N$. For (ii), we can let $n \to \infty$ in (57.5) using the Bounded-Convergence Theorem 21.1. For (iii), we have

$$|X_{T \wedge n} - X_0| = \left| \sum_{k=1}^{T \wedge n} (X_k - X_{k-1}) \right| \leqslant KT$$

and $\mathbf{E}(KT) < \infty$, so that the Dominated-Convergence Theorem 8.5 justifies letting $n \to \infty$ in (57.5) to obtain the answer we want.

Proof of (b). Apply (a) to X and to $(-X)$. □

(57.6) *Awaiting the almost inevitable.* In order to be able to apply result (iii) of Part (a) of the above theorem, we need ways of proving that (when true!) $\mathbf{E}(T) < \infty$. Then following announcement of the principle that '*whatever always stands a reasonable chance of happening will almost surely happen—sooner rather than later*' is often useful.

(57.7) LEMMA. *Suppose that T is a stopping time such that for some N in $\mathbf{N}$ and some $\varepsilon > 0$, we have, for every n in $\mathbf{N}$,*

$$\mathbf{P}(T \leqslant n + N \mid \mathscr{F}_n) > \varepsilon, \quad a.s.$$

Then $\mathbf{E}(T) < \infty$.

The proof is left as an exercise.

(57.8) THEOREM (Doob's Supermartingale Inequalities). *Let X be a non-negative supermartingale and T a stopping time. Then*

$$\mathbf{E}(X_T) \leqslant \mathbf{E}(X_0). \tag{57.9}$$

Moreover, for $c \geqslant 0$,

$$c\mathbf{P}\left\{\sup_k X_k > c\right\} \leqslant \mathbf{E}(X_0). \tag{57.10}$$

Proof. The result (57.9) is obtained by applying Fatou's Lemma to (57.2). The result (57.10) is then obtained by taking $T := \inf\{n : X_n > c\}$ in (57.9). □

58. The pre-T σ-algebra $\mathscr{F}_T$. Let T be a stopping time.

(58.1) DEFINITION (the pre-T σ-algebra $\mathscr{F}_T$). *For $F \subseteq \Omega$, we write $F \in \mathscr{F}_T$ if*

$$F \cap \{T \leqslant n\} \in \mathscr{F}_n \quad \textit{for every } n \in \mathbf{Z}^+ \cup \{\infty\},$$

or, equivalently, if

$$F \cap \{T = n\} \in \mathscr{F}_n \quad \textit{for every } n \in \mathbf{Z}^+ \cup \{\infty\}.$$

(58.2) *The intuitive meaning.* We regard the σ-algebra $\mathscr{F}_T$ as consisting of those events whose occurrence or non-occurrence can be decided from what our observer has seen up to and including time T. Note how well the following lemma therefore ties in with our intuition.

(58.3) LEMMA. *Let S and T be stopping times.*

(i) If X is an adapted process then $X_T \in \mathrm{m}\mathcal{F}_T$.
(ii) If $S \leqslant T$ then $\mathcal{F}_S \subseteq \mathcal{F}_T$.
(iii) $\mathcal{F}_{S \wedge T} = \mathcal{F}_S \cap \mathcal{F}_T$.
(iv) If $F \in \mathcal{F}_{S \vee T}$ then $F \cap \{S \leqslant T\} \in \mathcal{F}_T$.
(v) $\mathcal{F}_{S \vee T} = \sigma(\mathcal{F}_S, \mathcal{F}_T)$.

The proof is left as an easy exercise.

59. Optional sampling. This is an area in which one has to be careful. It is easy to devise fallacious quick 'proofs' of the following theorem that, one then realises, assume Part (ii) of the theorem. One is, for example, tempted to presuppose that $X_{T \wedge n}$ converges to X_T in $\mathcal{L}^1$. The fact that Part (ii) of the theorem is false in continuous time (though it is then still true for UI *martingales*) helps indicate the need for care. We give a proof that very clearly connects the so-called 'class (D) property' in Part (ii) with the existence of a Doob decomposition.

(59.1) THEOREM (Doob's Optional-Sampling Theorem). *Let S and T be stopping times with $0 \leqslant S \leqslant T \leqslant \infty$.*

(i) Let M be a UI martingale. Then

$$\mathbf{E}(M_T | \mathcal{F}_S) = M_S, \text{ a.s.} . \tag{59.2}$$

Moreover, M is of class (D) in that the family (M_T: T a stopping time) is UI.

(ii) Let X be a UI supermartingale, and let

$$X = X_0 + M - A \tag{59.3}$$

be its Doob decomposition (with M a martingale null at 0, and A a previsible increasing process null at 0). Then A is integrable in that $\mathbf{E}(A_\infty < \infty)$, and M is UI, whence X is of class (D) in that

$$\text{the family } (X_T: T \text{ a stopping time}) \text{ is UI.}$$

Moreover,

$$\mathbf{E}(X_T | \mathcal{F}_S) \leqslant X_S, \quad \text{a.s.}$$

Proof of Part (i) for the case when $0 \leqslant S \leqslant T \leqslant k$ for some $k \in \mathbf{N}$. Suppose that $0 \leqslant S \leqslant T \leqslant k$ for some $k \in \mathbf{N}$. Then M_T and M_S are in $\mathcal{L}^1$, because each is dominated by $|M_0| + \cdots + |M_k|$. Let $F \in \mathcal{F}_S$, and define the process $C := I_F I_{(S,T]}$, that is,

$$C_n(\omega) := \begin{cases} 1 & \text{if } \omega \in F \text{ and } S(\omega) < n \leqslant T(\omega), \\ 0 & \text{otherwise.} \end{cases}$$

Then (check!) C is previsible (and bounded and non-negative), whence

$$\mathbf{E}(C \bullet M)_k := \mathbf{E}(M_T - M_S; F) = 0,$$

and Part (i) follows for this case.

Completion of Proof of Part (i). We have, using the result just proved, the Tower Property (41)(i) and Theorem 50.1,

$$M_{T\wedge k} = \mathbf{E}(M_k|\mathscr{F}_{T\wedge k}) = \mathbf{E}(M_\infty|\mathscr{F}_k|\mathscr{F}_{T\wedge k}) = \mathbf{E}(M_\infty|\mathscr{F}_{T\wedge k}).$$

By Lévy's Upward Theorem 50.3, we have $\mathscr{L}^1$ convergence of this equation to

$$M_T = \mathbf{E}(M_\infty|\mathscr{G}), \quad \text{where} \quad \mathscr{G} := \sigma\left(\bigcup_k \mathscr{F}_{T\wedge k}\right).$$

Of course, $\mathscr{G} \subseteq \mathscr{F}_T$. Suppose that $F \in \mathscr{F}_T$. Then (check!) $F \cap \{T \leqslant k\} \in \mathscr{F}_{T\wedge k}$, so that $F \cap \{T < \infty\} \in \mathscr{G}$ and

$$\mathbf{E}(M_T; F \cap \{T < \infty\}) = \mathbf{E}(M_\infty; F \cap \{T < \infty\}).$$

Of course it is tautological that

$$\mathbf{E}(M_T; F \cap \{T = \infty\}) = \mathbf{E}(M_\infty; F \cap \{T = \infty\}).$$

Hence $\mathbf{E}(M_T; F) = \mathbf{E}(M_\infty; F)$, and we have proved (59.2).

The class (D) property of M now follows because of Theorem 44.1. □

Exercise. Prove that if $F \in \mathscr{F}_T$ then there exists $G \in \mathscr{G}$ such that $(G\backslash F) \cup (F\backslash G)$ is $\mathbf{P}$-null.

Proof of Part (ii). We have $\mathbf{E}(A_n) = \mathbf{E}(X_0) - \mathbf{E}(X_n)$, and since X is UI and therefore bounded in $\mathscr{L}^1$, we have

$$\mathbf{E}(A_\infty) = \uparrow\lim \mathbf{E}(A_n) < \infty.$$

Thus the process A is dominated by the element A_∞ of $\mathscr{L}^1$, so that A is UI. Since X is also UI, it follows that M is UI. We now know that the families

$$(M_T : T \text{ a stopping time}) \quad \text{and} \quad (A_T : T \text{ a stopping time})$$

are UI, the latter being dominated by A_∞. Hence

$$(X_T : T \text{ a stopping time})$$

is UI, and certainly each X_T is in $\mathscr{L}^1$. Next,

$$\begin{aligned} \mathbf{E}(X_T|\mathscr{F}_S) &= X_0 + E(M_T|\mathscr{F}_S) - \mathbf{E}(A_T|\mathscr{F}_S) \\ &\leqslant X_0 + M_S - \mathbf{E}(A_S|\mathscr{F}_S) \\ &= X_0 + M_S - A_S = X_S, \end{aligned}$$

and the theorem is proved. □

(59.4) (Riesz decomposition; potentials). *Again let X be a UI supermartingale. Then X has a unique Riesz* decomposition

$$X = Y + Z,$$

where Y is a martingale, and Z is a potential *that is, Z is a non-negative UI* *supermartingale with* $Z_\infty = 0$, *a.s.. In our discrete-parameter situation, Z is of class (D), and Z is the potential*

$$Z_n := \mathbf{E}(A_\infty - A_n | \mathscr{F}_n)$$

of our integrable previsible increasing process A.

The proof is an easy exercise.

(59.5) THEOREM. *Let X be a non-negative supermartingale, and let S and T be stopping times with* $S \leqslant T$. *Then*

$$\mathbf{E}(X_T | \mathscr{F}_S) \leqslant X_S, \quad a.s.$$

Note. Now there is *no* a.s. equality if X is a non-negative martingale.

Proof. As in (59.2),

$$\mathbf{E}(X_{T \wedge n} | \mathscr{F}_{S \wedge n}) \leqslant X_{S \wedge n}, \quad \text{a.s.},$$

and, by modifying the arguments for (59.3), we indeed have

$$\mathbf{E}(X_{T \wedge n} | \mathscr{F}_S) \leqslant X_{S \wedge n}, \quad \text{a.s.}$$

Now let $n \to \infty$, and use the conditional form of Fatou's Lemma. □

The following commutativity property is often useful.

(59.6) THEOREM. *For a stopping time T, define*

$$\mathbf{E}_T : L^1(\Omega, \mathscr{F}, \mathbf{P}) \to L^1(\Omega, \mathscr{F}_T, \mathbf{P})$$

(note the L^1 *rather than* $\mathscr{L}^1$*) by making* $\mathbf{E}_T(\xi)$ *the equivalence class containing* $\mathbf{E}(\xi | \mathscr{F}_T)$*, the distinction between* ξ *and its equivalence class already being blurred! Then, for stopping times S and T, we have*

$$\mathbf{E}_S \mathbf{E}_T = \mathbf{E}_T \mathbf{E}_S = \mathbf{E}_{S \wedge T}.$$

Proof. We make repeated use of Theorem 59.1 for UI martingales. Let $\xi \in \mathscr{L}^1(\Omega, \mathscr{F}, \mathbf{P})$ and let $\xi_n := \mathbf{E}(\xi | \mathscr{F}_n)$, a.s.. Let $\eta = \mathbf{E}(\xi | \mathscr{F}_T)$, a.s., so that, by (59.1), $\eta = \xi_T$, a.s.. Moreover, $\mathbf{E}(\eta | \mathscr{F}_n)$ and $\xi_{n \wedge T}$ are UI martingales, both with limiting value (a.s. equal to) ξ_T. Hence $\mathbf{E}(\eta | \mathscr{F}_n) = \xi_{n \wedge T}$, a.s. for all n, and, by (59.1) yet again, $\mathbf{E}(\eta | \mathscr{F}_S) = \xi_{S \wedge T}$. □

Note. In the continuous-parameter situation, we need a right-continuous version of ξ_T, and have to mirror the use of $\mathscr{L}^1$ rather than L^1 during the proof.

60. Exercises. There are a lot of exercises on this material in [W]. The following important exercises are not given there.

(*E60.41*) *Conditional independence.* Let $(\Omega, \mathcal{F}, \mathbf{P})$ be a probability triple, and let $\mathcal{A}, \mathcal{B}$ and $\mathcal{C}$ be sub-σ-algebras of $\mathcal{F}$. Show that the conditions

(i) $$\mathbf{P}(A \cap C \mid \mathcal{B}) = \mathbf{P}(A \mid \mathcal{B})\mathbf{P}(C \mid \mathcal{B}), \quad \text{a.s.} \quad (\forall A \in \mathcal{A}, \forall C \in \mathcal{C}).$$

(ii) $$\mathbf{P}(C \mid \sigma(\mathcal{A}, \mathcal{B})) = \mathbf{P}(C \mid \mathcal{B}), \quad \text{a.s.} \quad (\forall C \in \mathcal{C}),$$

are equivalent. When one (then each) of these conditions holds, we say that $\mathcal{A}$ *and* $\mathcal{C}$ *are conditionally independent given* $\mathcal{B}$.

The crucial application is to Markov-process theory, when $\mathcal{A}$ represents the Past, $\mathcal{B}$ the Present and $\mathcal{C}$ the Future.

(*E60.50*) *Martingales and differentiation.* Let $f \in \mathcal{L}^1([0,1], \mathcal{B}[0,1], \text{Leb})$. Define

$$f_n(x) := 2^n \int_{(k-1)2^{-n}}^{k2^{-n}} f(x)\,dx \quad \text{if } (k-1)2^{-n} \leqslant x < k2^{-n},$$

with (say) $f_n(1) := f(1)$. Prove that $f_n \to f$, a.e. and in $\mathcal{L}^1$. Use this result to complete Exercises E13.6b and E13.9e.

Hints

(*H60.41*) Suppose that (i) holds. We want to prove (ii). Now, sets of the form $A \cap B$, where $A \in \mathcal{A}$ and $B \in \mathcal{B}$, form a π-system that generates $\sigma(\mathcal{A}, \mathcal{B})$, so we need only prove, using (i), that

$$\mathbf{E}(\mathbf{P}(C \mid \mathcal{B}); A \cap B) = \mathbf{P}(A \cap B \cap C).$$

This and the 'converse' part are exercises in using the 'Taking out what is known' property (41)(j).

(*H60.50*) Take $\Omega := [0,1]$, $\mathcal{F} := \mathcal{B}[0,1]$ and $\mathbf{P} :=$ Leb. Define

$$\mathcal{F}_n := \sigma([(k-1)2^{-n}, k2^{-n}) : 1 \leqslant k < 2^n).$$

Then $f_n = \mathbf{E}(f \mid \mathcal{F}_n)$, and Lévy's Upward Theorem shows that $f_n \to f$, a.s. and in $\mathcal{L}^1$.

Hint for E13.6b. Suppose that $F \in \mathcal{F}$ satisfies $F \subseteq B$ and $\mathbf{P}(F) > 0$. Let $f := I_F$ and let $\varepsilon > 0$. Since $f_n \to f$, a.s., there will be some large n and some k such that

$$f_n(x) > 1 - \tfrac{1}{2}\varepsilon \quad \text{for } (k-1)2^{-n} \leqslant x < k2^{-n}.$$

But $(F + 2m\rho) \bmod 1$ is a subset of B, and, for $1 \leqslant i \leqslant 2^n$, we can, by suitable choice of m, show that

$$\mathbf{P}_*(B \cap [(i-1)2^{-n}, i2^{-n})) > (1-\varepsilon)2^{-n}.$$

5. CONTINUOUS-PARAMETER SUPERMARTINGALES

Regularisation: R-supermartingales

61. Orientation. You should not proceed with this Part 5 until you are very secure with the material of Part 4. *Everything* there will find applications here.

The essential first step, Doob's Regularity Theorem, helps establish that we can concentrate on right-continuous supermartingales relative to right-continuous filtrations. Right continuity of paths and filtrations will then allow us to transfer results from the discrete-parameter context of Part 4. We have to work with 'R-filtrations', filtrations satisfying the 'usual conditions', to obtain an adequate theory of stopping times.

(61.1) A guide to our notation. We shall begin by assuming given a 'rough' supermartingale $\{Y_t : t \in \mathbf{R}^+\}$ relative to a 'rough' filtration $\{\mathscr{G}_t : t \in \mathbf{R}^+\}$. This 'rough' setup is the 'obvious' generalisation of the discrete-parameter case; but it is not at all adequate. We shall show that, for almost all ω, the limit (through *rational* times)

$$X_t(\omega) := \lim_{q \downarrow\downarrow t} Y_q(\omega) := \lim_{\mathbf{Q} \ni q \to t, q > t} Y_q(\omega)$$

exists simultaneously for all t, and defines a modification of Y that is a right-continuous supermartingale relative to the 'usual augmentation' $\{\mathscr{F}_t : t \in \mathbf{R}^+\}$ of $\{\mathscr{G}_t\}$. We call the setup $\{(X_t, \mathscr{F}_t) : t \in \mathbf{R}^+\}$ the R-regularisation of $\{(Y_t, \mathscr{G}_t)\}$. If the map $t \mapsto Y_t$ is right-continuous from $[0, \infty)$ into $\mathscr{L}^1$, as it will be in all cases of interest, then X is a modification of Y.

The 'obvious' analogue of everything in Part 4 holds, except for the 'obvious' analogue of the Doob decomposition and the 'class (D) property of UI supermartingales' in Part (ii) of Theorem 59.1. The great (Doob–)Meyer Decomposition Theorem, which gives the correct generalisation of these things, is proved in in full generality in Chapter VI. The theory of stopping times explodes into a huge subject, many of the deeper parts of which are also studied in Chapter VI.

Note. We use 'rough' with the connotation of 'rough diamond'—one that can be 'polished'. Please note that 'raw' is a different technical term meaning 'not necessarily adapted'. In a sense, 'raw' means '*un*tameable'.

62. Some real-variable results. We collect here some elementary, but essential, real-variable results. We begin by defining the appropriate regularity property for the paths of supermartingales.

(62.1) DEFINITION (R-function on $\mathbf{R}^+$). *A function $x : \mathbf{R}^+ \to \mathbf{R}$ will be called*

an R-function *if*

$$x_t = \lim_{u\downarrow\downarrow t} x_u \quad \textit{for every } t \geqslant 0,$$

$$x_{t-} := \lim_{s\uparrow\uparrow t} x_s \quad \textit{exists finitely for every } t > 0.$$

The double-arrow notation used above will now be clarified.

(*62.2*) NOTATION (one-sided limits; $\downarrow\downarrow$, etc.). For a function $x: \mathbb{R}^+ \to \mathbb{R}$, we define

$$\limsup_{u\downarrow\downarrow t} x_u := \limsup_{u\to t, u>t} x_u = \inf_{v>t} \sup\{x_v : t < u \leqslant v\}.$$

The corresponding lim inf is defined in the obvious way, and the corresponding lim exists if and only if the lim sup and lim inf have the same finite value.

Of course, the $\uparrow\uparrow$ notation is used in a similar way.

Remarks. The French call an R-function càdlàg (*continu à droite et pourvu de limites à gauche*). Some writers use corlol (continuous on the right with limits on the left). In the first edition of this book, R-functions were called Skorokhod functions. Our current notation agrees with that in Volume 2.

We have already examined the difficulties caused by the fact that in studying processes with time-parameter set $\mathbb{R}^+$, we can only utilise σ-cylinders: essentially, we can only consider the behaviour of our process on a countable subset of times. For convenience, we work with the set $\mathbb{Q}^+$ of non-negative rational times. (In many ways, the set of dyadic-rational times would be more convenient!)

(*62.3*) IMPORTANT CONVENTION. *The symbol q as a subscript under a* lim *or* lim sup *or* lim inf *or* $\cap$ *stands for a* RATIONAL *number. We shall use s or u (or anything other than q) to signify a* real *number in this subscript context.*

(*62.4*) *More notation.* Let y be a function on $\mathbb{Q}^+$. We combine the notational conventions at (62.2) and (62.3) by writing, for example,

$$\limsup_{q\downarrow\downarrow t} y_q := \inf_{v>t} \sup\{y_q : q \in \mathbb{Q}^+, t < q \leqslant v\}.$$

Here the double arrow notation helps emphasise that if t happens to be rational then the value y_t is not relevant to the lim sup just described. The definition of analogous things will be obvious.

(*62.5*) DEFINITION (Regularisable function on $\mathbb{Q}^+$). *Let* $y: \mathbb{Q}^+ \to \mathbb{R}$. *We shall call* y regularisable *if*

$$\lim_{q\downarrow\downarrow t} y_q \quad \textit{exists finitely for every real } t \geqslant 0,$$

$$\lim_{q\uparrow\uparrow t} y_q \quad \textit{exists finitely for every real } t > 0,$$

(62.6) DEFINITION (Upcrossings; $U_N(y;[a,b])$ where $y:\mathbb{Q}^+\to\mathbb{R}$). *Let* $y:\mathbb{Q}^+\to\mathbb{R}$. *Let* $N\in\mathbb{N}$ *and let* $a,b\in\mathbb{R}$, *where* $a<b$. *We define the* number $U_N(y;[a,b])$ of upcrossings of $[a,b]$ by y during the interval $[0,N]$ *to be the supremum of* $k\in\mathbb{Z}^+$ *such that we can find rationals*

$$0\leqslant q_1<r_1<q_2<r_2<\cdots<q_k<r_k\leqslant N$$

with

$$y(q_i)<a,\qquad y(r_i)>b\qquad (1\leqslant i\leqslant k).$$

Remarks. We use $y(q)$ rather than y_q when typographically more convenient. Note that $U_N(y;[a,b])$ may well be ∞.

(62.7) THEOREM. *Let* $y:\mathbb{Q}^+\to\mathbb{R}$. *Then* y *is regularisable if and only if whenever* $N\in\mathbb{N}$ *and* $a,b\in\mathbb{Q}$ *with* $a<b$, *we have both*

$$\sup\{|y_q|:q\in\mathbb{Q}^+\cap[0,N]\}<\infty \tag{62.8}$$

and

$$U_N(y;[a,b])<\infty. \tag{62.9}$$

Proof. Now that you know the proof of Doob's Convergence Theorem 49.1, this is an easy exercise.

The 'if' part. Suppose that, whenever $N\in\mathbb{N}$ and $a,b\in\mathbb{Q}$ with $a<b$, statements (62.8) and (62.9) hold. Suppose for the purposes of contradiction that, for some t,

$$\limsup_{q\downarrow\downarrow t} y_q>\liminf_{q\downarrow\downarrow t} y_q. \tag{62.10}$$

If we choose a and b in $\mathbb{Q}$ so that $a<b$ and both a and b lie strictly between the lim inf and lim sup in (62.10) then, for $N>t$, we shall have $U_N(y;[a,b])=\infty$, contradicting (62.9). Hence (62.10) is false. We therefore see that for every $t\geqslant 0$,

$$\lim_{q\downarrow\downarrow t} y_q \text{ exists in } [\infty,\infty],$$

and (62.8) guarantees that this limit is finite. The proof for the $\uparrow\uparrow$ limits is similar.

The 'only if' part. If y is unbounded on $\mathbb{Q}^+\cap[0,N]$ for some $N\in\mathbb{N}$, we can choose $q(n)$ in $\mathbb{Q}^+\cap[0,N]$ such that $|y_{q(n)}|>n$. Let t be an accumulation point of the set $\{q(n)\}$. Then at least one of the limits

$$\lim_{q\downarrow\downarrow t} y_q,\quad \lim_{q\uparrow\uparrow t} y_q \tag{62.11}$$

must fail to exist in $\mathbb{R}$.

Suppose that, for some $a,b\in\mathbb{Q}$ with $a<b$ and for some $N\in\mathbb{N}$, $U_N(y;[a,b])=$

∞. Define $t := \inf\{r \in \mathbf{R}^+ : U_r(y; [a,b]) = \infty\}$, the definition of $U_r(y; [a,b])$ being obvious. Then at least one of the limits at (62.11) must fail to exist, the lim sup being at least b and the lim inf at most a. □

(62.12) COROLLARY. *If $\{Y_q : q \in \mathbf{Q}^+\}$ is an $\mathbf{R}$-valued stochastic process carried by $(\Omega, \mathscr{G}, \mathbf{P})$ then*

$$G := \{\omega : q \mapsto Y_q(\omega) \text{ is regularisable}\}$$

is an element of $\mathscr{G}$.

Proof. Theorem 62.7 allows us to exhibit G as a σ-cylinder, which is then automatically in $\mathscr{G}$. Do think this through.

(62.13) THEOREM. *Let $y: \mathbf{Q}^+ \to \mathbf{R}$ be a regularisable function. Then*

$$x_t := \lim_{q \downarrow\downarrow t} y_q \quad (t \in \mathbf{R}^+)$$

defines an R-function x.

63. Filtrations; supermartingales; R-processes; R-supermartingales. Let $(\Omega, \mathscr{G}, \mathbf{P})$ be a probability triple.

(63.1) DEFINITION (filtration $\{\mathscr{G}_t : t \in \mathbf{R}^+\}$; filtered space). *By a* filtration $\{\mathscr{G}_t : t \in \mathbf{R}^+\}$ on $(\Omega, \mathscr{G}, \mathbf{P})$, *we mean an increasing family of sub-σ-algebras of $\mathscr{G}$:*

$$\text{for } 0 \leqslant s \leqslant t, \quad \mathscr{G}_s \subseteq \mathscr{G}_t \subseteq \mathscr{G}_\infty := \sigma\Big(\bigcup_{u \in \mathbf{R}^+} \mathscr{G}_u\Big) \subseteq \mathscr{G}. \tag{63.2}$$

The setup $(\Omega, \mathscr{G}, \mathbf{P}; \{\mathscr{G}_t : t \in \mathbf{R}^+\})$ is then called a filtered space.

(63.3) DEFINITION (martingale, supermartingale, adapted process). *Let $(\Omega, \mathscr{G}, \mathbf{P}; \{\mathscr{G}_t : t \in \mathbf{R}^+\})$ be a filtered space. By a* martingale (relative to this setup), *we mean an $\mathbf{R}$-valued process $\{Y_t : t \in \mathbf{R}^+\}$ such that*

(i) $Y_t \in \mathscr{L}^1$ for every t;
*(ii) $\{Y_t\}$ is $\{\mathscr{G}_t\}$-*adapted *in that $Y_t \in \mathrm{m}\mathscr{G}_t$ for every t;*
(iii) for $0 \leqslant s \leqslant t$, $\mathbf{E}(Y_t | \mathscr{G}_s) = Y_s$, a.s.

For the definition of supermartingale *(respectively,* submartingale*), the '=' sign in (iii) is replaced by '$\leqslant$' (respectively '$\geqslant$').*

(63.4) LEMMA. *Let Y be a supermartingale relative to the filtered space $(\Omega, \mathscr{G}, \mathbf{P}; \{\mathscr{G}_t : t \in \mathbf{R}^+\})$. Let $t \in [0, \infty)$ and let $(q(n) : n \in -\mathbf{N})$ be a sequence of rationals with $q(n) \downarrow\downarrow t$ as $n \downarrow\downarrow -\infty$. Then*

$$\lim Y_{q(n)} \text{ exists a.s. and in } \mathscr{L}^1.$$

Proof. Simply apply the Lévy–Doob Downward Theorem 51.1 to the setup $(Y_{q(n)}, \mathcal{G}_{q(n)})$, noting that $\sup_{n \leqslant -1} \mathbf{E}(Y_{q(n)}) \leqslant \mathbf{E}(Y_t)$. □

(*63.5*) DEFINITION (R-process, R-supermartingale). *A process is called an* R-process *if all its sample functions are R-functions. By an* R-supermartingale, *we mean a process that is both an R-process and a supermartingale.*

64. Some important examples. We look at Brownian and Poisson examples, at hazard functions, and at an example of a UI martingale with a jump that is not in $\mathcal{L}^1$.

(*64.1*) **Example:** *pre-Brownian motion.* Consider the pre-Brownian motion

$$(\Omega, \mathcal{G}, \mathbf{P}; \{Y_t : t \geqslant 0\}) := (\mathbf{R}^T, \mathcal{B}^T, \mu; \{\pi_t : t \in T\}) \quad (T := [0, \infty))$$

in (32.5). For $0 \leqslant s \leqslant t$, $Y_t - Y_s$ is independent of any finite subfamily of $\{Y_r : r \leqslant s\}$, and therefore of the σ-algebra $\mathcal{G}_s := \sigma(Y_r : r \leqslant s)$. Hence

$$\mathbf{E}(Y_t - Y_s; \mathcal{G}_s) = \mathbf{E}(Y_t - Y_s) = 0, \quad \text{a.s.},$$

and so Y is a martingale relative to its natural filtration $\{\mathcal{G}_t\}$. Note that, since $\mathbf{E}[(Y_t - Y_s)^2] = t - s$, the map $t \mapsto Y_t$ is continuous into $\mathcal{L}^2$, and therefore into $\mathcal{L}^1$.

(*64.2*) **Example:** *compensated Poisson counting process.* Use Section 37 to construct a Poisson measure Λ on $[0, \infty)$ with intensity measure equal to the Lebesgue measure λ. Write $(\Omega, \mathcal{G}, \mathbf{P})$ for the carrier triple. Define $N_t := \Lambda[0, t]$, the number of Poisson occurrences during time interval $[0, t]$. Note that the function $t \mapsto N(t, \omega)$ is here already an R-function for every ω. For $0 \leqslant s \leqslant t$,

$$N_t - N_s = \Lambda(s, t]$$

is independent of $\mathcal{G}_s := \sigma(N_r : r \leqslant s)$, and $N_t - N_s$ has mean $t - s$. Hence $\{N_t - t\}$ is an R-martingale relative to the filtration $\{\mathcal{G}_t\}$. The process $\{N_t - t\}$ is called the compensated Poisson counting process. A serious study of 'compensators' is made in Chapter VI.

(*64.3*) **Example:** *hazard functions.* A more extended study of hazard or 'cumulative-risk' functions is made in Section VI.22.

Let $(\Omega, \mathcal{G}, \mathbf{P})$ be a probability triple, and let $T : \Omega \to (0, \infty)$ be a positive random variable. Let F be the distribution function of T: $F(t) := \mathbf{P}(T \leqslant t)$. Define the right-continuous process $A := I_{[T, \infty)}$ via

$$A_t(\omega) := \begin{cases} 1 & \text{if } t \geqslant T(\omega), \\ 0 & \text{if } t < T(\omega); \end{cases}$$

and put $\mathcal{G}_t := \sigma(A_s : s \leqslant t)$. Note that $\{\mathcal{G}_t\}$ is the smallest filtration relative to which T is a stopping time in that $\{T \leqslant t\} \in \mathcal{G}_t$ for every t.

Define the *hazard* or *comulative-risk* function

$$h(u) := \int_{(0,u]} \frac{dF(v)}{1 - F(v-)}.$$

(64.4) LEMMA. *The process M, where*

$$M_t := A_t - h(T \wedge t),$$

is a martingale relative to the filtration $\{\mathscr{G}_t\}$.

Proof. For $s \leqslant t$, the σ-algebra $\mathscr{G}_s$ is generated by the π-system consisting of Ω together with all sets of the form $\{T \leqslant r\}$, where $r \leqslant s$. Now, for $r \leqslant s$, we have

$$\mathbf{E}[M_t; T \leqslant r] = \mathbf{E}[1; T \leqslant r] - \mathbf{E}[h(T); T \leqslant r] = \mathbf{E}[M_s; T \leqslant r].$$

Next, we have

$$\begin{aligned}\mathbf{E}(M_t) &= \mathbf{P}(T \leqslant t) - \mathbf{E}[h(T); T \leqslant t] - \mathbf{E}[h(t); T > t] \\ &= F(t) - \int_{(0,t]} h(r)\, dF(r) - h(t)[1 - F(t)].\end{aligned}$$

However,

$$\begin{aligned}\int_{(0,t]} h(r)\, dF(r) &= \int_{(0,t]} dF(r) \int_{(0,r]} dF(v)[1 - F(v-)]^{-1} \\ &= \int_{(0,t]} dF(v)[1 - F(v-)]^{-1} \int_{[v,t]} dF(r) \\ &= \int_{(0,t]} dF(v)[1 - F(v-)]^{-1}[F(t) - F(v-)] \\ &= \int_{(0,t]} dF(v)[1 - F(v-)]^{-1}\{1 - F(v-) - [1 - F(t)]\} \\ &= \int_{(0,t]} dF(v) - h(t)[1 - F(t)] \\ &= F(t) - h(t)[1 - F(t)],\end{aligned}$$

so that $\mathbf{E}(M_t; \Omega) = \mathbf{E}(M_s; \Omega) = 0$. Hence, by the π-system characterisation of conditional expectation in Definition 39.1, we have $\mathbf{E}(M_t | \mathscr{G}_s) = M_s$, a.s.. □

(64.5) **Example.** Consider the previous example when T has the exponential distribution of rate 1, so that, for $t > 0$,

$$F(t) = 1 - e^{-t}, \quad h(t) = t, \quad M_t = A_t - (T \wedge t).$$

Let g be a deterministic function on $[0, \infty)$, let $G(t) := \int_0^s g(s)\,ds$, and define

$$Z_t := \int_0^t g(s)\,dM_s = g(T)I_{\{T \leqslant t\}} - G(T \wedge t)$$

$$= \begin{cases} g(T) - G(T) & \text{if } T \leqslant t, \\ -G(t) & \text{if } T > t. \end{cases}$$

Now choose g so that

$$G(t) = \frac{e^t - 1}{1 + t}, \qquad \text{whence} \quad g(t) = \frac{te^t + 1}{(1 + t)^2}.$$

(64.6) LEMMA. *The process Z is then a UI martingale such that $Z(T-)$ is not in $\mathscr{L}^1$.*

Proof. We have

$$|g(T) - G(T)| = \frac{|2 + T - e^T|}{(1 + T)^2} \leqslant 2 + \frac{e^T}{(1 + T)^2}.$$

Since

$$\mathbf{E}\frac{e^T}{(1 + T)^2} = \int_0^\infty \frac{e^s}{(1 + s)^2} e^{-s}\,ds = 1,$$

we see that the process $\{Z_t I_{\{T \leqslant t\}}\}$ is dominated in $\mathscr{L}^1$. Moreover,

$$\mathbf{E}(|Z_t I_{\{T > t\}}|; |Z_t| > K)$$

$$= \begin{cases} (1 + t)^{-1}(e^t - 1)e^{-t} & \text{if} (1 + t)^{-1}(e^t - 1) > K, \\ 0 & \text{otherwise.} \end{cases}$$

Thus both the processes $\{Z_t I_{\{T \leqslant t\}}\}$ and $\{Z_t I_{\{T > t\}}\}$ are UI, whence Z is UI.

Proof of the martingale property of Z, along the lines of the proof of Lemma 64.4, is left as an exercise. Note that

$$\mathbf{E}Z(T-) = -\mathbf{E}G(T-) = -\int_0^\infty \frac{e^s - 1}{1 + s} e^{-s}\,ds = -\infty. \qquad \square$$

Remarks. Examples similar to Z were studied by Dellacherie and by Doléans-Dade. We can obtain the same phenomenon in discrete time with $Z(T-1)$ instead of $Z(T-)$.

65. Doob's Regularity Theorem: Part 1. Now the full power of the Upcrossing Lemma is brought into play.

(65.1) THEOREM (Doob's Regularity Theorem: Part 1). *Let* $\{Y_t : t \in \mathbf{R}^+\}$ *be a supermartingale carried by the filtered space* $(\Omega, \mathscr{G}, \mathbf{P}; \{\mathscr{G}_t : t \in \mathbf{R}^+\})$. *Let*

$$G := \{\omega : \text{the map } q \mapsto Y_q(\omega) \text{ from } \mathbf{Q}^+ \text{ to } \mathbf{R} \text{ is regularisable}\}.$$

Then $G \in \mathscr{G}$ *and* $\mathbf{P}(G) = 1$. *For* $t \in \mathbf{R}^+$, *define*

$$X_t(\omega) := \begin{cases} \lim_{q \downarrow\downarrow t} Y_q(\omega) & \text{if } \omega \in G, \\ 0 & \text{if } \omega \notin G. \end{cases}$$

Then X is an R-process in that all sample paths of X are R-functions.

Proof. Look again at Theorem 62.7. Because there are only countable many triples (N, a, b) where $N \in \mathbf{N}$ and $a, b \in \mathbf{Q}$ with $a < b$, we need only show that, for fixed $N \in \mathbf{N}$ and fixed $a, b \in \mathbf{Q}$ with $a < b$, we have

$$\mathbf{P}(\sup\{|Y_q(\omega)| : q \in \mathbf{Q}^+ \cap [0, N]\} < \infty) = 1 \tag{65.2}$$

$$\mathbf{P}(U_N(Y|_{\mathbf{Q}^+}; [a, b]) < \infty) = 1, \tag{65.3}$$

where the number of upcrossings relates to the restriction of $t \mapsto Y_t(\omega)$ to the set $\mathbf{Q}^+ \cap [0, N]$.

Let $(D(m))$ be a sequence of finite subsets of $\mathbf{Q}^+ \cap [0, N]$, each containing 0 and N, and with $D(m) \uparrow \mathbf{Q}^+ \cap [0, N]$. Then, by Lemma 54.5 applied to $\{Y_q : q \in D(m)\}$, we have, for $c > 0$,

$$\begin{aligned} \mathbf{P}(\sup\{|Y_q(\omega)| : q \in \mathbf{Q}^+ \cap [0, N]\} > 3c) &= \uparrow \lim \mathbf{P}(\sup\{|Y_q(\omega)| : q \in D(m)\} > 3c) \\ &\leqslant 4\mathbf{E}(|Y_0|) + 3\mathbf{E}(|Y_N|), \end{aligned}$$

and (65.2) follows. By the Upcrossing Lemma 48.3, we find that

$$\mathbf{E}U_N(Y|_{\mathbf{Q}^+}; [a, b]) = \uparrow \lim \mathbf{E}U_N(Y|_{D(m)}; [a, b]) \leqslant \frac{E(|Y_n|) + |a|)}{b - a},$$

and (65.3) follows.

(65.4) **Example.** Suppose that $\Omega = \{+1, -1\}$, that $\mathbf{P}(\{+1\}) = \mathbf{P}(\{-1\}) = \frac{1}{2}$, that $\mathscr{G}_t = \{\varnothing, \Omega\}$ when $t \leqslant 1$ and that $\mathscr{G}_t = \mathscr{P}(\Omega)$ when $t > 1$. Suppose that, for $\omega \in \Omega$,

$$Y_t(\omega) = \begin{cases} 0 & \text{if } t \leqslant 1, \\ \omega & \text{if } t > 1. \end{cases}$$

Then Y is a martingale relative to the filtration $\{\mathscr{G}_t\}$, and

$$X_t(\omega) = \begin{cases} 0 & \text{if } t < 1, \\ \omega & \text{if } t \geqslant 1. \end{cases}$$

Note that X_1 is not $\mathscr{G}_1$-measurable, so that X is not a martingale relative to the filtration $\{\mathscr{G}_t\}$. Moreover, $\mathbf{P}(X_1 = Y_1) = 0$, so that X is not a modification of Y.

This example explains our next concerns.

66. Partial augmentation. We continue with the notation and assumptions of Theorem 65.1.

The set G in that theorem is an element of the σ-algebra $\mathcal{N}(\mathcal{G}_\infty)$ of **P**-trivial sets in $\mathcal{G}_\infty$, that is, sets in $\mathcal{G}_\infty$ of **P**-measure 0 or 1. The definition of X in Theorem 65.1 makes the following lemma obvious.

(*66.1*) LEMMA and DEFINITION (partial augmentation). *The process X is adapted to the filtration $\{\mathcal{H}_t\}$, where*

$$\mathcal{H}_t := \sigma(\mathcal{G}_{t+}, \mathcal{N}(\mathcal{G}_\infty)), \quad \textit{where} \quad \mathcal{G}_{t+} := \bigcap_{v>t} \mathcal{G}_v = \bigcap_{q>t} \mathcal{G}_q.$$

We call $\{\mathcal{H}_t\}$ the partial augmentation of $\{\mathcal{G}_t\}$.

(*66.2*) THEOREM (Doob's Regularity Theorem: Part 2). *The process X is a supermartingale relative to $\{\mathcal{H}_t\}$. Moreover, X is a modification of Y if and only if the map $t \mapsto Y_t$ is right-continuous into $\mathcal{L}^1$, that is, if and only if*

$$\lim_{v \downarrow\downarrow t} \mathbf{E}(|Y_v - Y_t|) = 0 \quad \textit{for every } t \geqslant 0.$$

Proof. For the moment, fix v and t with $v > t \geqslant 0$.

Suppose that $(q(n): n \in -\mathbf{N})$ is a sequence of rationals with $v > q(n) \downarrow\downarrow t$ as $n \downarrow\downarrow -\infty$. (For sequences, the $\downarrow\downarrow$ notation is understood to imply monotonicity: $q(n) \geqslant q(n-1) > t$.) We have

$$\mathbf{E}(Y_v | \mathcal{G}_{q(n)}) \leqslant Y_{q(n)}, \quad \text{a.s.}$$

Using the Lévy–Doob Downward Theorem 51.1 (for martingales!), we have

$$\mathbf{E}(Y_v | \mathcal{G}_{t+}) \leqslant X_t, \quad \text{a.s.},$$

whence, trivially,

$$\mathbf{E}(Y_v | \mathcal{H}_t) \leqslant X_t, \quad \text{a.s.} \tag{66.3}$$

Now suppose instead that $u \geqslant t$ and that $(q(n))$ is a sequence of rationals with $q(n) \downarrow\downarrow u$. From (66.3), we have

$$\mathbf{E}(Y_{q(n)} | \mathcal{H}_t) \leqslant X_t, \quad \text{a.s.} \tag{66.4}$$

However, we know from Lemma 63.4 and Theorem 65.1 that $Y_{q(n)} \to X_u$ in $\mathcal{L}^1$; and, by the $\mathcal{L}^1$ continuity of conditional expectations, we now have

$$\mathbf{E}(X_u | \mathcal{H}_t) \leqslant X_t, \quad \text{a.s.}$$

Hence X is a supermartingale relative to $\{\mathcal{H}_u\}$.

It now follows from Lemma 63.4 and the right-continuity of X that X is right-continuous in $\mathcal{L}^1$. Since we also know that if $q(n) \downarrow\downarrow t$ then $Y_{q(n)} \to X_t$ in $\mathcal{L}^1$, it follows that X is a modification of Y if and only if Y is right-continuous in $\mathcal{L}^1$. □

67. Usual conditions; R-filtered space; usual augmentation; R-regularisation. We continue the discussion in the last two sections.

The partial augmentation $\{\mathscr{H}_t\}$ of $\{\mathscr{G}_t\}$ does not allow a sufficiently rich class of stopping times. We need the so-called 'usual augmentation'.

(67.1) DEFINITION (usual conditions; R-filtered space). *A filtered space* $(\Omega, \mathscr{F}, \mathbf{P}; \{\mathscr{F}_t : t \in \mathbf{R}^+\})$ *is said to satisfy the usual conditions and then to be an* R-filtered space *if, in addition to the filtration property*

$$\text{for } 0 \leqslant s \leqslant t, \quad \mathscr{F}_s \subseteq \mathscr{F}_t \subseteq \mathscr{F}_\infty := \sigma\left(\bigcup_{u \in \mathbf{R}^+} \mathscr{F}_u \right) \subseteq \mathscr{F},$$

the following properties hold:

(i) the σ*-algebra* $\mathscr{F}$ *is* **P***-complete;*
(ii) $\mathscr{F}_0$ *contains all* **P***-null sets in* $\mathscr{F}$*;*
(iii) $\{\mathscr{F}_t\}$ *is right-continuous in that*

$$\mathscr{F}_t = \mathscr{F}_{t+} := \bigcap_{u > t} \mathscr{F}_u \quad \text{for all } t \geqslant 0.$$

(67.2) *Remark.* The 'usual conditions' described above are those standard in the literature. In some contexts, however, a more appropriate definition is obtained by replacing $\mathscr{F}$ by $\mathscr{F}_\infty$ in properties (i) and (ii). We stick to Definition 67.1, though the current Remark will haunt us in Markov-process theory.

(67.3) DEFINITION (usual augmentation, R-regularisation). *Let* $(\Omega, \mathscr{G}, \mathbf{P}, \{\mathscr{G}_t\})$ *be a filtered space. The* usual augmentation, *or* R-regularisation $(\Omega, \mathscr{F}, \mathbf{P}, \{\mathscr{F}_t\})$ *of this setup is the minimal enlargement ('enlargement' in that* $\mathscr{G} \subseteq \mathscr{F}$ *and* $\mathscr{G}_t \subseteq \mathscr{F}_t$ *for every* t*) that satisfies the usual conditions.*

(67.4) LEMMA. *The usual augmentation of* $(\Omega, \mathscr{G}, \mathbf{P}, \{\mathscr{G}_t\})$ *is obtained by making* $\mathscr{F}$ *the* **P***-completion of* $\mathscr{G}$*, and, with* $\mathscr{N}$ *denoting the collection of* **P***-null sets in* $\mathscr{F}$*, setting*

$$\mathscr{F}_t = \bigcap_{u > t} \sigma(\mathscr{G}_u, \mathscr{N}) = \sigma(\mathscr{G}_{t+}, \mathscr{N}). \tag{67.5}$$

If $t \geqslant 0$ *and* $F \in \mathscr{F}_t$ *then there exists* G *in* $\mathscr{G}_{t+}$ *such that*

$$F \Delta G := (F \setminus G) \cup (G \setminus F) \in \mathscr{N}.$$

Important Note. The equality on the last two terms of (67.5) needs proof because it is not true in general that if (Σ_n) is a decreasing sequence of σ-algebras on a set S with intersection Σ, and $\mathscr{S}$ is a σ-algebra on S, then $\bigcap \sigma(\Sigma_n, \mathscr{S}) = \sigma(\Sigma, \mathscr{S})$. See, for example, Section 4.12 of [W]. Of course, in (67.5), $\mathscr{N}$ consists of all **P**-null sets, and this makes (67.5) easy to prove. Go through the Exercise E79.67b of proving the lemma in full.

(67.6) PROPOSITION. *In the context of Parts 1 and 2 of the Regularity Theorem, X is an $\{\mathscr{F}_t\}$ supermartingale. The structure $\{(X_t, \mathscr{F}_t\}$ is called the* R-regularisation of $\{(Y_t, \mathscr{G}_t)\}$.

Proof. This is obvious because, for each t, $\mathscr{H}_t$ and $\mathscr{F}_t$ differ only by null sets, as was explained at the end of Lemma 67.4. □

We now examine what happens when our underlying filtration does satisfy the usual conditions.

(67.7) THEOREM (Doob's Regularity Theorem: Part 3). *Let $(\Omega, \mathscr{F}, \mathbf{P}, \{\mathscr{F}_t\})$ be an R-filtered space relative to which supermartingales etc. are defined. Let Y be a supermartingale. Then Y has an R-process modification Z if and only if the map $t \mapsto \mathbf{E}(Y_t)$ from $[0, \infty)$ to $\mathbf{R}$ is right-continuous, and then Z is an R-supermartingale.*

Proof. From the supermartingale property of Y, we have, for $u > t$,

$$\mathbf{E}(Y_u | \mathscr{F}_t) \leqslant Y_t, \quad \text{a.s.} \tag{67.8}$$

Now let X be constructed from Y as in Parts 1 and 2 of the Regularity Theorem. Let $u \downarrow\downarrow t$ in (67.8), and use the fact that $Y_u \to X_t$ in $\mathscr{L}^1$, to obtain

$$\mathbf{E}(X_t | \mathscr{F}_t) \leqslant Y_t, \quad \text{a.s.}$$

However, $\{X_t\}$ is $\{\mathscr{F}_t\}$-adapted, the usual augmentation of $\{\mathscr{F}_t\}$ being $\{\mathscr{F}_t\}$ itself. Hence

$$X_t \leqslant Y_t, \quad \text{a.s.}$$

But, if $t \mapsto \mathbf{E}(Y_t)$ is right-continuous then, since $Y_u \to X_t$ in $\mathscr{L}^1$, we have

$$\mathbf{E}(X_t) = \lim_{u\downarrow\downarrow t} \mathbf{E}(Y_u) = \mathbf{E}(Y_t), \tag{67.9}$$

and, on comparing (67.8) with (67.9), we see that $X_t = Y_t$, a.s. Thus, if the map $t \mapsto \mathbf{E}(Y_t)$ is right-continuous then X is an R-supermartingale modification of Y. The rest is trivial. □

The following lemma answers a frequently occurring question.

(67.10) LEMMA. *Suppose that Y is an R-supermartingale relative to a filtered space $(\Omega, \mathscr{G}, \mathbf{P}, \{\mathscr{G}_t\})$. Then Y is also an R-supermartingale relative to the usual augmentation $(\Omega, \mathscr{F}, \mathbf{P}, \{\mathscr{F}_t\})$.*

Proof. Let $0 \leqslant t < v$. Then, for $u(n) \downarrow\downarrow t$ with $u(n) \leqslant v$ for every n, we have

$$\mathbf{E}(Y_v | \mathscr{G}_{u(n)}) \leqslant Y_{u(n)}, \quad \text{a.s.},$$

whence, applying the Downward Theorem on the left-hand side and right continuity on the right-hand side, we obtain

$$\mathbf{E}(Y_v|\mathscr{G}_{t+}) \leqslant Y_t, \quad \text{a.s.}$$

Hence $\mathbf{E}(Y_v|\mathscr{F}_t) \leqslant Y_t$, a.s. □

68. A necessary pause for thought. We must not rush too quickly to claim that we can henceforth consider only R-supermartingales relative to R-filtered spaces.

Let us consider rather carefully the canonical pre-Brownian motion

$$(68.1) \qquad (\Omega, \mathscr{G}, \mathbf{P}; \{Y_t : t \geqslant 0\})$$

of (64.1). We define the natural filtration $\mathscr{G}_t := \sigma(Y_r : r \leqslant t)$ of Y. Then Y is a martingale relative to this filtration. Since the martingale Y is also continuous in $\mathscr{L}^1$, we can find a modification X of Y that is an R-supermartingale relative to the usual augmentation $(\mathscr{F}, \{\mathscr{F}_t\})$ of $(\mathscr{G}, \{\mathscr{G}_t\})$.

But $\mathscr{F}_t$ contains information about what happens just after time t, and various questions are raised. Have we destroyed the Markov property? We know that, for $t, u > 0$, $Y_{t+u} - Y_t$ is independent of $\mathscr{G}_t$, and it is easy to see that $X_{t+u} - X_t$ is independent of $\mathscr{G}_t$. But is $X_{t+u} - X_t$ independent of $\mathscr{F}_t$? More fundamentally, we can ask whether $\mathscr{F}_t$ really looks into the future, or whether it is true that $\mathscr{F}_t = \sigma(\mathscr{G}_t, \mathscr{N})$, where $\mathscr{N}$ is the collection of P-null sets in $\mathscr{F}$. We know, of course, that $\mathscr{F}_t = \sigma(\mathscr{G}_{t+}, \mathscr{N})$.

We now resolve these questions. It seems best to state the results first as they relate to the original setup (68.1).

(68.2) THEOREM. *For the canonical pre-Brownian motion setup at (68.1), with its natural filtration $\{\mathscr{G}_t\}$, the following results hold.*

(i) For $t \geqslant 0, \mathscr{U}_t := \sigma(Y_{t+u} - Y_t : u \geqslant 0)$ is independent of $\mathscr{G}_{t+}$.
(ii) For $t \geqslant 0, \mathscr{G}_{t+} \subseteq \sigma(\mathscr{G}_t, \mathscr{N}(\mathscr{G}_\infty))$, where $\mathscr{N}(\mathscr{G}_\infty)$ denotes the collection of **P***-null subsets in $\mathscr{G}_\infty$.*

We can interpret (i) as stating that looking ahead an infinitesimal amount by replacing $\mathscr{G}_t$ by $\mathscr{G}_{t+}$ does not destroy the crucial independence property. The result (ii) makes it clear that $\mathscr{G}_{t+}$ does not really look ahead of t: every $\mathscr{G}_{t+}$ set differs from some $\mathscr{G}_t$ set by a null set.

Proof of (i). It is clear from the independence properties of Y and from the construction of X in Part 1 of the Regularity Theorem that, for $t, u \geqslant 0$ and $\varepsilon > 0$,

$$X_{t+u+\varepsilon} - X_{t+\varepsilon} \text{ is independent of } \mathscr{G}_{t+(1/2)\varepsilon} \text{ and hence of } \mathscr{G}_{t+}.$$

Hence, for any G in $\mathscr{G}_{t+}$ and any bounded continuous function f on $\mathbf{R}$,

$$\mathbf{E}[f(X_{t+u+\varepsilon} - X_{t+\varepsilon}); G] = \mathbf{P}(G)\mathbf{E}[f(X_{t+u+\varepsilon} - X_{t+\varepsilon})].$$

We use the right-continuity of X and the Bounded-Convergence Theorem to deduce that

$$\mathbf{E}[f(X_{t+u}-X_t);G]=\mathbf{P}(G)\mathbf{E}[f(X_{t+u}-X_t)]. \tag{68.3}$$

The Monotone-Class Theorem shows that for each fixed G, the result (68.3) holds for every bounded Borel function f, whence $X_{t+u}-X_t$ is independent of $\mathscr{G}_{t+}$. Since Y is a modification of X, $Y_{t+u}-Y_t$ is independent of $\mathscr{G}_{t+}$.

But now, for $t,u,v\geqslant 0$, the variable $Y_{t+u+v}-Y_{t+u}$ is independent of $\mathscr{G}_{(t+u)+}$; and $\mathscr{G}_{t+}$ and $\sigma(Y_{t+u}-Y_t)$ are independent sub-σ-algebras of $\mathscr{G}_{(t+u)+}$; etc. Full proof that $\mathscr{U}_t$ is independent of $\mathscr{G}_{t+}$ is now left as an exercise.

Proof of (ii). Note that $\mathscr{G}_\infty=\sigma(\mathscr{G}_t,\mathscr{U}_t)$, so that $\mathscr{G}_\infty$ is generated by the π-system $\mathscr{I}_t$ of sets of the form $G_t\cap A_t$, where $G_t\in\mathscr{G}_t$ and $A_t\in\mathscr{U}_t$.

Let η be bounded $(\mathscr{G}_{t+})$-measurable, and let ξ be a version of $\eta-\mathbf{E}(\eta|\mathscr{G}_t)$. All that we need to show is that $\xi=0$, a.s.. Since ξ is $(\mathscr{G}_{t+})$-measurable, ξ is independent of $\mathscr{U}_t$. So, for $G_t\in\mathscr{G}_t$ and $A_t\in\mathscr{U}_t$, we have

$$\mathbf{E}(\xi;G_t\cap A_t)=\mathbf{P}(A_t)\mathbf{E}(\xi;G_t)=0,$$

the last equality holding because of the definition of conditional expectation. Hence ξ is orthogonal to the indicator function of any element of the π-system $\mathscr{I}_t$, and hence to the indicator function of every element of $\mathscr{G}_t$. Since ξ is $(\mathscr{G}_\infty)$-measurable, $\xi=0$, a.s. □

Theorem 68.2 has the following consequences for the R-regularisation $(X,\mathscr{F},\{\mathscr{F}_t\})$ of $(Y,\mathscr{G},\{\mathscr{G}_t\})$.

(68.4) THEOREM. *For the R-regularisation $(X,\mathscr{F},\{\mathscr{F}_t\})$ of the canonical pre-Brownian motion $(Y,\mathscr{G},\{\mathscr{G}_t\})$, the following statements are true.*

(i) *For $t\geqslant 0$, $\sigma(X_{t+u}-X_t:u\geqslant 0)$ is independent of $\mathscr{F}_t$.*
(ii) *For $t\geqslant 0$, $\mathscr{F}_t=\sigma(\mathscr{G}_t,\mathscr{N})$, where $\mathscr{N}$ is the collection of* **P**-*null elements of $\mathscr{F}$ (which here equals $\mathscr{F}_\infty$).*

Note that since X is a modification of Y, we have proved the following result.

(68.5) LEMMA. *There exists an R-process X with Wiener measure as its law.*

We shall see later that almost all paths of this process X are continuous.

Results such as *Theorem 68.4 are very much part of the reason that we can concentrate on R-supermartingales relative to R-filtrations.* We need considerable extensions of Theorem 68.4 later.

69. Convergence theorems for R-supermartingales. Right-continuity of paths allows us to prove the convergence theorem for continuous-parameter supermartingales in the same way as we proved the discrete-parameter result.

(69.1) THEOREM (Doob's Supermartingale-Convergence Theorem; compare Theorem 49.1). *Let X be an $\mathbf{R}$-supermartingale relative to a filtered space $(\Omega, \mathscr{G}, \mathbf{P}, \{\mathscr{G}_t\})$. Suppose further that X is bounded in $\mathscr{L}^1$: $\sup_t \mathbf{E}(|X_t|) < \infty$. Then*

$$X_\infty := \lim X_t \text{ exists (in } \mathbf{R}\text{)}, \quad \text{a.s.}$$

Proof. Since $t \mapsto X_t(\omega)$ is right-continuous,

$$\limsup_{t\to\infty} X_t = \limsup_{q\to\infty} X_q, \quad \liminf_{t\to\infty} X_t = \liminf_{q\to\infty} X_q.$$

If, therefore, $\lim_t X_t(\omega)$ did not exist in $[-\infty, \infty]$, we could find rationals a, b with $a < b$ such that

$$\liminf_{q\to\infty} X_q < a < b < \limsup_{q\to\infty} X_q,$$

whence the number $U_\infty(X|_{\mathbf{Q}^+}; [a,b])$ of upcrossings of $[a,b]$ by the restriction $X|_{\mathbf{Q}^+}$ would be infinite. However, by the Upcrossing Lemma and familiar arguments,

$$\mathbf{E}U_\infty(X|_{\mathbf{Q}^+}; [a,b]) \leqslant (b-a)^{-1}\left\{\sup_t \mathbf{E}(|X_t|) + |a|\right\} < \infty.$$

Thus X_∞ exists, a.s., in $[-\infty, \infty]$. That $X_\infty \in \mathbf{R}$, a.s. follows as usual by Fatou's Lemma. □

(69.2) THEOREM (Doob's Convergence Theorem: Part 2; compare Theorem 50.1). *Continue with the notation and assumptions of Theorem 69.1.*

(i) If $\{X_t : t \geqslant 0\}$ is further assumed to be UI then

$$X_t \to X_\infty \quad \text{in } \mathscr{L}^1,$$

and, for every t, $\mathbf{E}(X_\infty|\mathscr{G}_t) \leqslant X_t$, a.s., with a.s. equality if X is martingale.
(ii) If X is a martingale and $X_t \to X_\infty$ in $\mathscr{L}^1$ then X is UI.

Proof of (i) is an obvious modification of the proof of Theorem 50.1. Part (ii) is an immediate consequence of the $\mathscr{L}^1$ continuity of conditional expectations and the UI property established in Theorem 44.1.

(69.3) **Warning:** *in continuous time, an $\mathscr{L}^1$-convergent supermartingale need not be UI.* See Exercise 79.64 or part (v) of E79.66a.

(69.4) THEOREM (compare the Downward Theorem 51.1). *Suppose that we have an $\mathbf{R}$-supermartingale $(\Omega, \mathscr{G}, \mathbf{P}; \{(\mathscr{G}_t, X_t): t > 0\})$ with parameter set $(0, \infty)$ open at 0. Suppose further that $\sup_{t>0} \mathbf{E}(X_t) < \infty$. Then*

$$X_{0+} := \lim_{t\downarrow\downarrow 0} X_t \text{ exists a.s. and in } \mathscr{L}^1,$$

and, for every t, $\mathbf{E}(X_t|\mathscr{G}_{0+}) \leqslant X_0$, a.s.

Proof. Yet again, the existence of the limit in $[-\infty, \infty]$ follows from the Upcrossing Lemma. The rest follows closely the proof of Theorem 51.1. □

(*69.5*) THEOREM (Upward Theorem for martingales). *Suppose that* $(\Omega, \mathscr{F}, \mathbf{P}; \{\mathscr{F}_t : t \geqslant 0\})$ *satisfies the usual conditions. Let* $\xi \in \mathscr{L}^1(\Omega, \mathscr{F}, \mathbf{P})$. *Then there exists a UI R-martingale* $\{\xi_t : t \geqslant 0\}$ *with* $\xi_t = \mathbf{E}(\xi|\mathscr{F}_t)$, *a.s. As* $t \to \infty$, $\xi_t \to \mathbf{E}(\xi|\mathscr{F}_\infty)$, *a.s. and in* $\mathscr{L}^1$.

This is now an easy exercise.

70. Inequalities and $\mathscr{L}^p$ convergence for R-submartingales. Right continuity of paths also allows us to transfer the standard inequalities from the discrete-parameter case.

(*70.1*) THEOREM (Doob's Submartingale Inequality; compare Theorem 52.1). *Let Z be a non-negative R-submartingale relative to the filtered space* $(\Omega, \mathscr{G}, \mathbf{P}; \{\mathscr{G}_t\})$. *Then, for* $c > 0$ *and* $t \geqslant 0$,

$$c\mathbf{P}\left(\sup_{s \leqslant t} Z_s \geqslant c\right) \leqslant \mathbf{E}\left(Z_t ; \sup_{s \leqslant t} Z_s \geqslant c\right) \leqslant \mathbf{E}(Z_t).$$

Proof. Let $(D(m))$ be an increasing sequence of finite subsets of $[0, t]$ each containing the points 0 and t, and insist that the union of the $D(m)$ is dense in $[0, t]$. Then

$$\sup_{s \in [0,t]} Z_s(\omega) = \sup_m \sup_{s \in D(m)} Z_s(\omega),$$

so that

$$\left\{\sup_{s \in [0,t]} Z_s > c\right\} = \uparrow \lim_m \left\{\sup_{s \in D(m)} Z_s > c\right\}.$$

Note that we need '>' not '$\geqslant$' for this logic. Thus, for $\gamma > 0$,

$$\mathbf{P}\left\{\sup_{s \in [0,t]} Z_s > \gamma\right\} = \uparrow \lim_m \mathbf{P}\left\{\sup_{s \in D(m)} Z_s > \gamma\right\}$$

$$\leqslant \uparrow \lim_m \gamma^{-1} \mathbf{E}\left(Z_t ; \sup_{s \in D(m)} Z_s > \gamma\right)$$

$$= \gamma^{-1} \mathbf{E}\left(Z_t ; \sup_{s \in [0,t]} Z_s > \gamma\right).$$

Now let $\gamma \uparrow\uparrow c$. □

(*70.2*) THEOREM (Doob's $\mathscr{L}^p$ inequality; compare Theorem 52.3). *Let* $p > 1$ *and define q so that* $p^{-1} + q^{-1} = 1$. *Let Z be a nonnegative R-submartingale* *relative to some filtered space $(\Omega, \mathscr{G}, \mathbf{P}; \{\mathscr{G}_t\})$. Assume further that Z is bounded in $\mathscr{L}^p$. Define*

$$Z^* := \sup_{t \geq 0} |Z_t| = \sup_{t \geq 0} Z_t.$$

Then $Z^* \in \mathscr{L}^p$, and indeed

$$\|Z^*\|_p \leqslant q \sup_t \|Z_t\|_p. \tag{70.3}$$

The submartingale Z is therefore dominated by the element Z^* of $\mathscr{L}^p$. Also, $Z_\infty := \lim Z_t$ exists a.s. and in $\mathscr{L}^p$, and

$$\|Z_\infty\|_p = \sup_t \|Z_t\|_p = \lim_t \|Z_t\|_p.$$

If Z is of the form M, where M is an R-martingale bounded in $\mathscr{L}^p$, then $M_\infty := \lim M_t$ exists a.s. and in $\mathscr{L}^p$, and $\mathbf{E}(M_\infty | \mathscr{G}_t) = M_t$, a.s..

The proof is left as an exercise.

71. Martingale proof of Wiener's Theorem; canonical Brownian motion. We gave the Lévy–Ciesielski proof of Wiener's Theorem in Section I.6. Now, we give a martingale proof.

We know that the pre-Brownian motion Y has an R-process modification X that is a martingale relative to the usual augmentation $\{\mathscr{F}_t\}$ of the natural filtration of Y. Perhaps it is here most natural to work with the natural filtration $\{\mathscr{X}_t\}$ of X, defining martingales etc. with respect to $\{\mathscr{X}_t\}$. However, you can work with $\{\mathscr{F}_t\}$ if you wish: it does not matter.

(*71.1*) THEOREM. **P**-*almost all paths of X are continuous.*

Proof. The process X^4 is an R-submartingale, by the conditional form of Jensen's inequality. Hence, by the Submartingale Inequality, we have for $\varepsilon > 0$,

$$\varepsilon^4 \mathbf{P}\left(\sup_{s \leqslant \delta} |X_s| > \varepsilon\right) = \varepsilon^4 \mathbf{P}\left(\sup_{s \leqslant \delta} \{X_s^4\} > \varepsilon\right) \leqslant \mathbf{E}(X_\delta^4) = K\delta^2,$$

where $K = \mathbf{E}\xi^4$, ξ denoting a random variable with the standard normal distribution. (In fact, $K = 3$.) Thus if $D_n := \{k2^{-n} : 0 \leqslant k < 2^n\} \subset [0, 1]$, we have with $\delta_n := 2^n$ and $\varepsilon_n := n^{-1}$,

$$\mathbf{P}\left(\sup_{r \in D(n)} \sup_{s \leqslant \delta_n} |X_{r+s} - X_r| > \varepsilon_n\right) \leqslant 2^n \mathbf{P}\left(\sup_{s \leqslant \delta_n} |X_s - X_0| > \varepsilon_n\right)$$
$$\leqslant 2^n K \varepsilon_n^{-4} \delta_n^2 = 2^n K n^4 2^{-2n} = K n^4 2^{-n}.$$

However, $\sum K n^4 2^{-n}$ converges, so, by the First Borel–Cantelli Lemma, there exists a subset Ω_0 of Ω with $\mathbf{P}(\Omega_0) = 1$ such that, for $\omega \in \Omega_0$, there exists an $n_0(\omega)$

such that, for $n \geqslant n_0(\omega)$,

$$\sup_{r \in D(n)} \sup_{s \leqslant 2^{-n}} |X_{r+s}(w) - X_r(\omega)| \leqslant n^{-1},$$

whence

$$\sup_{r,s \in [0,1]: |r-s| \leqslant 2^{-n}} |X_{r+s}(\omega) - X_r(\omega)| \leqslant 3n^{-1}.$$

Thus, for $\omega \in \Omega_0$, $t \mapsto X_t(\omega)$ is (uniformly) continuous on $[0, 1]$. It is obvious now that **P**-almost-all paths of X are continuous on $[0, \infty)$. □

It is obvious that we can modify X so as to have *all* its paths continuous. (Take $X.(\omega) \equiv 0$ for bad ω.) By using exponential martingales instead of the X^4 submartingale, and a lot of ingenuity, one can refine this argument to obtain Lévy's Modulus-of-Continuity Theorem. See, for example, McKean [1].

(*71.2*) *Canonical Brownian motion.* Let $C := C(\mathbb{R}^+, \mathbb{R})$ be the space of all continuous functions from $[0, \infty)$ to $\mathbb{R}$. For $w \in C$ and $t \geqslant 0$, define

$$\pi_t(w) := w(t),$$

and define the σ-algebras

$$\mathscr{A}_t := \sigma(\pi_s : s \leqslant t) \quad (t > 0), \qquad \mathscr{A}_\infty := \sigma(\pi_s : s \geqslant 0).$$

If X is either the process of the last section or the process B of Section I.6 (in either case, with *all* paths made continuous) then Wiener measure **W** on $(C, \mathscr{B}_\infty)$ is given by the law of X:

$$\mathbf{W} = \mathbf{P} \circ X^{-1},$$

when we regard X as the measurable map $\omega \mapsto X.(\omega)$ from $(\Omega, \mathscr{F})$ to $(C, \mathscr{B}_\infty)$.

(*71.3*) DEFINITION (canonical Brownian motion started at 0) and LEMMA. *The setup*

$$(C, \mathscr{A}_\infty, \mathbf{W}; \pi_t, \mathscr{A}_t : t \geqslant 0)$$

is called canonical Brownian motion on $\mathbb{R}$ started at 0. *The Wiener measure on $(C, \mathscr{A}_\infty)$ is the unique probability measure on $(C, \mathscr{A}_\infty)$ such that, under* **W**,

(*i*) $\pi_0 = 0$, *a.s., and*
(*ii*) *whenever $t, u \geqslant 0$, $\pi_{t+u} - \pi_t$ is independent of $\mathscr{A}_t$ and has the $N(0, t-s)$ distribution.*

We know this. Now check out the following lemma.

(*71.4*) LEMMA. *For canonical Brownian motion, for every $t \geqslant 0$ the process*

$$\{\pi_{t+u} - \pi_t : u \geqslant 0\}$$

has law **W** *and is independent of $\mathscr{A}_t$.*

We know that such things are rather tiresome (and there are more such in the next section). However, you should be sure that you can prove such results as that just given. Of course, we more or less did this example during the proof of Theorem 68.2.

There are good reasons for *not* insisting that $X_0(\omega) = 0$ for *every* ω.

72. Brownian motion relative to a filtered space. Just when you thought that things were getting clean and tidy ... . Already during our discussion of Skorokhod embedding in Section I.7, we had to use non-canonical Brownian motion. We needed variables α, β etc. that are independent of our Brownian motion, and our canonical space will not carry these. So, more definitions....

(72.1) DEFINITION (Brownian motion relative to a filtered space). *Let* $(\Omega, \mathcal{G}, \mathbf{P}, \{\mathcal{G}_t\})$ *be a filtered space. By a* Brownian motion (on $\mathbf{R}$, starting at 0) relative to this setup, *we mean a process X such that*

(i) $X(0) = 0$, *a.s.;*
(ii) all paths of X are continuous;
(iii) whenever $t, u \geqslant 0$, $X_{t+u} - X_t$ *is independent of* $\mathcal{G}_t$ *and has the* $N(0, t-s)$ *distribution.*

(72.2) LEMMA. *If X is as in (72.1) then, for each* $t \geqslant 0$, *the process* $\{X_{t+u} - X_t : u \geqslant 0\}$ *has the Wiener measure as its law, and is independent of* $\mathcal{G}_t$. *Moreover, X is a Brownian motion relative to the usual augmentation of the setup* $(\Omega, \mathcal{G}, \mathbf{P}, \{\mathcal{G}_t\})$.

We essentially know all this.

We mention that Lévy's characterisation of Brownian motion (I.2.2) extends to this situation.

(72.3) LEMMA. *Let* $(\Omega, \mathcal{G}, \mathbf{P}, \{\mathcal{G}_t\})$ *be a filtered space. Let X be a continuous process adapted to this filtration such that* $X_0 = 0$ *almost surely. Then X is a Brownian motion (started at 0) relative to the setup* $(\Omega, \mathcal{G}, \mathbf{P}, \{\mathcal{G}_t\})$ *if and only if both* $\{X_t\}$ *and* $\{X_t^2 - t\}$ *are martingales relative to this setup.*

The proof is in Section IV.33.

Here is the result we need to begin to play the Skorokhod-embedding game of Section I.7 properly. (We also require the Strong Markov Theorem.)

(72.4) THEOREM. *Let* X^* *be a Brownian motion relative to* $(\Omega^*, \mathcal{G}^*, \mathbf{P}^*, \{\mathcal{G}_t^*\})$. *Let* $(\Omega^{**}, \mathcal{G}^{**}, \mathbf{P}^{**})$ *be a probability triple carrying a family* $\{\alpha_\lambda^{**} : \lambda \in \Lambda\}$ *of random variables.*

Let $\Omega := \Omega^* \times \Omega^{**}$, *with typical point* $\omega = (\omega^*, \omega^{**})$. *Define*

$$\mathcal{G}_t := \mathcal{G}_t^* \times \mathcal{G}^{**}, \qquad X_t(\omega) := X_t^*(\omega^*), \qquad \alpha_\lambda(\omega) := \alpha_\lambda^{**}(\omega^{**}),$$

and put $\mathbf{P} := \mathbf{P}^* \times \mathbf{P}^{**}$ *on* $\mathcal{G} := \mathcal{G}^* \times \mathcal{G}^{**}$. *Then*

(i) $\{X_t\}$ *is a Brownian motion relative to* $(\Omega, \mathcal{G}, \mathbf{P}; \{\mathcal{G}_t\})$;
(ii) *the family* $\{\alpha_\lambda : \lambda \in \Lambda\}$ *has the same* $\mathbf{P}$*-law as* $\{\alpha_\lambda^{**} : \lambda \in \Lambda\}$ *has* $\mathbf{P}^{**}$*-law*;
(iii) $\{X_t\}$ *and* $\{\alpha_\lambda : \lambda \in \Lambda\}$ *are independent families.*

Proof. For $t \geqslant 0$, *let*

$$\mathcal{A}_t := \sigma(X_{t+u} - X_t : u \geqslant 0).$$

Then $\mathcal{A}_t = \mathcal{A}_t^* \times \{\emptyset, \Omega^{**}\}$, with an obvious notation for $\mathcal{A}_t^*$, because $\mathcal{A}_t$ has no information about ω^{**}. For $A_t^* \in \mathcal{A}_t^*$, $G_t^* \in \mathcal{G}_t^*$ and $G^{**} \in \mathcal{G}^{**}$, we have

$$\begin{aligned}
&\mathbf{P}[(A_t^* \times \Omega^{**}) \cap (G_t^* \times G^{**})] \\
&\quad = \mathbf{P}[(A_t^* \cap G_t^*) \times G^{**}] && \text{(logic!)} \\
&\quad = \mathbf{P}^*(A_t^* \cap G_t^*)\mathbf{P}^{**}(G^{**}) && \text{(definition of } \mathbf{P}) \\
&\quad = \mathbf{P}^*(A_t^*)\mathbf{P}^*(G_t^*)\mathbf{P}^{**}(G^{**}) && (\mathbf{P}^*\text{-independence of } \mathcal{A}_t^* \text{ and } G_t^*) \\
&\quad = \mathbf{P}(A_t^* \times \Omega^{**})\mathbf{P}(G_t^* \times G^{**}) && \text{(definition of } \mathbf{P}).
\end{aligned}$$

That $\mathcal{A}_t$ and $\mathcal{G}_t$ are independent now follows from the π-system Lemma 22.2. The rest is easy. □

Stopping times

In continuous time, the theory of stopping times becomes a massive and very deep subject in its own right, many of the crucial parts of which we develop in Chapter VI. Here we concentrate on developing those results that we need in Chapters III and IV (the latter being on stochastic-integral theory for continuous processes).

73. Stopping time T, pre-T algebra $\mathcal{G}_T$, progressive process. Let $(\Omega, \mathcal{G}, \mathbf{P}, \{\mathcal{G}_t\})$ be filtered space. At the moment, we make no assumptions about usual conditions.

(73.1) DEFINITION (stopping time T, σ-algebra $\mathcal{G}_T$). *A map* $T : \Omega \to [0, \infty]$ *is called a* $\{\mathcal{G}_t\}$ stopping time *if*

(73.2) $$\{T \leqslant t\} := \{\omega : T(\omega) \leqslant t\} \in \mathcal{G}_t \textit{ for every } t \leqslant \infty.$$

We then define the pre-T σ-algebra $\mathcal{G}_T$ via

(73.3) $$\Lambda \in \mathcal{G}_T \quad \textit{if and only if} \quad \Lambda \cap \{T \leqslant t\} \in \mathcal{G}_t \textit{ for every } t \leqslant \infty.$$

(73.4) LEMMA. *The following results hold.*

(i) *If* $S \leqslant T$ *then* $\mathcal{G}_S \subseteq \mathcal{G}_T$.
(ii) $\mathcal{G}_{S \wedge T} = \mathcal{G}_S \cap \mathcal{G}_T$.

(iii) If $\Lambda \in \mathcal{G}_{S \vee T}$ *then* $\Lambda \cap \{S \leqslant T\} \in \mathcal{G}_T$.
(iv) $\mathcal{G}_{S \vee T} = \sigma(\mathcal{G}_S, \mathcal{G}_T)$.

(Compare Lemma 58.3—carefully!)

(73.5) $\{\mathcal{G}_{t+}\}$ *stopping times.* Note that $\{\mathcal{G}_{t+}\}$ is also a filtration, that T is a $\{\mathcal{G}_{t+}\}$ stopping time if and only if

$$\{T < t\} \in \mathcal{G}_t \quad \text{for every } t \leqslant \infty, \tag{73.6}$$

and that then, with $\mathcal{G}_{T+} := (\mathcal{G}_{\cdot+})_T$ in the obvious sense,

$$\begin{aligned} \Lambda \in \mathcal{G}_{T+} &\text{ if and only if } \Lambda \cap \{T \leqslant t\} \in \mathcal{G}_{t+}, \forall t, \\ &\text{ if and only if } \Lambda \cap \{T < t\} \in \mathcal{G}_t, \forall t. \end{aligned} \tag{73.7}$$

(73.8) LEMMA. *Let* $(S_n : n \in \mathbb{N})$ *be a sequence of* $\{\mathcal{G}_t\}$ *stopping times.*

(i) If $S_n \uparrow S$ *then* S *is a* $\{\mathcal{G}_t\}$ *stopping time.*
(ii) If $S_n \downarrow S$ *then* S *is a* $\{\mathcal{G}_{t+}\}$ *stopping time and* $\mathcal{G}_{S+} = \bigcap_n \mathcal{G}_{S_n+}$.

The proof is left as Exercise E79.73.

Note. Of course, if $S_n \uparrow S$, it need not be true that $\mathcal{G}_S = \sigma(\mathcal{G}_{S_n} : n \in \mathbb{N})$.

A very important part is played in the subject by a sequence

$$\text{Adapted} \supseteq \text{Progressive} \supseteq \text{Optional} \supseteq \text{Previsible}$$

of ever-more-restrictive notions of 'non-anticipating'. We now meet the second of these.

(73.9) DEFINITION (progressive process). *A process* $X = \{X_t : t \geqslant 0\}$ *(with values in an arbitrary measurable space* $(E, \mathcal{E})$*) is called* $\{\mathcal{G}_t\}$-progressive *if, for every* $t \geqslant 0$*, the restriction of* $(t, \omega) \mapsto X(t, \omega)$ *to* $[0, t] \times \Omega$ *is* $(\mathcal{B}[0, t] \times \mathcal{G}_t)$-*measurable.*

Note that a $\{\mathcal{G}_t\}$-*progressive process is automatically* $\{\mathcal{G}_t\}$-*adapted*—see Lemma 11.4.

An important example of a progressive process (in fact, of an optional process!) is provided by the following lemma.

(73.10) LEMMA. *A right-continuous adapted process with values in a metrisable space* $(E, \mathcal{B}(E))$ *is progressive.*

Proof. Fix $t \geqslant 0$. For $n \in \mathbb{N}$, define, for $s < t$,

$$Y^{(n)}(s, \omega) := X([k+1]2^{-n}t, \omega) \quad \text{if } k2^{-n}t \leqslant s < [k+1]2^{-n}t,$$

and put $Y^{(n)}(t, \omega) := X(t, \omega)$. Then $Y^{(n)}$ is trivially $(\mathcal{B}[0, t] \times \mathcal{G}_t)$-measurable, and $X = \lim Y^{(n)}$. □

You did of course notice that no analogue of Part (i) of Lemma 58.3 was given at (73.4). Here is the appropriate analogue.

(73.11) LEMMA. *If X is $\{\mathscr{G}_t\}$-progressive and T is a $\{\mathscr{G}_t\}$ stopping time then X_T is $\mathscr{G}_T$-measurable.*

Of course, $X_T(\omega)$ is defined to equal $X_{T(\omega)}(\omega)$ if $T(\omega) < \infty$. If, for example, X is a supermartingale such that X_∞ exists, we define $X_T(\omega) = X_\infty(\omega)$. In other cases, we would set $X_\infty(\omega)$ to be identically 0.

Proof. Fix $t \geqslant 0$. Define $\hat{\Omega}_t := \{\omega : T(\omega) \leqslant t\}$, and let $\hat{\mathscr{G}}_t$ be the σ-algebra of subsets of $\hat{\Omega}_t$ that are in $\mathscr{G}_t$. Define the map $\rho : \hat{\Omega}_t \to [0,t] \times \hat{\Omega}_t$ by $\rho(\omega) := (T(\omega), \omega)$, and define the map $\hat{X}_{(t)} : [0,t] \times \hat{\Omega}_t \to E$ by $\hat{X}_{(t)}(s,\omega) := X(s,\omega)$. Then we have the pictures

$$\hat{\Omega}_t \xrightarrow{\rho} [0,t] \times \hat{\Omega}_t \xrightarrow{\hat{X}_{(t)}} E,$$

$$\hat{\mathscr{G}}_t \xleftarrow{\rho^{-1}} \mathscr{B}[0,t] \times \hat{\mathscr{G}}_t \xleftarrow{\hat{X}_{(t)}^{-1}} \mathscr{E},$$

whence, for $\Gamma \in \mathscr{E}$,

$$\{\omega : X_T(\omega) \in \Gamma\} \cap \{T \leqslant t\} = (\hat{X}_{(t)} \circ \rho)^{-1}(\Gamma) \in \hat{\mathscr{G}}_t \subseteq \mathscr{G}_t.$$

74. First-entrance (début) times; hitting times; first-approach times: the easy cases. We now consider some important examples of stopping times.

(74.1) DEFINITION (first-entrance (début) times; hitting times). *If $\{X_t\}$ is a process with values in a measurable space $(E, \mathscr{E})$, and if $\Gamma \in \mathscr{E}$, we define*

$$D_\Gamma(\omega) := \inf\{t \geqslant 0 : X_t(\omega) \in \Gamma\},$$

$$H_\Gamma(\omega) := \inf\{t > 0 : X_t(\omega) \in \Gamma\},$$

with the usual convention that the infimum of the empty set is ∞. We call D_Γ the début of Γ for X *or* first-entrance time of X into Γ *and* H_Γ the hitting time of Γ by X.

The definition of hitting time may seem rather bizarre. However, it is the one that matters in potential theory.

Please do note carefully the hypotheses and conclusions of the following lemmas.

(74.2) LEMMA. *The first-entrance time D_F into a closed set F for a continuous $\{\mathscr{G}_t\}$ adapted process with values in a metric space (E, ρ) is a $\{\mathscr{G}_t\}$ stopping time. The hitting time H_F of F is a $\{\mathscr{G}_{t+}\}$ stopping time.*

Proof. Since $x \mapsto \rho(x, F)$ is continuous, $\omega \mapsto \rho(X_q(\omega), F)$ is $\mathscr{G}_q$-measurable, for

$q \in \mathbb{Q}^+$. But now, for $t \geqslant 0$, we have by path continuity,

$$D_F(\omega) \leqslant t \quad \text{if and only if} \quad \inf\{\rho(X_q(\omega), F) : q \in \mathbb{Q} \cap [0, t]\} = 0,$$

and the stopping-time property of D_F follows. Proof of the result for H_F is left to you. □

(74.3) LEMMA. *The first-entrance time D_G into an open set G for a right-continuous $\{\mathcal{G}_t\}$ adapted process with values in a topological space $(E, \mathcal{B}(E))$ is a $\{\mathcal{G}_{t+}\}$ stopping time.*

Proof. Right continuity of paths implies that

$$\{D_G < t\} = \bigcup_{q<t} \{X_q \in G\} \in \mathcal{G}_t,$$

so that D_G is a $\{\mathcal{G}_{t+}\}$ stopping time. Even if all paths of X are continuous, D_G need not be a $\{\mathcal{G}_t\}$ stopping time. (For example, suppose that X is real-valued, that for some ω, $X_t(\omega) \leqslant 1$ for $t \leqslant 1$, and that $X_1(\omega) = 1$. Let $G = (1, \infty)$. We cannot tell without looking slightly ahead of time 1 whether or not $D_G = 1$.)

(74.4) LEMMA and DEFINITION (first-approach time). *Let X be an adapted R-process with values in a metric space (E, ρ). Let K be a compact subset of E. Define the first-approach time A_K for K as*

$$A_K := \inf\{t \geqslant 0 : \text{either } X_t \text{ or } X_{t-} \text{ is in } K.\}$$

(We define $X_{0-} := X_0$.) Then A_K is a $\{\mathcal{G}_t\}$ stopping time.

Proof. We have $A_K \leqslant t$ if and only if $\inf\{\rho(X_q(\omega), K) : q \in \mathbb{Q} \cap [0, t]\} = 0$ or $X_t \in K$. □

75. Why 'completion' in the usual conditions has to be introduced. Now for the result that explains why we need the 'completion' part of the usual conditions. You are very strongly recommended to examine its proof, so that you start to appreciate the need for the 'transfinite' methods that find their proper form in proofs of the Début and Section Theorems.

(75.1) LEMMA. *Again suppose that X is an R-process with values in a separable metric space (E, ρ) and that K is a compact subset of E. Suppose that X is adapted to a filtration $\{\mathcal{F}_t\}$ that satisfies the usual conditions. Then D_K is a $\{\mathcal{F}_t\}$ stopping time.*

Proof. Let

$$S_1(\omega) := A_K := \inf\{t \geqslant 0 : \text{either } X_t \text{ or } X_{t-} \text{ is in } K\}.$$

The most intuitive feeling for ordinal numbers will suffice for our discussion.

Recall that one can count through countable ordinals as follows:

$$1, 2, 3, \ldots, \alpha, \alpha+1, \alpha+2, \ldots, 2\alpha, \ldots, 3\alpha, \ldots, 4\alpha, \ldots, \alpha^2 (= \alpha\alpha),$$
$$\ldots, \alpha^3, \ldots, \alpha^4, \ldots, \alpha^\alpha, \alpha^\alpha + 1, \ldots, \text{etc., etc.,}$$

where α is the first infinite ordinal. Each countable ordinal is either the successor $\eta + 1$ of some η or a *limit* ordinal, which is the supremum of ordinals less than it. Define S_η for countable ordinals η as follows:

$$S_{\eta+1} := \inf\{t \geqslant S_\eta : \text{either } X_t \text{ or } X_{t-} \text{ is in } K\},$$
$$S_\eta := \uparrow \lim_{\gamma \uparrow\uparrow \eta} S_\gamma \quad \text{if } \eta \text{ is a limit ordinal.}$$

The argument used to prove Lemma 74.4 shows that each S_η is a $\{\mathscr{F}_t\}$ stopping time.

The process X may approach K at S_β, and 'jump away at the last minute', in which case $S_{\beta+1} > S_\beta$; but it can only make countably many such jumps away. (It is easy to prove that if the sum of an arbitrary sequence of non-negative terms is finite then only countable many terms can be non-zero.) We have $D_K(\omega) = S_{\delta(\omega)}(\omega)$, where $\delta(\omega)$ is the first—necessarily countable—ordinal such that $S_{\delta(\omega)}(\omega) = S_{\delta(\omega)+1}(\omega)$. The problem is that the number of countable ordinals, the number of possible values of $\delta(\omega)$, is uncountable (in much the same way that the number of finite ordinals is infinite).

The probability measure $\mathbf{P}$ now comes into play. Let

$$c_\eta := \mathbf{E}\exp(-S_\eta), \qquad c := \inf c_\eta,$$

the infimum being over all countable ordinals. For $n \in \mathbf{N}$, we can find $\eta(n)$ such that $c_{\eta(n)} < c + n^{-1}$. Let $\eta(\infty)$ be the *countable* ordinal $\lim \eta(n)$. Then $\eta(\infty)$ is independent of ω, and, since $c_{\eta(\infty)} = c$, we have

$$S_{\eta(\infty)} = \sup S_{\eta(n)}, \quad \text{a.s.}(\mathbf{P}).$$

It is now clear that D_K is almost surely equal to the $\{\mathscr{F}_t\}$ stopping time $S_{\eta(\infty)}$. □

It is not at all easy to *prove that* D_K need not be $\mathscr{G}_\infty$-measurable if $\{\mathscr{G}_t\}$ is the unaugmented natural filtration of X, but Dellacherie [2] succeeded in doing so.

(75.2) *The effect of usual augmentation on stopping times.* Proofs of Strong Markov Theorems rely heavily on the following result.

(75.3) THEOREM (Dynkin). *Let T be a $\{\mathscr{F}_t\}$ stopping time, where $\{\mathscr{F}_t\}$ is the usual augmentation of $\{\mathscr{G}_t\}$. Then there exists a $\{\mathscr{G}_{t+}\}$ stopping time S such that $\mathbf{P}(S = T) = 1$. Furthermore, $\mathscr{F}_T$ is the smallest σ-algebra containing $\mathscr{G}_{S+}$ and all $\mathbf{P}$-null sets in $\mathscr{F}$.*

Proof. For $n \in \mathbf{N}$, define

(75.4) $$T^{(n)}(\omega) := (k+1)2^{-n} \quad \text{if } k2^{-n} \leqslant t < (k+1)2^{-n},$$

with $T^{(n)}(\omega) := \infty$ if $T(\omega) = \infty$. Then

$$\Lambda_{k,n} := \{\omega : T^{(n)}(\omega) = k2^{-n}\} \in \mathscr{F}_{k2^{-n}} \quad (k \in \mathbb{N} \cup \{\infty\}),$$

so that we can find $\Lambda^*_{k,n}$ in $\mathscr{G}_{k2^{-n}+}$ such that $\Lambda^*_{k,n} = \Lambda_{k,n}$, a.s. (in that their indicator functions agree, a.s.). Define

$$R^{(n)}(\omega) := \begin{cases} k2^{-n} & \text{on} \quad \Lambda^*_{k,n} \setminus \left[\bigcup_{j<k} \Lambda^*_{j,n}\right] \quad (k \in \mathbb{N}), \\ \infty & \text{on} \quad \Omega \setminus \left[\bigcup_{j \in \mathbb{N}} \Lambda^*_{j,n}\right]. \end{cases}$$

Then $R^{(n)}$ is a $\{\mathscr{G}_{t+}\}$ stopping time. Put $S^{(n)} := \inf_{m \leq n} R^{(m)}$. Then $S^{(n)}$ is a $\{\mathscr{G}_{t+}\}$ stopping time. It is easily checked that $S^{(n)} = T^{(n)}$ a.s.(**P**), and we know from Lemma 73.8 that $S := \downarrow \lim S^{(n)}$ is a $\{\mathscr{G}_{t+}\}$ stopping time that clearly satisfies $S = T$, a.s.(**P**). Proof of the statement about $\mathscr{F}_T$ is now an easy exercise. □

76. Début and Section Theorems. The Début and Section Theorems are the fundamental 'measurability' results for stopping-time theory. The proper techniques for deriving these results involve *Choquet capacitability theory* and earlier methods of Sierpinski and others. We owe the theory principally to Sierpinski, Choquet, Ray, Doob, Hunt, Dynkin, Meyer and Dellacherie. Dellacherie and Meyer [1] is the definitive account. It is rather hard going, and it might be advisable to read (say) the appendix to Dynkin [1] first. Perhaps the best thing to do is to take the Début and Section Theorems on trust for a while until you understand how such things are used.

(76.1) THEOREM (Début Theorem). *Let X be a progressive process relative to a filtration $\{\mathscr{G}_t\}$, and with values in some topological space E (with its Borel σ-algebra $\mathscr{B}(E)$). Then, for $B \in \mathscr{B}(E)$, the equation*

$$D_B(\omega) := \inf\{t \geq 0 : X_t(\omega) \in B\}$$

defines an $\{\mathscr{F}_t\}$ stopping time, where $\{\mathscr{F}_t\}$ is the usual augmentation of $\{\mathscr{G}_t\}$.

For a proof, see T.IV.50 of Dellacherie and Meyer [1]. (Note that $I_B \circ X$ is a progressive process with values in $\{0, 1\}$.)

(76.2) THEOREM (Section Theorem: first draft). *Let X be a right-continuous process, $\{\mathscr{G}_t\}$-adapted, with values in some complete separable metric space E. Let B be a Borel subset of E. Then, for $\varepsilon > 0$, there exists a $\{\mathscr{G}_t\}$ stopping time T such that*

(i) $X_{T(\omega)}(\omega) \in B$ on $\{T < \infty\}$;
(ii) $\mathbf{P}(T < \infty) \geq \mathbf{P}(D_B < \infty) - \varepsilon$.

If $\{\mathscr{G}_t\}$ satisfies the usual conditions then we often can, and do, take $T = D_K$, where K is some 'large' compact subset of B.

The hypotheses of Theorem 76.2 are not the natural ones—hence the 'first draft' in the title. It is important to note that *Theorem 76.2 would be false if X were only assumed progressive: the correct measurability requirement on X is that it be 'optional'*. We look further into these matters in Chapter VI.

The significance of the Début Theorem will already be clear to you. The Section Theorem may not convey much as the moment. Let us therefore give you now an illustration of its use. We shall take for granted the strong Markov property of a certain process; this is proved in Chapter III.

(*76.3*) *Example.* For $n \in \mathbf{N}$, let $X_n = \{X_n(t): t \geqslant 0\}$ be a Markov chain with two states $\{1, -1\}$ and Q-matrix

$$\begin{pmatrix} -q_n & q_n \\ q_n & -q_n \end{pmatrix}.$$

Assume that the chains X_n are independent of one another. Let E be the multiplicative group $E = \{-1, 1\}^{\mathbf{N}}$ with the obvious product topology, so that E is homeomorphic to the Cantor set. Let

$$\mathbf{X}(t) := (X_1(t), X_2(t), \ldots).$$

Then $\mathbf{X}$ is a strong Markov process on E; more precisely, if $\{\mathscr{F}_t\}$ denotes the usual augmentation of the natural filtration of $\mathbf{X}$, and if T is an $\{\mathscr{F}_t\}$ stopping time, then the process $\{\mathbf{X}(T+t)/\mathbf{X}(T)\}$ is independent of $\mathscr{F}_T$ and has the same law as $\{\mathbf{X}(t)/\mathbf{X}(0)\}$. (You should consider the adjustments to be made when T can be ∞).

If the q_n grow sufficiently rapidly, then one can prove by local-time techniques that

$$\mathbf{P}\left(\sup_{x \in E} D_{\{x\}} < \infty \right) = 1,$$

so that, almost surely, by some finite random time, $\mathbf{X}$ will have visited all points of E.

What we do now is investigate the 'robustness' of the Strong Law of Large Numbers (SLLN) when, as we now assume,

$$q_n = 1 \quad \text{for all } n.$$

Then

$$\mathbf{P}[X_n(t) = X_n(0)] = \tfrac{1}{2}(1 + e^{-2t}), \quad \forall (n, t),$$

and one can easily use the SLLN to prove that, for fixed t,

$$\mathbf{P}\left[\limsup n^{-1} \sum_{k \leqslant n} X_k(t) = e^{-2t} \limsup n^{-1} \sum_{k \leqslant n} X_k(0) \right] = 1.$$

In particular, if T is a stopping time then, for fixed t,

$$\mathbf{P}\left[\limsup n^{-1}\sum_{k\leqslant n} X_k(T+t)=e^{-2t}\limsup n^{-1}\sum_{k\leqslant n} X_k(T)\,\middle|\, T<\infty\right]=1. \tag{76.4}$$

Suppose henceforth that

$$\mathbf{P}[X_n(0)=\pm 1]=\tfrac{1}{2},\quad \forall n.$$

Let

$$\Gamma := \left\{x\in E: n^{-1}\sum_{k\leqslant n} x_k \to 0 \text{ as } n\to\infty\right\}.$$

Then, by the SLLN, for fixed t,

$$\mathbf{P}[\mathbf{X}(t)\in\Gamma]=1,$$

so, by Fubini's Theorem,

$$\mathbf{P}[\mathbf{X}(t)\in\Gamma \text{ for } \textit{almost} \text{ all } t]=1. \tag{76.5}$$

However, if T is a stopping time then, by (76.4),

$$\mathbf{P}[\mathbf{X}(T+t)\notin\Gamma \mid \mathbf{X}(T)\notin\Gamma]=1 \quad \forall t\geqslant 0,$$

whence, *a fortiori*,

$$\mathbf{P}[X(t)\in\Gamma \text{ for almost all } t;\ \mathbf{X}(T)\notin\Gamma,\ T<\infty]=0. \tag{76.6}$$

On comparing (76.5) with (76.6), we see that, for any stopping time T,

$$\mathbf{P}[T<\infty;\mathbf{X}(T)\notin\Gamma]=0.$$

By the Section Theorem,

$$\mathbf{P}[\mathbf{X}(t)\in\Gamma \text{ for } \textit{all } t\geqslant 0]=\mathbf{P}(D_{E\setminus\Gamma}=\infty)=1.$$

Thus, when $q_n=1$ for all n, the SLLN is preserved at *all* times. We saw earlier that if the q_n grow rapidly then there will be a random set of measure zero of times on which the SLLN fails in every possible way. This kind of discussion takes us into the area of 'quasi' potential theory—see, for example, Fukushima [2] and Lyons [3].

77. Optional Sampling for R-supermartingales under the usual conditions. Let $(\Omega,\mathscr{F},\mathbf{P},\{\mathscr{F}_t\})$ satisfy the usual conditions. Martingales, supermartingales, stopping times etc. are defined relative to $\{\mathscr{F}_t\}$.

(77.1) THEOREM (approximation from above). *Let X be an R-supermartingale, and let T be a stopping time. For $n\in\mathbb{N}$, let*

$$\mathbb{D}_n^+ := \{k2^{-n}: k\in\mathbb{Z}^+\}$$

be the set of non-negative dyadic rationals of order (less than or equal to) n. Fix

$t \geqslant 0$. *Define*

$$(77.2) \qquad T^{(n)}(\omega) := \inf\{q \in \mathbb{D}_n^+ : q > T(\omega)\}, \quad t^{(n)} := \inf\{q \in \mathbb{D}_n^+ : q > t\}.$$

Then $T^{(n)}$ is a stopping time relative to $\{\mathscr{F}_q : q \in \mathbb{D}_n^+\}$, $T^{(n)} \downarrow T$ and $\mathscr{F}(T^{(n)}) \downarrow \mathscr{F}(T)$. Moreover,

$$(77.3) \qquad X(T^{(n)} \wedge t^{(n)}) \to X(T \wedge t), \quad \text{a.s. and in } \mathscr{L}^1.$$

In particular, $X(T \wedge t) \in \mathscr{L}^1$.

Proof. The fact that $\mathscr{F}(T^{(n)}) \downarrow \mathscr{F}(T)$ follows from Lemma 73.8. Applying the argument used at (59.2) to the finite discrete-parameter set $\mathbb{D}_{n+1}^+ \cap [0, t+1]$, we obtain

$$\mathbf{E}[X(T^{(n)} \wedge t^{(n)}) | \mathscr{F}(T^{(n+1)} \wedge t^{(n+1)})] \leqslant X(T^{(n+1)} \wedge t^{(n+1)}), \quad \text{a.s.,}$$

and

$$\mathbf{E}X(T^{(n)} \wedge t^{(n)}) \leqslant \mathbf{E}X(0).$$

The result (77.3) now follows from the Downward Theorem 51.1. (Our new labelling reverses the 'direction' of the ns!) □

(77.4) THEOREM (stopped supermartingales are supermartingales). *Let X be an R-supermartingale, and let T be a stopping time. Then $X^T := \{X_{T \wedge t} : t \geqslant 0\}$ is an R-supermartingale. Of course, X^T is an R-martingale if X is an R-martingale.*

Proof. Fix $0 \leqslant s \leqslant t$. Define $T^{(n)}$ and $t^{(n)}$ as in Theorem 77.1, and define $s^{(n)}$ analogously. By discrete-parameter theory, we have for $n, r \in \mathbb{N}$ with $r \geqslant n$,

$$\mathbf{E}[X(T^{(n)} \wedge t^{(n)}) | \mathscr{F}(s^{(r)})] \leqslant X(T^{(n)} \wedge s^{(r)}), \quad \text{a.s.}$$

Letting $r \uparrow \infty$ and using the Downward Theorem for martingales, we have

$$\mathbf{E}[X(T^{(n)} \wedge t^{(n)}) | \mathscr{F}_s] \leqslant X(T^{(n)} \wedge s), \quad \text{a.s.}$$

But we may now use the result (77.3) on the left-hand side and right-continuity of paths on the right-hand side to obtain

$$\mathbf{E}[X(T \wedge t) | \mathscr{F}_s] \leqslant X(T \wedge s), \quad \text{a.s.}$$

as required. □

(77.5) THEOREM (Doob's Optional-Sampling Theorem for UI R-supermartingales). *Suppose that X is an R-supermartingale, and that either X is UI, so that $\{X_t : t \geqslant 0\}$ is a UI family, or X is non-negative. Let S and T be stopping times with $S \leqslant T$. Then $X_T \in \mathscr{L}^1$, and*

$$\mathbf{E}(X_\infty | \mathscr{F}_T) \leqslant X_T, \quad \text{a.s.,} \qquad \mathbf{E}(X_T | \mathscr{F}_S) \leqslant X_S, \quad \text{a.s.,}$$

with a.s. equality in both places if X is a UI R-martingale. In particular, a UI *martingale M is of class (D) in that the family*

$$\{M_T : T \text{ a stopping time}\}$$

is UI.

Remark. You have already been warned that the 'obvious' analogue of the Doob decomposition for supermartingales is false in continuous time. See Exercise E79.77a. We therefore cannot proceed as we did in discrete time by deriving the supermartingale result from the martingale result. We combine approximation with the discrete-parameter result for supermartingales.

Proof. From the discrete-parameter results (59.1) and (59.5) for $\mathbb{D}^+_{n+1}$, we have

$$\mathbf{E}[X(T^{(n)})|\mathscr{F}(T^{(n+1)})] \leqslant X(T^{(n+1)}), \quad \text{a.s.,}$$

and

$$\mathbf{E}X(T^{(n)}) \leqslant \mathbf{E}X(0).$$

Hence, by the Downward Theorem, $X(T^{(n)}) \to X(T)$ in $\mathscr{L}^1$ as well as almost surely (by right-continuity). Next, we have

$$\mathbf{E}[X(\infty)|\mathscr{F}(T^{(n)})] \leqslant X(T^{(n)}), \quad \text{a.s.,}$$

whence

$$\mathbf{E}[X(\infty)|\mathscr{F}(T)] \leqslant X(T), \quad \text{a.s.,}$$

We also have, for $n, r \in \mathbb{N}$ with $r \geqslant n$,

$$\mathbf{E}[X(T^{(n)})|\mathscr{F}(S^{(r)})] \leqslant X(S^{(r)}), \quad \text{a.s.,}$$

whence, as previously, on letting $r \uparrow \infty$ and then $n \uparrow \infty$, we obtain the desired result. □

The martingale case of the above Optional-Sampling Theorem is used repeatedly in stochastic-integral theory. The following little result will also be found useful in that subject.

(77.6) THEOREM. *Let $\{M_t : 0 \leqslant t \leqslant \infty\}$ be a progressive process such that for each (finite or infinite) stopping time T, we have $E(|M_T|) < \infty$ and $\mathbf{E}(M_T) = 0$. Then M is a UI martingale.*

Proof. Let $t \geqslant 0$ and $F \in \mathscr{F}_t$. Define

$$T(\omega) := \begin{cases} t & \text{if } \omega \in F, \\ \infty & \text{if } \omega \in F^c := \Omega \backslash F. \end{cases}$$

Then (check!) T is a stopping time, and

$$\mathbf{E}(M_T) = \mathbf{E}(M_T; F) + \mathbf{E}(M_\infty; F^c) = 0,$$
$$\mathbf{E}(M_\infty) = \mathbf{E}(M_\infty; F) + \mathbf{E}(M_\infty; F^c) = 0,$$

whence $E(M_\infty; F) = \mathbf{E}(M_t; F)$, and so, $M_t = \mathbf{E}(M_\infty | \mathscr{F}_t)$, a.s. □

Do the exercise of extending the commutativity property in Theorem 59.6 to the following continuous time.

(77.7) THEOREM. *Continue to assume the usual conditions. Let S and T be stopping times. Then*

$$\mathbf{E}_S \mathbf{E}_T = \mathbf{E}_T \mathbf{E}_S = \mathbf{E}_{S \wedge T},$$

where E_T *denotes the conditional expectation map*

$$\mathbf{E}(\cdot | \mathscr{F}_T): L^1(\Omega, \mathscr{F}, \mathbf{P}) \to L^1(\Omega, \mathscr{F}_T, \mathbf{P}).$$

See the note following Theorem 59.6.

78. Two important results for Markov-process theory. We apologise for 'floating in and out of the usual conditions'. We shall see that there are reasons for so doing.

(78.1) THEOREM. *The following results hold.*

(i) Let Z be an R-process on a triple $(\Omega, \mathscr{G}, \mathbf{P})$. *Set*

$$U(\omega) := \inf\{t \geqslant 0 : Z_t(\omega) = 0 \text{ or } Z_{t-} = 0\},$$

so that U is the time of first approach by Z to 0. *Then U is* $\mathscr{G}$*-measurable and*

$$G := \{\omega : Z_\cdot(\omega) \equiv 0 \text{ on } [U, \infty)\} \in \mathscr{G}.$$

(ii) Let X be a non-negative R-supermartingale relative to a setup $(\Omega, \mathscr{G}, \mathbf{P}, \{\mathscr{G}_t : t \geqslant 0\})$. *Set*

$$T(\omega) := \inf\{t \geqslant 0 : X_t(\omega) = 0 \quad \text{or} \quad X_{t-} = 0\}.$$

Then

$$\mathbf{P}(X \equiv 0 \text{ on } [T, \infty)) = 1.$$

Proof of (i). That U is $\mathscr{G}$-measurable is obvious from the fact that we could take every $\mathscr{G}_t$ to be $\mathscr{G}$ in Lemma 74.4. Then, with ε, q_1 and q_2 rational, we have

$$G = \bigcap_{\varepsilon > 0} \bigcup_{q_1 \geqslant 0} \bigcap_{q_2 > q_1 + \varepsilon} \{\omega : q_1 \leqslant U(\omega) < q_1 + \varepsilon; Z_{q_2}(\omega) = 0\}.$$ □

Proof of (ii). We know that X remains a supermartingale relative to the usual augmentation $(\Omega, \mathscr{F}, \mathbf{P}, \{\mathscr{F}_t : t \geqslant 0\})$ of $(\Omega, \mathscr{G}, \mathbf{P}, \{\mathscr{G}_t : t \geqslant 0\})$, and that T is an $\{\mathscr{F}_t\}$ stopping time. For $n \in \mathbb{N}$, set $S_n := \inf\{t : X_t < n^{-1}\}$. Then S_n is a stopping time with $S_n \leqslant T$. For any rational $q \geqslant 0$, S_n and $T+q$ are stopping times, and $S_n \leqslant T+q$. Hence, for every n, $EX(T+q) \leqslant EX(S_n) \leqslant n^{-1}$, whence $\mathbf{P}[X(T+q)=0]=1$. The rest is obvious. □

The next result has been very influential in shaping the development of Markov-process theory, particularly in connection with the problem of determining the 'correct' hypotheses for that theory.

(78.2) THEOREM (Meyer). *Let $\{\mathscr{F}_t\}$ satisfy the usual conditions. For each $n \in \mathbb{N}$, let X^n be an R-supermartingale relative to $\{\mathscr{F}_t\}$. Suppose that the sequence $(X^n : n \in \mathbb{N})$ is increasing in that*

$$X^n(t,\omega) \leqslant X^{n+1}(t,\omega), \quad \forall (n,t,\omega).$$

Define

$$X(t,\omega) := \sup_n X^n(t,\omega) \leqslant \infty.$$

Then $\mathbf{P}$-almost all paths of the $(\infty,\infty]$-valued process X are R-functions.

The original proof in Meyer [1] remains the simplest. There is a nice, more sophisticated, proof in Getoor [1].

79. Exercises. Substantial hints for most of the exercises are given at the end of the section.

(*E79.63*) Let B be a path-continuous $BM(\mathbb{R})$. Show that $X_t := e^{\frac{1}{2}t} \cos B_t$ and $Y_t := e^{\frac{1}{2}t} \sin B_t$ define martingales X and Y relative to the natural filtration of B. Thus (X, Y) is a martingale in $\mathbb{R}^2$ and $X_t^2 + Y_t^2 = e^t$. This is behaviour very different from that of continuous martingales on $\mathbb{R}$, which, as we shall see, all resemble Brownian motion on $\mathbb{R}$.

(*E79.64*) *Empirical distributions.* Let $n \in \mathbb{N}$. Let $X_1, X_2, \ldots, X_n$ be independent random variables each with the uniform distribution on $[0,1]$. For $0 \leqslant t < 1$, define

$$G_n(t) := n^{-1} \# \{k \leqslant n : X_k \leqslant t\}, \quad A_n(t) := n^{\frac{1}{2}}[G_n(t) - t].$$

Let $\mathscr{G}_n(t) := \sigma(G_n(s) : s \leqslant t)$. Prove that

$$M_n(t) := \frac{A_n(t)}{1-t}, \quad B_n(t) := A_n(t) + \int_0^t \frac{A_n(s)}{1-s}\, ds,$$

$$V_n(t) := B_n(t)^2 - G_n(t),$$

define martingales M_n, B_n and V_n with parameter set $[0,1)$ relative to $\{\mathscr{G}_n(t):$

$t \in [0, 1)\}$. Which of these martingales are UI on $[0, 1)$? Note that if we set $M(t) := 0$ and $\mathscr{G}_t = \sigma(X_1, \ldots, X_n)$ for $1 \leqslant t < \infty$ then M_n is a supermartingale $\mathscr{L}^1$ convergent at ∞, but is not UI.

Note. It is intuitively clear that, as $n \to \infty$, we should have, in some sense,

$$G_n(t) \to t, \quad B_n \to B, \quad A_n \to A$$

(on parameter set $[0, 1)$), where B is a Brownian motion and A is a Brownian bridge.

(*E79.66a*) Test your understanding of the regularisation results by proving the following 'L-analogues'. Part (iii) is particularly instructive. We shall say that $x : \mathbb{R}^+ \to \mathbb{R}$ is an *L-function* if

$$x_t = \lim_{s \uparrow\uparrow t} x_s \quad (\forall t > 0), \qquad x_{t+} := \lim_{u \downarrow\downarrow t} x_u \quad \text{exists in } \mathbb{R} \quad (\forall t \geqslant 0).$$

(i) If $y : \mathbb{Q}^+ \to \mathbb{R}$ is regularisable, then

$$z_t := \begin{cases} \lim_{q \uparrow\uparrow t} y_q & \text{if } t > 0, \\ y_0 & \text{if } t = 0 \end{cases}$$

defines an L-function z.

(ii) If $\{Y_t : t \in \mathbb{R}^+\}$ is a supermartingale carried by $(\Omega, \mathscr{G}, \mathbf{P}; \{\mathscr{G}_t : t \in \mathbb{R}^+\})$ then the set G of those ω for which $q \mapsto Y_q(\omega)$ is regularisable is in $\mathscr{G}$, and $\mathbf{P}(G) = 1$. If we set

$$Z_t(\omega) := \begin{cases} \lim_{q \uparrow\uparrow t} Y_q & \text{if } t > 0, \\ y_0 & \text{if } t = 0 \end{cases}$$

then Z is an L-process.

(iii) Define $\mathscr{G}_{t-} := \sigma(\mathscr{G}_s : s < t)$ for $t > 0$ and $\mathscr{G}_{0-} := \mathscr{G}_0$, and set $\mathscr{J}_t := \sigma(\mathscr{G}_{t-}, \mathscr{N}(\mathscr{G}_\infty))$, where $\mathscr{N}(\mathscr{G}_\infty)$ is the collection of $\mathbf{P}$-null sets in $\mathscr{G}_\infty$. Then Z is adapted to $\{\mathscr{J}_t\}$. Moreover, Z is a supermartingale relative to $\{\mathscr{J}_t\}$.

(iv) Z is a modification of Y if the map $t \mapsto Y_t$ is left-continuous into $\mathscr{L}^1$.

(v) In contrast to the case of R-regularisation, Z can be a modification of Y even though $t \mapsto Y_t$ is not left-continuous into $\mathscr{L}^1$. Convince yourself of this by taking an example in which

$$Y_t = B\left(\frac{t}{1-t} \wedge \tau\right) \quad (t < 1), \qquad Y_t := -1 \quad (t \geqslant 1),$$

where B is a $\mathrm{BM}_0(\mathbb{R})$ and $\tau := \inf\{u : B_u = -1\}$.

(*E79.66b*) Let $\{\mathscr{G}_t\}$ be any filtration, and define $\mathscr{H}_t := \sigma(\mathscr{G}_s : s < t)$ for $t > 0$. Prove that, for every t, $\mathscr{H}_{t+} = \mathscr{G}_{t+}$.

(*E79.67a*) Let $(\Omega, \mathscr{F}, \mathbf{P})$ be a triple in which $\mathscr{F}$ is $\mathbf{P}$-complete. Let $\mathscr{N}$ be the collection of all $\mathbf{P}$-null sets in $\mathscr{F}$. Let $\mathscr{K}$ be a sub-σ-algebra of $\mathscr{F}$. Prove that

$$L := \sigma(\mathscr{K}, \mathscr{N}) = \{U \in \mathscr{F} : U \Delta K \in \mathscr{N} \text{ for some } K \in \mathscr{K}\} =: R.$$

Note that, since $\mathscr{K} \cup \mathscr{N} \subseteq R \subseteq L$, you need only prove that R is a σ-algebra.

(*E79.67b*) Prove Lemma 67.4.

(*E79.71a*) *Likelihood-ratio martingales.* Let $(\Omega, \mathscr{G}, \{\mathscr{G}_t\})$ be a filtered space. Suppose that $\mathbf{P}$ and $\mathbf{Q}$ are probability measures on $(\Omega, \mathscr{G})$ such that $\mathbf{Q}$ is absolutely continuous relative to $\mathbf{P}$ on every $\mathscr{G}_t$ (though not necessarily on all of $\mathscr{G}$). Define M_t to be a version of the Radon–Nikodým derivative:

$$M_t := \left(\frac{d\mathbf{Q}}{d\mathbf{P}} \text{ on } \mathscr{G}_t\right), \quad \text{a.s.}$$

Prove that M is a $(\mathbf{P}, \{\mathscr{G}_t\})$ martingale.

(i) Let $(\Omega, \mathscr{G}, \{\mathscr{G}_t\}) = (C, \mathscr{A}, \{\mathscr{A}_t\})$, the canonical filtered space for continuous processes on $\mathbb{R}$. Let $\mathbf{P}$ be the Wiener law of BM_0, and let $\mathbf{Q}$ be the law of Brownian motion with drift c starting at 0. Prove that

$$\left(\frac{d\mathbf{Q}}{d\mathbf{P}} \text{ on } \mathscr{G}_t\right) = \exp(c\pi_t - \tfrac{1}{2}c^2 t).$$

(ii) Now let Ω be the space of all right-continuous functions $w: \mathbb{R}^+ \to \mathbb{Z}^+$ such that $w(0) = 0$ and that w is constant except for a series of upward jumps of size 1. Let

$$\pi_t(w) := w(t), \qquad \mathscr{G}_t := \sigma(\pi_s : s \geq 0), \qquad \mathscr{G}_t := \sigma(\pi_s : s \leq t).$$

Let λ and μ be positive constants. Let $\mathbf{P}$ (respectively $\mathbf{Q}$) be the law of the Poisson process of rate λ (respectively μ). Calculate $d\mathbf{Q}/d\mathbf{P}$ on $\mathscr{G}_t$.

(*E79.71b*) *Projecting onto smaller filtrations.* Let $(\Omega, \mathscr{G}, \mathbf{P}; \{\mathscr{G}_t\})$ be a filtered space carrying a supermartingale Y. Suppose that $\{\mathscr{H}_t\}$ is a smaller filtration in that $\mathscr{H}_t \subseteq \mathscr{G}_t$ for every t. Prove that

$$Z_t := \mathbf{E}(Y_t \mid \mathscr{H}_t), \quad \text{a.s.},$$

defines a supermartingale Z relative to $\{\mathscr{H}_t\}$.

You have already seen at an intuitive level how this applies to give Azéma's martingale in Section 37. Here is another application.

Let $X_t := B_t + ct$, where B is a $\mathrm{BM}_0(\mathbb{R})$ and $c > 0$. Let $\{\mathscr{H}_t\}$ be the natural filtration of B (equivalently of X). Let $\sigma := \sup\{t : X_t = 0\}$, with the convention that $\sup \emptyset = 0$. Prove that $Z_t := \mathbf{P}(\sigma > t \mid \mathscr{H}_t)$ defines a supermartingale Z relative to $\{\mathscr{H}_t\}$. Obtain the explicit formula $Z_t = \exp(-2cX_t^+)$, a.s. Deduce that

$$(*) \qquad \mathbf{P}(\sigma \in dt) = c(2\pi t)^{-\frac{1}{2}} \exp(-\tfrac{1}{2}c^2 t)\, dt.$$

Give an immediate direct proof of (*): '*And the last shall be first*'. (*Note.* Excursion theory allows one to write down formulae analogous to (*) in very general situations.)

(*E79.71c*) This exercise gives a simple proof of Lévy's descriptions of local time in (I.14.7) and (37.9). Suppose that Λ is a Poisson measure on $[0, \infty) \times (0, \infty)$ with intensity measure Leb $\times\, \nu$, where $\nu \neq 0$, and where

$$c(r) := \lim_{\varepsilon \downarrow 0} \frac{\nu(\varepsilon, \infty)}{\nu(r\varepsilon, \infty)} \quad \text{exists for } \tfrac{1}{2} \leqslant r \leqslant 1$$

and defines a strictly increasing continuous function c on $[\frac{1}{2}, 1]$. Set

$$M(t, \varepsilon) := \nu(\varepsilon, \infty)^{-1} \Lambda([0, t] \times (\varepsilon, \infty)) - t.$$

Show that $M(\cdot, \varepsilon)$ is a martingale, and that, for $\frac{1}{2} \leqslant r < 1$, it is almost surely true that $M(\cdot, r^n) \to 0$ as $n \to \infty$, uniformly on compact t-intervals. Deduce that

almost surely, $M(t, \varepsilon) \to 0$ uniformly on compact t-intervals.

(*E79.71d*) Read Section IV.2 of Volume 2, and make the sketched proof of Lévy's quadratic-variation result in that section rigorous. (Volume 2 does give rigorous proofs of much more general results.)

(*E79.73*) Prove Lemma 73.8.

(*E79.77a*) Let M be a continuous martingale such that

$$\mathbf{P}\left(\sup_t M_t = \infty, \inf_t M_t = \infty\right) = 1.$$

Define

$$T(0) := 0, \qquad T(n) := \inf\{t > T(n-1) : |M_t - M_{T(n-1)}| = 1\}.$$

Prove that the law of $\{M_{T(n)} : n \in \mathbf{Z}^+\}$ is the (Markovian!) law of Simple Random Walk.

(*E79.77b*) Let M be a continuous martingale such that

$$\mathbf{P}(\sup M_t \geqslant 1) = \mathbf{P}(\inf M_t \leqslant -1) = 1.$$

Define $U_t := \sup_{s \leqslant t} M_s$, $L_t := \inf_{s \leqslant t} M_s$ and

$$T := \inf\{r : U_r - L_r = 1\}.$$

What is the distribution of M_T? (This question arose, and was solved, in work by LCGR on financial problems. We now have many solutions.)

(*E79.77c*) *The Helms–Johnson example.* This is a first look at one of the most celebrated counterexamples in the subject, one to which we return in Sections III.31, IV.14 and VI.33.

Suppose that X is a UI supermartingale, and that

(i) $$X = M - A,$$

where M is a martingale and A a process with non-decreasing paths. Show that $\mathbf{E}(A_\infty) < \infty$, and deduce that M is UI. Prove that X *is of class (D)* in that the family

$$\{X_T : T \text{ a stopping time}\}$$

is UI.

Let $\mathbf{B}$ be a $BM_0(\mathbb{R}^3)$, and define the Helms–Johnson process

(ii) $$X_t := 1/|\mathbf{B}_{t+1}|.$$

Prove that X is a UI supermartingale relative to its own natural filtration, but that X is not of class (D), so that X does not have a 'Doob decomposition' as (i). *Hints.* Show that

(iii) $$\mathbf{E}(X_t^2) = 1/(t+1).$$

If $T_n := \inf\{t : X_t \geqslant n\}$ then, by Corollary I.18.3 (which we assume),

(iv) $$\mathbf{P}(T_n < \infty \,|\, X_0) = \min(n^{-1}X_0, 1).$$

(*E79.77d*) *More on likelihood-ratio martingales.* Let $(\Omega, \mathscr{F}, \mathbf{P}; \{\mathscr{F}_t\})$ satisfy the usual conditions. Martingales, a.s., etc. are defined relative to this setup. Let $\mathbf{Q}$ be a probability measure on $(\Omega, \mathscr{F})$ such that $\mathbf{Q} \ll \mathbf{P}$ on every $\mathscr{F}_t$. Choose a right-continuous martingale M with $d\mathbf{Q}/d\mathbf{P} = M_t$ on every $\mathscr{F}_t$. Show that if T is an almost surely finite stopping time such that $\{M_{T\wedge t} : t \geqslant 0\}$ is UI then $\mathbf{Q} \ll \mathbf{P}$ on $\mathscr{F}_T$ and

$$\frac{d\mathbf{Q}}{d\mathbf{P}} = M_T, \quad \text{a.s. on } \mathscr{F}_T.$$

Prove Reuter's Theorem that if $\mathbf{X}$ is a Brownian motion on $\mathbb{R}^n$ with constant drift vector $\mathbf{c}$ and started at 0, and if $T := \inf\{t : |\mathbf{X}_t| = 1\}$ is the hitting time of the unit sphere, then T and $\mathbf{X}_T$ are independent variables. Why does this not conflict with intuition?

Hints for selected exercises

(*H79.63*) We know that $\exp(i\theta B_t + \frac{1}{2}\theta^2 t)$ is a martingale with values in $\mathbb{C}$. Now take $\theta = 1$.

(*H79.64*) For $0 \leqslant t \leqslant u \leqslant 1$,

$$\mathbf{E}(G_n(u) | \mathscr{G}_n(t)) = G_n(t) + [1 - G_n(t)]\frac{u-t}{1-t}.$$

This rearranges to say that M_n is a martingale. The real reason that B_n is a martingale is the SDE

$$dB_n = (1-t)\,dM_n.$$

But we can prove the property directly: for $0 \leqslant t \leqslant u \leqslant 1$,

$$\mathbf{E}(B_n(u)|\mathscr{G}_n(t)) = \frac{A_n(t)(1-u)}{1-t} + \int_0^t \frac{A_n(s)}{1-s}\,ds + \int_t^u \frac{A_n(t)}{1-t}\,ds.$$

(Think briefly on how to justify this step.) There is an SDE reason for the martingale property of V_n too, but you can obtain this directly using the variance of a binomial distribution.

Of course, $M_n(1-) = 0$, a.s., so M_n cannot be UI on $[0,1)$: if it were, we would have $M_n(t) = \mathbf{E}(M_n(1-)|\mathscr{G}_t) = 0$, a.s. We know that $\mathbf{E}\{B_n(t)^2\} = \mathbf{E}G_n(t) = t$, so that B_n is $\mathscr{L}^2$ bounded and therefore UI. A bit more calculation shows that V_n is also $\mathscr{L}^2$ bounded.

(*H79.66a*) (Hint for Part (iii), which is the tricky bit.) Fix $u > 0$. For $0 \leqslant r < u$, with $r \in \mathbb{Q}$, write

$$V_r := Y_r - \mathbf{E}(Y_u|\mathscr{G}_r).$$

Check that V is a non-negative supermartingale with parameter set $\mathbb{Q} \cap [0,u)$. Hence, for $0 \leqslant q < t < r < u$, with $q \in \mathbb{Q}$, we have

$$\mathbf{E}(V_{u-}|\mathscr{G}_q) = \mathbf{E}\left(\liminf_{r\uparrow\uparrow u} V_r|\mathscr{G}_q\right) \leqslant \liminf \mathbf{E}(V_r|\mathscr{G}_q) \leqslant V_q, \quad \text{a.s.}$$

But, by the Upward Theorem, as $r \uparrow\uparrow u$,

$$Y_r = V_r + \mathbf{E}(Y_u|\mathscr{G}_r) \to Y_{u-} = V_{u-} + \mathbf{E}(Y_u|\mathscr{G}_{u-}), \quad \text{a.s.,}$$

whence

$$\mathbf{E}(Y_{u-}|\mathscr{G}_q) = \mathbf{E}(V_{u-}|\mathscr{G}_q) + \mathbf{E}(Y_u|\mathscr{G}_q) \leqslant V_q + \mathbf{E}(Y_u|\mathscr{G}_q) = Y_q.$$

Now let $q \uparrow\uparrow t$ to get the required result that Z is a $\{\mathscr{J}_t\}$ supermartingale.

(*H79.67b*) For $u > t$, $\mathscr{F}_u$ must extend $\sigma(\mathscr{G}_u, \mathscr{N})$, whence

$$\mathscr{F}_t \supseteq \mathscr{H}_t := \bigcap_{u>t} \sigma(\mathscr{G}_u, \mathscr{N}).$$

However, $\{\mathscr{H}_t\}$ satisfies the usual conditions, so that $\mathscr{H}_t = \mathscr{F}_t, \forall t$.

Suppose now that $H \in \mathscr{H}_t$ and that $u(n) \downarrow\downarrow t$. By Exercise 79.67a, we can find $G_{u(n)}$ in $\mathscr{H}_{u(n)}$ such that $N_{u(n)} := H \Delta G_{u(n)} \in \mathscr{N}$. Now set $G := \bigcap G_{u(n)}$ and $N := \bigcup N_{u(n)}$. Then $G \in \mathscr{G}_{t+}$ and $N \in \mathscr{N}$, and, since you can easily check that $H \Delta G \subseteq N$, the proof is finished.

(*H79.71a*) For every t, let $M_t = d\mathbf{Q}/d\mathbf{P}$ on $\mathscr{G}_t$, a.s. Then, for $s < t$, we have, for

$G_s \in \mathcal{G}_s \subseteq \mathcal{G}_t$,

$$\mathbf{E}(M_t; G_s) = \mathbf{Q}(G_s) = \mathbf{E}(M_s; G_s),$$

so that M is a martingale.

(i) If $p(t; x, y)$ is the transition density function for Brownian motion, and $q_c(t; x, y)$ that for Brownian motion with drift c, then

$$q_c(t; x, y) = \exp[c(y-x) - \tfrac{1}{2}c^2 t]\, p(t; x, y),$$

and, for $n \in \mathbf{N}$, $0 = t_0 < t_1 < \cdots < t_n = t$, and $x_0, x_1, \ldots, x_n \in \mathbf{R}$ with $x_0 = 0$,

$$\mathbf{Q}\left(\bigcap_{i=1}^{n} \{\pi_{t_i} \in dx_i\}\right) = \prod_{i=1}^{n} q_c(t_i - t_{i-1}; x_{i-1}, x_i)\, dx_i = \cdots.$$

(ii) For $j, n \in \mathbf{Z}^+$, we have

$$p_\mu(t; j, j+n) = \exp(-\mu t)\frac{(\mu t)^n}{n!} = \exp[(\lambda - \mu)t]\left(\frac{\mu}{\lambda}\right)^n p_\lambda(t; j, j+n);$$

and the answer is $\exp[(\lambda - \mu)t](\mu/\lambda)^{\pi_t}$.

(*H79.71b*) The 'theory' part is just the Tower Property of conditional expectations: for $0 \leqslant s \leqslant t$,

$$\begin{aligned} \mathbf{E}(Z_t \mid \mathcal{H}_s) &= \mathbf{E}(Y_t \mid \mathcal{H}_t \mid \mathcal{H}_s) = \mathbf{E}(Y_t \mid \mathcal{H}_s) \\ &= \mathbf{E}(Y_t \mid \mathcal{G}_s \mid \mathcal{H}_s) \leqslant \mathbf{E}(Y_s \mid \mathcal{H}_s) = Z_s, \quad \text{a.s.} \end{aligned}$$

As regards the example, $Y_t := I_{\sigma > t}$ is decreasing, and is obviously a supermartingale relative to the filtration with $\mathcal{G}_t = \mathcal{H}_\infty$ for every t. Hence Z is a supermartingale. That $Z_t = \exp(-2cX_t^+)$ you know from Section I.9. The formula (*) now follows from

$$\mathbf{P}(\sigma > t) = \mathbf{E}\mathbf{E}(Y_t \mid \mathcal{H}_t) = \mathbf{E}(Z_t)$$

and some integration.

Now for the immediate proof. Let $\tilde{B}$ be the BM_0 defined as $\tilde{B}(t) = tB(1/t)$. Then

$$\begin{aligned} \sigma &= \sup\{t : B(t) + ct = 0\} = \sup\{t : t\tilde{B}(1/t) + ct = 0\} \\ &= \sup\{t : \tilde{B}(1/t) = -c\} = 1/\inf\{t : \tilde{B}(t) = -c\}. \end{aligned}$$

(*H79.71c*) Since $\mathbf{E}\{M(t,\varepsilon)^2\} = \nu(\varepsilon, \infty)^{-1} t$, the fact that $M(\cdot, r^n) \to 0$ uniformly on compact t-intervals follows from the Submartingale Inequality and the Borel–Cantelli Lemma. Interpolating for ε between r^{n+1} and r^n for r very close to 1, by using monotonicity, is left to you.

(*H79.73*) If $S_n \uparrow S$ then $\{S \leqslant t\} = \bigcap \{S_n \leqslant t\}$.

If $S_n \downarrow S$ then $\{S < t\} = \bigcup \{S_n < t\}$.

If $S_n \downarrow S$ and $G \in \bigcap \mathcal{G}_{S_n+}$ then, for every n and every t,

$$G \cap \{S_n < t\} \in \mathcal{G}_t,$$

whence $G \cap \{S < t\} = \bigcup (G \cap \{S_n < t\}) \in \mathcal{G}_t$.

(*H.79.77b*) Simon Harris found this direct solution. Suppose that $0 < x < y < 1$. If $M_T \in [x, y]$ then

(i) M hits $y - 1$ before it hits y; and
(ii) after hitting $y - 1$, it hits x before it hits $x - 1$.

Hence, by the logic of the previous question,

$$\mathbf{P}(M_T \in [x, y]) \leqslant y(y - x).$$

On the other hand, if the following three conditions hold:

(i) M hits $y - 1$ before it hits x;
(ii) after hitting $y - 1$, M hits 0 before it hits $x - 1$;
(iii) after hitting $y - 1$ and then 0, M hits y before $y - 1$;

then $M_T \in [x, y]$. Hence

$$P(M_T \in [x, y]) \geqslant \frac{x}{1 + x - y} \frac{y - x}{1 - x}(1 - y).$$

The rest is easy.

(*H.79.77c*) If X is a UI supermartingale satisfying (i), then X is bounded in $\mathcal{L}^1$, and, since

$$\mathbf{E}(A_t) = \mathbf{E}(A_0) + \mathbf{E}(X_t - X_0),$$

we must have $\mathbf{E}(A_\infty) < \infty$. But now X and A are UI, whence M is also UI, and, by Theorem 77.5, M is of class (D). Since A is trivially of class (D), it follows that X is of class (D).

In the example,

$$\mathbf{E}(X_t^2) = \int_0^\infty r^2 (2\pi t)^{-3/2} \exp\left(\frac{-r^2}{2t}\right) 4\pi r^2 \, dr = \frac{1}{t + 1},$$

so that X is bounded in $\mathcal{L}^2$, and so is UI. Clearly, $X_\infty = 0$, a.s., so that, if X is of class (D) then, for any sequence (T_n) of stopping times with $T_n \uparrow \infty$, we shall have $X(T_n) \to 0$ in $\mathcal{L}^1$. But, for the given sequence of stopping times, we have

$$\liminf \mathbf{E} X(T_n) \geqslant \liminf n \mathbf{P}(T_n < \infty) \geqslant E(X_0) > 0.$$

(*H.79.77d*) Let $F \in \mathcal{F}_T$. Then

$$F \cap \{T \leqslant t\} \in \mathcal{F}_{T \wedge t} = \mathcal{F}_t \cap \mathcal{F}_T,$$

and so

$$\mathbf{Q}(F \cap \{T \leqslant t\}) = \mathbf{E}(M_t; F \cap \{T \leqslant t\}) = \mathbf{E}(M_{T \wedge t}; F \cap \{T \leqslant t\}).$$

Now let $t \uparrow \infty$ to get

$$\mathbf{Q}(F \cap \{T < \infty\}) = \mathbf{E}(M_T : F \cap \{T < \infty\}).$$

For the proof of Reuter's Theorem see IV.39.6 in Volume 2.

6. PROBABILITY MEASURES ON LUSIN SPACES

This Part consists of two main themes: 'weak' convergence of measures, and existence of regular conditional probabilities. These themes are linked via their use of *inner regularity of measures relative to compact sets.* In the Daniell–Kolmogorov Theorem 31.1, we were able to utilize this inner regularity by making the assumption that the state-space E was Lusin. Now we are going to make this assumption on the sample space Ω. The product sample space from the DK Theorem is useless for this purpose, but we have now learnt that we can work with the space of continuous paths or that of R-paths, both of which spaces are Lusin (and even Polish).

We present—fully, and in the simplest way we could devise—all the 'theory' of 'weak convergence', and how it applies to the space of *continuous* paths, the only case that we shall need. See 'Pointers to the main results' below. For the way in which the theory applies to the space of R-functions with the Skorokhod topology, you will have to see the excellent books by Billingsley [2] Parthasarathy [1] and Ethier and Kurtz [1].

Let $X_1, \ldots, X_n$ be independent identically distributed real-valued random variables each with mean 0 and variance 1. Let μ_n be the law of

$$Y_n := n^{-1/2}(X_1 + \cdots + X_n).$$

The Central Limit Theorem tells us that μ_n converges 'weakly' to the law μ of a standard normal $\mathrm{N}(0, 1)$ random variable. If, for example, each X is $+1$ or -1 with probability $\frac{1}{2}$ each, and if $\mathbf{A}$ denotes the set of algebraic numbers, then

$$\mu_n(\mathbf{A}) = \mathbf{P}(Y_n \in \mathbf{A}) = 1 \not\to 0 = \mu(\mathbf{A}).$$

For which Borel sets do we have $\mu_n(B) \to \mu(B)$? How do we formulate 'weak convergence' generally so that it will apply, for example, to the Donsker Invariance Principle of Section I.8 and allow us to derive interesting consequences (look ahead to (84.7)) from it? These are among the questions that we shall address. We use weak convergence to obtain *existence* of solutions to martingale problems in Section V.23 of Volume 2.

Regular conditional probabilities provide an important language for probability theory and an important technique for martingale problems etc. See the proof that solutions of martingale problems are Markovian in Section V.21 in Volume 2.

We begin by recalling the Stone–Weierstrass Theorem and the Riesz Representation Theorem, results that we shall also need in Markov-process theory. For the case where J is a compact metrisable space, we shall put in explicit form (also useful later) the fact that the set $\Pr(J)$ of probability measures on $(J, \mathscr{B}(J))$ is again a compact metrisable space. Results for the case in which we are most interested, that of probability measures on a Polish space S (a space homeomorphic to a complete separable metric space), will be deduced by embedding S as a Borel subset of a compact metrisable space K. Therefore the only property of S that we need is that S is a Lusin space (a space homeomorphic to a Borel subset of a compact metrisable space!). Since Lusin spaces occur frequently, we use the 'Lusin' hypothesis—but the main reason for doing so is that it makes things *easier*!

The vexed question of terminology. There have always been conflicts of terminology in this area. Probabilists have always used 'weak convergence' for something close to functional analysts' 'weak* convergence' (and very different from functional analysts' 'weak convergence'). Since our treatment depends crucially on the fact that, in functional analysts' language, 'the unit ball in the dual space is weak* compact', and since we rely on functional analysis more in this part of the chapter than elsewhere in the book, we use terminology consistent with that of functional analysis while 'doing the work'. Having made everything clear (we hope), we shall then, from a clearly signalled point on, regress to probabilists' terminology.

Note on functional analysis. Through no fault of their own, many research students these days are less familiar with functional analysis than students were when the first edition of this book was published. We therefore include a more systematic linking of our results with those in the functional-analysis texts. We emphasize that one does not know *a priori* that, for a Polish space S, the $C_b(S)$ topology on $\Pr(S)$ is metrisable. This is why we are forced to use nets rather than sequences.

Pointers to the main results. The 'weak', or $C_b(S)$, topology on the set $\Pr(S)$ of probability measures on a Lusin space S is defined in (83.1). This topology is shown to be metrisable in (83.7). Prohorov's sufficient condition (also necessary when S is Polish) for conditional compactness of subsets of $\Pr(S)$ is given in (83.10). The Continuous-Mapping Principle is given in (84.2); and Skorokhod's Representation Theorem (which gives a clear picture of the Continuous-Mapping Theorem and of much else) is found in Section 86. It is shown in Section 82 that, with the topology of uniform convergence on compacts, the space $W = C([0,\infty);\mathbb{R})$ is Polish, with the usual algebra of σ-cylinders as its Borel σ-algebra. Prohorov's theorem on 'weak' compactness is translated into practicable form for $\Pr(W)$ in Section 85. Finally, the relation between 'weak' convergence and convergence of finite-dimensional distributions for W is explained in Section 87.

'Weak convergence'

80. $C(J)$ and $\Pr(J)$ when J is compact Hausdorff. *Let J be a compact Hausdorff space.* This is a standard setting in functional analysis. We recall various fundamental results, for which D&S (that is, Dunford and Schwartz [1]), remains a superb reference. (Nowadays, you have to read 'sphere' there as 'ball'.)

Let $C(J)$ denote the Banach algebra of continuous (and necessarily bounded) real-valued functions on J with the usual supremum norm.

(80.1) THEOREM *(Stone–Weierstrass Theorem, D&S IV.6.16). Let A be a subalgebra of $C(J)$ that contains constant functions and separates points of J: for $x \in J$, there exist elements f and g in A such that $f(x) \neq g(x)$. Then A is dense in $C(J)$.*

(80.2) DEFINITION ($\Pr(J)$, inner regularity *for a single measure). Let* $\Pr(J)$ *denote the set of probability measures on $(J, \mathscr{B}(J))$. An element μ of* $\Pr(J)$ *is called* inner regular *if, for every $B \in \mathscr{B}(J)$,*

$$\mu(B) = \sup\{\mu(K) : K \text{ compact}, K \subseteq B\}.$$

(80.3) THEOREM (Riesz Representation Theorem, D&S IV.6.3). *Let ϕ be a linear increasing functional $\phi: C(J) \to \mathbb{R}$ such that $\phi(\mathbf{1}) = 1$. Then there exists a unique inner regular element μ of* $\Pr(J)$ *such that*

$$\phi(f) = \mu(f) = \int_J f \, d\mu.$$

Of course, $\mathbf{1}$ denotes the constant function equal to 1 on J; and 'ϕ is increasing' means that $f \leqslant g$ on J implies that $\phi(f) \leqslant \phi(g)$. The hypotheses on ϕ force ϕ to be a bounded linear functional of norm 1: $\phi \in C(J)^*$.

(80.4) Discussion. What is going on here? Why the mysterious inner regularity? The answer is that the smallest σ-algebra on J with respect to which all continuous functions are measurable, the so-called *Baire σ-algebra on J*, may well be smaller than $\mathscr{B}(J)$. A probability measure on the Baire σ-algebra has a unique extension to an inner regular element of $\Pr(J)$. *Does this matter to probabilists*? Yes, it does. These ideas can be used in a very illuminating alternative approach to the proof of the Daniell–Kolomogorov Theorem and

to a study of its limitations and how to deal with them. Nelson [1] deserves much of the credit for this. See also Meyer [1] and Tjur [1].

(80.5) *The C(J), or weak*, topology on C(J)**. The $C(J)$ topology on the space $C(J)^*$ of all bounded linear functionals on the Banach space $C(J)$ is obtained by making sets of the form

(80.6) $$\{\phi \in C(J)^* : |\phi(f_i) - \phi_0(f_i)| < \varepsilon, 1 \leqslant i \leqslant n\}$$

a basis for neighbourhoods of the point ϕ_0 in $C(J)^*$. Note that this topology is automatically Hausdorff.

We can study general topology in much the same way as we studied elementary metric topology provided that we replace convergence of sequences by convergence of *nets* (generalised sequences). See D&S I.7. A *directed set D* is a partially ordered set for which every finite subset has an upper bound in D. A net is a family $(x_\alpha : \alpha \in D)$ parametrised by some directed set D. If $(x_\alpha : \alpha \in D)$ is a net of points in a topological space E, we say that $x_\alpha \to x$ if, for every open set G containing x, there exists α_0 in D such that $x_\alpha \in G$ whenever $\alpha \geqslant \alpha_0$.

The $C(J)$ topology of $C(J)^*$ may therefore be described by saying that a net (ϕ_α) of elements of $C(J)^*$ converges to the element ϕ of $C(J)^*$ if and only if

(80.7) $$\phi_\alpha(f) \to \phi(f) \quad \textit{for all } f \textit{ in } C(J).$$

For $f \in C(J)$, the statement that $\phi_\alpha(f) \to \phi(f)$ means that, given $\varepsilon > 0$, there exists an element α_0 in D such that

$$|\phi_\alpha(f) - \phi(f)| < \varepsilon \quad \text{whenever } \alpha_0 \leqslant \alpha.$$

(*80.8*) THEOREM (Alaoglu, D&S V.4.2). *The unit ball*

$$\{\phi \in C(J)^* : \|\phi\| \leqslant 1\}$$

in $C(J)^*$ is compact in the $C(J)$ topology.

In effect, this is just Tychonov's Theorem. We can identify the set of inner regular probability measures on $\Pr(J)$ with the closed set of elements $\phi \in C(J)^*$ such that ϕ is increasing and maps **1** to 1. Hence the inner regular elements of $\Pr(J)$ form a compact set in the $C(J)$ topology.

81. $C(J)$ and $\Pr(J)$ when J is compact metrisable. *In this section, we assume that J is a compact metrisable space.* Things become much simpler. It is difficult to overstate the importance of the following result.

(*81.1*) THEOREM. *Every element of* $\Pr(J)$ *is inner regular.*

Note. The Baire and Borel σ-algebras on J agree.

Proof. Let $\mu \in \Pr(J)$. Let $\mathscr{A}$ be the class of B in $\mathscr{B}(J)$ such that, for every $\varepsilon > 0$, there exist a compact set K with $K \subseteq B$ and an open set G with $G \supseteq B$ such that

$$(81.2) \qquad \mu(B \backslash K) < \varepsilon, \quad \mu(G \backslash B) < \varepsilon.$$

It is immediate that the complement of a set in $\mathscr{A}$ is again in $\mathscr{A}$. Suppose that B_1 and B_2 are in $\mathscr{A}$, and let $B := B_1 \cap B_2$. Let $\varepsilon > 0$. For $i = 1, 2$, choose a compact K_i and an open G_i with

$$\mu(B_i \backslash K_i) < \tfrac{1}{2}\varepsilon, \quad \mu(G_i \backslash B_i) < \tfrac{1}{2}\varepsilon.$$

Then if $K := K_1 \cap K_2$ and $G := G_1 \cap G_2$, we have (81.2). By this stage we know that $\mathscr{A}$ is an algebra.

Now let (B_n) be an increasing sequence of elements of $\mathscr{A}$ with union B. For each n, choose a compact $K_n \subseteq B_n$ and an open G_n with $G_n \supseteq B_n$ with

$$\mu(B_n \backslash K_n) < \varepsilon 2^{-n-1}, \quad \mu(G_n \backslash B_n) < \varepsilon 2^{-n}.$$

Then $G := \bigcup G_n$ is open and $\mu(G \backslash B) < \varepsilon$. The set $L := \bigcup K_n$ satisfies $\mu(B \backslash L) < \frac{1}{2}\varepsilon$, but L need not be compact. However, for some N, $K := \bigcup_{n \leq N} K_n$ is compact with $\mu(K) > \mu(L) - \frac{1}{2}\varepsilon$; and we have proved (81.2). We now know that $\mathscr{A}$ is a σ-algebra.

If K is a closed (hence compact) subset of J then K is the intersection of the sequence (G_n) of open sets:

$$K = \bigcap G_n, \quad G_n := \{x \in J : \rho(x, K) < n^{-1}\},$$

where ρ is the metric on J. Hence every closed set is in $\mathscr{A}$, and $\mathscr{A} = \mathscr{B}(J)$. □

(*81.3*) THEOREM. *$C(J)$ is separable, and* $\Pr(J)$ *in its $C(J)$ topology is compact metrizable.*

Proof. Let (x_n) be a countable dense subset of J (why is there such a set?), and define

$$h_k(x) := \rho(x, x_k).$$

Note that the functions h_k separate points of J.

Let A be the collection of functions in $C(J)$ that are finite sums of the form

$$q1 + \sum q(k_1, \ldots, k_r; n_1, \ldots, n_r) h_{k_1}^{n_1} \cdots h_{k_r}^{n_r},$$

where q and the $q(\cdot, \cdot, \ldots)$ are rational constants. Then the closure of A is an algebra containing all constant functions and separating points of J. By the Stone–Weierstrass Theorem, A is dense in $C(J)$. Since A is countable, $C(J)$ is a separable metric space.

Let (f_n) be a countable dense subset of $C(J)$. Consider the map

$$(81.4) \qquad \Pr(J) \ni \mu \mapsto (\mu(f_1), \mu(f_2), \ldots) \in V := \prod [-\|f_n\|, \|f_n\|].$$

This map is clearly one-one (why?). Moreover, for a net (μ_α) in $\Pr(J)$,

$$\mu_\alpha(f)\to\mu(f),\ \forall f\in C(J), \quad \text{if and only if} \quad \mu_\alpha(f_n)\to\mu(f_n),\ \forall n.$$

Hence the map is (81.4) is a homeomorphism, and $\Pr(J)$ is homeomorphic to a compact subset of the metrizable space V. Note that the Riesz Representation Theorem is still necessary to obtain the fact that the image of $\Pr(J)$ in V is closed. □

82. Polish and Lusin spaces. We begin by recalling the definitions.

(82.1) DEFINITION (Polish space; Lusin space). *Let S be a topological space. Then S is called a* Polish space *if the topology of S arises from a metric with respect to which S is complete. The space S is called a* Lusin space *if S is homeomorphic to a Borel subset of a compact metric space J.*

(82.2) *The space $(W, \mathscr{A})$.* By far the most important example for us is the case when S is the space

$$W := C([0,\infty),\mathbb{R})$$

of all continuous functions on $\mathbb{R}$. This is the path space for (1-dimensional) Brownian motion and diffusions.

(82.3) LEMMA. *In the topology of uniform convergence on compact sets, W is a Polish space, and the σ-algebra of σ-cylinders,*

$$\mathscr{A} := \sigma(\pi_t : t\geqslant 0), \qquad \pi_t(w) := w(t) \quad (w\in W),$$

is the Borel σ-algebra $\mathscr{B}(W)$ on W.

Proof. A suitable metric on W is given by

$$\rho(w_1,w_2) := \sum 2^{-n}\rho_n(w_1,w_2)[1+\rho_n(w_1,w_2)]^{-1},$$

where

$$\rho_n(w_1,w_2) := \sup_{t\in[0,n]} |w_1(t)-w_2(t)|.$$

That W is complete and separable follows from the corresponding result for $C([0,n])$.

That $\mathscr{A}\subseteq\mathscr{B}(W)$ follows because each π_t is a continuous map from W to $\mathbb{R}$.

Next note that if $w_1\in W$ then

$$\rho_n(w,w_1) = \sup_{q\in\mathbb{Q}\cap[0,n]} |\pi_q(w)-\pi_q(w_1)|,$$

so that each ρ_m, and therefore also ρ, is $\mathscr{A}$-measurable. If F is a closed subset of W and $\{w_n\}$ is a countable dense subset of F then

$$F = \{w \in W : \inf_n \rho(w, w_n) = 0\},$$

and $F \in \mathscr{A}$. Thus $\mathscr{A} = \mathscr{B}(W)$.

(82.4) LEMMA. *If S is a Lusin space then every probability measure on S is inner regular.*

Proof. This is an immediate consequence of Theorem 81.1. A probability measure μ on $(S, \mathscr{B}(S))$ has a canonical extension to a probability measure on $(J, \mathscr{B}(J))$ with $\mu(J \backslash S) = 0$. □

(82.5) THEOREM. *A topological space S is Polish if and only if it is homeomorphic to a G_δ subset (countable intersection of open sets) of a compact metric space J. In particular, every Polish space is a Lusin space.*

The 'if' part is not necessary for us. You can find it in Section 6, No. 1, Theorem 1 of Bourbaki [1]. Our proof of the 'only if' part is an extended version of one found there.

Proof of the 'only if' part. Let S be our Polish space. We prove that S may be embedded as a G_δ subset of the compact metrisable space $J := [0, 1]^{\mathbf{N}}$. This result is so important for us that we provide every detail of the proof.

Step 1: Put $\hat{\rho} := \rho/(1 + \rho)$. Then $\hat{\rho}$ is also a metric under which S is complete and separable, and $0 \leqslant \hat{\rho} \leqslant 1$. Choose a countable dense subset $\{x_n : n \in \mathbf{N}\}$ of S, and let α be the map $\alpha: S \to J$ defined as follows:

$$\alpha(x) := (\hat{\rho}(x, x_1), \hat{\rho}(x, x_2), \ldots).$$

Let us prove that α is a homeomorphism of S to $\alpha(S)$. We need only show that if $(x(n))$ is a sequence of elements of S and $x \in S$, then the statements

$$x(n) \to x, \tag{82.6}$$

$$\hat{\rho}(x(n), x_k) \to \hat{\rho}(x, x_k) \quad \text{for every } k \tag{82.7}$$

are equivalent. Since each $\hat{\rho}(\cdot, x_k)$ is continuous on S, it is immediate that (82.6) implies (82.7). Suppose now that (82.7) holds. Since

$$\hat{\rho}(x(n), x) \leqslant \hat{\rho}(x(n), x_k) + \hat{\rho}(x_k, x),$$

we have from (82.7)

$$\limsup \hat{\rho}(x(n), x) \leqslant 2\hat{\rho}(x, x_k), \quad \forall k.$$

Now let $x_k \to x$ to see that $\hat{\rho}(x(n), x) \to 0$.

Step 2: For the moment fix $x \in S$. Let d be a metric giving the topology of J.

Since α^{-1} is continuous on $\alpha(S)$ at $\alpha(x)$, for $\varepsilon > 0$, we can find $\delta(\varepsilon) > 0$ such that $y \in S$ and $d(\alpha(x), \alpha(y)) < \delta(\varepsilon)$ imply that $\hat{\rho}(x, y) < \varepsilon$. In particular, if $n \in \mathbf{N}$, then, taking $\varepsilon = (2n)^{-1}$ and $\delta = \min(\delta(\varepsilon), \varepsilon)$, we see that if $B_{J,d}(\alpha(x), \delta)$ is the open ball in J of d-radius δ, then $B_{J,d}(\alpha(x), \delta)$ has d-diameter at most $1/n$ and $\alpha(S) \cap B_{J,d}(\alpha(x), \delta)$ has $\hat{\rho}$-diameter at most $1/n$.

Step 3: Now think of S as identified with $\alpha(S)$ sitting on J. For $n \in \mathbf{N}$ and $x \in \bar{S}$, the closure of S in J, put x in U_n if x has a J-neighbourhood $N_{x,n}$ of d-diameter less than $1/n$ such that the $\hat{\rho}$-diameter of $S \cap N_{x,n}$ is also less than $1/n$. We have already proved at Step 2 that $U_n \supseteq S$. Now we claim that U_n is open in $\bar{S}$. So suppose that $x \in U_n$ with $N_{x,n}$ as above. Any $z \in \bar{S}$ sufficiently close to x in the d-metric will also belong to $N_{x,n}$, and we can take $N_{z,n} = N_{x,n}$. Hence U_n *is* open in $\bar{S}$.

Suppose that $x \in \bigcap_n U_n$. For each n, pick a point x_n of S (remember that $x \in \bar{S}$!) in $\bigcap_{k \leqslant n} N_{x,k}$. Then $d(x, x_n) \leqslant 1/n$, so that $x_n \to x$ in (J, d). But also, for $r \geqslant n$, the points x_r and x_n are in $N_{x,n}$, so that $\hat{\rho}(x_r, x_n) < 1/n$. Hence (x_r) is a Cauchy sequence in $(S, \hat{\rho})$, a complete metric space. Hence, for some $x_0 \in S, x_n \to x_0$ in $(S, \hat{\rho})$. But, since α is a homeomorphism, we must also have $x_n \to x_0$ in J. Thus $x = x_0 \in S$. Since U_n is open in $\bar{S}$, we must have $U_n = \bar{S} \cap V_n$, where V_n is open in J. Hence

$$S = \bigcap_n U_n = \bar{S} \cap \left(\bigcap_n V_n \right).$$

To show that S is a G_δ in J, we need only show that $\bar{S}$ is a G_δ in J; and this is obvious because

$$\bar{S} = \bigcap_n \{ y \in J : d(y, \bar{S}) < 1/n \}. \qquad \square$$

83. The $C_b(S)$ topology of Pr(S) when S is a Lusin space; Prohorov's Theorem. *Throughout this important section. S denotes a Lusin space, so that S is a Borel subset of a compact metric space* (J, ρ).

(*83.1*) DEFINITION (the $C_b(S)$ topology of Pr(S)). *We denote the space of bounded continuous functions on S by* $C_b(S)$. *We denote by* Pr(S) *the set of probability measures on* $(S, \mathscr{B}(S))$. *If* (μ_α) *is a net of elements of* Pr(S) *and* $\mu \in \Pr(S)$, *we say that* μ_α converges to μ in the $C_b(S)$ topology *if*

$$\mu_\alpha(f) \to \mu(f) \quad \text{for all } f \in C_b(S).$$

A basis of neighbourhoods of the point $\mu_0 \in \Pr(S)$ *is therefore provided by sets of the form*

$$\{ \mu \in \Pr(S) : |\mu(f_i) - \mu_0(f_i)| < \varepsilon_i, \quad 1 \leqslant i \leqslant n \},$$

where $n \in \mathbf{N}$, *each* $f_i \in C_b(S)$ *and each* $\varepsilon_i > 0$

The most obvious way to guarantee $C_b(S)$ convergence of *sequences* of elements in $\Pr(S)$ is as follows.

(83.2) LEMMA. *Suppose that* (X_n) *is a sequence of* $(S, \mathscr{B}(S))$*-valued random variables on some probability triple* $(\Omega, \mathscr{F}, \mathbf{P})$, *and that* $X_n \to X$, *a.s. Then the law* μ_n *of* X_n *converges to the law* μ *of* X *in the* $C_b(S)$ *topology. The same conclusion holds if we only have* $X_n \to X$ *in probability in that, for every* $\varepsilon > 0$,

$$\mathbf{P}[\rho(X_n, X) > \varepsilon] \to 0 \quad \text{as } n \to \infty.$$

Proof. First assume almost sure convergence. Then, for $f \in C_b(S)$, we have

$$\mu_n(f) = \mathbf{E} f(X_n) \to \mathbf{E} f(X) = \mu(f).$$

If we assume convergence in probability then we shall have almost sure convergence along any sufficiently fast subsequence, etc. □

(83.3) *Example.* Let $S = C([0,1], \mathbb{R})$ with the supremum-norm topology. Let $S^{(n)}$ be the normalized random walk in Section I.8. Then, by (I.8.3), we see that the law of $S^{(n)}$ converges in the $C_b(S)$ topology to the law of Brownian motion with parameter set $[0,1]$.

Now, back to the theory! If $(x_\alpha : \alpha \in D)$ is a net in $\mathbb{R}$, we define

$$\limsup x_\alpha := \inf_{\alpha_0 \in D} \sup \{x_\alpha : \alpha \geqslant \alpha_0\},$$

and we define the lim inf analogously. We have $x_\alpha \to x$ if and only if

$$\limsup x_\alpha = \liminf x_\alpha = x.$$

(83.4) THEOREM. *The following three conditions on a net* (μ_α) *of elements of* $\Pr(S)$ *are equivalent:*

(83.5) (i) $\mu_\alpha \to \mu$ *in the* $C_b(S)$ *topology;*
(83.5) (ii) $\limsup \mu_\alpha(F) \leqslant \mu(F)$ *for every closed* $F \subseteq S$;
(83.5)(iii) $\liminf \mu_\alpha(G) \geqslant \mu(G)$ *for every open* $G \subseteq S$;

If therefore (83.5)(i) holds and $B \in \mathscr{B}(S)$ *satisfies* $\mu(\partial B) = 0$, *where* ∂B *is the frontier (closure\interior) of* B, *then* $\mu_\alpha(B) \to \mu(B)$.

Proof. The equivalence of (83.5)(ii) and (83.5)(iii) is trivial.

Now suppose that (83.5)(i) holds, and that F is closed in S. For each n, the function f_n defined by

$$f_n(x) = \max(0, 1 - n\rho(x, F))$$

is an element of $C_b(S)$, and $f_n \downarrow I_F$ (the indicator function of F) as $n \uparrow \infty$. For each n,

$$\limsup_\alpha \mu_\alpha(I_F) \leqslant \lim_\alpha \mu_\alpha(f_n) = \mu(f_n),$$

so that (83.5)(ii) follows on letting $n \uparrow \infty$.

To finish the proof, we need to show that (83.5)(ii) implies (83.5)(i); and for this, we follow Billingsley [2]. Assume (83.5)(ii). We first show that

$$\limsup \mu_\alpha(f) \leqslant \mu(f), \quad \forall f \in C_b(S). \tag{83.6}$$

By replacing f by a suitable linear combination $af + b$, where $a > 0$, we see that it is enough to prove (83.6) when $0 < f < 1$. Pick such an f, and define

$$F_i := \{s \in S : f(s) \geqslant i/k\} \quad (0 \leqslant i \leqslant k),$$

where k is a temporarily fixed positive integer. Then $F_i \downarrow$ as $i \uparrow$, and

$$k^{-1}(i-1) \leqslant f < k^{-1} i \quad \text{on } F_{i-1} \backslash F_i,$$

so that

$$k^{-1} \sum_{i=1}^{k} (i-1)\mu(F_{i-1} \backslash F_i) \leqslant \mu(f) \leqslant k^{-1} \sum_{i=1}^{k} i\mu(F_{i-1} \backslash F_i).$$

By partial summation, we find that

$$k^{-1} \sum_{i=1}^{k} \mu(F_i) \leqslant \mu(f) \leqslant k^{-1} + k^{-1} \sum_{i=1}^{k} \mu(F_i).$$

Thus, since each F_i is closed and (83.5)(ii) holds,

$$\begin{aligned} \mu(f) \geqslant k^{-1} \sum \mu(F_i) \geqslant k^{-1} \sum \limsup \mu_\alpha(F_i) &\geqslant \limsup k^{-1} \sum \mu_\alpha(F_i) \\ &\geqslant \limsup \mu_\alpha(f) - k^{-1}. \end{aligned}$$

Since k is arbitrary, (83.6) follows, and, by applying (83.6) both to f and to $-f$, (83.5)(i) follows. □

(83.7) THEOREM. *For $\mu \in \Pr(S)$, let $\hat{\mu}$ be the extension of μ to an element of* $\Pr(J)$ *with $\hat{\mu}(J \backslash S) = 0$. Then the map $\mu \mapsto \hat{\mu}$ is a homeomorphism of* $\Pr(S)$ *with its $C_b(S)$ topology to the subset $\{v : v(S) = 1\}$ of* $\Pr(J)$ *with its $C_b(J)$ topology. Hence the $C_b(S)$ topology of* $\Pr(S)$ *is metrisable.*

Proof. We must show that *if (μ_α) is a net in* $\Pr(S)$ *and $\mu \in \Pr(S)$ then the conditions*

$$\mu_\alpha(f) \to \mu(f), \quad \forall f \in C_b(S), \tag{83.8}$$

$$\hat{\mu}_\alpha(f) \to \hat{\mu}(f), \quad \forall f \in C_b(J) \tag{83.9}$$

are equivalent. Since the restriction to S of an element of $C_b(J)$ is automatically in $C_b(S)$, it is obvious that (83.8) implies (83.9).

Now assume that (83.9) holds. A closed subset F of S is of the form $S \cap Y$, where Y is closed in J. By Theorem 83.4,

$$\limsup \mu_\alpha(F) = \limsup \hat{\mu}_\alpha(Y) \leqslant \hat{\mu}(Y) = \mu(F),$$

so that, again by Theorem 83.4 the result (83.8) holds. □

Of course the metrisability of Pr(S) is a relief: we can return to working with sequences. '*Let the ungodly fall into their own nets together; and let me ever escape them*' (Book of Psalms).

(*83.10*) THEOREM (Prohorov's Theorem) and DEFINITION (tightness). *A sufficient condition for a subset H of* Pr(S) *to be conditionally compact (that is, for its closure to be compact) in the $C_b(S)$ topology is that H be* tight *in the following sense:*

(*83.11*) *for each $\varepsilon > 0$ there exists a compact subset K_ε of S such that*

$$\mu(K_\varepsilon) > 1 - \varepsilon, \quad \forall \mu \in H.$$

If S is Polish then this tightness condition is also necessary.

It is the 'sufficiency' part that is important for us.

Proof of sufficiency. Suppose that (83.11) holds. Since Pr(J) and Pr(S) are metrisable, conditional compactness is the same as conditional sequential compactness. Further, we know from Theorem 81.3 that every subset of Pr(J) is conditionally sequentially compact. It now follows from Theorem 83.7 that we need only show that *if $\mu_n \in H$ and $\hat{\mu}_n \to \nu$ in* Pr(J) *then $\nu(S) = 1$*. This is, however, almost obvious: from Theorem 83.4, we have

$$\nu(K_\varepsilon) \geqslant \limsup_n \hat{\mu}_n(K_\varepsilon) \geqslant 1 - \varepsilon.$$

Hence $\nu(S) = 1$, as required. □

Proof of necessity when S is Polish. Let S be Polish. Let ρ be a metric on S such that (S, ρ) is complete with countable dense set $\{x_n\}$. Let V be a compact subset of Pr(S) in the $C_b(S)$ topology, and let $\varepsilon > 0$ be given.

For each $r \in \mathbb{N}$, the open subsets G_r^n of S, where

$$G_r^n := \bigcup_{j \leqslant n} B_\rho(x_j, 1/r), \quad \text{where } B_\rho(y, \delta) := \{x \in S : \rho(x, y) < \delta\},$$

satisfy $G_r^n \uparrow S$ as $n \uparrow \infty$, so that

$$U_r^n := \{\mu \in V : \mu(G_r^n) > 1 - \varepsilon 2^{-r}\} \uparrow V \quad \text{as } n \uparrow \infty.$$

However, it is clear from the result (83.5)(iii) that

$$\{\mu \in V : \mu(G_r^n) \leqslant 1 - \varepsilon 2^{-r}\}$$

is closed in V, whence U_r^n is open in V. Since V is compact, $U_r^{n(r)} = V$ for some $n(r)$, so that

$$\mu(G_r^{n(r)}) > 1 - \varepsilon 2^{-r}, \quad \forall \mu \in V.$$

Now put

$$K := \bigcap \overline{G_r^{n(r)}},$$

the 'bar' signifying closure in S. Then $\mu(K) > 1 - \varepsilon$. Moreover, K is closed in S, and therefore complete under ρ. For every r,

$$K \subseteq \bigcup_{j \leq n(r)} B_\rho(x_j, 2/r),$$

so that K is totally bounded. Because K is complete and totally bounded for the metric ρ, K is compact (D&S, I.6.14). The proof is complete. □

84. Some useful convergence results. Again, let S be a Lusin space.

(*84.1*) LEMMA. *Suppose that h is a measurable function from $(S, \mathscr{B}(S))$ to $(\hat{S}, \mathscr{B}(\hat{S}))$ where $\hat{S}$ is another Lusin space, with metric $\hat{\rho}$. Then the set of points D_h at which h is discontinuous is in $\mathscr{B}(S)$.*

Proof. With δ and ε denoting positive rationals, we have

$$D_h = \bigcup_{\varepsilon} \bigcap_{\delta} A_{\varepsilon,\delta},$$

where $A_{\varepsilon,\delta}$ is the *open* subset of S consisting of those x in S for which there exist y, z in S such that $\rho(x, y) < \delta, \rho(x, z) < \delta$ and $\hat{\rho}(h(y), h(z)) > \varepsilon$. □

(84.2) LEMMA (Continuous-Mapping Principle). *Let h and D_h be as in Lemma 84.1. Suppose that (μ_n) is a sequence in* Pr (*S*) *with $\mu_n \to \mu$, and that $\mu(D_h) = 0$. Then*

(84.3) $\qquad \mu_n \circ h^{-1} \to \mu \circ h^{-1}$ *in the $C_b(\hat{S})$ topology of* Pr ($\hat{S}$).

Proof. Let Γ be a closed set in $\hat{S}$. Let F be the closure in S of $h^{-1}(\Gamma)$. Then

$$h^{-1}(\Gamma) \subseteq F \subseteq h^{-1}(\Gamma) \cup D_h.$$

From (83.5)(ii),

$$\limsup \mu_n \circ h^{-1}(\Gamma) \leq \limsup \mu_n(F) \leq \mu(F) \leq \mu \circ h^{-1}(\Gamma),$$

so that, by Lemma 83.4, the result (84.3) holds. □

(*84.4*) *$C_b(\mathbf{R})$ convergence in* Pr ($\mathbf{R}$). Let μ be an element of Pr ($\mathbf{R}$), and introduce the associated distribution function $F(x) := \mu(-\infty, x]$. A point $a \in \mathbf{R}$ is called an *atom* of μ (or of F) if

$$\mu(\{a\}) = F(a) - F(a-) > 0,$$

or equivalently, if F is discontinuous at a. Since μ can have at most n atoms of mass $1/n$, the number of atoms of μ is countable, so that the set of non-atoms of μ is dense in $\mathbf{R}$.

(84.5) LEMMA. *Let* (μ_n) *be a sequence of elements of* Pr(**R**), *and let* $\mu \in \text{Pr}(\mathbf{R})$. *Introduce the associated distribution functions* F_n *and* F. *Then the following conditions are equivalent:*

(84.6) (i) $\mu_n \to \mu$ *in the* $C_b(\mathbf{R})$ *topology of* Pr(**R**);

(84.6)(ii) $F_n(x) \to F(x)$ *at every non-atom of* F.

(Skorokhod representation for $C_b(\mathbf{R})$ convergence in Pr(**R**)). *Moreover, if (84.6)(ii) holds then we can find a probability triple* $(\Omega, \mathcal{F}, \mathbf{P})$ *carrying* $(\mathbf{R}, \mathcal{B})$*-valued random variables* X_n *with law* μ_n *and* X *with law* μ *such that* $X_n \to X$ *almost surely.*

For the general Skorokhod representation for $C_b(S)$ convergence in Pr(S), with S a Lusin space, see Section 86.

Proof that (84.6)(i) implies (84.6)(ii). This is an immediate application of the last sentence of Lemma (83.5), taking B there to be $(\infty, x]$.

Proof that (84.6)(ii) implies (84.6)(i). It is clearly enough to prove the last sentence of the Lemma. Suppose that (84.6)(ii) holds.

Let $(\Omega, \mathcal{F}, \mathbf{P}) = ([0,1], \mathcal{B}([0,1]), \text{Leb})$. For $\omega \in \Omega$, define

$$X^+(\omega) := \sup\{x : F(x) \leqslant \omega\} = \inf\{x : F(x) > \omega\},$$
$$X^-(\omega) := \sup\{x : F(x) < \omega\} = \inf\{x : F(x) \geqslant \omega\};$$

and make the analogous definitions for $X_n^{\pm}$. If $z > X^-(\omega)$ then $F(z) \geqslant \omega$, so that, by right-continuity of F, we have $F(X^-(\omega)) \geqslant \omega$, and

$$X^-(\omega) \leqslant c \quad \text{implies that} \quad \omega \leqslant F(X^-(\omega)) \leqslant F(c).$$

We now see that $X^-(\omega) \leqslant c$ if and only if $\omega \leqslant F(c)$, whence $\mathbf{P}[X^-(\omega) \leqslant c] = F(c)$.

If $\omega < F(c)$ then $X^+(\omega) \leqslant c$, so that

$$F(c) = \mathbf{P}[\omega < F(c)] \leqslant \mathbf{P}[X^+(\omega) \leqslant c].$$

But, since $X^- \leqslant X^+$, we must have

$$\mathbf{P}[X^+(\omega) \leqslant c] \leqslant \mathbf{P}[X^-(\omega) \leqslant c] = F(c),$$

and it is now clear that equality must hold throughout, so that *both* X^- *and* X^+ *have law* μ. Since, for every rational c, we have

$$\mathbf{P}(X^- \leqslant c < X^+) = \mathbf{P}(X^- \leqslant c) - \mathbf{P}(X^+ \leqslant c) = 0,$$

it is clear that X^+ *and* X^- *are almost surely equal.*

Fix $\omega \in \Omega$. Let z be a non-atom of F with $z > X^+(\omega)$. Then $F(z) > \omega$, so that (by (84.6)(ii)), for large n, we shall have $F_n(z) > \omega$ and $X_n^+(\omega) \leqslant z$. Hence

$$\limsup X_n^+(\omega) \leqslant z.$$

But we can choose non-atoms z with $z \downarrow\downarrow X^+(\omega)$ to get

$$\limsup X_n^+(\omega) \leqslant X^+(\omega).$$

Finally, since $\liminf X_n^-(\omega) \geqslant X^-(\omega)$ follows similarly and $X^+ = X^-$ almost surely, we have, with X_n [respectively, X] denoting either of X_n^+, X_n^- [respectively, X^+, X^-],

$$X_n \to X, \quad \text{a.s.} \qquad \square$$

(84.7) *The arcsine law.* Consider the situation in (83.3) in which $S := C([0,1];\mathbb{R})$ with the supremum norm, and in which $S^{(n)}$ is the normalised random walk of Section I.8. Let μ_n be the law of $S^{(n)}$ on $(S, \mathscr{B}(S))$ and μ the Wiener law of BM_0. For $w \in S$, define

$$h(w) := \mathrm{Leb}\,\{s: 0 \leqslant s \leqslant 1; w(s) \geqslant 0\}.$$

Then (why?) h is a monotone limit of continuous functions on S, and so is Borel from S to $[0,1]$. If $w_k \to w$ in S, then

$$\begin{aligned}\{s \leqslant 1 : w(s) > 0\} &\subseteq \liminf \{s \leqslant 1 : w_k(s) > 0\}\\ &\subseteq \limsup \{s \leqslant 1 : w_k(s) > 0\}\\ &\subseteq \{s \leqslant 1 : w(s) \geqslant 0\}.\end{aligned}$$

Hence

$$h(w) \leqslant \liminf h(w_k) \leqslant \limsup h(w_k) \leqslant h(w) + \int_0^1 I_{\{0\}}(w(t))\,dt.$$

Fubini's Theorem shows that

$$\int_S \mu(dw) \int_0^1 I_{\{0\}}(w(t))\,dt = \int_0^1 \mu\{w : w(t) = 0\}\,dt = 0,$$

so that $\mu(D_h) = 0$. Hence, by Lemmas 84.2 and 84.5, we have, for $0 \leqslant u \leqslant 1$,

$$\mu_n\{w : h(w) \leqslant u\} \to \mu\{w : h(w) \leqslant u\} = (2/\pi) \arcsin u^{1/2}.$$

The last equality is Lévy's arcsine law, which is proved in Section III.23, and, in a more illuminating way by excursion theory, in Section VI.53.

85. Tightness in $\Pr(W)$ when W is the path-space $W := C([0,\infty);\mathbb{R})$. The Arzelà–Ascoli Theorem allows us to translate Prohorov's Theorem into 'practicable' terms when $S = W$ with the Fréchet topology in Section 82.

(85.1) THEOREM (Arzelà–Ascoli Theorem, D&S, IV.6.7). *A subset Γ of W is conditionally compact if and only if the following two conditions hold:*

(85.2(i) $\qquad \sup\{|w(0)| : w \in \Gamma\} < \infty;$

(85.2)(ii) $\qquad$ *for each* $N \in \mathbb{N}$, $\displaystyle \limsup_{\delta \downarrow 0}\ {}_{w \in \Gamma} \Delta(\delta, N; w) = 0,$

where

$$\Delta(\delta, N; \omega) := \sup\{|w(t_1) - w(t_2)| : t_1, t_2 \in [0, N]; |t_1 - t_2| < \delta\}$$

And here is how it is applied.

(85.3) THEOREM. *A subset H of* Pr(W) *is conditionally compact if and only if the following two conditions hold:*

(85.4) (i) $$\lim_{a\to\infty} \sup_{\mu \in H} \mu\{w : |w(0)| > a\} = 0;$$

(85.4)(ii) *for every* $\varepsilon > 0$ *and* $N \in \mathbf{N}$, $$\lim_{\delta \downarrow 0} \sup_{\mu \in H} \mu\{w : \Delta(\delta, N, w) > \varepsilon\} = 0.$$

Only the 'if' part matters to us.

Proof of 'if' part. By Prohorov's Theorem, we must show that if $\eta > 0$ is given then we can find a conditionally compact subset Γ of W with $\mu(\Gamma) > 1 - \eta, \forall \mu \in H$.

So suppose that conditions (85.4) hold and that $\eta > 0$. Choose a so that if

$$A := \{w : |w(0)| \leqslant a\}$$

then $\mu(A) > 1 - \frac{1}{2}\eta, \forall \mu \in H$. Choose $\delta = \delta(n, N)$ such that if

$$A_{n,N} := \{w : \Delta(\delta, N; w) \leqslant 1/n\},$$

then $\mu(A_{n,N}) > 1 - \eta 2^{-(n+N+2)}, \forall \mu \in H$. Put

$$\Gamma := A \cap \bigcap_{n,N} A_{n,N}.$$

Then $\mu(\Gamma) > 1 - \eta$, and since Γ satisfies the conditions (85.2), Γ is conditionally compact in W. □

Martingale methods will provide the best way of establishing the conditions (85.4) in the cases that concern us. But moment criteria and the deep Garsia–Rodemich–Rumsey inequality provide very important ways of establishing these conditions in other contexts. Here, as an example, is a result motivated by Kolmogorov's criterion (I.25.2) for path continuity.

(85.5) THEOREM. *Let H be a subset of* Pr(W) *such that (85.4)(ii) holds and that, for every* $N \in \mathbf{N}$, *there exist constants* γ_N, δ_N *and* C_N *in* $(0, \infty)$ *such that*

$$\sup_{\mu \in H} \int_W |w(t_2) - w(t_1)|^{\gamma_N} \mu(dw) \leqslant C_N |t_2 - t_1|^{1+\delta_N} \quad (\forall t_1, t_2 \in [0, N]).$$

Then H is conditionally compact in Pr(W).

86. The Skorokhod representation of $C_b(S)$ convergence on Pr(S). The following theorem gives a nice (and useful) way of thinking about $C_b(S)$ convergence on Pr(S).

(86.1) THEOREM (Skorokhod). *Suppose that S is a Lusin space, that μ_n $(n\in\mathbb{N})$ and μ are elements of* Pr(S), *and that $\mu_n \to \mu$ in the $C_b(S)$ topology. Then there exists a triple $(\Omega, \mathcal{F}, \mathbf{P})$ carrying $(S, \mathcal{B}(S))$-valued random variables X_n with law μ_n and X with law μ such that $X_n \to X$ almost surely.*

(86.2) OBSERVATION. *If S is some fixed Lusin space for which the theorem is true (for arbitrary μ_n and μ in* Pr(S)*) then the theorem remains true when S is replaced by an element $\tilde{S}$ of $\mathcal{B}(S)$.*

Proof of Observation. This is obvious from the proof of Theorem 83.7. □

Proof of Theorem. We already know from Lemma 84.5 that the theorem is true when $S = \mathbb{R}$. In particular, it is true when S is the Cantor set $C \subset [0, 1]$. Now, the space $\{0, 1\}^{\mathbb{N}}$ is homeomorphic to C via the map

$$(\varepsilon_n) \mapsto 2\sum \varepsilon_n 3^{-n},$$

and it is therefore clear that $C^{\mathbb{N}}$ is homeomorphic to C.

By now, we know that the theorem is true when $S = C^{\mathbb{N}}$. We also know that we need only prove the theorem when S is a compact metrisable space (J, d). However, the map from J to $[0, \frac{1}{2}]^{\mathbb{N}}$ defined by

$$x \mapsto (\tfrac{1}{2}d(x, x_k)[1 + d(x, x_k)]^{-1} : k\in\mathbb{N}),$$

where $\{x_n\}$ is a dense subset of J, is a homeomorphism of J to a compact (and therefore Borel) subset of $[0, \frac{1}{2}]^{\mathbb{N}}$. Hence all we need do is deduce the result for $S = [0, \frac{1}{2}]^{\mathbb{N}}$ from the known result of $C^{\mathbb{N}}$.

So, let $S := [0, \frac{1}{2}]^{\mathbb{N}}$, and let $\mu_n \to \mu$ in the $C_b(S)$ topology of Pr(S). For each k in $\mathbb{N}$, there can be at most countably many real numbers a such that $\mu\{x\in S : x_k = a\} > 0$. Hence, if $\mathbb{D}$ denotes the set of dyadic rationals then we can find r in $[0, \frac{1}{2}]$ such that

$$\mu\{x\in S : x_k + r \in [0, 1]\backslash\mathbb{D}, \forall k\in\mathbb{N}\} = 1.$$

Now let c be the continuous standard Cantor functions $c : [0, 1] \to [0, 1]$, and let

$$\gamma(t) := \inf\{u\in[0, 1] : c(u) \geq t\}, \quad t\in[0, 1].$$

Then $\gamma : [0, 1] \to C$, γ is Borel, γ is continuous at points of $[0, 1]\backslash\mathbb{D}$, and $c \circ \gamma = \text{id}$ on $[0, 1]$. The map $\tau : S \to C^{\mathbb{N}}$, where

$$\tau(x) := (\gamma(x_k + r) : k\in\mathbb{N}),$$

is Borel-measurable (look at the inverse image of special cylinders!), and the set of its discontinuities has μ-measure 0. Hence, by Lemma 84.2,

$$\mu_n \circ \tau^{-1} \to \mu \circ \tau^{-1} \quad \text{in the } C_b(C^{\mathbb{N}}) \text{ topology of } \Pr(C^{\mathbb{N}}).$$

Since our theorem is true when $S = C^{\mathbb{N}}$, we can find $(\Omega, \mathscr{F}, \mathbf{P})$ carrying $(C, \mathscr{B}(C))$-valued random variables Y_n with law $\mu_n \circ \tau^{-1}$ and Y with law $\mu \circ \tau^{-1}$ such that

$$Y_n \to Y \quad \text{in } C^{\mathbb{N}}, \quad \text{almost surely.}$$

Now consider the map $\phi: C^{\mathbb{N}} \to [-1,1]^{\mathbb{N}}$ defined by

$$\phi(y) := (c(y_k) - r : k \in \mathbb{N}).$$

Note that $\phi \circ \tau = \text{id}$ on S. Since ϕ is continuous, $\phi(Y_n) \to \phi(Y)$ almost surely. If V is an $(S, \mathscr{B}(S))$ random variable with law μ, then $\tau(V)$ has law $\mu \circ \tau^{-1}$, just like Y, and $\phi(\tau(V)) = V$. Hence the random variable $\phi(Y)$ has law $\hat{\mu}$, the obvious extension of μ from S to $[-1,1]^{\mathbb{N}}$. If we now throw away any ω for which either some $Y_n(\omega)$ or $Y(\omega)$ is not in S, we have completed the requisite construction. □

Note. We did not need to consider, for example, whether the image of S under τ is Borel in $C^{\mathbb{N}}$.

87. Weak convergence versus convergence of finite-dimensional distributions. We now revert to probabilists' terminology.

(87.1) TERMINOLOGY (weak convergence, convergence in law). *Let S be a Lusin space and let μ_n $(n \in \mathbb{N})$ and μ be elements of* $\Pr(S)$. *We say that*

$$\mu_n \text{ converges weakly to } \mu,$$

and write

$$\mu_n \Rightarrow \mu,$$

if and only if

$$\mu_n \textit{ converges to } \mu \textit{ in the } C_b(S) \textit{ topology.}$$

If X_n, carried by $(\Omega, \mathscr{F}_n, \mathbf{P}_n)$ and with law μ_n, and X, carried by $(\Omega, \mathscr{F}, \mathbf{P})$ and with law μ, are $(S, \mathscr{B}(S))$-valued random variables, and if also $\mu_n \Rightarrow \mu$, then we say that

$$X_n \text{ converges in law to } X.$$

For $W = C(\mathbb{R}^+, \mathbb{R})$ and for each finite subset U of $\mathbb{R}^+$, we have the restriction map $\pi_U: W \to \mathbb{R}^U$ as in our study of the DK Theorem.

(87.2) DEFINITION (convergence of finite-dimensional distributions). *Let W be the path-space $W = C(\mathbb{R}^+, \mathbb{R})$. Let μ_n $(n \in \mathbb{N})$ and μ be elements of* $\Pr(W)$. *We say that*

the finite-dimensional distributions of μ_n converge to those of μ

if, for every finite subset U of $\mathbb{R}^+$,

$$\mu_n \circ \pi_U^{-1} \Rightarrow \mu \circ \pi_U^{-1}$$

(in the $C_b(\mathbb{R}^U)$ topology of $\Pr(\mathbb{R}^U)$ *).*

The following result might clarify certain things.

(87.3) LEMMA. *Preserve the meaning of W, μ_n and μ. Then $\mu_n \Rightarrow \mu$ if and only if both of the following conditions hold:*

(87.4) (i) *the finite-dimensional distributions of μ_n converge to those of μ;*
(87.4)(ii) *the family $(\mu_n : n \in \mathbb{N})$ is tight.*

Proof of 'only if' part. Suppose that $\mu_n \Rightarrow \mu$. Since each map π_U is continuous, convergence of the finite-dimensional distributions follows from Lemma 84.2. The tightness follows from the 'necessity' part of Prohorov's Theorem, W being Polish. □

Proof of 'if' part. Suppose that (87.4)(i) and (87.4)(ii) hold. Then, by the 'sufficiency' part of Prohorov's Theorem, (μ_n) is conditionally sequentially compact. But if $\mu_{n(k)} \Rightarrow \nu$, then the finite-dimensional distributions of $\mu_{n(k)}$ converge to those of ν; and, since μ and ν therefore have the same finite-dimensional distributions, they are equal. □

Now do Exercise E91.87.

Regular conditional probabilities

Here we prove the existence of regular conditional probabilities under topological assumptions. (Recall that we know from Section 43 that regular conditional probabilities do not always exist.) We also discuss the Markov property of Brownian motion using the language of regular conditional probabilities: this is meant to prepare you for the next chapter on Markov processes.

We are even going to start switching *now* to 'Markov-process' notation. It seems to be impossible to study Markov processes rigorously without a baroque extravaganza of σ-algebras and filtrations: one uses symbols with 'degrees' such as $\mathscr{F}^\circ$ for '*uncompleted*' σ-algebras (rather like those we have hitherto called $\mathscr{G}$), $\mathscr{F}^\mu$ for the completion of $\mathscr{F}^\circ$ with respect to a certain $\mathbf{P}^\mu$ measure, etc.

So, we are going to work with a basic σ-algebra $\mathscr{F}^\circ$ and a sub-σ-algebra $\mathscr{G}^\circ$.

88. Some preliminaries. Let Ω be a set.

(88.1) DEFINITION (countably generated σ-algebra; atom of a σ-algebra). *Let $\mathscr{G}^\circ$ be a σ-algebra on Ω. Then $\mathscr{G}^\circ$ is said to be* countably generated *if there exists* *a sequence* $G_1, G_2, \ldots$ *of elements of* $\mathscr{G}^\circ$ *such that* $\mathscr{G}^\circ = \sigma\{G_1, G_2, \ldots\}$. *For* $\omega \in \Omega$, *the* atom $A(\omega)$ of $\mathscr{G}^\circ$ containing ω *is defined to be the set*

$$A(\omega) := \bigcap\{G \in \mathscr{G}^\circ : \omega \in G\}.$$

Exercises

(i) Prove that if S is a compact metrisable space then $\mathscr{B}(S)$ is countably generated. Deduce that the same holds for a Lusin space S.
(ii) Find a sub-σ-algebra of $\mathscr{B}[0,1]$ that is not countably generated—there is an obvious one.
(iii) In general, $A(\omega)$ need not be an element of $\mathscr{G}^\circ$. Prove that $A(\omega) \in \mathscr{G}^\circ$ if $\mathscr{G}^\circ$ is countably generated. This is done later in this section.

The proof of the main theorem in the next section requires a slight modification of the Riesz Representation Theorem 80.3. Suppose that Ω is a compact metrizable space. Suppose that $\mathscr{C}$ is a countable dense subset of $C(\Omega)$ such that $\mathbf{1} \in \mathscr{C}$ and $\mathscr{C}$ is a vector space over the rational field $\mathbb{Q}$. For example, $\mathscr{C}$ could be the algebra A used in the proof of Theorem 81.3. Suppose that $\phi : \mathscr{C} \to \mathbb{R}$ is $\mathbb{Q}$-linear, increasing, and satisfies $\phi(\mathbf{1}) = 1$. Let us prove that for $h \in C(\Omega)$,

$$\phi^*(h) := \inf\{\phi(g) : g \in \mathscr{C}, g \geqslant h\} = \sup\{\phi(f) : f \in \mathscr{C}, f \leqslant h\}.$$

[For, by adding to h a suitable multiple of $\mathbf{1}$, we can suppose that $h \geqslant 1$ on Ω. Then, for rational $\varepsilon > 0$, we can find f and g in $\mathscr{C}$ such that

$$(1-\varepsilon)h \leqslant f \leqslant h \leqslant g \leqslant (1+\varepsilon)h,$$

whence $f \geqslant (1-\varepsilon)(1+\varepsilon)^{-1} g$.] By applying the Riesz theorem to ϕ^*, we find that there is a unique probability measure μ on $(\Omega, \mathscr{B}(\Omega))$ such that

$$\phi(f) = \int_\Omega f(\omega)\mu(d\omega), \quad \forall f \in \mathscr{C}.$$

89. The main existence theorem. We recall the definition of regular conditional probability within the theorem.

(*89.1*) THEOREM (Doob, …, Kuratowski, …). *Let* $(\Omega, \mathscr{F}^\circ, \mathbf{P})$ *be a probability triple in which* Ω *is a Lusin space and* $\mathscr{F}^\circ = \mathscr{B}(\Omega)$. *Then there exists a* regular conditional probability $(\mathbf{P}|\mathscr{G}^\circ)$ of $\mathbf{P}$ given $\mathscr{G}^\circ$, *that is, a function*

$$(\mathbf{P}|\mathscr{G}^\circ) : \mathscr{F}^\circ \times \Omega \to [0,1]$$

such that

(*i*) *for each* $F \in \mathscr{F}^\circ$, *the function* $\omega \mapsto (\mathbf{P}|\mathscr{G}^\circ)(F, \omega)$ *is a version of* $\mathbf{P}(F|\mathscr{G}^\circ)$;
(*ii*) *for every* ω, *the map* $F \mapsto \mathbf{P}(F|\mathscr{G}^\circ)(F, \omega)$ *is a probability measure on* $\mathscr{F}^\circ$.

The stochastic process $\{[\mathbf{P}|\mathcal{G}^\circ](F,\cdot):F\in\mathcal{F}^\circ\}$ is, modulo indistinguishability, the unique modification of the process $\{\mathbf{P}(F|\mathcal{G}^\circ):F\in\mathcal{F}^\circ\}$ with the regularity properties (i) and (ii).

Assume further that $\mathcal{G}^\circ$ is a countably generated sub-σ-algebra of $\mathcal{F}^\circ$. Then $(P|\mathcal{G}^\circ)$ has the further properties:

(iii) $\{\omega:(\mathbf{P}|\mathcal{G}^\circ)(G,\omega)=I_G(\omega),\forall G\in\mathcal{G}^\circ\}$ is a set in $\mathcal{G}^\circ$ of $\mathbf{P}$-measure 1.
(iv) if $A(\omega)$ denotes the atom of $\mathcal{G}^\circ$ containing ω then $\{\omega:(\mathbf{P}|\mathcal{G}^\circ)(A(\omega),\omega)=1\}$ is in $\mathcal{G}^\circ$ and

$$\mathbf{P}\{\omega:(\mathbf{P}|\mathcal{G}^\circ)(A(\omega),\omega)=1\}=1.$$

Proof when Ω is a compact metrisable space. Assume that Ω is a compact metrisable space. Let $\mathcal{C}$ be a countable dense subset of $C(\Omega)$ containing **1** and such that $\mathcal{C}$ is a vector space over $\mathbb{Q}$. For each $f\in\mathcal{C}$, choose and fix some version of $\mathbf{E}(f|\mathcal{G}^\circ)$ and write

$$\phi_\omega(f):=\mathbf{E}(f|\mathcal{G}^\circ)(\omega).$$

The set Γ of 'good' ω for which the statements

$$\phi_\omega(q_1f_1+q_2f_2)=q_1\phi_\omega(f_1)+q_2\phi_\omega(f_2),\quad \forall q_1,q_2\in\mathbb{Q},\ \forall f_1,f_2\in\mathcal{C},$$
$$\phi_\omega(f_1)\leqslant\phi_\omega(f_2)\quad \text{whenever } f_1,f_2\in\mathcal{C} \text{ and } f_1\leqslant f_2,$$
$$\phi_\omega(\mathbf{1})=1,$$

are simultaneously true is in $\mathcal{G}^\circ$, and elementary properties of conditional expectations show that it has probability 1. For $\omega\in\Gamma$, the map ϕ_ω on $\mathcal{C}$ is $\mathbb{Q}$-linear and increasing, with $\phi_\omega(\mathbf{1})=1$, so that from the discussion in Section 88, there exists a probability measure $(\mathbf{P}|\mathcal{G}^\circ)(\cdot,\omega)$ such that

$$\phi_\omega(f)=\mathbf{E}(f|\mathcal{G}^\circ)(\omega)=\int f(\tilde\omega)(\mathbf{P}|\mathcal{G}^\circ)(d\tilde\omega,\omega),\quad \forall f\in\mathcal{C}.$$

For $\omega\notin\Gamma$, define $(\mathbf{P}|\mathcal{G}^\circ)(\cdot,\omega):=\nu$, for some arbitrary but fixed element ν of $\Pr(\Omega)$.

It is now trivial that, for fixed ξ,

$$\mathbf{E}(\xi|\mathcal{G}^\circ)=\int\xi(\tilde\omega)(\mathbf{P}|\mathcal{G}^\circ)(d\tilde\omega,\omega),\quad \text{a.s. } (\mathbf{P},\mathcal{G}^\circ),$$

first for each ξ in $C(\Omega)$ by uniform convergence, then for each $\xi\in b\mathcal{F}^\circ$ by monotone-class arguments, then for each ξ in $\mathcal{L}^1(\Omega,\mathcal{F}^\circ,\mathbf{P})$ by truncation.

Now we must prove properties (iii) and (iv) under the assumption that $\mathcal{G}^\circ$ is generated by a countable sequence $G_1,G_2,\ldots$. Thus $\mathcal{G}^\circ$ is *a fortiori* generated by the countable π-system $\mathcal{K}$ consisting of all finite intersections of the G_i. Now let

$$\Omega_1:=\{\omega:(\mathbf{P}|\mathcal{G}^\circ)(K,\omega)=I_K(\omega),\forall K\in\mathcal{K}\}\in\mathcal{G}^\circ,$$

where, as usual, I_K denotes the indicator function of K. Since $\mathcal{K}$ is countable and, obviously,

$$\mathbf{E}(I_K|\mathscr{G}^\circ) = I_K, \quad \text{a.s.}\,(\mathbf{P},\mathscr{G}^\circ),$$

it follows that $\mathbf{P}(\Omega_1) = 1$. For each $\omega \in \Omega_1$, the set of those $G \in \mathscr{G}^\circ$ on which the *measures*

$$G \mapsto (\mathbf{P}|\mathscr{G}^\circ)(G,\omega), \quad G \mapsto I_G(\omega)$$

agree is a d-system; and, since this d-system includes the π-system $\mathscr{K}$, it includes $\sigma(\mathscr{K}) = \mathscr{G}^\circ$, by Dynkin's Lemma. Hence, for $\omega \in \Omega_1$,

$$(\mathbf{P}|\mathscr{G}^\circ)(G,\omega) = I_G(\omega), \quad \forall G \in \mathscr{G}^\circ.$$

Now let $A(\omega)$ be the atom of G° containing ω. Then, for $\omega \in \Omega_1$, we can conclude that

$$(\mathbf{P}|\mathscr{G}^\circ)(A(\omega),\omega) = \mathbf{I}_{A(\omega)}(\omega) = 1,$$

once we know that $A(\omega) \in \mathscr{G}^\circ$. But you have already done this exercise by showing that

$$A(\omega) = \left(\bigcap_{G_n \ni \omega} G_n \right) \cap \left(\bigcap_{G_n^c \ni \omega} (G_n^c) \right),$$

where $\mathscr{G}^\circ = \sigma(G_1, G_2, \ldots)$ as before and $G_n^c := \Omega \backslash G_n$.

All further details are left to you. □

Proof when Ω is a Lusin space. Assume that Ω is a Lusin space. Thus Ω is in $\mathscr{B}(J)$ for some compact metrisable space J. Regard $\mathbf{P}$ as extended to $(J, \mathscr{B}(J))$ in the obvious way, and apply the 'compact' result already obtained to the triple $(J, \mathscr{B}(J), \mathbf{P})$ with subalgebra $\mathscr{H}^\circ := \sigma(\mathscr{G}^\circ)$ on J. On the $\mathbf{P}$-null, $\mathscr{H}^\circ$-measurable set of ω^* in J for which $(\mathbf{P}|\mathscr{H}^\circ)(\Omega, \omega^*) \neq 1$, redefine $(\mathbf{P}|\mathscr{H}^\circ)(\cdot, \omega^*) := \nu$, where ν is an arbitrary but fixed element of $\Pr(J)$ with $\nu(\Omega) = 1$. Set

$$(\mathbf{P}|\mathscr{G}^\circ)(F,\omega) := (\mathbf{P}|\mathscr{H}^\circ)(F,\omega) \quad (F \in \mathscr{F}^\circ, \omega \in \Omega),$$

and we are finished. □

See Dellacherie and Meyer [1], Stroock and Varadhan [1] and Parthasarathy [1] for other accounts of essentially the same proof. Stroock and Varadhan illuminate the relevance of tightness to the DK Theorem by relating that theorem to Theorem 89.1.

Remark. The theory of regular conditional probabilities is sometimes called *la théorie de désintegration des mesures.*

90. Canonical Brownian Motion CBM ($\mathbb{R}^N$); Markov property of $\mathbf{P}^x$ laws. We end this chapter (except for an important set of exercises) by revising our view of canonical Brownian motion in the light of the ideas we have studied.

On no account skip this section. It is important for your later understanding.

We now use the basic notation:

(90.1) (i) $\Omega := C(\mathbf{R}^+, \mathbf{R}^N), \quad X_t(\omega) := X(t,\omega) := \omega(t) \quad (t \geq 0; \omega \in \Omega);$

(90.1) (ii) $\mathcal{F}^\circ := \sigma(X_s : s \geq 0), \quad \mathcal{F}_t^\circ := \sigma(X_s : s \leq t) \quad (t \geq 0);$

(90.1)(iii) $$p(t; x, y) := (2\pi t)^{-N/2} \exp\left(-\frac{\|y - x\|^2}{2t}\right) \quad (t > 0; x, y \in \mathbf{R}^N),$$

(90.1)(iv) $$P(t; x, dy) := \begin{cases} p(t; x, y)\, dy & \text{if } t > 0, \\ \varepsilon_x(dy) & \text{if } t = 0. \end{cases}$$

Here ε_x is the unit mass at x. Thus $p(t; x, y)$ is the Brownian transition-density function, and $P(t; x, B)$ is the transition function from x into Borel set B in time t.

By Wiener's Theorem, for each x in $\mathbf{R}^n$, there exists a unique measure $\mathbf{P}^x$ on $(\Omega, \mathcal{F}^0)$ such that for $n \in \mathbf{N}$, for $0 \leq t_1 \leq \cdots \leq t_n$ and for $x_1, \ldots, x_n \in \mathbf{R}^N$,

(90.2) $$\mathbf{P}^x\left(\bigcap_{k=1}^n \{X(t_k) \in dx_k\}\right) = \prod_{k=1}^n P(t_k - t_{k-1}; x_{k-1}, dx_k),$$

where $t_0 := 0$ and $x_0 := x$. Equation (90.2) makes rigorous sense when integrated over a Borel subset of $(\mathbf{R}^N)^n$. Let $\mathbf{E}^x$ be the expectation associated with $\mathbf{P}^x$.

It is obvious from (90.2) that, for a special cylinder set F, the map

(90.3) $$x \mapsto \mathbf{P}^x(F) \text{ is Borel-measurable on } \mathbf{R}^n.$$

The class of all $F \in \mathcal{F}^\circ$ for which (90.3) holds is clearly a d-system; and, since it contains the π-system of special cylinders, we see that

(90.4) $$x \mapsto \mathbf{P}^x(F) \text{ is } \mathcal{B}(\mathbf{R}^N)\text{-measurable for every } F \in \mathcal{F}^\circ.$$

Space shifts $(\sigma_x : x \in \mathbf{R}^N)$. For $x \in \mathbf{R}^N$ and $\omega \in \Omega$, define

(90.5) $$\sigma_x : \Omega \to \Omega, \quad (\sigma_x \omega)(t) := \omega(t) + x \quad (t \geq 0).$$

Then σ_x is a continuous map from Ω to Ω if we use the usual topology of uniform convergence on compacts. Moreover,

$$\mathbf{P}^x = \mathbf{P}^0 \circ \sigma_x^{-1}, \quad \mathbf{E}^x \xi = \mathbf{E}^0 \xi \circ \sigma_x \quad (\xi \in b\mathcal{F}^\circ).$$

It is now clear that

(90.6) *the map $x \mapsto \mathbf{P}^x$ is continuous from $\mathbf{R}^N$ to* $\Pr(\Omega)$ *with its weak, $C_b(\Omega)$, topology.*

Towards the Markov property. There are two equivalent ways of formulating the conditional independence of past and future given the present, which is the

Markov property:

(90.7) $\mathbf{P}(\text{past and future} \mid \text{present}) = \mathbf{P}(\text{past} \mid \text{present})\mathbf{P}(\text{future} \mid \text{present});$

(90.8) $\mathbf{P}(\text{future} \mid \text{past and present}) = \mathbf{P}(\text{future} \mid \text{present}).$

See Exercise E60.41. We are going to concentrate on the latter formulation, but make it more precise by using the language of regular conditional probabilities. We also build the time-homogeneity property into our formulation.

Time shifts $(\theta_t : t \geqslant 0)$. For $t \geqslant 0$, define the map

(90.9) $$\theta_t : \Omega \to \Omega, \quad (\theta_t \omega)(u) := \omega(t+u) \quad (u \geqslant 0).$$

If Λ denotes an event then $\theta_t^{-1}(\Lambda)$ denotes that event shifted through time t: thus if $\Lambda = \{X_h \in B\}$ then $\theta_t^{-1}(\Lambda) = \{X_{t+h} \in B\}$. If η is a function on Ω, we write $\theta_t \eta$ for $\eta \circ \theta_t$:

(90.10) $$\theta_t \eta(\omega) := (\theta_t \eta)(\omega) := \eta(\theta_t \omega).$$

Note that, since Ω is a Polish space and $\mathscr{F}^\circ = \mathscr{B}(\Omega)$, and since also $\mathscr{F}_t^\circ$ is countably generated (why?), a regular conditional probability $(\mathbf{P}^x \mid \mathscr{F}_t^\circ)$ must exist for $x \in \mathbb{R}^N$. One of the most intuitive statements of the time-homogeneous Markov property is that $(\mathbf{P}^x \mid F_t^\circ) \circ \theta_t^{-1}$ is (indistinguishable from) $\mathbf{P}^{X(t)}$. This is part of our next theorem. Because we are considering CBM $(\mathbb{R}^N)$ already equipped with all its $\mathbf{P}^x$ laws, we do not need here the existence theorem on regular conditional probabilities. However, we shall need that theorem later, in the Stroock–Varadhan theory of martingale problems. Roughly speaking, we shall need to say there that CBM $(\mathbb{R}^N)$ with just one given law (say the $\mathbf{P}^0$ law) can sense the family of $\mathbf{P}^x$ laws.

(*90.11*) THEOREM (Markov properties of CBM $(\mathbb{R}^N)$). *The following results hold.*

(i) For every $x \in \mathbb{R}^N$ *and every* $t \geqslant 0$, *we have (modulo indistinguishability)*

(90.12) $$(\mathbf{P} \mid \mathscr{F}_t^\circ) \circ \theta_t^{-1} = \mathbf{P}^{X(t,\omega)} \text{ on } \mathscr{F}^\circ.$$

(ii) For $x \in \mathbb{R}^N$ *and* $t \geqslant 0$, *and for* $\xi \in \mathscr{F}_t^\circ$ *and* $\eta \in \mathrm{b}\mathscr{F}^\circ$, *we have*

(90.13) $$\mathbf{E}^x[\xi \theta_t \eta] = \mathbf{E}^x[\xi \mathbf{E}^{X(t)} \eta].$$

The same result holds if $\xi \in (\mathrm{m}\mathscr{F}_t^\circ)^+$ *and* $\eta \in (\mathrm{m}\mathscr{F}^\circ)^+$.

Discussion and proof. Both parts of the theorem describe the way in which the Markov property knits together the various $\mathbf{P}^x$ laws. The formula (90.13) is the most useful statement of the Markov property for doing calculations. You will soon acquire fluency in its use. At first sight, (90.13) looks rather complicated; and, even though it will look more complicated when written out in full, it is worth expanding its statement for clarity. It says that

$$\int_{\omega\in\Omega} \xi(\omega)\eta(\theta_t\omega)\mathbf{P}^x(d\omega) = \int_{\omega\in\Omega} \xi(\omega)\left(\int_{\tilde{\omega}\in\Omega} \eta(\tilde{\omega})\mathbf{P}^{X(t,\omega)}(d\tilde{\omega})\right)\mathbf{P}^x(d\omega).$$

A number of measurability and other questions will have occurred to you in connection with the statement of the theorem. Indeed, once one is clear what the theorem means, its proof is almost obvious!

The result (90.4) extends by monotone-class arguments to yield the fact that

(90.14) $\qquad$ for $\eta\in b\mathscr{F}^\circ$, the map $x\mapsto\mathbf{E}^x\eta$ is $\mathscr{B}(\mathbf{R}^N)$-measurable.

We know that $\omega\mapsto X(t,\omega)$ is $\mathscr{F}_t^\circ$-measurable, so that

(90.15) $\qquad$ for $\eta\in b\mathscr{F}^\circ$, the map $\omega\mapsto\mathbf{E}^{X(t,\omega)}(\eta)$ is $\mathscr{F}_t^\circ$-measurable.

Because $F\mapsto\mathbf{P}^{X(t)}(F)$ is already a measure on $\mathscr{F}^\circ$ and because $\mathscr{F}^\circ$ is countably generated, to prove part (i), we need only prove that, for fixed $F\in\mathscr{F}^\circ$,

(90.16) $$\mathbf{P}^x(\theta_t^{-1}F\,|\,\mathscr{F}_t^\circ) = \mathbf{P}^{X(t)}(F), \quad \text{a.s.}\,(\mathbf{P}^x, \mathscr{F}_t^\circ),$$

a statement about ordinary (as opposed to regular) conditional probabilities. To prove (90.16), we must show that, for $G\in\mathscr{F}_t^\circ$, we have

(90.17) $$\mathbf{P}^x(G\cap\theta_t^{-1}F) = \mathbf{E}^x(\mathbf{P}^{X(t)}(F); G).$$

It is enough to prove this when F and G are special cylinders. To avoid lots of integrals, we assume formally that

$$G = \{X(t)\in dy\}\cap\bigcap_{i=1}^{j}\{X(s_i)\in dx_i\}, \qquad F = \bigcap_{k=1}^{r}\{X(u_k)\in dz_k\}.$$

Then

$$\mathbf{P}^{X(t,\omega)}(F) = \prod_{k=1}^{r} P(u_k - u_{k-1}; z_{k-1}, dz_k) \quad (u_0 := 0, z_0 := X(t,\omega))$$

and

$$\mathbf{E}^x(P^{X(t)}(F); G) = \left\{\prod_{i=1}^{j} P(s_i - s_{i-1}; x_{i-1}, dx_i)\right\} P(t - s_j; x_j, dy)\mathbf{P}^y(F),$$

and (check!) this agrees with the left-hand side of (90.17).

When $\xi = I_G$ and $\eta = I_F$, (90.13) reduces to (90.17). Monotone-class arguments round everything off. □

91. Exercises. In these exercises, we use the probabilistic notation and terminology explained in Section 87. For numerous exercises on weak convergence of measures on $(\mathbf{R}, \mathscr{B}(\mathbf{R}))$, see [W].

(*E91.82*) Prove that if S is a G_δ subset (countable intersection of open sets) of a compact metric space J, then S is Polish. (*Hint.* Begin with the case when S is an open subset.) Deduce that the set $\mathbb{R}\backslash\mathbb{Q}$ of irrational numbers, with its usual topology, is Polish.

(*E91.83a)* The space $\Pr(S)$ shares many of the properties of S. We know that if J is compact metrisable then $\Pr(J)$ is compact metrizable. Show that if G is open in J and $\eta\in\mathbb{R}$, then $\{\mu\in\Pr(J):\mu(G)>\eta\}$ is open in $\Pr(J)$. Deduce that if S is a G_δ subset of J then $\{\mu\in\Pr(J):\mu(S)=1\}$ is a G_δ in $\Pr(J)$: in other words, if S is Polish then $\Pr(S)$ is Polish. Use the Monotone-Class Theorem to prove that the set of f in $b\mathscr{B}(J)$ for which $\mu\mapsto\mu(f)$ is $\mathscr{B}(\Pr(J))$ measurable coincides with $b\mathscr{B}(J)$. Hence, if S is a Borel subset of J then $\{\mu\in\Pr(J):\mu(S)=1\}$ is Borel in $\Pr(J)$: in other words, if S is a Lusin space, then so is $\Pr(S)$.

It should be noted that if S is Polish then there is a natural metric, the *Prohorov metric*, under which $\Pr(S)$ is complete and separable. See, for example, Ethier and Kurtz [1].

(*E91.83b*) *Weak convergence of empirical distributions.* Let $X_1, X_2, \ldots$ be independent identically distributed real-valued random variables carried by $(\Omega, \mathscr{F}, \mathbf{P})$. Let μ be the common law of the X_k. For $\omega\in\Omega$, let $\mu_n(\omega)$ be the empirical distribution

$$\mu_n(\omega)(B) := n^{-1}\#\{k\leqslant n: X_k(\omega)\in B\} \quad (B\in\mathscr{B}).$$

Prove that $\{\omega:\mu_n(\omega)\Rightarrow\mu\}\in\mathscr{F}$ and $\mathbf{P}(\mu_n\Rightarrow\mu)=1$.

(*E91.83c*) Let $S=C([0,1];\mathbb{R})$. For $\lambda<1$ and $w\in S$, define

$$\{\tau(\lambda)w\}(t) := \lambda^{-1}w(\lambda^2 t). \quad t\in[0,1].$$

Thus $\tau(\lambda): S\to S$. Fix $c<1$, and, for $w\in S$, define a probability measure $\mu_n(w)$ on $(S,\mathscr{B}(S))$ via

$$\mu_n(w)(\mathscr{B}) := n^{-1}\#\{k\leqslant n:\tau(c^n)w\in\mathscr{B}\}.$$

Let $\mathbf{P}$ be the (Wiener measure on $(S, \mathscr{B}(S))$. We know that $\tau(c)$ preserves $\mathbf{P}$: $\mathbf{P}\circ\tau(c)^{-1}=\mathbf{P}$ on $(S,\mathscr{B}(S))$. Prove that if f is a bounded measurable function from $(S,\mathscr{B}(S))$ to $\mathbb{R}$ such that $f(\tau(c)w)=f(w)$ for every w then f is constant a.s. ($\mathbf{P}$). It now follows from the Ergodic Theorem that $\mu_n(w)(f)\to\mathbf{E}(f)$, a.s. It therefore follows (why?) that $\mathbf{P}(\mu_n\Rightarrow\mathbf{P})=1$.

(*E91.86*) The Continuous-Mapping Principle (84.2) was used in the proof of the Skorokhod Representation Theorem in Section 86. Even though it is a circular argument, convince yourself that the Skorokhod result greatly illuminates Lemma 84.2.

Now use the Skorokhod Representation Theorem to prove the following result, which we need in Section V.23 of Volume 2 and which is often used in the literature.

Let $\{g_k : k \geqslant 0\}$ be a uniformly bounded sequence of functions on a Lusin space S, and suppose that $\mu_k \in \Pr(S)$ and $\mu_k \Rightarrow \mu$. If $\{g_k : k \geqslant 0\}$ is equicontinuous at each point of S and $g_k \to g$ pointwise then

$$\int_S g_k \, d\mu_k \to \int_S g \, d\mu.$$

(*E91.87*) The purpose of this exercise is to give an example in which we have convergence of finite-dimensional distributions *without* convergence in law. (Clearly, tightness, must fail in such a case.) 'Almost anything will do', but we give a very concrete example.

Let h be the 'tent function'

$$h(x) := \begin{cases} 1 - |x| & \text{if } |x| \leqslant 1, \\ 0 & \text{if } |x| > 1. \end{cases}$$

Let U be a random variable carried by some $(\Omega, \mathscr{F}, \mathbf{P})$ and uniformly distributed on $[\frac{1}{3}, \frac{2}{3}]$. For $\omega \in \Omega$ and $t \in [0, 1]$, define for $n \in \mathbf{N}$,

$$X_n(t, \omega) := h(3^n\{t - U(\omega)\}), \qquad X(t, \omega) := 0.$$

Regard X_n and X as $C[0, 1]$-valued random variables. Prove that, for every t, $X_n(t) \to X(t)$ almost surely, so that the finite-dimensional distributions of X_n converge to those of X. Prove that X_n does not converge in law to X.

(*E91.90*). Let X be CBM ($\mathbf{R}^N$) under the $\mathbf{P}^0$ measure. Prove that a regular conditional probability of $\mathbf{P}^0$ given $\sigma(X(1))$ will agree with $\mathbf{Q}$ on $\mathscr{F}_1^\circ$, where $\mathbf{Q}$ is the law of the Brownian bridge from 0 to $X(1)$ in time 1.

Some hints

(*H91.82*) For an open subset G of J define

$$\rho_G(x, y) := \rho(x, y) + \left| \frac{1}{\rho(x, G^c)} - \frac{1}{\rho(y, G^c)} \right|,$$

where $G^c := J \backslash G$. For $S = \bigcap G(n)$, define

$$\rho_S(x, y) = \sum 2^{-n} \frac{\rho_{G(n)}(x, y)}{1 + \rho_{G(n)}(x, y)}.$$

(*H91.83b*) Think of $\mathbf{R}$ as an open subset of $J := [-\infty, \infty]$. Choose a countable dense subset $\{f_r\}$ of $C(J)$. For each r,

$$\frac{1}{n}\sum_{k=1}^{n} f_r(X_k) \to \mu(f_r), \quad \text{a.s.},$$

by the Strong Law.

(*H91.83c*) If $f(w) = f(\tau(c)w)$ for every w then, for every n,

$$f(w) = f(\tau(c^n)w), \qquad \text{so that} \quad f \in m\sigma\{w(s): s \leqslant c^n\}.$$

But (see part (ii) of Theorem 68.2) $\bigcap \sigma\{w(s): s \leqslant c^n\}$ is **P**-trivial, so f is a.s. constant.

CHAPTER III

Markov Processes

As explained in the Preface to this edition, this chapter remains very much as it was originally: the only significant difference is that the functional analysis—in Hille–Yosida theory and in Ray's Theorem—is given much fuller treatment. There is a sense, therefore, in which the chapter is caught in a 'time warp'; but we hope and trust that it will still serve as a useful introduction to Markov processes. The Preface advised further reading. The table of contents is a good guide to what this chapter contains.

1. TRANSITION FUNCTIONS AND RESOLVENTS

1. What is a (continuous-time) Markov process? The first two sections are intended to explain and motivate something of what follows. Details may be left vague—at least for the time being!

Informally, a Markov process models the motion of a particle that moves around in a measurable space $(E, \mathscr{E})$ in a memoryless way. As carrier set-up, we need a filtered space $(\Omega, \{\mathscr{F}_t\})$, on which, for every $t \geqslant 0$, an $\mathscr{F}_t$-measurable random variable X_t, which gives the position of our particle at time t, is defined. It proves necessary to introduce a probability law $\mathbf{P}^x$ for each point x in E. (In the Feller–Dynkin context that we meet first, $\mathbf{P}^x$ denotes the law of the process when it starts at x. In the Ray context, this idea has to be modified somewhat.) The Markov property ties together the various $\mathbf{P}^x$ laws. It is all very similar to what we have just studied for the Brownian case in Section 90 of Chapter II. The following definition gives the idea of a Markov process and a precise definition of a transition function. The '$P_t(x, E) \leqslant 1$' condition allows for the possible death of our particle.

(*1.1*) DEFINITION (Markov process, transition function). *A* Markov process

$$X = (\Omega, \{\mathscr{F}_t : t \geqslant 0\}, \{X_t : t \geqslant 0\}, \{P_t : t \geqslant 0\}, \{\mathbf{P}^x : x \in E\})$$

with state-space $(E, \mathscr{E})$ *is an E-valued stochastic process adapted to* $\{\mathscr{F}_t\}$ *such that, for* $0 \leqslant s \leqslant t, f \in b\mathscr{E}$ *and* $x \in E$,

$$\mathbf{E}^x[f(X_{s+t}) | \mathscr{F}_s] = (P_t f)(X_s), \quad \mathbf{P}^x \text{ a.s.,} \tag{1.2}$$

where $\{P_t\}$ *is a* transition function on $(E, \mathscr{E})$, *a family of kernels* $P_t: E \times \mathscr{E} \to [0,1]$ *such that*

(1.3) (i) *for* $t \geqslant 0$ *and* $x \in E$, $P_t(x, \cdot)$ *is a measure on* $\mathscr{E}$ *with* $P_t(x, E) \leqslant 1$;
(1.3) (ii) *for* $t \geqslant 0$ *and* $\Gamma \in \mathscr{E}$, $P_t(\cdot, \Gamma)$ *is* $\mathscr{E}$*-measurable;*
(1.3)(iii) *for* $s, t \geqslant 0$, $x \in E$ *and* $\Gamma \in \mathscr{E}$,

$$P_{t+s}(x, \Gamma) = \int_E P_s(x, dy) P_t(y, \Gamma).$$

Equation (1.3)(iii) is called the Chapman–Kolmogorov equation. We can equally think of the transition function as inducing/being a family $\{P_t\}$ of positive bounded operators of norm less than or equal to 1 on $b\mathscr{E}$, with

$$P_t f(x) := (P_t f)(x) = \int_E P_t(x, dy) f(y),$$

in which case the Chapman–Kolmogorov equation becomes the semigroup property

$$P_s P_t = P_{s+t} \quad (s, t \geqslant 0).$$

Much of the interest of the subject arises from the interplay between the analysis of transition functions and the sample-path description of the Markov process. The starting point is sometimes the one, sometimes the other. For example, a diffusion X is often given to us 'pathwise' as the solution of some stochastic differential equation; and, except in a few trivial cases, there will be no closed-form expression for the transition function $\{P_t\}$ of X. (Often, there will be no need to know about $\{P_t\}$.) On the other hand, in Markov-chain theory, we begin with a semigroup satisfying some minimal regularity properties; this time, it is the process that is hard to get hold of, and indeed the existence and properties of this process will need deep and difficult results, which it is the aim of this chapter to develop.

The most important results of the chapter say that if a transition semigroup $\{P_t\}$ is given then, under certain mild regularity conditions, there will exist on some probability space a Markov process X with R-paths (in E or perhaps in a suitably-topologised compactification of E) and such that the strong Markov property holds:

$$\mathbf{E}^x[f(X_{S+t}) | \mathscr{F}_S] = (P_t f)(X_S), \quad \mathbf{P}^x \text{ a.s.},$$

whenever S is a finite stopping time. The chapter will explain numerous methods (time-substitution, Feynman–Kac formula etc.) that may be applied to this good version X.

2. The finite-state-space Markov chain. To illustrate certain concepts, we start with the simplest possible case. A Markov process whose state-space (before compactification!) is a countable set is called a *Markov chain*. (But beware: some authors use 'chain' to signify 'with discrete *time* parameter'.) To keep things really simple, we shall in this section assume that E is a finite set, and we shall use the notations

$$p_{ij}(t) := P_t(i, \{j\}), \quad P(t) = \{p_{ij}(t) : i, j \in E\}.$$

We shall assume that $\{P_t\}$ is *honest* in that $P_t(i, E) = 1$ for all i and t. A transition semigroup is now a semigroup of $E \times E$ matrices. That some regularity is necessary is obvious if we consider the case when E has two points and

$$P(0) = \begin{pmatrix} 1 & 0 \\ 0 & 1 \end{pmatrix}, \quad P(t) = \begin{pmatrix} \frac{1}{2} & \frac{1}{2} \\ \frac{1}{2} & \frac{1}{2} \end{pmatrix} \quad (t > 0);$$

then no chain X with transition function $\{P_t\}$ can have a right-continuous version. Some basic regularity assumption is needed: we consider only semigroups that are standard in that

(2.1) $$p_{ij}(t) \to \delta_{ij} \quad (t \downarrow 0).$$

It can be shown (see later in this chapter) that the condition (2.1) implies that

(2.2) $$p'_{ij}(0) := q_{ij} \text{ exists for } i, j \in E.$$

The matrix $Q := \{q_{ij} : i, j \in E\}$ is the *Q-matrix* or *infinitesimal generator* of $\{P_t\}$, and has the properties

(2.3) $$q_{ij} \geqslant 0 \quad (i \neq j), \qquad \sum_k q_{ik} = 0 \quad (i \in E).$$

Starting from the transition function, how would we construct an associated Markov process/chain X? One remark is that, for an initial distribution μ of X_0, the finite-dimensional distributions of X are determined via

(2.4) $$\mathbf{E}^\mu \prod_{i=0}^{n} f_i(X_{t_i}) = \mu(f_0 P_{s_1} f_1 P_{s_2} f_2 \cdots P_{s_n} f_n),$$

where $0 = t_0 < t_1 < \cdots < t_n$ and $s_j := t_j - t_{j-1}$. It is not hard to check (using the semigroup property) that the finite-dimensional distributions are consistent in the sense of the Daniell–Kolmogorov Theorem, whence, by that theorem (II.31.1), there exists a process X with the required law. *But can we suppose that X has R-paths and the strong Markov property?* If not, we can do nothing with X. The italicised question is not trivial; of course, that the answer is 'Yes' follows from theory developed below.

Could we start from the process? We could indeed. Let $q_i := -q_{ii}$, as usual. The well-known 'jump–hold' construction of a Markov chain starts with a discrete time chain $\{Y(n)\}$ with $Y(0)$ having initial distribution μ and transition matrix J, where

$$J_{ij} := \begin{cases} q_{ij}/q_i & \text{if } i \neq j, \\ 0 & \text{if } i = j. \end{cases}$$

(We assume $q_i \neq 0$ for each i.) Let $\{V_r : r \in \mathbb{Z}^+\}$ be a family of independent positive variables each exponentially distributed with rate 1. Set

$$A_n := \sum_{r=0}^{n} q(Y_r)^{-1} V_r, \qquad \tau_t := \inf\{n : A_n > t\}, \qquad X_t := Y(\tau_t).$$

Then X has R-paths (we use the discrete topology on E). Moreover, it is true that X is Markovian with Q-matrix Q; but how do we *prove* the Markov property, that the Q-matrix really is Q, etc.? Can we prove that X has the strong Markov property?

The reader who tries to answer these questions will see that there are fundamental and non-trivial issues even in this simple example; we need an adequate theory of Markov processes to take us clear of these fundamental (but not very exciting) questions.

Before abandoning this example, we can learn from it important elements of the structure of a general Markov process. Since $P'(0)$ exists and is equal to Q, it follows that, for $t \geqslant 0$,

(2.5) $$P'(t) = \lim_{\varepsilon \downarrow 0} \varepsilon^{-1}[P(t+\varepsilon) - P(t)] = \lim_{\varepsilon \downarrow 0} \varepsilon^{-1} P(t)\{P(\varepsilon) - I\} = P(t)Q;$$

and, solving this equation, we find that

(2.6) $$P(t) = \exp(tQ);$$

the semigroup is here *generated* by its generator Q in a very simple way. Next, we define the *resolvent* $\{R_\lambda : \lambda > 0\}$ of the semigroup by

(2.7) $$R_\lambda := \int_0^\infty e^{-\lambda t} P_t \, dt.$$

This is just the (componentwise) Laplace transform of the semigroup; or we may regard it (more helpfully) as follows:

(2.8) $$(\lambda R_\lambda)_{ij} = \int_0^\infty \lambda e^{-\lambda t} p_{ij}(t) \, dt = \mathbf{P}(X_T = j \mid X_0 = i),$$

where T is a random variable independent of X with the exponential distribution of rate λ. In view of (2.6), it is immediate that

(2.9) $$R_\lambda = (\lambda - Q)^{-1},$$

and various simple algebraic properties follow immediately, most notably the resolvent equation

(2.10) $$R_\lambda - R_\mu = (\mu - \lambda) R_\lambda R_\mu \quad (\lambda, \mu > 0).$$

The structure we have just described for a finite-state-space Markov chain is indeed, when suitably interpreted, a feature of all Feller–Dynkin processes: the *(infinitesimal) generator* of a semigroup is just its derivative at 0, the resolvent is given by (2.9), and the semigroup is then found by inverting the Laplace transform as in (2.7). Sense can be made of the exponential formula (2.6). The resolvent equation holds in complete generality; and we shall see that resolvents are 'smoother', and in many ways more fundamental, than semigroups. Technical problems arise in the general situation because the derivative with respect to t of $P_t f$ does not exist for all f, so the generator is defined only on a subspace of $b\mathcal{E}$,....

The Hille–Yosida Theorem and Ray's marvellous extension of it will allow us to cope with these analytical problems. Once we have the semigroup, and hence a crude Daniell–Kolmogorov version of our desired process, we find that there are always enough supermartingales around to allow us to 'smooth' our process by Doob's regularisation theorems.

3. Transition functions and their resolvents. Let $\{P_t\}$ be a transition function on $(E, \mathcal{E})$. We shall say that $\{P_t\}$ is *honest* ('conservative' and 'strictly Markovian' are often used) if

$$P_t(x, E) = 1, \quad \forall t, x.$$

The possibility $P_t(x, E) < 1$ must be allowed in the theory for reasons that will become clear later. The intuitive significance is that $1 - P_t(x, E)$ represents the probability that our Markov particle has 'died' before or at time t. It is convenient (even when $\{P_t\}$ is honest, for we need licence to kill an honest process!) to adjoin a *coffin state* ∂ to E producing an extended state-space $E_\partial := E \cup \partial$. Let $\mathcal{E}_\partial := \sigma(\mathcal{E}, \partial)$ be the smallest σ-algebra on E_∂ extending $\mathcal{E}$ and containing $\{\partial\}$. The transition function $\{P_t\}$ extends to an honest transition function $\{P_t^{+\partial}\}$ on $(E_\partial, \mathcal{E}_\partial)$ in the obvious way: for $t \geqslant 0$,

(3.1) (i) $$P_t^{+\partial}(x, \partial) := 1 - P_t(x, E) \quad (x \in E, t \geqslant 0),$$

(3.1) (ii) $$P_t^{+\partial}(\partial, \cdot) := \varepsilon_\partial, \quad \text{the unit mass at } \partial,$$

(3.1)(iii) $$P_t^{+\partial}(\cdot, \cdot) := P_t(\cdot, \cdot) \quad \text{on } E \times \mathcal{E}.$$

It is profitable to think about our (unextended) transition function $\{P_t\}$ on $(E, \mathcal{E})$ in another way. It is easy to see that the equation

$$P_t f(x) = \int_E P_t(x, dy) f(y)$$

sets up a one-to-one correspondence between transition functions on $(E, \mathcal{E})$ and *sub-Markov semigroups* on $b\mathcal{E}$. A (one-parameter) sub-Markov semigroup $\{P_t : t \geqslant 0\}$ on $b\mathcal{E}$ is a family of bounded linear operators on $b\mathcal{E}$ such that

(3.2) (i) $$P_t : b\mathcal{E} \to b\mathcal{E},$$

(3.2) (ii) $$0 \leqslant f \leqslant 1 \Rightarrow 0 \leqslant P_t f \leqslant 1,$$

(3.2)(iii) $$P_{s+t} = P_s P_t,$$

(3.2)(iv) $$f_n \downarrow 0 \Rightarrow P_t f_n \downarrow 0, \quad \forall t.$$

In this formulation, $\{P_t\}$ is honest if and only if $P_t 1 = 1, \forall t$.

(3.3) *'Normal' transition functions.* In all cases of interest, $\mathscr{E}$ contains all singleton sets $\{x\}$. We say that our transition function $\{P_t\}$ is *'normal'* if $P_0(x, \cdot) = \varepsilon_x(\cdot)$, the unit mass at x, for all x in E. In semigroup language, this means that $P_0 = I$, the identity map on $b\mathscr{E}$.

(3.4) *Remark.* Note that (3.2)(iii) implies that $P_0^2 = P_0$ for every transition function $\{P_t\}$. In the theory of Feller processes, which we develop first, we shall have the 'normal' situation: $P_0 = I$. In the theory of Ray processes, the condition $P_0 = I$ can fail.

(3.5) *Example.* We have already met the Brownian transition function on $(\mathbf{R}^n, \mathscr{B}(\mathbf{R}^n))$ defined as follows:

$$P_t(x, \Gamma) := \int_\Gamma p(t; x, y)\, dy \quad (t > 0; x \in \mathbf{R}^n; \Gamma \in \mathscr{B}(\mathbf{R}^n)),$$

where p is the Brownian transition-density function:

$$p(t; x, y) := (2\pi t)^{-n/2} \exp(-|y - x|^2/2t),$$
$$P_0(x, \cdot) := \varepsilon_x(\cdot).$$

(3.6) *Example.* Let I be a countable set and let $\mathscr{I}$ be the set of all subsets of I. Let $\{p_{ij}(t) : t \geqslant 0; i, j \in I\}$ be a *transition matrix function* on I, so that the following three conditions are satisfied:

(3.7) (i) $$p_{ij}(t) \geqslant 0, \qquad \forall i, j, t,$$

(3.7) (ii) $$\sum_{j \in I} p_{ij}(t) \leqslant 1, \qquad \forall i, t,$$

(3.7)(iii) $$p_{ij}(s + t) = \sum_{k \in I} p_{ik}(s) p_{kj}(t), \quad \forall i, j, s, t.$$

Then

$$P_t(i, J) := \sum_{j \in J} p_{ij}(t) \quad (t \geqslant 0; i \in I; J \in \mathscr{I})$$

defines a transition function on $(I, \mathscr{I})$, and all such transition functions arise in this way.

The only transition functions on countable sets of any interest are those that satisfy the continuity requirement:

(3.7)(iv) $$\lim_{t \downarrow 0} p_{ij}(t) = p_{ij}(0) = \delta_{ij}.$$

These are what Chung [1] calls 'standard' transition functions. We are going to take the view that for transition functions on countable sets, (3.7iv) comes as part of the definition. It is interesting that the full weight of Ray theory (or something like it) is needed to handle Markov chains. As we shall see, the true state-space of the Markov process is generally much larger than I. (We reserve the symbol E for *that* space.)

(3.8) *Resolvents.* Suppose now that $\{P_t\}$ is a *measurable* transition function on the measurable space $(E, \mathscr{E})$, so that, in addition to the conditions (1.3), we have the measurability requirement:

(3.9) $\forall \Gamma \in \mathscr{E}$, the map $(x, t) \mapsto P_t(x, \Gamma)$ is $(\mathscr{E} \times \mathscr{B}[0, \infty))$-measurable from $E \times [0, \infty)$ to $\mathbb{R}$.

For $\lambda > 0$, we can then define a map $R_\lambda: \mathrm{b}\mathscr{E} \to \mathrm{b}\mathscr{E}$ as follows: for $x \in E$,

$$R_\lambda f(x) := \int_{[0,\infty)} e^{-\lambda t} P_t f(x)\, dt = \int_E R_\lambda(x, dy) f(y), \tag{3.10}$$

where

$$R_\lambda(x, \Gamma) := \int_{[0,\infty)} e^{-\lambda t} P_t(x, \Gamma)\, dt.$$

(Trivial applications of monotone-class theorems and of Fubini's theorem are now made without comment.)

Then $\{R_\lambda : \lambda > 0\}$ is a *sub-Markovian resolvent* on $\mathrm{b}\mathscr{E}$:

(3.11) (i) $R_\lambda: \mathrm{b}\mathscr{E} \to \mathrm{b}\mathscr{E}$;
(3.11) (ii) $0 \leqslant f \leqslant 1 \Rightarrow 0 \leqslant \lambda R_\lambda f \leqslant 1$;
(3.11)(iii) the resolvent equation holds:

$$R_\lambda - R_\mu + (\lambda - \mu) R_\lambda R_\mu = 0;$$

(3.11)(iv) $f_n \downarrow 0 \Rightarrow R_\lambda f_n \downarrow 0 \, (\forall \lambda)$.

Of course we have the following characterisation of the *honest* (or strictly Markovian) situation:

$$P_t 1 = 1, \quad \forall t \Leftrightarrow \lambda R_\lambda 1 = 1, \quad \forall \lambda.$$

Terminology. R_λ is often called the *λ-potential operator* associated with $\{P_t\}$. We shall call R_λ the *λ-resolvent* of $\{P_t\}$.

(3.12) *Note.* If S and T are independent $[0, \infty)$-valued random variables with the exponential distributions of rate λ and μ respectively, then, for $\lambda \neq \mu$,

$$\mathbf{E} P_S f(x) = \lambda R_\lambda f(x), \quad \mathbf{P}(S + T \in du) = \lambda\mu \frac{e^{-\lambda u} - e^{-\mu u}}{\mu - \lambda} du.$$

This gives a probabilistic interpretation of the resolvent equation since

$$\mathbf{E}(P_{S+T}f) = \mathbf{E}\mathbf{E}(P_S P_T f \mid S) = \mathbf{E}P_S \mu R_\mu f = \lambda\mu R_\lambda R_\mu f.$$

(*3.13*) *Exercise.* Prove that for BM($\mathbb{R}$), and for $\lambda > 0$,

$$R_\lambda f(x) = \int_{\mathbb{R}} r_\lambda(x, y) f(y)\, dy, \qquad \text{where } r_\lambda(x, y) = \gamma^{-1} \exp[-\gamma|y - x|],$$

γ denoting $(2\lambda)^{1/2}$. (*Hint.* For $x > 0$ and $\gamma > 0$,

$$\begin{aligned} I &:= \int_0^\infty (2\pi t)^{-1/2} \exp[-\tfrac{1}{2}\gamma^2 t - \tfrac{1}{2}x^2 t^{-1}]\, dt \\ &= 2(2\pi\gamma)^{-1/2} x^{1/2} e^{-\gamma x} \int_0^\infty \exp[-\tfrac{1}{2}\gamma x (s - s^{-1})^2]\, ds, \end{aligned}$$

by putting $t = xs^2/\gamma$. But now the map $s \mapsto u(s) := s - s^{-1}$ maps $(0, \infty)$ one–one onto $(-\infty, \infty)$ and the inverse map $u \mapsto s(u)$ satisfies

$$s(u) = u + s(-u), \qquad \text{whence} \quad s'(u) + s'(-u) = 1.$$

Hence

$$\begin{aligned} I &= 2(2\pi\gamma)^{-1/2} x^{1/2} e^{-\gamma x} \int_0^\infty \exp[-\tfrac{1}{2}\gamma x u^2]\, du \\ &= \gamma^{-1} e^{-\gamma x}. \end{aligned}$$

4. Contraction semigroups on Banach spaces. In many respects, resolvents are more fundamental than transition functions. In the theory of Ray processes, we shall construct transition functions from resolvents. The starting-point for this construction will be the Hille–Yosida theorem on contraction semigroups of operators on Banach spaces. That theorem is proved in Section 5. Here we introduce contraction semigroups and their resolvents and infinitesimal generators, and explain how these concepts are related. The formal equations of the theory, namely

$$P_s P_t = P_{s+t}, \quad P_t = \exp(t\mathcal{G}), \quad P_0' = \mathcal{G},$$

$$R_\lambda = \int_0^\infty e^{-\lambda t} P_t\, dt = (\lambda - \mathcal{G})^{-1}$$

are known to us, but we must now translate them into rigorous mathematics. (But see Note (4.16) below.)

(*4.1*) DEFINITION (strongly continuous contraction semigroup (SCCSG)). *Let B_0 be a Banach space. A family $\{P_t : t \geq 0\}$ of bounded linear operators $P_t : B_0 \to B_0$ is called a (one-parameter)* strongly continuous contraction semigroup (SCCSG) *if the following conditions hold*:

(4.2) (i) for $f \in B_0$, $\|P_t f - f\| \to 0$ as $t \downarrow 0$ (strong continuity)

(4.2) (ii) $\|P_t\| \leqslant 1$ for $t \geqslant 0$ (contraction property);

(4.2)(iii) $P_s P_t = P_{s+t}$ for $s, t \geqslant 0$ (semigroup property).

(By 'strongly continuous', we therefore mean 'of class C_0' in the terminology of Hille and Phillips [1].)

Suppose that $\{P_t : t \geqslant 0\}$ is an SCCSG on B_0. Then, for $t \geqslant s \geqslant 0$, and $f \in B_0$,

$$\|P_t f - P_s f\| = \|P_s(P_{t-s} f - f)\| \leqslant \|P_{t-s} f - f\|,$$

from which it follows that the map $t \mapsto P_t f$ is continuous from $[0, \infty)$ into B_0. We may therefore define the resolvent $\{R_\lambda : \lambda > 0\}$ of $\{P_t : t \geqslant 0\}$ via

(4.3)
$$R_\lambda f := \int_0^\infty e^{-\lambda t} P_t f \, dt,$$

the integral being the limit (in the (strong) topology of B_0) of approximating Riemann sums.

(4.4) DEFINITION (contraction resolvent). *Let B be a Banach space, and let $\{R_\lambda : \lambda > 0\}$ be a family of bounded linear operators $R_\lambda : B \to B$. We call $\{R_\lambda : \lambda > 0\}$* a contraction resolvent *if*

(4.5) (i) $\|\lambda R_\lambda\| \leqslant 1$ *for* $\lambda > 0$, *and*
(4.5)(ii) *the resolvent equation holds:*

$$R_\lambda - R_\mu + (\lambda - \mu) R_\lambda R_\mu = 0 \quad (\lambda, \mu > 0).$$

We already know how to prove that if $\{P_t : t \geqslant 0\}$ is an SCCSG on B_0, then its resolvent defined in (4.3) is a contraction resolvent on B_0. But since then

$$\lambda R_\lambda f - f = \int_0^\infty e^{-t} (P_{t/\lambda} f - f) \, dt,$$

it is clear that the resolvent of an SCCSG is a strongly continuous contraction resolvent (SCCR) in the sense of the following definition.

(4.6) DEFINITION (strongly continuous contraction resolvent (SCCR)). *By a* strongly continuous contraction resolvent (SCCR) *on a Banach space B_0, we mean a contraction resolvent $\{R_\lambda : \lambda > 0\}$ on B_0 with the additional property*

(4.7) $\|\lambda R_\lambda f - f\| \to 0$ as $\lambda \to \infty$ (strong continuity).

The Hille–Yosida theorem gives the 'Tauberian' converse to the 'Abelian' result we have just seen by showing that

(4.8) *an SCCR is the resolvent of an SCCSG.*

Suppose now that $\{R_\lambda : \lambda > 0\}$ is a contraction resolvent on a Banach space B. Since

$$R_\mu = R_\lambda[I + (\lambda - \mu)R_\mu],$$

it is clear that

(4.9) the range $R_\lambda B$ of R_λ is a space $\mathscr{R}$ independent of λ.

Since, for $g \in B$,

$$(\lambda R_\lambda - I)R_\mu g = \frac{\mu}{\lambda - \mu} R_\mu g - \frac{\lambda R_\lambda}{\lambda - \mu} g,$$

we see that for $f \in \mathscr{R}$, and therefore for f in the closure $\bar{\mathscr{R}}$ of $\mathscr{R}$, $\lambda R_\lambda f \to f$ as $\lambda \to \infty$. Indeed, it is now obvious that

$$B_0 := \{h \in B : \lambda R_\lambda h \to h \text{ as } \lambda \to \infty\} = \bar{\mathscr{R}}. \tag{4.10}$$

(4.11) DEFINITION ((infinitesimal) generator of an SCCSG). *Let B_0 be a Banach space and let $\{P_t : t \geqslant 0\}$ be an SCCSG on B_0. The* (infinitesimal) generator *$\mathscr{G}$ of $\{P_t : t \geqslant 0\}$ is the (generally unbounded) operator $\mathscr{G}: \mathscr{D}(\mathscr{G}) \to B_0$ defined as follows. We write $f \in \mathscr{D}(\mathscr{G})$ if, for some g in B_0, we have*

$$\| \varepsilon^{-1}(P_\varepsilon f - f) - g \| \to 0 \quad \text{as } \varepsilon \downarrow 0,$$

and we then define $\mathscr{G} f$ to equal g.

Let $\{P_t : t \geqslant 0\}$ be an SCCSG on B_0. We are now going to (formulate and) prove that, for $\lambda > 0$,

(4.12) *the operators R_λ and $(\lambda - \mathscr{G})$ are inverses.*

For $g \in B_0$, we have as $\varepsilon \downarrow 0$,

$$\varepsilon^{-1}(R_\lambda g - e^{-\lambda\varepsilon} P_\varepsilon R_\lambda g) = \varepsilon^{-1} \int_0^\varepsilon e^{-\lambda s} P_s g \, ds \to g; \tag{4.13}$$

and it is clear that

(4.14) *for $\lambda > 0$ and $g \in B_0$, $R_\lambda g \in \mathscr{D}(\mathscr{G})$ and $(\lambda - \mathscr{G})R_\lambda g = g$.*

For $f \in \mathscr{D}(\mathscr{G})$, and $t > 0$,

$$\varepsilon^{-1}(P_{t+\varepsilon} f - P_t f) = P_t \varepsilon^{-1}(P_\varepsilon f - f) \to P_t \mathscr{G} f,$$

and it is easy to obtain the results

$$P_t f \in \mathscr{D}(\mathscr{G}), \qquad \frac{d}{dt} P_t f = P_t \mathscr{G} f = \mathscr{G} P_t f,$$

$$P_t f - f = \int_0^t P_s \mathscr{G} f \, ds = \int_0^t \mathscr{G} P_s f \, ds.$$

On taking Laplace transforms of the last equation, we obtain

$$R_\lambda f - \lambda^{-1} f = \lambda^{-1} R_\lambda \mathscr{G} f;$$

in other words,

(4.15) if $f \in \mathscr{D}(\mathscr{G})$, *then* $R_\lambda(\lambda - \mathscr{G})f = f$.

(*4.16*) *Note.* In this section, we have used without proof properties of strong Riemann integrals. If you want the full theory of these, see Hille and Phillips [1] or Dynkin [2].

We end this section with a useful lemma that sometimes enables us to calculate the precise domain $\mathscr{D}(\mathscr{G})$ for an SCCSG $\{P_t : t \geq 0\}$.

(*4.17*) LEMMA (Dynkin, Reuter). *Suppose that $\mathscr{C}$ is an extension of $\mathscr{G}$. Thus suppose that $\mathscr{C}$ is a linear map from $\mathscr{D}(\mathscr{C})$, with $\mathscr{D}(\mathscr{G}) \subseteq \mathscr{D}(\mathscr{C}) \subseteq B_0$, into B_0 and that $\mathscr{C} f = \mathscr{G} f$, $\forall f \in \mathscr{D}(\mathscr{G})$. Suppose also that, for $f \in \mathscr{D}(\mathscr{C}), \mathscr{C} f = f \Rightarrow f = 0$. Then $\mathscr{C} = \mathscr{G}$; or, equivalently, $\mathscr{D}(\mathscr{C}) = \mathscr{D}(\mathscr{G})$.*

Proof. Suppose that $f \in \mathscr{D}(\mathscr{C})$. Put $g := f - \mathscr{C} f$. Then $h := R_1 g \in \mathscr{D}(\mathscr{G})$ and

$$h - \mathscr{C} h = h - \mathscr{G} h = g = f - \mathscr{C} f,$$

so that

$$f - h = \mathscr{C}(f - h)$$

and $f = h \in \mathscr{D}(\mathscr{G})$. □

5. The Hille–Yosida Theorem. Here now is the route from resolvent to semigroup.

(*5.1*) THEOREM (Hille–Yosida). *Let $\{R_\lambda : \lambda > 0\}$ be a strongly continuous contraction resolvent family on B_0. Then there exists a unique strongly continuous contraction semigroup (SCCSG) $\{P_t : t \geq 0\}$ on B_0 such that*

$$\int_{[0,\infty)} e^{-\lambda t} P_t f \, dt = R_\lambda f, \quad \forall \lambda > 0, \forall f \in B_0. \tag{5.2}$$

Indeed, if we define

$$G_\lambda := \lambda(\lambda R_\lambda - I), \tag{5.3}$$

$$P_{t,\lambda} := \exp(t G_\lambda) = e^{-\lambda t} \sum_{n=0}^{\infty} (\lambda t)^n (\lambda R_\lambda)^n / n!, \tag{5.4}$$

then, for each f in B_0,

$$P_t f = \lim_{\lambda \to \infty} P_{t,\lambda} f, \quad \forall t \geq 0. \tag{5.5}$$

(5.6) *Preliminaries.* In the last section, we obtained the fact (4.12) that if $\{R_\lambda : \lambda > 0\}$ and $\mathscr{G}$ are the resolvent and generator of an already given SCCSG $\{P_t : t \geqslant 0\}$, then R_λ and $\lambda - \mathscr{G}$ are inverse operators.

In our current situation, we are given an SCCR $\{R_\lambda : \lambda > 0\}$ as data. No semigroup $\{P_t : t \geqslant 0\}$ is yet available. Even so, we are guided by (4.12).

We know that the range $R_\lambda B_0$ of R_λ is a space $\mathscr{R}$ independent of λ, and that $\mathscr{R}$ is dense in B_0 (since $\lambda R_\lambda f \to f$ as $\lambda \to \infty$). If $h \in B_0$ and $R_\lambda h = 0$ for some λ, then

$$R_\mu h = \{I + (\lambda - \mu) R_\mu\} R_\lambda h = 0$$

for every μ; and, since $\mu R_\mu h \to h$ as $\mu \to \infty$, we must have $h = 0$. Thus the map $R_\lambda : B_0 \to \mathscr{R}$ is a bijection. On combining this fact with the resolvent equation, we see that there is a uniquely defined operator $\mathscr{G}$ with domain $\mathscr{D}(\mathscr{G}) := \mathscr{R}$ such that

$$(\lambda - \mathscr{G})^{-1} = R_\lambda$$

in that (4.14) and (4.15) hold for our new situation.

The operator G_λ in (5.3) is bounded. What makes the result (5.5) a particularly satisfying interpretation of $P_t = \exp(t\mathscr{G})$ is the fact that, for $f \in B_0$,

(5.7) $$f \in \mathscr{D}(\mathscr{G}) \quad \text{if and only if} \quad g := \lim_{\lambda \to \infty} G_\lambda f \text{ exists,}$$
$$\text{and then } \mathscr{G} f = g.$$

Proof of (5.7). Suppose first that $f \in \mathscr{D}(\mathscr{G})$. Then

$$G_\lambda f = \lambda R_\lambda \mathscr{G} f \to \mathscr{G} f.$$

Suppose conversely that $G_\lambda f \to g$. Then, for fixed $\mu > 0$, the resolvent equation shows that, as $\lambda \to \infty$,

$$R_\mu G_\lambda f = \lambda \left(\frac{\mu R_\mu - \lambda R_\lambda}{\lambda - \mu} \right) f \to \mu R_\mu f - f.$$

Since we also have $R_\mu G_\lambda f \to R_\mu g$, it must be the case that $f = R_\mu(\mu f - g)$, whence $f \in \mathscr{D}(\mathscr{G})$ and $\mathscr{G} f = g$. □

Proof of Theorem 5.1. Since G_λ is a bounded operator, it is well known that

(5.8) (i) $P_{s,\lambda} P_{t,\lambda} = P_{s+t,\lambda}$;
(5.8) (ii) $\lim_{h \downarrow\downarrow 0} h^{-1}(P_{h,\lambda} - I) = G_\lambda$ (uniform operator topology);
(5.8)(iii) $P_{t,\lambda} - I = \int_0^t P_{s,\lambda} G_\lambda \, ds$.

Since $\| \lambda R_\lambda \| \leqslant 1$, it is clear from (5.4) that

(5.8)(iv) $$\| P_{t,\lambda} \| \leqslant 1.$$

Recall that R_λ and R_μ commute, whence G_λ and G_μ commute and $P_{t,\lambda}$ commutes with $P_{s,\mu}$. Thus we may calculate, with $P(t,\lambda)$ standing for $P_{t,\lambda}$ when convenient,

$$P_{t,\lambda}f - P_{t,\mu}f = \sum_{k=1}^{n} P\left(\frac{k-1}{n}t,\lambda\right)P\left(\frac{n-k}{n}t,\mu\right)\left[P\left(\frac{t}{n},\lambda\right)-P\left(\frac{t}{n},\mu\right)\right]f,$$

so that, by (5.8)(iv),

$$\|P_{t,\lambda}f - P_{t,\mu}f\| \leqslant n\left\|\left[P\left(\frac{t}{n},\lambda\right)f-f\right]-\left[P\left(\frac{t}{n},\mu\right)f-f\right]\right\|.$$

Letting $n\to\infty$ and using (5.8)(ii), we find that

$$\|P_{t,\lambda}f - P_{t,\mu}f\| \leqslant t\|G_\lambda f - G_\mu f\|.$$

It now follows from (5.7) that, for $f\in\mathscr{D}(\mathscr{G})$, the limit

(5.9) $$P_t f := \lim_{\lambda\to\infty} P_{t,\lambda}f$$

exists uniformly over compact t-intervals, so that $t\mapsto P_t f$ is continuous for $f\in\mathscr{D}(\mathscr{G})$. Since $\mathscr{D}(\mathscr{G})$ is dense in B_0, the limit in (5.9) exists for each f in B_0, and $t\mapsto P_t f$ is continuous for $f\in B_0$. That (5.2) holds now follows from the fact that

$$\int_0^\infty e^{-\lambda t}P_{t,\mu}f\,dt = (\lambda - G_\mu)^{-1}f \to R_\lambda f \quad (\mu\to\infty).$$

Exercise. Prove this by using the resolvent equation to show that, with $\gamma := \lambda\mu(\lambda+\mu)^{-1}$,

$$(\lambda - G_\mu)^{-1} = (\lambda+\mu)^{-2}\mu^2 R_\gamma + (\lambda+\mu)^{-1}I.$$

The proof of the Hille–Yosida Theorem is now complete. □

(5.10) LEMMA. *The operator $\mathscr{G}$ does generate $\{P_t\}$ in the sense that $\{P_t\}$ is uniquely determined by $(\mathscr{G},\mathscr{D}(\mathscr{G}))$.*

Proof. For f in B_0, we can (for each $\lambda>0$) determine $R_\lambda f$ as the *unique* solution in $\mathscr{D}(\mathscr{G})$ of the equation

$$(\lambda - \mathscr{G})R_\lambda f = f.$$

The function $t\mapsto P_t f$ is then uniquely specified by the fact that its Laplace transform is $R_\lambda f$. Of course, in this special 'semigroup' situation, inversion of the Laplace transform may be effected directly via (5.5). □

Note. There is no need to worry about the uniqueness theorem for Laplace transforms in the Banach-space context—just apply an element of the dual space and use the real-variable result.

(5.11) *Limitations of the HY Theorem.* There are two reasons why the HY theorem may not be entirely appropriate. The first is that $\mathscr{D}(\mathscr{G})$ may be too *small* for many purposes. This is certainly the case in Markov-chain theory, where we need to introduce a 'natural' generator extending $\mathscr{G}$. The second reason is (of course!) that $\mathscr{D}(\mathscr{G})$ may be too *large*. This is the case in diffusion theory in dimension $n \geqslant 2$, where the 'differential generator', a contraction of $\mathscr{G}$, is more tractable and often contains all the relevant information.

2. FELLER–DYNKIN PROCESSES

6. Feller–Dynkin (FD) semigroups. *Until further notice, suppose that E is a locally compact Hausdorff space with countable base (LCCB) and that* $\mathscr{E} = \mathscr{B}(E)$. It is well known that if E is not compact, then we can adjoin to E a point ∂ so that $E_\partial := E \cup \partial$ is compact metrisable. Thus ∂ is the point at infinity in the one-point compactification of E. The notation is meant to indicate that ∂ can be used as a coffin state. If E is compact, make ∂ a point isolated from E. In either case, E is σ-compact and Polish.

We write:

$C(E)$ for the space of all ($\mathbb{R}$-valued) continuous functions on E;
$C_b(E)$ for the space of bounded continuous functions on E;
$C_0(E)$ for the space of (bounded) continuous functions on E which vanish at infinity;
$C_\kappa(E)$ for the space of continuous functions on E with compact support.

As an extension of the Riesz representation theorem (II.80.3) we have the following result. Again see Theorem IV.6.3 of Dunford and Schwartz [1].

(6.1) THEOREM. *A bounded linear functional φ on $C_0(E)$ may be written uniquely in the form*

$$\varphi(f) = \mu(f) := \int_E f(x)\mu(dx)$$

where μ is a signed measure on E of finite total variation.

By a *sub-Markov kernel V* on $(E, \mathscr{E})$, we mean a map $V: E \times \mathscr{E} \to [0, 1]$ such that

(i) $\forall x \in E$, $V(x, \cdot)$ is a subprobability measure on $(E, \mathscr{E})$ so that $V(x, E) \leqslant 1$;
(ii) $\forall \Gamma \in \mathscr{E}$, $V(\cdot, \Gamma)$ is $\mathscr{E}$-measurable.

Exercise. Derive the following theorem from Theorem 6.1 by using the Monotone-Class Theorem II.3.1.

(6.2) THEOREM. *Suppose that $V: C_0(E) \to \mathrm{b}\mathscr{E}$ is a (bounded) linear operator* *that is sub-Markov in the sense that* $0 \leqslant f \leqslant 1$ *implies* $0 \leqslant Vf \leqslant 1$. *Then there exists a unique sub-Markov kernel (also denoted by) V on* $(E, \mathscr{E})$ *such that*

$$Vf(x) = \int V(x, dy) f(y), \quad \forall f \in C_0(E), \forall x \in E.$$

Hence V has a canonical extension (via the integral) to a map $V: \mathrm{b}\mathscr{E} \to \mathrm{b}\mathscr{E}$.

Every author has his or her own definition of 'Feller semigroup', so be careful when moving from book to book. The modern trend is to mean by the *Feller property* of a transition function on $(E, \mathscr{E})$ the property

$$P_t: C_b(E) \to C_b(E), \quad \forall t \geqslant 0, \tag{6.3}$$

and by the *strong Feller property* the property

$$P_t: \mathrm{b}\mathscr{E} \to C_b(E), \quad \forall t \geqslant 0. \tag{6.4}$$

There are good reasons for using the 'Feller' label for all kinds of subtle modifications of these statements.

To avoid causing still further terminological clashes, let us give a new (and perfectly just) name to a favourite class of semigroups.

(6.5) DEFINITION (Feller–Dynkin semigroup). *A* Feller–Dynkin (FD) semi-group *is a strongly continuous, sub-Markov semigroup* $\{P_t : t \geqslant 0\}$ *of linear operators on* $C_0(E)$:

(6.6) (i) $$P_t: C_0(E) \to C_0(E);$$

(6.6) (ii) $$\forall f \in C_0(E), \quad 0 \leqslant f \leqslant 1 \Rightarrow 0 \leqslant P_t f \leqslant 1;$$

(6.6)(iii) $$P_s P_t = P_{s+t}, \quad \forall s, t \geqslant 0; \quad P_0 = I, \text{ the identity on } C_0(E);$$

(6.6)(iv) $$\forall f \in C_0(E), \quad \| P_t f - f \| \to 0 \quad \text{as } t \downarrow 0.$$

(Here then we have a situation to which the HY theorem applies with $B = B_0 = C_0(E)$.)

It follows easily from Theorem 6.2 that to any FD semigroup there corresponds a *'Feller–Dynkin' transition function* on $(E, \mathscr{E})$.

The following lemma is very important for verifying conditions (6.6) in practice.

(6.7) LEMMA. *If* $\{P_t : t \geqslant 0\}$ *is a sub-Markov semigroup on* $C_0(E)$ *satisfying (6.6)(i)–(iii) then (6.6)(iv) is implied by the apparently weaker condition*

(6.6)(iv)* $$\forall f \in C_0(E), \quad \forall x \in E, \quad P_t f(x) \to f(x) \quad \text{as } t \downarrow 0.$$

Proof. If $f \in C_0(E)$ and $x \to y$ in E then, by the Dominated-Convergence Theorem,

$$(R_\lambda f)(x) := \int_0^\infty e^{-\lambda t} P_t f(x)\, dt \to \int_0^\infty e^{-\lambda t} P_t f(y)\, dt = R_\lambda f(y).$$

It is therefore clear that $R_\lambda: C_0(E) \to C_0(E)$, and $\{R_\lambda : \lambda > 0\}$ is a contraction resolvent on $C_0(E)$. We know from the Hille–Yosida Theorem that the common domain B_0 of strong continuity of $\{P_t : t \geqslant 0\}$ and $\{R_\lambda : \lambda > 0\}$ is given by $B_0 = \overline{R_\lambda C_0(E)}$ for every λ. We need to prove that $B_0 = C_0(E)$. If $B_0 \neq C_0(E)$ then, by the Hahn–Banach theorem, we can find a non-trivial linear functional φ on $C_0(E)$ such that φ annihilates B_0. If μ is the signed measure that represents φ in the Riesz theorem, we shall have

$$\int \lambda R_\lambda f(x)\mu(dx) = \varphi(\lambda R_\lambda f) = 0$$

for every f in $C_0(E)$ and every $\lambda > 0$. However,

$$\lambda R_\lambda f(x) = \int_0^\infty e^{-s} P_{s/\lambda} f(x)\, ds \to f(x) \quad (\lambda \to \infty)$$

by the assumption (6.6)(iv)*. Hence $\varphi(f) = 0$ for every f in $C_0(E)$, contradicting the fact that φ is non-trivial. Hence B_0 does equal $C_0(E)$. □

You might like to try to prove Lemma 6.7 directly, without using Hille–Yosida machinery.

Let $\{P_t\}$ be an FD transition function on $(E, \mathscr{E})$. Let $\mathscr{G}$ be the (strong) generator of the FD semigroup $\{P_t\}$. Then, for $f \in \mathscr{D}(\mathscr{G})$,

$$\mathscr{G} f(x) = \lim_{t \downarrow\downarrow 0} t^{-1} \left[\int_E P_t(x, dy) f(y) - f(x) \right].$$

Let $f \in \mathscr{D}(\mathscr{G}) \subseteq C_0(E)$ and let f attain its supremum (as it must) at the point x. Then if $f(x) \geqslant 0$, we must have $\mathscr{G} f(x) \leqslant 0$. (If $\{P_t\}$ is honest then we will have $\mathscr{G} f(x) \leqslant 0$ irrespective of the sign of $f(x)$.)

This fact motivates

(6.8) LEMMA (Dynkin's Maximum Principle). *Suppose that $\mathscr{C}: \mathscr{D}(\mathscr{C}) \to C_0(E)$ is a linear operator extending $\mathscr{G}$. Suppose that if $f \in \mathscr{D}(\mathscr{C})$ and f attains its maximum at x and $f(x) \geqslant 0$, then $\mathscr{C} f(x) \leqslant 0$. Then $\mathscr{C} = \mathscr{G}$.*

Proof. By Lemma 4.17, we need only prove that $f \in \mathscr{D}(\mathscr{C})$ and $\mathscr{C} f = f$ imply that $f = 0$. So suppose that $f \in \mathscr{D}(\mathscr{C})$ and $\mathscr{C} f = f$. Let f attain its maximum at x. If $f(x) \geqslant 0$ then $\mathscr{C} f(x) \leqslant 0$, so that $f(x) = \mathscr{C} f(x) = 0$. By applying the same argument to $-f$, we see that $f = 0$. □

(6.9) Generator of Brownian motion. Let $E = \mathbb{R}^n$. Let $P_t(x, dy)$ denote the transition function of CBM ($\mathbb{R}^n$). See Example 1.6. For $f \in C_0(\mathbb{R}^n)$, set

$$P_t f(x) := \int P_t(x, dy) f(y) = \mathbf{W}^x f(X_t).$$

Here X is CBM ($\mathbb{R}^n$) and $\mathbf{W}^x$ is the Wiener measure corresponding to starting position x. The fact that $P_t: C_0 \to C_0$ (where $C_0 = C_0(\mathbb{R}^n)$) is easily established by analysis, and is an immediate consequence of the already established fact that $x \mapsto \mathbf{W}^x$ is continuous from $\mathbb{R}^n$ to $\Pr(W)$. The fact that

$$\lim_{t \downarrow 0} P_t f(x) = f(x) \quad (f \in C_0(\mathbb{R}^n), x \in \mathbb{R}^n)$$

is also easy to establish analytically, and is probabilistically obvious because of the (right)-continuity of X_t at 0. Thus $\{P_t\}$ is an FD semigroup on $C_0(\mathbb{R}^n)$.

The natural domain in $C_0 = C_0(\mathbb{R}^n)$ of the operator $\frac{1}{2}\Delta$ (Δ being Laplace's operator) is defined to be

$$\mathscr{D}(\tfrac{1}{2}\Delta) := \{f \in C_0 : \tfrac{1}{2}\Delta f \text{ exists and } \tfrac{1}{2}\Delta f \in C_0\}.$$

In a moment, we shall prove that if $n = 1$ then $\mathscr{G} = \frac{1}{2}\Delta$. The situation in dimension $n \geqslant 2$ is more complicated. The operator $\mathscr{G}$ is the *closure* of $\frac{1}{2}\Delta$: thus $f \in \mathscr{D}(\mathscr{G})$ if and only if there exist functions f_n in $\mathscr{D}(\frac{1}{2}\Delta)$ and a function g in C_0 such that $\|f_n - f\| \to 0$ and $\|\frac{1}{2}\Delta f_n - g\| \to 0$; and then $\mathscr{G}f = g$. We examine this case later. The moral is that *for dimension* $n \geqslant 2$, *infinitesimal generators are not really the right things to look at.* The Stroock–Varadhan theory tells us how we should view things.

Now *consider the 1-dimensional case.* From (3.13), it follows that

$$R_\lambda f(x) = \int_{\mathbb{R}} r_\lambda(x, y) f(y)\, dy \quad (\lambda > 0, f \in C_0),$$

where

$$r_\lambda(x, y) := \gamma^{-1} \exp(-\gamma |y - x|), \qquad \gamma := (2\lambda)^{1/2}.$$

Fix $\lambda > 0$. Suppose that $h \in \mathscr{D}(\mathscr{G})$, so that $h = R_\lambda f$ for some f in C_0. (In the present context, we have $B = B_0 = C_0$.) Then

$$h'(x) = \int_{\mathbb{R}} \gamma r_\lambda(x, y) \operatorname{sgn}(y - x) f(y)\, dy$$

where

$$\operatorname{sgn} x := \begin{cases} 1 & \text{if } x > 0, \\ -1 & \text{if } x < 0, \\ 0 & \text{if } x = 0. \end{cases}$$

On differentiating again, we find that

$$\lambda h - \tfrac{1}{2}h'' = f = \lambda h - \mathscr{G}h.$$

Hence $\frac{1}{2}\Delta$ is an extension of $\mathscr{G}$. By Dynkin's maximal principle (or by direct application of Lemma 4.17), $\mathscr{G} = \frac{1}{2}\Delta$.

7. The existence theorem: canonical FD processes. *Let E continue to denote an LCCB and let $\mathscr{E} := \mathscr{B}(E)$. Suppose that $\{P_t\}$ is an FD semigroup on $C_0 := C_0(E)$. We shall show that there exists a strong Markov, E_∂-valued* R*-process X with* *transition function* $\{P_t\}$. (Strictly speaking, the transition function of X is $\{P_t^{+\partial}\}$, but, conventionally, we say that $\{P_t\}$ is the transition function, and $\{P_t^{+\partial}\}$ the extended transition function, of X.) We then obtain Dynkin's simple and extremely illuminating formula for the (strong) generator $\mathscr{G}$ of $\{P_t\}$.

The technique used for establishing the existence of X is the same as that which we used for CBM($\mathbb{R}$). We first construct the Daniell–Kolmogorov (DK) process Y associated with $\{P_t\}$ and then obtain X by smoothing the paths of Y via the regularity theorem for supermartingales.

(7.1) *Use of the DK Theorem.* Let $\Omega := E_\partial^{[0,\infty)}$, the space of all functions ω from $[0,\infty)$ to E_∂. For $t \geqslant 0$, let Y_t be the coordinate projection mapping Ω to E_∂ via $Y_t(\omega) := \omega(t)$. Set

$$\mathscr{G}^\circ := \sigma\{Y_s : s \geqslant 0\}, \qquad \mathscr{G}_t^\circ := \sigma\{Y_s : s \leqslant t\}.$$

For every probability measure μ on $(E_\partial, \mathscr{E}_\partial)$, the DK theorem guarantees the existence of a unique probability measure $\mathbf{P}^\mu$ on $(\Omega, \mathscr{G}^\circ)$ such that, for $n \in \mathbb{N}$, $0 \leqslant t_1 \leqslant t_2 \leqslant \cdots \leqslant t_n$ and $x_0, x_1, \ldots, x_n \in E_\partial$,

(7.2)
$$\mathbf{P}^\mu[Y(0) \in dx_0; Y(t_1) \in dx_1; \ldots; Y(t_n) \in dx_n] = \mu(dx_0) P_{t_1}^{+\partial}(x_0, dx_1) \cdots P_{t_n - t_{n-1}}^{+\partial}(x_{n-1}, dx_n).$$

The semigroup (Chapman–Kolmogorov) property guarantees the required consistency, and the fact that E_∂ is compact metric gives more than adequate topological structure.

We write

(7.3)
$$\mathbf{P}^x := \mathbf{P}^{\varepsilon_x}, \qquad \varepsilon_x \text{ being the unit mass at } x.$$

We can verify by the usual monotone-class methods (Exercise!) that the map $x \mapsto \mathbf{P}^x(\Lambda)$ is $\mathscr{E}$-measurable for every Λ in $\mathscr{G}^\circ$. (Problems concerning the weak continuity of the map $x \mapsto \mathbf{P}^x$ are discussed in Section 13). Hence, for $\eta \in b\mathscr{G}^\circ$ and $t \geqslant 0$, the map

$$\omega \mapsto \mathbf{E}^{Y(t,\omega)}\eta$$

is $\mathscr{G}_t^\circ$-measurable, where $\mathbf{E}^x$ (respectively $\mathbf{E}^\mu$) denotes the expectation corresponding to $\mathbf{P}^x$ (respectively $\mathbf{P}^\mu$).

The Markov property knitting the laws $\{\mathbf{P}^x : x \in E_\partial\}$ can now be expressed: for $\eta \in b\mathscr{G}^\circ, \mu \in \Pr(E_\partial)$ and $t \geqslant 0$,

(7.4)
$$\mathbf{E}^\mu[\eta \circ \theta_t \mid \mathscr{G}_t^\circ] = \mathbf{E}^{Y(t)}\eta, \quad \text{a.s.}(\mathbf{P}^\mu).$$

Here, of course, θ_t is the time-shift map:

$$\theta_t : \Omega \to \Omega, \qquad \theta_t\omega(s) := \omega(t+s).$$

See Section II.90 for (7.4). (The 'Brownian' proof obviously transfers!)

In particular, we have, for $f \in C_0$ and $s, t \geqslant 0$,

(7.5) (i)
$$\mathbf{E}^\mu[f \circ Y_{s+t} \mid \mathscr{G}_t^\circ] = P_s f(Y_t), \quad \text{a.s.}(\mathbf{P}^\mu)$$

and this, together with the fact that

(7.5)(ii) $$\mathbf{P}^\mu \circ Y_0^{-1} = \mu,$$

completely determines all the laws $\mathbf{P}^\mu$.

(7.6) *Path regularisation.* Suppose that h is of the form $R_1 g$, where $g \in C_0^+$ (the set of non-negative elements in C_0). Then

(7.7) *h is 1-super-median for $\{P_t\}$: by definition, this means that*

$$0 \leqslant e^{-s} P_s h \leqslant h, \quad \forall s \geqslant 0.$$

Proof of (7.7).

$$e^{-s} P_s R_1 g = e^{-s} P_s \int_0^\infty e^{-u} P_u g \, du = \int_s^\infty e^{-v} P_v g \, dv \leqslant R_1 g. \qquad \square$$

Hence, for every μ,

$$\mathbf{E}^\mu[e^{-(s+t)} h(Y_{s+t}) | \mathscr{G}_t^\circ] = e^{-(s+t)} P_s h(Y_t) \leqslant e^{-t} h(Y_t),$$

so that

(7.8) *$e^{-t} h(Y_t)$ is a supermartingale relative to $(\mathscr{G}_t^\circ, \mathbf{P}^\mu)$.*

Hence (see Section II.65.1).

(7.9) *a.s.($\mathbf{P}^\mu$), the following statement holds: the limit $\lim_{\mathbf{Q} \ni q \downarrow\downarrow t} e^{-q} h(Y_q)$ exists for all t and defines an R-function of t.*

Now let $g_0, g_1, g_2, \ldots$ be a countable dense subset of C_0^+ with $g_0 > 0$ on E. Put $h_n = R_1 g_n$ and let $\mathscr{H} := \{h_0, h_1, h_2, \ldots\}$. Then $h_0 > 0$ on E and $\mathscr{H}$ separates points of E_∂ because $\mathscr{H} - \mathscr{H}$ is dense in $\mathscr{D}(\mathscr{G})$ and $\mathscr{D}(\mathscr{G})$ is dense in C_0. Since $\mathscr{H}$ is countable, it follows that for every μ,

(7.10) *a.s.($\mathbf{P}^\mu$), (7.9) holds for all $h \in \mathscr{H}$.*

But the map $x \mapsto (h_0(x), h_1(x), \ldots)$ (with $h(\partial) = 0, \forall h \in \mathscr{H}$) is a homeomorphism of E_∂ onto a subset of $\mathbf{R}^\infty$. Hence we can conclude that, a.s.($\mathbf{P}^\mu$),

(7.11) *$X_t := \lim_{\mathbf{Q} \ni q \downarrow\downarrow t} Y_q$ exists, $\forall t$, and defines an R-process X.*

It is worth explaining things in a little more detail. Let Ω_0 be the set of ω in Ω for which the limit $X_t(\omega)$ exists for every t and defines an R-map $t \mapsto X_t(\omega)$. Then $\Omega_0 \in \mathscr{G}^\circ$ and $\mathbf{P}^\mu(\Omega_0) = 1$ for all $\mu \in \Pr(E_\partial)$. For $\omega \in \Omega \backslash \Omega_0$, define $X_t(\omega) := \partial, \forall t$. Then X is an R-process and X_t is $\mathscr{G}^\circ$-measurable for each t.

The crucial result that, *for each μ, X is a ($\mathbf{P}^\mu$) modification of Y:*

(7.12) $$\mathbf{P}^\mu[X_t = Y_t] = 1, \quad \forall t, \forall \mu,$$

must be established directly, since we cannot appeal to Theorem II.67.7 in the absence of the usual conditions.

Proof of (7.12). For $f_1, f_2 \in C_0(E)$ and $\mathbb{Q} \ni q \downarrow\downarrow t$,

$$\begin{aligned}\mathbf{E}^\mu[f_1(Y_t)f_2(X_t)] &= \lim \mathbf{E}^\mu[f_1(Y_t)f_2(Y_q)] \\ &= \lim \mathbf{E}^\mu[f_1(Y_t)P_{q-t}f_2(Y_t)] = \mathbf{E}^\mu[f_1(Y_t)f_2(Y_t)].\end{aligned}$$

By monotone-class arguments, $\mathbf{E}^\mu f(Y_t, X_t) = \mathbf{E}^\mu f(Y_t, Y_t)$ for $f \in \mathrm{b}(\mathscr{E}_\partial \times \mathscr{E}_\partial)$, and (7.12) follows. □

Finally, note that, since $h_0 := R_1 g_0 > 0$ on E, we can conclude from Theorem II.78.1 that, for every μ, it is true a.s.($\mathbf{P}^\mu$) that

(7.13) *$\forall t$, if either X_{t-} or $X_t = \partial$, then $X_u = \partial, \forall u \geqslant t$.*

According to that theorem, the statement (7.13) corresponds to a $\mathscr{G}^\circ$-measurable set.

(7.14) *Canonical FD processes.* The DK theorem has served its purpose. The clumsy space $\Omega := E_\partial^{[0,\infty)}$ is no longer needed. As in the switch from $\mathbf{R}^{[0,\infty)}$ to C for CBM($\mathbb{R}$) in Section II.71, we can now tidy things up.

(7.15) *Let Ω now denote the space of R-paths:*

$$\omega: [0, \infty) \to E_\partial,$$

such that if either $\omega(t-)$ or $\omega(t) = \partial$ then $\omega(u) = \partial, \forall u \geqslant t$. By convention, each ω in Ω is extended to a map $\omega: [0, \infty] \to E_\partial$ by setting $\omega(\infty) := \partial$.

Note. It is important that we do *not* require the existence of the limit $\lim_{t\uparrow\uparrow\infty} \omega(t)$ for ω in Ω.

For $\omega \in \Omega$ and $t \geqslant 0$, define

(7.16) (i) $$X_t(\omega) := \omega(t),$$

(7.16) (ii) $$\mathscr{F}^\circ := \sigma\{X_s : 0 \leqslant s < \infty\} = \sigma\{X_s : 0 \leqslant s \leqslant \infty\},$$

(7.16)(iii) $$\mathscr{F}_t^\circ := \sigma\{X_s : s \leqslant t\}.$$

(7.17) THEOREM (Dynkin, Kinney, Blumenthal). *For $\mu \in \Pr(E_\partial)$, there exists a unique probability measure $\mathbf{P}^\mu$ on $(\Omega, \mathscr{F}^\circ)$ such that, for $n \in \mathbf{N}, 0 \leqslant t_1 \leqslant t_2 \leqslant \cdots \leqslant t_n$ and $x_0, x_1, \ldots, x_n \in E_\partial$,*

(7.18) $$\begin{aligned}&\mathbf{P}^\mu[X(0) \in dx_0; X(t_1) \in dx_1; \ldots; X(t_n) \in dx_n] \\ &= \mu(dx_0)P_{t_1}^{+\partial}(x_0, dx_1) \cdots P_{t_n - t_{n-1}}^{+\partial}(x_{n-1}, dx_n).\end{aligned}$$

This very important theorem merely reinterprets the results obtained above. The new $\mathbf{P}^\mu$ is the old '$\mathbf{P}^\mu$ law of X'.

The set-up

$$X = (\Omega, \mathscr{F}^\circ, \{X_t : 0 \leqslant t \leqslant \infty\}, \{\mathbf{P}^\mu : \mu \in \Pr(E_\partial)\})$$

is called the *canonical FD process* associated with the FD semigroup $\{P_t\}$. Of course, X has the same simple Markov properties as the process Y. In particular, for $s, t \geqslant 0, \xi \in b\mathscr{F}_t^\circ$ and $f \in C_0$,

(7.19) $$\mathbf{E}^\mu[\xi f(X_{t+s})] = \mathbf{E}^\mu[\xi P_s f(X_t)].$$

This formula will allow us to utilize the smoothness of the semigroup $\{P_t\}$ and the right-continuity of $\{X_t\}$ to show that X (unlike Y) has the *strong* Markov property.

(7.20) *Lifetime.* The random time

$$\zeta := \inf\{t : X(t) = \partial\} = \inf\{t : X(t-) = \partial \text{ or } X(t) = \partial\}$$

is called the *lifetime* of X.

(7.21) *Note that if $t < \zeta(\omega)$ then the set of values $\{X(s, \omega) : s \leqslant t\}$ is precompact in E.*

8. Strong Markov property: preliminary version. Recall that an $\{\mathscr{F}_{t+}^\circ\}$ stopping time is a map $T : \Omega \to [0, \infty]$ such that

$$\{\omega : T(\omega) \leqslant t\} \in \mathscr{F}_{t+}^\circ, \quad \forall t \in [0, \infty).$$

(Here $\mathscr{F}_{\infty+}^\circ := \mathscr{F}_\infty^\circ := \mathscr{F}^\circ$.) Equivalently, T is a map $T : \Omega \to [0, \infty]$ such that

(8.1) (i) $$\{\omega : T(\omega) < t\} \in \mathscr{F}_t^\circ, \quad \forall t \in [0, \infty].$$

For such a T, we define $\mathscr{F}_{T+}^\circ$ to be the σ-algebra of sets Λ in $\mathscr{F}^\circ$ for which

(8.1)(ii) $$\Lambda \cap \{\omega : T(\omega) < t\} \in \mathscr{F}_t^\circ, \quad \forall t \in [0, \infty].$$

For $0 \leqslant t \leqslant \infty$, define $\theta_t : \Omega \to \Omega$ as usual:

(8.2) (i) $$\theta_t \omega(s) := \omega(t + s), \quad \forall s,$$

where, of course, $\infty + s = s + \infty = \infty, \forall s$. If T is a map from Ω to $[0, \infty]$, define

(8.2) (ii) $$\theta_T \omega = \theta_{T(\omega)} \omega.$$

Recall that, for a function η on Ω, we write $\theta_T \eta$ for $\eta \circ \theta_T$ and that, for example,

(8.2)(iii) $$(\xi \theta_T \eta)(\omega) := \xi(\omega) \eta(\theta_{T(\omega)} \omega).$$

(8.3) THEOREM (Strong Markov Theorem; Dynkin, Yuškevič, Blumenthal). *Let T be an $\{\mathscr{F}_{t+}^\circ\}$ stopping time. Then $\forall \mu \in \Pr(E_\partial), \forall \eta \in b\mathscr{F}^\circ$,*

(8.4) $$\mathbf{E}^\mu[\theta_T \eta \mid \mathscr{F}_{T+}^\circ] = \mathbf{E}^{X(T)}[\eta], \quad a.s.(\mathbf{P}^\mu).$$

Equivalently, $\forall \mu \in \Pr(E_\partial), \forall \xi \in b\mathscr{F}_{T+}^\circ, \forall \eta \in b\mathscr{F}^\circ$,

(8.5) $$\mathbf{E}^\mu[\xi \theta_T \eta] = \mathbf{E}^\mu[\xi \mathbf{E}^{X(T)} \eta].$$

Notes

(i) We have already mentioned that (8.5) expresses the strong Markov theorem in a form ideally suited to applications. Certain slight variants of (8.5) are sometimes required. Thus (8.5) will obviously hold if $\xi \in \mathrm{m}^+\mathscr{F}^\circ_{T+}$ (the set of non-negative $\mathscr{F}^\circ_{T+}$ measurable functions from Ω to $[0, \infty]$) and $\eta \in \mathrm{m}^+\mathscr{F}^\circ$.

(ii) The début and section theorems make it clear that *Theorem 8.3 needs to be extended to take account of completions of σ-algebras* before it is of any real use for discontinuous processes. See Section 9 for the appropriate extension.

Proof of Theorem 8.3. As we used to do for martingales, put

$$T^{(n)}(\omega) := \begin{cases} k2^{-n} & \text{if } (k-1)2^{-n} \leqslant T(\omega) < k2^{-n},\ k \in \mathbf{N}, \\ \infty & \text{if } T(\omega) = \infty. \end{cases}$$

Suppose that $\Lambda \in \mathscr{F}^\circ_{T+}$. Then

$$\Lambda_{n,k} := \{\omega : T^{(n)}(\omega) = k2^{-n}\} \cap \Lambda \in \mathscr{F}^\circ_{k2^{-n}}.$$

Thus, applying the simple Markov property (7.19) with $\xi_{n,k}$ as the indicator function of $\Lambda_{n,k}$, we find that, for $\mu \in \Pr(E_\partial)$ and $f \in C_0$,

$$\begin{aligned} \mathbf{E}^\mu[f \circ X(T^{(n)} + s); \Lambda] &= \sum_{k \leqslant \infty} \mathbf{E}^\mu[f \circ X(k2^{-n} + s); \Lambda_{n,k}] \\ &= \sum \mathbf{E}^\mu[P_s f(X_{k2^{-n}}); \Lambda_{n,k}] \\ &= \mathbf{E}^\mu[(P_s f) \circ X(T^{(n)}); \Lambda]. \end{aligned}$$

Keep μ and s fixed and let $n \to \infty$. *By right-continuity of paths,*

$$X(T^{(n)} + s) \to X(T + s), \qquad X(T^{(n)}) \to X(T).$$

Since $f \in C_0$, we have $P_s f \in C_0$ *by the Feller–Dynkin property*, so

$$f \circ X(T^{(n)} + s) \to f(X_{T+s}), \qquad P_s f \circ X(T^{(n)}) \to P_s f(X_T).$$

Hence, by the Dominated-Convergence Theorem,

$$\mathbf{E}^\mu[f(X_{T+s}); \Lambda] = \mathbf{E}^\mu[P_s f(X_T); \Lambda], \tag{8.6}$$

and monotone-class arguments give

$$\mathbf{E}^\mu[\xi f(X_{T+s})] = \mathbf{E}^\mu[\xi P_s f(X_T)], \quad \forall \xi \in \mathrm{b}\mathscr{F}^\circ_{T+}. \tag{8.7}$$

Now consider the expression

$$V := \mathbf{E}^\mu[\xi f(X_{T+s}) g(X_{T+s+u})],$$

where $f, g \in C_0$ and $\xi \in \mathrm{b}\mathscr{F}^\circ_{T+}$. We can apply (8.7) with $T + s$ playing the role of T and $\xi f(X_{T+s})$ playing the role of ξ to obtain

$$V = \mathbf{E}^\mu[\xi f(X_{T+s}) P_u g(X_{T+s})].$$

Now we can apply (8.7) again with 'T as T, and ξ as ξ' but with $f(x)$ replaced by $f(x)(P_u g)(x)$ to find that

$$V = \mathbf{E}^{\mu}[\xi \times (\{P_s(f \times P_u g)\} \circ X_T)].$$

The brackets are meant to help clarify the structure, but you can ignore as many as you wish!

You can now check that we have just established (8.5) in the case when

$$\eta(\omega) = f(X_s(\omega))g(X_{s+u}(\omega)),$$

and you can see that the case when

$$\eta = f_1(X_{s_1})f_2(X_{s_2})\cdots f_n(X_{s_n}) \quad (f_1, f_2, \ldots, f_n \in C_0) \tag{8.8}$$

can be established similarly. Now let $\mathscr{H}$ be the algebra of functions η that are sums of products of the form (8.8) and apply the Monotone-Class Theorem II.3.2 to obtain the general case of (8.5). You should check that ∂ causes no trouble in this proof. □

9. Strong Markov property: full version; Blumenthal's 0–1 Law. The Strong Markov Theorem (8.3) is inadequate because it only applies to $\{\mathscr{F}^{\circ}_{t+}\}$-stopping times, whereas, for example, the début of a compact set for an R-process is not an $\{\mathscr{F}^{\circ}_{t+}\}$-stopping time (see Section II.75). We therefore need the extension to be described in this section.

For each μ in $\Pr(E_\partial)$, we now define

(9.1) $(\Omega, \mathscr{F}^{\mu}, \{\mathscr{F}^{\mu}_t\})$ *to be the usual* $\mathbf{P}^{\mu}$ *augmentation of* $(\Omega, \mathscr{F}^{\circ}, \{\mathscr{F}^{\circ}_t\})$.

First, then, $\mathscr{F}^{\mu}$ *is the* $\mathbf{P}^{\mu}$ *completion of* $\mathscr{F}^{\circ}$. This means that $\Lambda \in \mathscr{F}^{\mu}$ if and only if there exist $\Lambda_{1,\mu}, \Lambda_{2,\mu}$ in $\mathscr{F}^{\circ}$ with

$$\Lambda_{1,\mu} \subseteq \Lambda \subseteq \Lambda_{2,\mu}, \qquad \mathbf{P}^{\mu}(\Lambda_{1,\mu}) = \mathbf{P}^{\mu}(\Lambda_{2,\mu});$$

and then we set $\mathbf{P}^{\mu}(\Lambda) := \mathbf{P}^{\mu}(\Lambda_{1,\mu})$. Further, for $t \geqslant 0$, $\mathscr{F}^{\mu}_t$ *is the smallest* σ*-algebra on* Ω *extending* $\mathscr{F}^{\circ}_{t+}$ *and containing all* $\mathbf{P}^{\mu}$ *null sets in* $\mathscr{F}^{\mu}$.

Now put

$$\mathscr{F} := \bigcap_{\mu} \mathscr{F}^{\mu}, \qquad \mathscr{F}_t := \bigcap_{\mu} \mathscr{F}^{\mu}_t, \tag{9.2}$$

the intersections being taken over all μ in $\Pr(E_\partial)$. Although for each μ,

$(\Omega, \mathscr{F}^{\mu}, \{\mathscr{F}^{\mu}_t\}, \mathbf{P}^{\mu}\}$ satisfies the usual conditions (see Section II.67),

$(\Omega, \mathscr{F}, \{\mathscr{F}_t\}, \mathbf{P}^{\mu})$ does *not* satisfy the usual conditions (except in trivial cases).

However, *we do have* $\mathscr{F}_{t+} = \mathscr{F}_t$, because $\mathscr{F}^{\mu}_{t+} = \mathscr{F}^{\mu}_t (\forall \mu)$.

(9.3) THEOREM (Début Theorem, see II.76). *For* $B \in \mathscr{B}(E_\partial)$, *set*

$$D_B := \inf\{t \geqslant 0 : X_t \in B\},$$

$$H_B := \inf\{t > 0 : X_t \in B\}.$$

Then D_B and H_B are $\{\mathcal{F}_t^\mu\}$ stopping times for every μ, so that D_B and H_B are $\{\mathcal{F}_t\}$ stopping times.

(9.4) THEOREM (Strong Markov Theorem for X, definitive form). *Let T be an $\{\mathcal{F}_t\}$ stopping time; then, for $\mu \in \Pr(E_\partial), \xi \in b\mathcal{F}_T, \eta \in b\mathcal{F}$, we have*

(9.5) (i) $$\mathbf{E}^\mu[\eta \circ \theta_T | \mathcal{F}_T] = \mathbf{E}^{X(T)}[\eta], \quad a.s.(\mathbf{P}^\mu),$$

(9.5)(ii) $$\mathbf{E}^\mu[\xi \theta_T \eta] = \mathbf{E}^\mu[\xi \mathbf{E}^{X(T)} \eta].$$

It is necessary first to give careful thought to the technical problem of what Theorem 9.4 *means*, because many measurability properties are implicit in its statement. (We are sympathetic up to a point if you regard the whole business of completions as an unavoidable and artificial nuisance in a subject, probability theory, which is fundamentally a branch of applied mathematics. We shall therefore try to deal with this area in as succinct a way as possible.)

Routine applications of monotone-class arguments are skipped in the following discussion.

We know from Lemma II.75.3 that

(9.6) *for $\mu \in \Pr(E_\partial)$, there exists an $\{\mathcal{F}_{t+}^\circ\}$ stopping time $T(\mu)$ such that*

$$\mathbf{P}^\mu[T(\mu) = T] = 1.$$

Further, it is easily shown that, for $\mu \in \Pr(E_\partial)$, we can find a function η_μ in $b\mathcal{F}^\circ$ with $\mathbf{P}^\mu[\eta_\mu = \eta] = 1$. Then $\eta \circ \theta_T$ is, a.s.($\mathbf{P}^\mu$), equal to the $\mathcal{F}^\circ$-measurable function $\eta_\mu \circ \theta_{T(\mu)}$, so that $\eta \circ \theta_T$ is $\mathcal{F}^\mu$-measurable. Hence

(9.7) *$\eta \circ \theta_T$ is $\mathcal{F}$-measurable.*

Thus the conditional expectation $\mathbf{E}^\mu[\eta \circ \theta_T | \mathcal{F}_T]$ can be interpreted by reference either to the 'carrier' triple $(\Omega, \mathcal{F}, \mathbf{P}^\mu)$ or to $(\Omega, \mathcal{F}^\mu, \mathbf{P}^\mu)$.

Next, we prove that

(9.8) *$\forall \Lambda \in \mathcal{F}$, the map $x \mapsto \mathbf{P}^x(\Lambda)$ is universally ($\mathcal{E}_\partial^*$) measurable on E_∂.*

(Recall that $\mathcal{E}_\partial^*$, the *universal completion of* $\mathcal{E}_\partial$, is defined as

(9.9) $$\mathcal{E}_\partial^* := \bigcap [\mathcal{E}_\partial^\nu : \nu \in \Pr(E_\partial)],$$

where $(E_\partial, \mathcal{E}_\partial^\nu)$ is the ν-completion of $(E_\partial, \mathcal{E}_\partial)$.)

Proof of (9.8). If $\nu \in \Pr(E_\partial)$ and $\Lambda \in \mathcal{F} (\subseteq \mathcal{F}^\nu)$, we can find $\Lambda_{1,\nu}$ and $\Lambda_{2,\nu}$ in $\mathcal{F}^\circ$ with

$$\Lambda_{1,\nu} \subseteq \Lambda \subseteq \Lambda_{2,\nu} \quad \text{and} \quad \mathbf{P}^\nu(\Lambda_{1,\nu}) = \mathbf{P}^\nu(\Lambda_{2,\nu}).$$

But it is clear from the definition of $\mathbf{P}^\nu$ on $\mathcal{F}^\circ$ that, for $k = 1$ or 2,

$$\mathbf{P}^\nu(\Lambda_{k,\nu}) = \int \mathbf{P}^x(\Lambda_{k,\nu}) \nu(dx).$$

Hence

$$\mathbf{P}^x(\Lambda_{1,\nu}) \leqslant \mathbf{P}^x(\Lambda) \leqslant \mathbf{P}^x(\Lambda_{2,\nu}), \quad \forall x,$$

and

$$\int \mathbf{P}^x(\Lambda_{1,\nu})\nu(dx) = \int \mathbf{P}^x(\Lambda_{2,\nu})\nu(dx).$$

Since $x \mapsto P^x(\Lambda_{k,\nu})$ is $\mathscr{E}$-measurable for $k=1,2$, it follows that $x \mapsto \mathbf{P}^x(\Lambda)$ is $\mathscr{E}^\nu$-measurable. Since ν is arbitrary, (9.8) follows. □

The above proof of (9.8) yields the intuitively obvious result:

(9.10) $$\forall \Lambda \in \mathscr{F}, \forall \nu \in \Pr(E), \quad \mathbf{P}^\nu(\Lambda) = \int_{E_\partial} \mathbf{P}^x(\Lambda)\nu(dx).$$

We know (II.73.11) that if S is an $\{\mathscr{F}^\circ_{t+}\}$ stopping time then

(9.11) *X_S is $\mathscr{F}^\circ_{S+}$-measurable from Ω to $(E_\partial, \mathscr{E}_\partial)$.*

By some obvious further arguments based on (9.6), we can show that

(9.12) *X_T is $\mathscr{F}_T$-measurable from Ω to $(E_\partial, \mathscr{E}^*_\partial)$.*

By composition, it follows from (9.8) and (9.12) that

(9.13) *$\forall \Lambda \in \mathscr{F}$, the map $\omega \mapsto \mathbf{P}^{X(T(\omega),\omega)}(\Lambda)$ is $\mathscr{F}_T$-measurable.*

Finally, (9.13) implies that

(9.14) $$\forall \eta \in b\mathscr{F}, \quad \mathbf{E}^{X(T)}[\eta] \in b\mathscr{F}_T.$$

The measurability implications of Theorem 9.4 are now clear. Of course, the proof of the theorem is now trivial from (9.6) and the 'algebraic' Strong Markov Theorem (8.3).

(9.15) THEOREM (Blumenthal's 0–1 Law). *If $\Lambda \in \mathscr{F}_0$ then $\forall x \in E_\partial, \mathbf{P}^x(\Lambda) = 0$ or 1.*

Proof. Apply the Strong Markov Theorem with $T=0, \xi = I_\Lambda$ and $\eta = I_\Lambda$. Since $\mathbf{P}^x[X(0)=x]=1, \forall x$,

$$\mathbf{E}^x[I_\Lambda] = \mathbf{E}^x[I_\Lambda \theta_0 I_\Lambda] = \mathbf{E}^x[I_\Lambda \mathbf{E}^x I_\Lambda] = (\mathbf{E}^x[I_\Lambda])^2.$$ □

(9.16) COROLLARY. *If T is an $\{\mathscr{F}_t\}$ stopping time then $\forall x \in E_\partial, \mathbf{P}^x[T=0]=0$ or 1.*

Proof. $\{T=0\} \in \mathscr{F}_0$. □

(9.17) *Example.* Let $x \in E, B \in \mathscr{E}$. Prove that

either $\mathbf{P}^x[H_B=0]=1$, in which case x is called regular for B,
or $\mathbf{P}^x[H_B=0]=0$, in which case x is called irregular for B.

It is often difficult to decide which alternative obtains as classical examples like *Lebesgue's thorn* (see Section 7.11 of Itô and Mckean [1]) demonstrate. The connection between the present concept of 'regular' and that of a 'regular' boundary point in the Dirichlet problem is described in Section I.22.

(9.18) *Almost surely.* A statement S about points ω in Ω will be said to hold almost surely (a.s.) if $\Lambda := \{\omega : S(\omega) \text{ is true}\} \in \mathcal{F}$ and $\mathbf{P}^\mu(\Lambda) = 1, \forall \mu$. Thus 'a.s.' means 'a.s.($\mathbf{P}^\mu$), $\forall \mu$'.

Exercises

(9.10) Let X be CBM($\mathbf{R}^3$). Let $V \subset \mathbf{R}^3$ and let b be a point of $\mathbf{R}^3$ such that b is the tip of a cone that lies entirely within V. Prove that b is regular for V (for X). Why is it obvious that if L is a line in $\mathbf{R}^3$ then no point b is regular for L (for X)?

(9.20) Let X be a canonical FD process, and let $x \in E$. Set

$$U_x := \inf\{t > 0 : X_t \neq x\}.$$

Prove that

$$\mathbf{P}^x[U_x > s + t] = \mathbf{P}^x[U > s]\mathbf{P}^x[U > t]$$

and deduce that $\mathbf{P}^x[U_x > t] = \exp(-q_x t)$ for some q_x with $0 \leqslant q_x \leqslant \infty$. Explain why if X is honest and has *continuous* paths then $q_x = 0$ or ∞.

10. Some fundamental martingales; Dynkin's formula. It should not surprise you to learn that, at various levels of sophistication, the Strong Markov Theorem can be presented as just a corollary of the optional stopping theorem for martingales. This does not matter too much to us now, since we already have the strong Markov theorem.

What does matter is that it is very advantageous to regard some of the traditional consequences of the Strong Markov Theorem as martingale results. This will be a recurring theme in this book. For now, let us adopt a martingale approach (guided by Meyer's book [3]) to Dynkin's formula and Blumenthal's Quasi-left-continuity Theorem.

X continues to denote our FD process. Let us write

$$\text{(10.1)} \qquad \mathrm{b}\mathcal{E}_0^* := \mathrm{b}\mathcal{E}_\partial^* \cap \{f : f(\partial) = 0\}, \qquad \mathrm{m}^+\mathcal{E}_0^* := \mathrm{m}^+\mathcal{E}_0^* \cap \{f : f(\partial) = 0\}.$$

Recall that $\mathrm{m}^+\mathcal{E}_\partial^*$ denotes the set of $\mathcal{E}_\partial^*$-measurable functions from E_∂ to $[0, \infty]$. For f in $\mathrm{b}\mathcal{E}_0^*$ (or $\mathrm{m}^+\mathcal{E}_0^*$), we have

$$\text{(10.2)} \qquad P_t f(x) = \int_E P_t(x, dy) f(y) = \mathbf{E}^x f(X_t)$$

for x in E (and, by convention, for $x = \partial$ with $P_t f(\partial) = 0$). Then $P_t : \mathrm{b}\mathcal{E}_0^* \to \mathrm{b}\mathcal{E}_0^*$.

For an $\{\mathscr{F}_t\}$ stopping time T, and for $\lambda \geqslant 0$, we set

$$(10.3)\qquad P_T^\lambda f(x) := \mathbf{E}^x[e^{-\lambda T} f(X_T)], \quad P_T f(x) := \mathbf{E}^x[f(X_T)]:$$

here again, we allow $f \in \mathrm{b}\mathscr{E}_0^*$ or $f \in \mathrm{m}^+\mathscr{E}_0^*$. In particular, $P_t^\lambda = \mathrm{e}^{-\lambda t} P_t$. We have $P_T^\lambda : \mathrm{b}\mathscr{E}_0^* \to \mathrm{b}\mathscr{E}_0^*$. If B is a Borel subset of E, we write

$$(10.4)\qquad P_B \text{ for } P_{H_B}, \quad H_B := \inf\{t > 0 : X_t \in B\}.$$

You will appreciate that it is the necessity of completion in the Début Theorem that forces our present concern with universally measurable functions.

It follows from (10.2) that, for $g \in C_0$ and $\lambda > 0$,

$$(10.5)\qquad R_\lambda g(x) := \mathbf{E}^x \int_0^\infty e^{-\lambda t} g(X_t)\, dt.$$

By appeal to the general form of Fubini's Theorem, we can extend (10.5) to the case when $g \in \mathrm{b}\mathscr{E}_0^*$ or $\mathrm{m}^+\mathscr{E}_0^*$.

(*10.6*) *Exercise* (simple proof of Dynkin's formula). Deduce from the Strong Markov Theorem that if T is an $\{\mathscr{F}_t\}$ stopping time then

$$P_T^\lambda P_t^\lambda = P_{T+t}^\lambda$$

(but show that $P_T^\lambda P_t^\lambda \neq P_t^\lambda P_T^\lambda$ in general). Hence obtain Dynkin's formula: for $g \in C_0, \lambda > 0, x \in E$,

$$(10.7)\qquad R_\lambda g(x) = \mathbf{E}^x \int_0^T e^{-\lambda t} g(X_t)\, dt + P_T^\lambda R_\lambda g(x).$$

Of course, $P_T^\lambda R_\lambda g(x)$ means $(P_T^\lambda R_\lambda g)(x)$.

The alternative proof we now give of Dynkin's formula (10.7) is the key to many of the deepest results in the subject.

For the moment, fix $g \in C_0$ and $\lambda > 0$, and put

$$(10.8)\qquad \eta := \int_0^\infty e^{-\lambda s} g(X_s)\, ds \in \mathrm{b}\mathscr{F}^\circ.$$

Then $R_\lambda g(x) = \mathbf{E}^x \eta, \forall x$. Since

$$\eta = \int_0^t e^{-\lambda s} g(X_s)\, ds + e^{-\lambda t} \eta \circ \theta_t,$$

we can use the *simple* Markov property to find the following:

(10.9) *for every* x,

$$t \mapsto \int_0^t e^{-\lambda s} g(X_s)\, ds + e^{-\lambda t} R_\lambda g(X_t),$$

is an R-modification of the UI martingale $t \mapsto \mathbf{E}^x[\eta \mid \mathscr{F}_t]$. By the Optional-Stopping

Theorem, if T is an $\{\mathscr{F}_t\}$ stopping time (and hence an $\{\mathscr{F}_t^{\varepsilon_x}\}$ stopping time),

$$\mathbf{E}^x \int_0^T e^{-\lambda s} g(X_s)\,ds + \mathbf{E}^x[e^{-\lambda T} R_\lambda g(X_T)] = R_\lambda g(x);$$

in other words, Dynkin's formula (10.7) holds.

Now pick $f \in \mathscr{D}(\mathscr{G})$ and $\lambda > 0$, and apply (10.9) to $g = (\lambda - \mathscr{G})f$. We see that (10.10) *for* $\lambda > 0, f \in \mathscr{D}(\mathscr{G})$ *and* $x \in E$,

$$C_t^{\lambda,f} := e^{-\lambda t} f(X_t) - f(X_0) + \int_0^t e^{-\lambda s}(\lambda - \mathscr{G}) f \circ X_s\,ds$$

defines a UI R-martingale $C^{\lambda,f}$ *relative to* $(\{\mathscr{F}_t\}, \mathbf{P}^x)$. By the Optional-Stopping Theorem, *if* T *is an* $\{\mathscr{F}_t\}$ *stopping time then,*

$$\text{(10.11)} \qquad \mathbf{E}^x\, e^{-\lambda T} f(X_T) - f(x) = \mathbf{E}^x \int_0^T e^{-\lambda s}(\mathscr{G} - \lambda) f \circ (X_s)\,ds.$$

If $\mathbf{E}^x(T) < \infty$ *for some* x, *we can let* $\lambda \downarrow\downarrow 0$ *to obtain (for such* x*):*

$$\text{(10.12)} \qquad \mathbf{E}^x f(X_T) - f(x) = \mathbf{E}^x \int_0^T \mathscr{G} f(X_s)\,ds.$$

The formula (10.12) is also called *Dynkin's formula.* Since (10.7) and (10.11) are the same and (10.12) is an immediate corollary, we shall mean by 'Dynkin's formula' any or all of (10.7), (10.11) and (10.12).

It is easy to verify the following directly:

(10.13) *for* $f \in \mathscr{D}(\mathscr{G})$,

$$C_t^f := f(X_t) - f(X_0) - \int_0^t \mathscr{G} f \circ X_s\,ds$$

defines a martingale relative to $(\{\mathscr{F}_t\}, \mathbf{P}^x)$ *for all* x. *This corresponds to the analytical fact that, for* $f \in \mathscr{D}(\mathscr{G})$,

$$P_t f - f - \int_0^t P_s \mathscr{G} f\,ds = 0.$$

(*10.14*) *Example.* Let X be CBM($\mathbb{R}$) so that $\mathscr{G} = \frac{1}{2}d^2/dx^2$ on its natural domain in $C_0(\mathbb{R})$. Fix $b > 0$ and $\lambda > 0$. We can certainly find f in $\mathscr{D}(\mathscr{G})$ with

$$f(x) = \exp[x(2\lambda)^{1/2}] \quad \text{for } -\infty < x \leqslant b.$$

Apply (10.11) with $x = 0$ and $T = H_b$ to find that

$$\mathbf{E}^0 e^{-\lambda H_b} f(b) - f(0) = 0,$$

since $(\mathscr{G} - \lambda) f = 0$ on $(-\infty, b)$. Hence

$$\mathbf{E}^0 e^{-\lambda H_b} = \exp[-b(2\lambda)^{1/2}],$$

in agreement with our earlier findings. □

11. Quasi-left-continuity. We shall return to *obviously* applicable ideas very shortly (and there are a lot of applications coming up soon). First, it is convenient to prove Theorem 11.1, which (precisely) asserts that X is *quasi-left-continuous* (qlc).

Later in this chapter, we shall see how the modification of the qlc property needed for Ray processes clarifies the role of branch-points. However, it is only when we begin to consider the modern 'general theory of processes' in Volume 2 that we find what 'qlc' is really about. Still, the historical order is a good one to follow when learning.

(*11.1*) THEOREM (Blumenthal's qlc Theorem). *Let* (T_n) *be a strictly increasing sequence of stopping times with limit* T. *Then*

$$X(T_n) \to X(T), \quad \textit{a.s. on } \{T < \infty\}.$$

Note. 'Stopping time' here means, of course, '$\{\mathscr{F}_t\}$ stopping time'.

Proof. It is enough to prove the theorem when $T \leqslant c$ for some non-random constant c. (For the general case, we can then replace T_n by $T_n \wedge (c - c/n)$ and finally let $c \uparrow\uparrow \infty$ through a countable sequence.) So assume that $T \leqslant c$ for some c.

Since X has R-paths, $\lim X(T_n)$ exists and equals $X(T-)$. Thus we must prove that, a.s., $X_T = X_{T-}$.

Define (but note that this is *not* the fundamental definition of $\mathscr{F}_{T-}$; see Volume 2):

$$\mathscr{F}_{T-} := \sigma(\mathscr{F}_{T_n} : n = 1, 2, 3, \dots). \tag{11.2}$$

Fix x and, for the moment, fix f in $\mathscr{D}(\mathscr{G})$. By the Martingale-Convergence Theorem (Theorem II.69.2) we have

$$\mathbf{E}^x[f(X_T) | \mathscr{F}_{T_n}] \to \mathbf{E}^x[f(X_T) | \mathscr{F}_{T-}], \quad \text{a.s.}(\mathbf{P}^x). \tag{11.3}$$

But, by (10.13) and the Optional-Stopping Theorem (using the condition $T \leqslant c$ for justification),

$$\mathbf{E}^x[f(X_T) | \mathscr{F}_{T_n}] = f(X_{T_n}) + \mathbf{E}^x\left[\int_{T_n}^{T} \mathscr{G}f \circ X_s \, ds \,\middle|\, \mathscr{F}_{T_n}\right], \quad \text{a.s.}(\mathbf{P}^x). \tag{11.4}$$

Now we can choose a subsequence $(n(k))$ with

$$\mathbf{E}^x\left[\left(\int_{T_{n(k)}}^{T} \mathscr{G}f \circ X_s \, ds\right)^2\right] \leqslant 2^{-3k}$$

so that (by the Borel–Cantelli Lemma and the contraction property of conditional expectations), a.s.$(\mathbf{P}^x)$,

$$\mathbf{E}^x\left[\int_{T_{n(k)}}^{T} \mathscr{G}f \circ X_s \, ds \,\middle|\, \mathscr{F}_{T_{n(k)}}\right] \leqslant 2^{-k}$$

for all large k (greater than $k_0(\omega)$). Hence, letting n tend to ∞ through $(n(k))$

in (11.4), we obtain

$$\text{(11.5)} \qquad \mathbf{E}^x[f(X_T)|\mathscr{F}_{T-}] = f(X_{T-}), \quad \text{a.s.}(\mathbf{P}^x).$$

Since $\mathscr{D}(\mathscr{G})$ is separable and dense in C_0, it now follows that (for our fixed x) (11.5) is true if $f \in C_0$. Hence, for $f \in C_0$,

$$\mathbf{E}^x[\{f(X_T) - f(X_{T-})\}^2 | \mathscr{F}_{T-}] = f(X_{T-})^2 - 2f(X_{T-})^2 + f(X_{T-})^2 = 0,$$

a.s.($\mathbf{P}^x$). The rest is trivial. □

Not only is X quasi-left-continuous, but the filtration $\{\mathscr{F}_t\}$ is quasi-left-continuous.

(*11.6*) THEOREM (Meyer). *The filtration $\{\mathscr{F}_t\}$ is qlc: if (T_n) is an increasing sequence of stopping times with limit T then*

$$\mathscr{F}_T = \sigma(\mathscr{F}_{T_n} : n = 1, 2, 3, \ldots).$$

See Theorem VI.18.2 in Volume 2.

12. Characteristic operator. A point x of E is called *absorbing* if either of the following two equivalent conditions holds:

(i) $\mathbf{P}^x[X(t) = x, \forall t] = 1$;
(ii) $P_t(x, \{x\}) = 1, \quad \forall t$.

(*12.1*) LEMMA (Dynkin). *Let $x \in E$ and let d be a metric giving the topology of E. If x is not absorbing then, for all sufficient small $\eta > 0$,*

$$\mathbf{E}^x V_{\eta,x} < \infty, \quad \textit{where} \quad V_{\eta,x} := \inf\{t : d(x, X_t) \geqslant \eta\}.$$

So as not to interrupt things, we defer the proof of this lemma to (12.4).

We now define Dynkin's *characteristic operator* $\mathscr{C}$ of X. If x is absorbing, define

$$\mathscr{C}f(x) := 0, \quad \forall f \in C_0.$$

If x is not absorbing, define

$$\mathscr{C}f(x) := \lim_{\eta \downarrow\downarrow 0} \frac{\mathbf{E}^x[f \circ X(V_{\eta,x})] - f(x)}{\mathbf{E}^x V_{\eta,x}},$$

if the limit exists. The domain $\mathscr{D}(\mathscr{C})$ is defined to be the set of those f in C_0 for which $\mathscr{C}f(x)$ exists for every x and for which $\mathscr{C}f(\cdot) \in C_0$.

(*12.2*) THEOREM (Dynkin's Characteristic-Operator Theorem for FD processes). *We have*

$$\mathscr{G} = \mathscr{C}.$$

Proof. It is clear from the definition of $\mathscr{C}$ that $\mathscr{C}$ satisfies Dynkin's Maximum Principle (6.8). Hence it is sufficient to show that $\mathscr{C}$ extends $\mathscr{G}$. However, this fact is an immediate consequence of Dynkin's formula (10.12) with $T = V_{\eta,x}$, since $\mathscr{G}f$ is continuous at x. □

Dynkin's splendid theorem has all sorts of important consequences. Let us first see what it has to say for CBM($\mathbb{R}^n$).

(*12.3*) *Example.* Let X be CBM($\mathbb{R}^n$). Choose d to be the Euclidean metric on $\mathbb{R}^n$. Since $B_t^2 - t$ is a martingale if B is a BM$_0$($\mathbb{R}$), it is 'obvious' that

$$\mathbf{E}^x V_{\eta,x} = n^{-1}\eta^2.$$

Exercise. Give *rigorous* proof by the optional-stopping theorem.

Further, as we have seen before, the $\mathbf{P}^x$ distribution of the variable $X(V_{\eta,x})$ is the uniform probability distribution $\mu_{\eta,x}$ (say) on the sphere $S_{\eta,x} := \{y : d(x,y) = \eta\}$. Now, if $f \in \mathscr{D}(\frac{1}{2}\Delta)$, then the Gauss–Green Theorem shows that

$$\lim_{\eta \downarrow\downarrow 0} n\eta^{-2}\left[\int_{S_{\eta,x}} f(y)\mu_{\eta,x}(dy) - f(x)\right] = \tfrac{1}{2}\Delta f(x),$$

so that $f \in \mathscr{D}(\mathscr{C})$ and $\mathscr{C}f = \frac{1}{2}\Delta f$. Hence $\mathscr{G}(=\mathscr{C})$ is an extension of $\frac{1}{2}\Delta$. We already know that $\mathscr{G} = \frac{1}{2}\Delta$ if $n = 1$. See Section 7.2 of Itô and McKean [1] for the fact that if $n \geqslant 2$ then $\mathscr{G}$ is the *closure* of $\frac{1}{2}\Delta$ and a proper extension of $\frac{1}{2}\Delta$.

(*12.4*) *Proof of Lemma 12.1.* Suppose that x is not absorbing. Set

$$B_\varepsilon(x) := \{y : d(x,y) < \varepsilon\}.$$

Then, for some $\varepsilon > 0, t > 0$ and $\alpha > 0$,

$$P_t^{+\partial}(x, E_\partial \backslash \bar{B}_\varepsilon(x)) > \alpha,$$

where $\bar{B}_\varepsilon(x)$ is the closure of $B_\varepsilon(x)$ in E_∂. Let G be the open set $G := E_\partial \backslash \bar{B}_\varepsilon(x)$. Let (h_n) be a sequence of continuous functions on E_∂ increasing to the indicator function of G. Then $P_t^{+\partial} h_n \uparrow P_t^{+\partial}(\cdot, G)$, so that

$$\{y : P_t^{+\partial}(y, G) > \alpha\} = \bigcup_n \{y : P_t^{+\partial} h_n(y) > \alpha\}$$

is open. Hence, for some positive η, which we can and do suppose to be less than ε,

$$P_t^{+\partial}(y, E_\partial \backslash \bar{B}_\varepsilon(x)) > \alpha, \quad \forall y \in B_\eta(x).$$

An obvious use of the simple Markov property now shows that

$$\mathbf{P}^x[X_{kt} \in B_\eta(x), \forall k \leqslant n] \leqslant (1-\alpha)^n,$$

and it is an elementary consequence that

$$\mathbf{E}^x V_{\eta,x} \leqslant t\alpha^{-1}.$$ □

13. Feller–Dynkin diffusions. We now assume that $E=\mathbb{R}^n$, but E could equally well be an n-dimensional C^∞ manifold. Recall that the *lifetime* ζ of our process X is defined as follows:

$$\zeta(\omega):=\inf\{t: X_t(\omega)=\partial\}.$$

By an *FD diffusion* on $\mathbb{R}^n$, we mean an FD process X with the following additional properties:

(13.1) (i) the paths $t\mapsto X_t(\omega)$ are *continuous* on $[0,\zeta)$;
(13.1)(ii) the domain $\mathscr{D}(\mathscr{G})$ of the generator $\mathscr{G}$ of X contains $C_\kappa^\infty:=C_\kappa^\infty(\mathbb{R}^n)$, the space of infinitely differentiable functions of compact support.

Let X be an FD diffusion. Then the restriction $\mathscr{L}$ (say) of $\mathscr{G}$ to C_κ^∞ satisfies the following conditions:

(13.2) (i) $\mathscr{L}$ is a linear map from C_κ^∞ to C_0;
(13.2) (ii) $\mathscr{L}$ is *local*: if functions f and g in C_κ^∞ agree in some neighbourhood of a point x, then $\mathscr{L}f(x)=\mathscr{L}g(x)$;
(13.2)(iii) $\mathscr{L}$ satisfies the *maximum principle*: if f in C_κ^∞ attains its maximum at x and $f(x)\geqslant 0$, then $\mathscr{L}f(x)\leqslant 0$.

The property (13.2)(i) is obvious, and (13.2)(iii) is already familiar to us. Since X has *continuous* paths, it is clear from the definition of $\mathscr{C}$ that $\mathscr{C}$ is local. The property (13.2)(ii) now follows because $\mathscr{L}\subseteq\mathscr{G}=\mathscr{C}$.

The three properties (13.2) imply the following theorem.

(*13.3*) THEOREM (Dynkin). *The restriction $\mathscr{L}$ of $\mathscr{G}$ to C_κ^∞ is a second-order elliptic operator of the form*

$$\mathscr{L}f(x)=\frac{1}{2}\sum_i\sum_j a_{ij}(x)\partial_i\partial_j f(x)+\sum_i b_i(x)\partial_i f(x)-c(x)f(x),$$

where ∂_i denotes $\partial/\partial x_i$ and

(13.4) (i) *$\forall i,j$, the functions $a_{ij}(\cdot)$, $b_i(\cdot)$ and $c(\cdot)$ are continuous;*
(13.4) (ii) *$\forall x$, the matrix $\{a_{ij}(x): 1\leqslant i,j\leqslant n\}$ is non-negative definite symmetric;*
(13.4)(iii) *$\forall x$, $c(x)\geqslant 0$.*

Proof. Note that it follows from the local and maximum-principle properties of $\mathscr{L}$ that $\mathscr{L}$ satisfies the Local-Maximum Principle: if f in $\mathscr{D}(\mathscr{L})$ has a *local* maximum at x and $f(x)\geqslant 0$, then $\mathscr{L}f(x)\leqslant 0$.

For x in E, we can find φ in C_κ^∞ with $\varphi=1$ in a neighbourhood of x. For such a φ, we can define $c(x)=-\mathscr{G}\varphi(x)$; this defines $c(x)$ independently of the particular φ chosen.

For fixed $x=(x_1,x_2,\dots,x_n)$ in E, set

$$b_i(x)=\mathscr{G}\varphi_i(x),$$

where $\varphi_i\in C_\kappa^\infty$ and $\varphi_i(y)=y_i-x_i$ near x, and

$$a_{ij}(x):=\mathscr{G}(\varphi_i\varphi_j)(x).$$

Then c, b_i and a_{ij} are continuous. For $\lambda_1, \lambda_2, \ldots, \lambda_n \in \mathbb{R}$, the function h with

$$h(y) := -\left[\sum_i \lambda_i \varphi_i(y)\right]^2$$

has a local maximum at x, so that

$$\sum_i \sum_j a_{ij}(x)\lambda_i\lambda_j = -\mathscr{L}h(x) \geqslant 0.$$

Hence the symmetric matrix $\{a_{ij}(x): 1 \leqslant i, j \leqslant n\}$ is non-negative definite.

Now, if $f \in C_\kappa^\infty$, Taylor's formula gives (for y near x)

$$f(y) = \psi(y) + o(|y - x|^2),$$

where

$$\psi(y) := f(x)\varphi(y) + \sum_i \partial_i f(x)\varphi_i(y) + \tfrac{1}{2}\sum_i \sum_j \partial_i \partial_j f(x)\varphi_i(y)\varphi_j(y).$$

(Recall that $\varphi(y) = 1$ near x.) Note that

$$\mathscr{L}\psi(x) = -c(x)f(x) + \sum_i b_i(x)\partial_i f(x) + \tfrac{1}{2}\sum_i \sum_j a_{ij}(x)\partial_i \partial_j f(x).$$

For $\varepsilon > 0$, a function in C_κ^∞ defined near x by

$$y \mapsto f(y) - \psi(y) - \varepsilon|y - x|^2$$

has a local maximum at x. Hence

$$\mathscr{L}f(x) - \mathscr{L}\psi(x) - \varepsilon\sum_i a_{ii} \leqslant 0.$$

You can see why $\mathscr{L}f(x) = \mathscr{L}\psi(x)$, so the proof is complete. □

Suppose given a second-order elliptic operator $\mathscr{L}$ from $C_\kappa^\infty(\mathbb{R}^n)$ of $C_0(\mathbb{R}^n)$ of the form described in Theorem 13.3 (equivalently, satisfying the conditions (13.2)). Suppose that $\{P_t\}$ is an FD semigroup on $\mathbb{R}^n$ with generator $\mathscr{G}$ extending $\mathscr{L}$ and that X is the canonical FD process associated with $\{P_t\}$. We now prove that, a.s., X has *continuous* paths up to time ζ, so that (ignoring null sets) X is an FD diffuson.

(*13.5*) THEOREM (Dynkin, Kinney). *Let* $\mathscr{S}: C_\kappa^\infty \to C_0$ *be an elliptic operator of the type described in Theorem 13.3. Suppose that* $\{P_t\}$ *is an FD semigroup with generator extending* $\mathscr{L}$ *and that* X *is the associated FD process. Then, a.s., the paths of* X *are continuous on* $[0, \zeta)$.

Proof. To avoid annoyances, we give the proof when $\{P_t\}$ is further assumed to be honest, so $\zeta = \infty$, a.s. (Since then $P_t 1 = 1$, $\forall t$, it is easy to see that $c = 0$. However, the condition '$c = 0$' does not imply honesty, because it does not preclude explosion in which X reaches infinity 'continuously' in a finite time. More about explosion, and more about the case when $c \neq 0$, later.)

Let K be a closed ball in $\mathbb{R}^n$ and let G be an open ball containing K. It is well known that there exists $f \in C_K^\infty$ with $f = 1$ on $K, f = 0$ on $\mathbb{R}^n \backslash G, 0 \leqslant f \leqslant 1$ everywhere. Since $c = 0$, we have $\mathscr{L}f = 0$ on K. For $x \in K$, we have

$$P_t(x, \mathbb{R}^n \backslash G) \leqslant f(x) - P_t f(x) = -\int_0^t P_s \mathscr{L} f(x)\, ds.$$

Since $\| P_s \mathscr{L} f - \mathscr{L} f \| \to 0$ $(s \downarrow 0)$, it is now clear that

$$\sup_{x \in K} t^{-1} P_t(x, \mathbb{R}^n \backslash G) \to 0 \quad (t \downarrow\downarrow 0). \tag{13.6}$$

We now wish to prove that, for each compact K, for $\varepsilon, u > 0$ and for $x \in K$,

$$\mathbf{P}^x \left[\bigcup_{k=0}^{n-1} \{ |X(ku/n) - X((k+1)u/n)| > 3\varepsilon; X_s \in K, \forall s \leqslant u \} \right] \to 0. \tag{13.7}$$

as $n \uparrow \infty$. The theorem will then be obvious. (See (7.21).) The probability in (13.7) is dominated by

$$n \sup_{y \in K} P_{u/n}(y, \mathbb{R}^n \backslash B_{3\varepsilon}(y)),$$

where $B_\varepsilon(x)$ denotes the open ball of radius ε around x. Hence we need only show that

$$t^{-1} \sup_{y \in K} P_t(y, \mathbb{R}^n \backslash B_{3\varepsilon}(y)) \to 0 \quad (t \downarrow\downarrow 0).$$

This is an immediate consequence of (13.6). For let $x_1, x_2, \ldots, x_r$ in K be such that $B_\varepsilon(x_1), B_\varepsilon(x_2), \ldots, B_\varepsilon(x_r)$ cover K. Apply (13.6) to the case where $K = \bar{B}_\varepsilon(x_k)$ and $G = B_{2\varepsilon}(x_k)$. Let $\eta > 0$ be given. Then there exists $\delta_k > 0$ such that, $\forall t < \delta_k$,

$$\sup_{y \in B_\varepsilon(x_k)} t^{-1} P_t(y, \mathbb{R}^n \backslash B_{3\varepsilon}(y)) \leqslant \sup_{y \in B_\varepsilon(x_k)} t^{-1} P_t(y, \mathbb{R}^n \backslash B_{2\varepsilon}(x_k)) < \eta.$$

Now take $\delta := \min(\delta_1, \delta_2, \ldots, \delta_r)$, etc. □

Wiener's Theorem is obviously a corollary of Theorem 13.5.

Two problems remain.

(i) Does there exist an FD semigroup with generator $\mathscr{G}$ extending $\mathscr{L}$?
(ii) If so, is there only one such semigroup?

See Section V.22 for the answers.

(13.8) *The weak-continuity problem for* $\{\mathbf{P}^x\}$. It would be wrong to leave the present theoretical discussion of the Feller property without mentioning the connection with weak continuity. (This connection will be clearly apparent in our later treatment of Stroock–Varadhan theory. References for the special case of diffusions will be given at that stage.)

If X is an honest FD diffusion for which $P_t: C_b(\mathbb{R}^n) \to C_b(\mathbb{R}^n)$, then the map

$x \mapsto \mathbf{P}^x$ *is continuous from* $\mathbb{R}^n$ *to* $\Pr(W)$, *where* W *is the space of continuous paths in* $\mathbb{R}^n$. For all FD diffusions, we can make the same statement, provided we make a suitable slight adjustment of the concept of weak convergence.

It has been mentioned earlier that the theory of weak convergence is very highly developed in the case when W is replaced by the space D of R-paths (with values in a compact metric space E_∂) with the Skorokhod J_1 topology. This provides the appropriate setting for studying weak continuity of $\{\mathbf{P}^x\}$ for general FD semigroups. See Skorokhod's classic paper [2], and Billingsley [3], Aldous [1] and Ethier and Kurtz [1] for interesting work.

14. Characterisation of continuous real Lévy processes. Let X be a continuous 1-dimensional Lévy process, so that X has stationary independent increments. Then X is a continuous Markov process with shift-invariant transition function $\{P_t\}$:

$$P_t(x, x+\Gamma) = P_t(0, \Gamma) \quad (t \geqslant 0, x \in \mathbb{R}, \Gamma \in \mathscr{B}).$$

We shall use Dynkin's characteristic-operator formula to prove Lévy's theorem that

$$X_t = \sigma B_t + \mu t$$

for some Brownian motion B and constants σ and μ.

It is strictly elementary to show that $\{P_t\}$ has the FD property; and, since X is continuous, X is strong Markov. We do not yet know that the domain of the generator of X extends C_κ^∞.

Write

$$\mathscr{D} := \{f \in C_0 : f'' \in C_0\}.$$

Note that, for $f \in \mathscr{D}$, $f' \in C_0$ because

$$f(x+1) - f(x) = f'(x) + \tfrac{1}{2} f''(x + \theta_x) \quad \text{for some } \theta_x \text{ in } (0, 1).$$

We shall prove that every element of $\mathscr{D}$ belongs to the domain $\mathscr{D}(\mathscr{C})$ of the characteristic operator $\mathscr{C}$ of X and that there exist constants $\sigma \in \mathbb{R}^+$ and $\mu \in \mathbb{R}$ such that

$$\mathscr{C} f = \tfrac{1}{2}\sigma^2 f'' + \mu f', \quad \forall f \in \mathscr{D}.$$

Since the operator $\frac{1}{2}\sigma^2 d^2/dx^2 + \mu d/dx$ with domain $\mathscr{D}$ is exactly the generator of the FD semigroup of $\sigma B_t + \mu t$, the desired result follows.

We shall assume that, for $a < x < b$,

$$0 < \mathbf{P}^x[H_a < H_b] = 1 - \mathbf{P}^x[H_b < H_a] < 1.$$

where $H_y := \inf\{t : X_t = y\}$. (The remaining cases are easily shown to be trivial in that, for them, $X_t = X_0 + \mu t$ for some $\mu \in \mathbb{R}$.)

For $h > 0$, put

$$\tau_0^h := 0, \quad \tau_{n+1}^h := \inf\{t > \tau_n^h : |X(t) - X(\tau_n^h)| = h\}.$$

Then

$$\{X(\tau_n^h): n = 0, 1, 2, \ldots\}$$

is just a simple (Bernoulli) random walk. From standard elementary results on gambler's ruin (see Sections 1–3 of Chapter XIV of Feller [1]), we can show that there exist constants $\gamma \in \mathbb{R}$ and $\beta \in \mathbb{R}^{++}$ such that, for $a < x < b$,

$$\text{(14.1)} \qquad \mathbf{P}^x[H_a < H_b] = \frac{e^{\gamma b} - e^{\gamma x}}{e^{\gamma b} - e^{\gamma a}} \quad \left(:= \frac{b-x}{b-a} \quad \text{if } \gamma = 0 \right),$$

$$\text{(14.2)} \qquad \mathbf{E}^x[H_a \wedge H_b] = \frac{\beta}{\gamma} \frac{(b-x)(e^{\gamma x} - e^{\gamma a}) + (x-a)(e^{\gamma x} - e^{\gamma b})}{e^{\gamma a} - e^{\gamma b}}$$

$$(:= \tfrac{1}{2}\beta(b-x)(x-a) \quad \text{if } \gamma = 0).$$

To prove (14.1), (14.2), first obtain $\mathbf{P}^x[H_a < H_b]$ and $\mathbf{E}^x[H_a \wedge H_b]$ (in terms of $\mathbf{E}^0[H_{-h} \wedge H_h]$) when $x - a$ and $b - x$ are both multiples of h, and employ obvious monotonicity properties in 'letting $h \downarrow\downarrow 0$'. You will find that β can be defined as

$$\text{(14.3)} \qquad \beta := \lim_{h \downarrow\downarrow 0} 2h^{-2} \mathbf{E}^0[H_{-h} \wedge H_h] > 0;$$

and it is indeed the existence of the limit at (14.3) rather than the much more informative (14.2) that we really require.

Recall that $\mathscr{C}f(x)$ is defined as

$$\text{(14.4)} \qquad \mathscr{C}f(x) := \lim_{\eta \downarrow\downarrow 0} \frac{\mathbf{E}^x[f \circ X(\tau_1^\eta)] - f(x)}{\mathbf{E}^x[\tau_1^\eta]},$$

provided the limit exists. But, from (14.1) and (14.2), we find that

$$\mathscr{C}f(x) = \frac{\gamma}{\beta} \lim_{\eta \downarrow\downarrow 0} \frac{(e^{\gamma\eta} - 1) f(x - \eta) + (1 - e^{-\gamma\eta}) f(x + \eta) - (e^{\gamma\eta} - e^{-\gamma\eta}) f(x),}{e^{\gamma\eta} + e^{-\gamma\eta} - 2},$$

and now it follows (by Taylor series expansion) that, for $f \in \mathscr{D}$, we have $f \in \mathscr{D}(\mathscr{C})$ and

$$\mathscr{C}f = \tfrac{1}{2}\sigma^2 f'' + f', \qquad \text{where } \sigma^2 := 2\beta^{-1}, \quad \mu := -\gamma\beta^{-1}.$$

The proof of Lévy's Theorem is complete. □

Exercise. Explain why the 1-dimensional result just proved implies the n-dimensional case (I.28.12) of Lévy's result.

For left-invariant diffusions on Lie groups, see Section V.35 in Volume 2.

15. Consolidation. We can profitably draw together a few threads. You will notice that the first sentence of the next section reads: 'Let $X = (X_t, \Omega, \{\mathscr{F}_t\}, \mathbf{P}^x : x \in E_\partial)$ be an FD process with transition function $\{P_t\}$.' Let us revise what is involved in this statement. The space E is an LCCB and the space $E_\partial := E \cup \partial$ is compact metrisable. We can regard $\{P_t\}$ as a semigroup on $C_0(E)$ satisfying (6.6)(i)–(iv)—but note particularly the significance of (6.6)(iv)*—or as the corresponding transition function derived via (6.2). We can therefore also regard P_t as a map $P_t : b\mathscr{E} \to b\mathscr{E}$. The extended transition function $\{P_t^{+\partial}\}$ is an FD transition function on E_∂. Since we have so far met only *canonical* FD processes, we take X to be the canonical process described in Section 7. The process X has *R-paths*, and if either $X(s-,\omega) = \partial$ or $X(s,\omega) = \partial$ then $X(t,\omega) = \partial$ ($\forall t \geqslant s$). The *lifetime* ζ of X is defined as $\zeta := \inf\{t : X_t = \partial\}$. The σ-algebra $\mathscr{F}_t^\circ$ is as in (7.16), and $\mathscr{F}_t$ is as in (9.2). We note particularly the *Début Theorem* 9.3. The process X is *strong Markov* relative to the filtration $\{\mathscr{F}_t\}$ in the sense described in Theorem 9.4, and is *quasi-left-continuous* in the sense of Theorem 11.1. Moreover (Theorem 11.6), the filtration $\{\mathscr{F}_t\}$ is also qlc. The resolvent of $\{P_t\}$ (or of X) is defined in (3.10). *Dynkin's formula* (10.7), which 'decomposes the resolvent at a stopping time T' is extremely important. Finally, the *generator* $\mathscr{G}$ of X may be defined via either (4.11) or Dynkin's characteristic-operator formula (12.2) (which are equivalent for our FD process X).

3. ADDITIVE FUNCTIONALS

16. PCHAFs; λ-excessive functions; Brownian local time. Let

$$X = (X_t, \Omega, \{\mathscr{F}_t\}, \mathbf{P}^x : x \in E_\partial)$$

be an FD process with transition function $\{P_t\}$. (You will see that the FD property is not really used in our results on additive functionals, so that we can (and shall) apply these results in more general contexts when these arise. We have a clear idea of what an FD process is, and it provides a good enough context to be getting on with.)

Let c be a measurable function from E to $[0,\infty)$. (Take $c(\partial) = 0$ by convention.) Define

$$(16.1) \qquad A_t(\omega) := \int_0^t c \circ X_s(\omega)\, ds, \quad \text{or} \quad A_t := \int_0^t c(X_s)\, ds \quad \text{for short.}$$

Expressions of this type occur as cost functions in control theory, and represent occupation times when c is an indicator function. In practice, we often wish to find the $\mathbf{P}^x$ distribution of A_t for each x. For this purpose we use the *Feynman–Kac (FK) formula*, first developed for quantum-theoretic applications. We prove the FK formula in Section 19.

A second major application of additive functionals is as compensators of certain potentials. For example, if A_t is defined as in (16.1) and $h(x) := \mathbf{E}^x A_\infty$, assumed everywhere finite, then

$$(16.2) \qquad \mathbf{E}(A_\infty \mid \mathscr{F}_t) = A_t + h(X_t)$$

expresses the Doob decomposition of the supermartingale $h(X_t)$ (you should check that this *is* a supermartingale!) More commonly, we first discount time to ensure convergence, forming, for example,

$$\varphi(x) := \mathbf{E}^x \int_0^\infty e^{-\lambda t} f(X_t)\, dt := R_\lambda f(x).$$

The analogue of (16.2) now is

$$\mathbf{E}\left[\int_0^\infty e^{-\lambda u} f(X_u)\, du \,\middle|\, \mathscr{F}_t\right] = \int_0^t e^{-\lambda u} f(X_u)\, du + e^{-\lambda t} R_\lambda f(X_t)$$

expressing the Doob decomposition of the supermartingale $e^{-\lambda t} R_\lambda f(X_t)$. To say that $e^{-\lambda t} R_\lambda f(X_t)$ is a supermartingale amounts to saying that $g := R_\lambda f \geqslant e^{-\lambda t} P_t g$ for all $t \geqslant 0$. It is *almost* the case that if ψ is a function such that $\psi \geqslant e^{-\lambda} P_t \psi$ for all $t \geqslant 0$ then $\psi = R_\lambda f$ for some f; the exact statement is the Volkonskii–Šur–Meyer Theorem 16.7.

The third major use of additive functionals is as random clocks to time-change Markov processes; we saw a very important use of this technology in I.5.13, and will explore the technique thoroughly in Sections 22–25, and again in Volume 2.

(*16.3*) DEFINITION. (PCHAF) A perfect continuous homogeneous additive functional (PCHAF) of X *is an* $\{\mathscr{F}_t\}$*-adapted process* A *such that, for some set* Ω_0 *in* $\mathscr{F}$ *with* $\mathbf{P}^x(\Omega_0) = 1$, $\forall x$, *the following properties hold for every* ω *in* Ω_0:

(*i*) $t \mapsto A_t(\omega)$ *is continuous, non-decreasing and* $A_0(\omega) = 0$;
(*ii*) $\forall s, \forall t,\ A_{s+t}(\omega) = A_s(\omega) + A_t(\theta_s \omega)$;
(*iii*) $A(\omega)$ *is constant on* $[\zeta(\omega), \infty)$.

Terminological note. In Dynkin [2], 'perfect' refers to the 'adapted' property of A. In Blumenthal and Getoor [1], 'perfect' refers to the fact that the 'exceptional' set $\Omega \backslash \Omega_0$ in (ii) can be chosen independently of s and t. Dynkin calls this 'strict homogeneity'. We win either way. Everyone would agree as to what a PCHAF is.

If, for example, the function c in (16.1) is (non-negative and) bounded then (16.1) defines a PCHAF A. (Some boundedness condition on c is needed to keep A_t finite.) It is true that all PCHAFs are 'limits' of such 'integral' PCHAFs, and this is reflected in the important Existence and Uniqueness Theorem 16.7.

(*16.4*) DEFINITION (uniformly λ-excessive). *Let* $\lambda > 0$ *be fixed. An* $\mathscr{E}^*$*-measurable function* f *from* E *to* $\mathbb{R}$ *is called* uniformly λ-excessive *if*

(*i*) f *is bounded;*
(*ii*) f *is* λ*-super-median (see Section 7):*

$$0 \leqslant e^{-\lambda t} P_t f(x) \leqslant f(x), \quad \forall t > 0,\ \forall x \in E;$$

(*iii*) $\|e^{-\lambda t} P_t f - f\| \to 0$ *as* $t \downarrow 0$.

The norm in (iii) is of course the supremum norm. We take $f(\partial)=0$ by convention.

(16.5) *Example.* For any FD process, $R_\lambda h$ is uniformly λ-excessive for $h\in C_0^+$.

(16.6) *Example.* Let X be CBM($\mathbb{R}$). Then

$$f(x):=\gamma^{-1}\exp(-\gamma|x|),\quad \gamma:=(2\lambda)^{1/2},$$

defines a uniformly λ-excessive function f. (See Exercise 16.12.)

(16.7) THEOREM (Volkonskii, Šur, Meyer). *Fix $\lambda>0$. Let f be a uniformly λ-excessive function on E. Then there exists a unique PCHAF A of X such that f is the λ-potential of A:*

$$f(x)=\mathbf{E}^x\int_0^\infty e^{-\lambda t}\,dA_t. \tag{16.8}$$

Uniqueness means that if B is another PCHAF for which (16.8) holds then

$$\mathbf{P}^x[A_t=B_t,\forall t]=1,\ \forall x.$$

Note that our conventions force both sides of (16.8) to equal 0 when $x=\partial$.

Before proving Theorem 16.7, let us look at Examples 16.5 and 16.6. In the case of (16.5),

$$A_t=\int_0^t h(X_s)\,ds.$$

(16.9) *Brownian local time at* 0. Let A be the unique PCHAF of CBM ($\mathbb{R}$) with λ-potential f as in (16.6). Now f was chosen so that

$$f(x)=\mathbf{E}^x[\exp(-\lambda H_0)]f(0)=P^\lambda_{H_0}f(x)$$

in the notation of Section 10. By 'Dynkin's formula' (see Exercise 16.14),

$$\mathbf{E}^x\int_{H_0}^\infty e^{-\lambda t}\,dA_t=P^\lambda_{H_0}f(x)=f(x)=\mathbf{E}^x\int_0^\infty e^{-\lambda t}\,dA_t,$$

so that

$$\mathbf{P}^x[A(H_0)=0]=1.$$

It is now obvious (see Exercise 16.14) that (a.s.) *A grows only when X is at* 0:

$$A_t=\int_0^t I_{\{0\}}(X_s)\,dA_s. \tag{16.10}$$

It is further obvious from the 'uniqueness' part of Theorem 16.7 that, *up to constant multiples, A is the only PCHAF satisfying* (16.10).

(16.11) *Remarks.* In dimension $n\geqslant 2$, we cannot define Brownian local time at a point (why?), but we can define the local time spent on certain *sets*.

Exercises. These exercises provide good practice in the use of θ_t operators.

(*16.12*) Prove that the function f of Example 16.6 is uniformly λ-excessive for $X = \text{CBM}(\mathbb{R})$. *Hint.* Use probabilistic (as opposed to analytical) reasoning to show that f is λ-super-median.

Solution. Since f is in $C_0(\mathbb{R})$ and X is FD, the only point that needs proof is that f is λ-super-median. Since, as we have already seen,

$$f(x) = \mathbf{E}^x[\exp(-\lambda H_0)]f(0),$$

it is enough to prove that

$$g(x) := \mathbf{E}^x[\exp(-\lambda H_0)]$$

defines a λ-super-median function g. Now, for $t \geqslant 0$,

$$H_0 \leqslant \inf\{s > t : X_s = 0\} = t + H_0 \circ \theta_t.$$

Hence, with $\eta := \exp(-\lambda H_0)$, we have

$$\begin{aligned} g(x) = \mathbf{E}^x[\eta] &\geqslant e^{-\lambda t}\mathbf{E}^x[\theta_t \eta] = e^{-\lambda t}\mathbf{E}^x[\mathbf{E}^{X(t)}\eta] \\ &= e^{-\lambda t}\mathbf{E}^x[g(X_t)] = e^{-\lambda t}P_t g(x), \end{aligned} \qquad \square$$

as required.

(*16.13*) Though we are at present primarily interested in the 'Brownian local time' case, this exercise covers the general situation. Show that if A is a PCHAF (of some FD process X) with finite λ-potential f then, for a stopping time T,

$$\mathbf{E}^x \int_T^\infty e^{-\lambda t}\, dA_t = P_T^\lambda f(x).$$

This generalises Dynkin's formula (10.7), which becomes the special case when A is an integral PCHAF of the form (16.1).

Solution. We calculate

$$\int_T^\infty e^{-\lambda t}\, dA_t = \int_T^\infty \lambda e^{-\lambda t}(A_t - A_T)\, dt = e^{-\lambda t}\int_0^\infty \lambda e^{-\lambda s}(A_{T+s} - A_T)\, ds$$

$$= e^{-\lambda t}\int_0^\infty \lambda e^{-\lambda s} A_s \circ \theta_T\, ds = e^{-\lambda T}\theta_T \eta,$$

where

$$\eta := \int_0^\infty e^{-\lambda s}\, dA_s = \int_0^\infty \lambda e^{-\lambda s} A_s\, ds.$$

By the Strong Markov Theorem,

$$\mathbf{E}^x \int_T^\infty e^{-\lambda t}\, dA_t = \mathbf{E}^x[e^{-\lambda T}\theta_T \eta] = \mathbf{E}^x[e^{-\lambda T}\mathbf{E}^{X(T)}\eta]$$

$$= \mathbf{E}^x[e^{-\lambda T} f(X_T)] = P_T^\lambda f(x),$$

as required. □

Note. On taking $T = u$, we find that

$$f(x) = \mathbf{E}^x \int_0^\infty e^{-\lambda t}\, dA_t \geqslant \mathbf{E}^x \int_u^\infty e^{-\lambda s}\, dA_s = e^{-\lambda u} P_u f(x),$$

so that f must be λ-super-median. Because of the Monotone-Convergence Theorem, we can see that f is even λ-*excessive* in that $e^{-\lambda u}P_u f(x) \uparrow f(x)$ as $u \downarrow 0$; but it need *not* be true that f is *uniformly* λ-excessive.

(*16.14*) Now return to the 'Brownian local time' case and prove (16.10).

Solution. Since $\mathbf{P}^y[A(H_0) = 0] = 1$, $\forall y$, we have

$$\mathbf{P}^x[A(t + H_0 \circ \theta_t) - A(t) = 0] = \mathbf{E}^x \mathbf{P}^{X(t)}[A(H_0) = 0] = 1.$$

Thus, for every x,

$$\mathbf{P}^x[A(q + H_0 \circ \theta_q) = A(q), \quad \forall q \in \mathbf{Q}^+] = 1.$$

The rest is easy. □

17. Proof of the Volkonskii–Šur–Meyer Theorem. If the λ-excessive function f were of the form $f = R_\lambda g$, we would recover g by taking $g = (\lambda - \mathscr{G})f$. The idea in general is to approximate $\lambda - \mathscr{G}$ by $n(I - P_{1/n}^\lambda)$, where we recall $P_t^\lambda := e^{-\lambda t}P_t$. Thus we define

$$g_n := n(f - P_{1/n}^\lambda f) \geqslant 0,$$

and then define

$$f_n := R_\lambda g_n = n \int_0^{1/n} P_t^\lambda f\, dt.$$

It is easy and important that $f_n \uparrow f$ *uniformly* on E. Then

$$f_n(x) = R_\lambda g_n(x) = \mathbf{E}^x \int_0^\infty e^{-\lambda t} g_n(X_s)\, ds = \mathbf{E}^x \int_0^\infty e^{-\lambda t}\, dA_n(t),$$

where

$$A_n(t) := \int_0^t g_n(X_s)\, ds.$$

Put

$$C_n(t) := \int_0^t e^{-\lambda s}\, dA_n(s) = \int_0^t e^{-\lambda s} g_n(X_s)\, ds,$$

so that

(17.1) $$f_n(x) = \mathbf{E}^x[C_n(\infty)].$$

For $n \geqslant m$, and with the shorthand

$$g_{m,n} := g_n - g_m \leqslant g_n, \quad f_{m,n} := f_n - f_m \geqslant 0,$$

we make the estimate (with $\|f\|$ and $\|f - f_m\|$ denoting supremum norms)

$$0 \leqslant \mathbf{E}^x([C_n(\infty) - C_m(\infty)]^2)$$

$$= 2\mathbf{E}^x \int_{s=0}^{\infty} \int_{t=0}^{\infty} e^{-\lambda s} g_{m,n}(X_s)\, e^{-\lambda(s+t)} g_{m,n}(X_{s+t})\, ds\, dt$$

$$= 2\mathbf{E}^x \int_s \int_t e^{-\lambda(2s+t)} g_{m,n}(X_s) P_t g_{m,n}(X_s)\, ds\, dt$$

(because of the simple Markov property)

$$= 2\mathbf{E}^x \int_{s=0}^{\infty} e^{-2\lambda s} g_{m,n}(X_t) f_{m,n}(X_s)\, ds \quad (\text{since } f_{m,n} = R_\lambda g_{m,n})$$

$$\leqslant 2\mathbf{E}^x \int_{s=0}^{\infty} e^{-2\lambda s}\, g_n(X_s) f_{m,n}(X_s)\, ds$$

$$\leqslant 2\|f - f_m\| \mathbf{E}^x \int_{s=0}^{\infty} e^{-\lambda s}\, g_n(X_s)\, ds = 2\|f - f_m\|\, f_n(x)$$

$$\leqslant 2\|f\|\, \|f - f_m\|.$$

(The 'Fubini' operations are justified because $\|g_{m,n}\| \leqslant (m+n)\|f\| < \infty$.) But

(17.2) $$\mathbf{E}^x[C_n(\infty) | \mathscr{F}_t] = C_n(t) + e^{-\lambda t} f_n(X_t), \quad \text{a.s.}(\mathbf{P}^x),$$

so that, for each x,

(11.3) $$M_{m,n}(t) := C_n(t) - C_m(t) + e^{-\lambda t} f_n(X_t) - e^{-\lambda t} f_m(X_t)$$

defines a martingale relative to $(\{\mathscr{F}_t\}, \mathbf{P}^x)$ with terminal value $C_n(\infty) - C_m(\infty)$. For the moment, *assume* that the martingale $M_{m,n}$ has R-paths. Set

$$M^*_{m,n} := \sup |M_{m,n}(t)|.$$

Then, by Doob's L^2 inequality (II.70.2),

$$\mathbf{E}^x M^*_{m,n} \leqslant [\mathbf{E}^x(M^{*2}_{m,n})]^{1/2} \leqslant (8\|f\|\, \|f - f_m\|)^{1/2}.$$

If we choose a sequence $n(k)$ such that $\sum \|f - f_{n(k)}\|^{1/2}$ converges then (Borel–

Cantelli), almost surely (a.s.($\mathbf{P}^x$), $\forall x$),

$$\sum[C_{n(k+1)}(t) - C_{n(k)}(t) + e^{-\lambda s} f_{n(k+1)}(X_t) - e^{-\lambda t} f_{n(k)}(X_t)]$$

converges *uniformly* over $t \in [0, \infty]$. But

$$\sup_t |e^{-\lambda t} f_{n(k+1)}(X_t) - e^{-\lambda t} f_{n(k)}(X_t)| \leqslant \| f_{n(k+1)} - f_{n(k)} \| \leqslant \| f - f_{n(k)} \|,$$

and, since $\sum \| f - f_{n(k)} \| < \infty$, we deduce that, almost surely,

$$C(t) := \lim C_{n(k)}(t)$$

exists uniformly over $t \in [0, \infty]$ and defines a continuous increasing process C. It is easy to check that

$$A(t) := \int_0^t e^{\lambda s} dC_s$$

satisfies (a.s.)

$$A(t) = \lim A_{n(k)}(t),$$

the limit existing uniformly over compact intervals. Hence A is a PCHAF. That

$$f(x) = \mathbf{E}^x[C(\infty)] = \mathbf{E}^x \int_0^\infty e^{-\lambda t}\, dA(t)$$

follows from (17.1) because $C_n(\infty) \to C(\infty)$ in $\mathscr{L}^2$ and hence in $\mathscr{L}^1$.

(17.4) *R-property of* $M_{m,n}$. In the cases that will concern us, f_m and f_n are *continuous* functions on E, and then $M_{m,n}$ inherits the R-property from X. If we assume only that f_m and f_n are *Borel* (as will happen if f is Borel), then we shall see in Volume 2 that the section theorem implies that $M_{m,n}$ is an R-process. However, the R-property of $M_{m,n}$ can be established in general: f is *nearly Borel*. See Theorem II.2.12 of Blumenthal and Getoor [1].

(17.5) *Connection with the Meyer decomposition.* We have

$$e^{-\lambda t} f(X_t) = \mathbf{E}^x[C(\infty) | \mathscr{F}_t] - C(t).$$

This is (see Volume 2) the Meyer decomposition of the 'regular class (D) potential' $t \mapsto e^{-\lambda t} f(X_t)$ on $(\Omega, \{\mathscr{F}_t\}, \mathbf{P}^x)$ as the 'potential' of the continuous integrable increasing process C. The uniqueness part of the Meyer decomposition theorem implies the *uniqueness of* A asserted in Theorem 16.7. Though it is not difficult to prove the uniqueness of A directly (you should be able to sense how to do it from the proof of the existence part of Theorem 16.7), it is best thought of in terms of martingale theory, so we take the uniqueness result for granted for now. □

18. Killing. We now construct a process $\hat{X}$ that represents 'X killed at rate dA_t'. (A now denotes some fixed PCHAF of X.) The intuitive idea is that $\hat{X}$ agrees with X up to time $\hat{\zeta} \leqslant \zeta$ and $\hat{X}(t) = \partial$, $\forall t \geqslant \hat{\zeta}$, where

$$\hat{\mathbf{P}}[\hat{\zeta} > t | X] = `\prod_{s \leqslant t} [1 - dA(s)]' = e^{-A(t)}. \tag{18.1}$$

Set

$$M_t := M(t) := e^{-A(t)},$$

so that M is a PCHMF of X. The 'M' in PCHMF stands of course for *multiplicative* and reflects the property

$$M_{s+t}(\omega) = M_s(\omega) M_t(\theta_s \omega). \tag{18.2}$$

The type of construction required is standard and straightforward, but even the most intuitive ideas *look* tedious in probability theory. Blumenthal and Getoor [1] do their killing as quickly and humanely as possible, and we shall follow them. Needless to say, they do a much more thorough post-mortem, if you want all the gory details.

The situation is rather curious because one normally works with σ-algebras $\{\hat{\mathscr{F}}_t^\circ\}$ for $\hat{X}$ that are, so to speak, 'half-completed'. (See Blumenthal and Getoor [1].) We shall keep things 'algebraic' by assuming that A (equivalently, M) is $\{\mathscr{F}_t^\circ\}$-adapted (as it will be in all cases of interest to us). We can then work with $\{\mathscr{F}_t^\circ\}$ and Borel functions on E instead of $\{\mathscr{F}_t\}$ and universally-measurable functions on E. We further assume that we have 'weeded out' (rejected) a null set so that for *every* ω, $M_0(\omega) = 1$ and $t \mapsto M_t(\omega)$ is continuous non-increasing.

Let $\hat{\Omega} := \Omega \times [0, \infty]$ and let $\hat{\omega} = (\omega, \xi)$ denote the typical point of $\hat{\Omega}$. Let $\mathscr{R}$ be the Borel σ-algebra on $[0, \infty]$ and set $\hat{\mathscr{F}}^\circ := \mathscr{F}^\circ \times \mathscr{R}$. Define $\hat{\zeta}(\hat{\omega}) := \zeta(\omega) \wedge \xi$, and put

$$\hat{X}_t(\hat{\omega}) := \begin{cases} X_t(\omega) & \text{if } t < \xi, \\ \partial & \text{if } t \geqslant \xi. \end{cases}$$

Define $\hat{\theta}_t \hat{\omega} := (\theta_t \omega, (\xi - t) \vee 0)$. Then

$$\hat{X}_t \circ \hat{\theta}_h = \hat{X}_{t+h}, \quad \forall t, \forall h. \tag{18.3}$$

Let $\hat{\Omega}_t := \Omega \times (t, \infty] \in \hat{\mathscr{F}}^\circ$, and, for $\hat{\Lambda} \in \hat{\mathscr{F}}^\circ$, put $\hat{\Lambda}$ in $\hat{\mathscr{F}}_t^\circ$ if there exists Λ in $\mathscr{F}_t^\circ$ such that

$$\hat{\Lambda} \cap \hat{\Omega}_t = \Lambda \times (t, \infty].$$

Then

(18.4) *$\{\hat{\mathscr{F}}_t^\circ\}$ is a filtration of $(\hat{\Omega}, \hat{\mathscr{F}}^\circ)$ and $\hat{X}$ is $\{\hat{\mathscr{F}}_t^\circ\}$-adapted.*

For $\omega \in \Omega$, define a probability measure α_ω on $[0, \infty]$ by setting $\alpha_\omega(\{0\}) := 0$ and

$$\alpha_\omega(t, \infty] := M_t(\omega) \quad (t \in [0, \infty]).$$

If $\hat{\Lambda} \in \hat{\mathscr{F}}^\circ = \mathscr{F}^\circ \times \mathscr{R}$, then, for each ω,

$$\hat{\Lambda}^\omega := \{\xi \in [0, \infty] : (\omega, \xi) \in \hat{\Lambda}\} \in \mathscr{R},$$

and we can easily check by monotone-class arguments that $\omega \mapsto \alpha_\omega(\hat{\Lambda}^\omega)$ is $\mathscr{F}^\circ$-measurable. Define

$$\hat{\mathbf{P}}^x(\hat{\Lambda}) := \mathbf{E}^x[\alpha_\omega(\hat{\Lambda}^\omega)] \quad (x \in E_\partial). \tag{18.5}$$

Then $\hat{\mathbf{P}}^x$ is a probability measure on $(\hat{\Omega}, \hat{\mathscr{F}}^\circ)$ for each x in E_∂. Recall that we use $\mathrm{b}\mathscr{E}_0$ to denote the space of bounded $\mathscr{E}_\partial$-measurable functions f on $\mathscr{E}_\partial$ such that $f(\partial) = 0$.

(*18.6*) THEOREM. *$\hat{X} = (\hat{\Omega}, \hat{\mathscr{F}}^\circ, \hat{\mathscr{F}}_t^\circ, \hat{X}_t, \hat{\theta}_t, \hat{\mathbf{P}}^x)$ is a Markov process on E with transition semigroup $\{\hat{P}_t\}$, where*

$$\hat{P}_t f(x) := \mathbf{E}^x[M_t f(X_t)] \quad (f \in \mathrm{b}\mathscr{E}_0, x \in E). \tag{18.7}$$

Note. You will appreciate that the notation $(\hat{\Omega}, \hat{\mathscr{F}}^\circ, \{\hat{\mathscr{F}}_t^\circ\}, \{\hat{X}_t\}, \ldots)$ would be just too unwieldy.

Clarification of Theorem 18.6. Since we have not previously met a sextuple like $\hat{X}$ except in the case of canonical FD processes, a word of clarification is necessary. We are not going to spell out all the axiomatics. (See Blumenthal and Getoor [1].) It is clear that, for example, statements (18.3) and (18.4) form part of the definition of the statement that the sextuple $\hat{X}$ forms a Markov process. But only one property really matters:

$$\hat{\mathbf{E}}^x[f(\hat{X}_{s+t}) | \hat{\mathscr{F}}_s^\circ] = \hat{P}_t f(\hat{X}_s) \quad (f \in \mathrm{b}\mathscr{E}_0). \tag{18.8}$$

Proof of (18.8). Let $\hat{\Lambda} \in \hat{\mathscr{F}}_s^\circ$, so that $\hat{\Lambda} \cap \hat{\Omega}_s = \Lambda \times (s, \infty]$ for some Λ in $\mathscr{F}_s^\circ$. Then, since $f(\partial) = 0$, you can check that

$$\begin{aligned} \hat{\mathbf{E}}^x[f(\hat{X}_{s+t}); \hat{\Lambda}] &= \mathbf{E}^x[f(X_{s+t}) M_{s+t}; \Lambda] \\ &= \mathbf{E}^x[(M_s I_\Lambda) \theta_s(M_t[f \circ X_t])] \\ &= \mathbf{E}^x[(M_s I_\Lambda) \hat{P}_t f(X_s)] = \hat{\mathbf{E}}^x[\hat{P}_t f(\hat{X}_s); \hat{\Lambda}]. \end{aligned}$$ □

(*18.9*) *Illustrative examples.* The Feynman–Kac formula will allow us to calculate the infinitesimal generator of $\{\hat{P}_t\}$ in important cases.

(*18.10*) *Example.* We shall see that (modulo certain technical qualifications) if X is an FD diffusion with 'differential' generator $\mathscr{L}$ satisfying

$$\mathscr{L} f(x) = \tfrac{1}{2} \sum \sum a_{ij}(x) \partial_i \partial_j f(x) + \sum b_i(x) \partial_i f(x), \tag{18.11}$$

and

$$A(t) := \int_0^t c(X_s)\, ds \quad (c \text{ non-negative}) \tag{18.12}$$

then $\hat{X}$ is a 'locally' FD diffusion with differential generator

$$\hat{\mathscr{L}}f(x) = \mathscr{L}f(x) - c(x)f(x). \tag{18.13}$$

The significance of the operation of killing is that, in connection with the big problems mentioned in Section 13, we can restrict attention to the case $c(x) = 0$, $\forall x$.

(*18.14*) *Killing at constant rate.* Perhaps the most frequently used application of killing occurs when we take c to be a constant function: $c(x) = \lambda\ (\geqslant 0)$, $\forall x$. (X now denotes an arbitrary FD process.) The killing operation in this case simply involves constructing a variable ξ that, under each $\mathbf{P}^x$, is independent of X and has the exponential distribution of rate λ. ($\xi = \infty$ if $\lambda = 0$). We then put

$$\hat{X}_t := \begin{cases} X_t & (t < \xi), \\ \partial & (t \geqslant \xi). \end{cases}$$

Note that our convention that A be constant on $[\zeta, \infty]$ leads us to take $A_t = (\lambda t) \wedge \zeta$ instead of $A_t = \lambda t$, but it does not matter: 'After the first death, there is no other.' We have

$$\hat{P}_t = P_t^\lambda := e^{-\lambda t} P_t, \quad \hat{\mathscr{G}} = \mathscr{G} - \lambda.$$

The Feynman–Kac formula (Section 19) localizes this idea.

(*18.15*) *Exercise.* Let G be an open set, and let $F := E_\partial \backslash G$. Define

$$X_t^G := \begin{cases} X_t & (t < H_F), \\ \partial & (t \geqslant H_F). \end{cases}$$

Thus X^G is the process X killed on first leaving G. Formulate and prove the result that X^G is Markov, noting especially how the *terminal* time property

$$H_F - t = H_F \circ \theta_t \quad \text{on } \{H_F > t\} \tag{18.16}$$

is used in your proof. Explain how X^G corresponds to the $\hat{X}$ process obtained by killing X in accordance with the right-continuous multiplicative functional M, where

$$M_t := \begin{cases} 1 & (t < H_F), \\ 0 & (t \geqslant H_F). \end{cases}$$

See III.3.7 of Blumenthal and Getoor [1].

19. The Feynman–Kac formula. We shall examine three approaches to the Feynman–Kac formula: one analytical, one via Markov-process theory and (inevitably) one via martingale theory. Inevitably too, the martingale approach is the one that best corresponds to our intuition.

Analytical approach. We first consider a simple situation with hypotheses that are much too strong for certain applications. Let X be an FD process with transition semigroup $\{P_t\}$ acting on C_0 and with generator $\mathscr{G}$. Let v be a bounded continuous non-negative function on E and set

$$A_t := \int_0^t v(X_s)\,ds, \qquad M_t := e^{-A(t)}.$$

We prove that (note that our notation is consistent with our previous notation if v is a constant function)

$$P_t^v f(x) := \mathbf{E}^x[M_t f(X_t)] \tag{19.1}$$

defines an FD semigroup $\{P_t^v\}$ with generator $\mathscr{G}^v$ satisfying

$$\mathscr{D}(\mathscr{G}^v) = \mathscr{D}(\mathscr{G}), \qquad \mathscr{G}^v f(x) = \mathscr{G} f(x) - v(x) f(x). \tag{19.2}$$

In fact, since $f \mapsto vf$ is a bounded operator (which we denote by v), it is a standard piece of semigroup theory that $\mathscr{G}^v$, considered as *defined* by (19.2), generates an FD semigroup $\{P_t^v\}$ with

$$P_t^v f = \lim_{n \uparrow\uparrow \infty} (e^{-tv/n} P_{t/n})^n f \quad (\text{in } C_0), \tag{19.3}$$

where, of course,

$$(e^{-tv/n} f)(x) := e^{-tv(x)/n} f(x).$$

(19.4) *Exercise.* Convince yourself that (19.3) is the analytical counterpart to (19.1). Do not worry about rigour, because we are now going to *prove* something much better.

Markov-process approach. Let us reformulate (19.2) in terms of resolvents:

$$R_\lambda^v = R_\lambda - R_\lambda v R_\lambda^v. \tag{19.5}$$

We can prove (19.5) directly and under wide conditions. It is important that we can drop the continuity requirement on v. So let us assume that *v is non-negative and $\mathscr{E}$-measurable and that $R_\lambda v(x) < \infty, \forall x$.* As for X, we do not need anything as strong as the FD property (though you can assume it for now). The only hypothesis really needed on X is that $(t, \omega) \mapsto X(t, \omega)$ be measurable relative to the respective σ-algebras $\mathscr{B}[0, \infty) \times \mathscr{F}$ and $\mathscr{E}$.

We use the property

$$A_{s+u} - A_s = A_u \circ \theta_s$$

and the simple Markov property of X to calculate, for $f \in b\mathscr{E}_0$,

$$R_\lambda f(x) - R_\lambda^v f(x) = \mathbf{E}^x \int_0^\infty e^{-\lambda t - A(t)} f(X_t)(e^{A(t)} - 1)\,dt$$

$$= \mathbf{E}^x \int_0^\infty dt\, e^{-\lambda t - A(t)} f(X_t) \int_0^t ds\, v(X_s) e^{A(s)}$$

$$= \mathbf{E}^x \int_0^\infty ds\, e^{-\lambda s} v(X_s) \int_0^\infty du \exp(-\lambda u - A_u \circ \theta_s) f(X_u \circ \theta_s)$$

$$= \mathbf{E}^x \int_0^\infty e^{-\lambda s} v(X_s) R_\lambda^v f(X_s)\, ds = R_\lambda v R_\lambda^v f(x).$$

(*19.6*) *Exercise.* Prove the easier, but less useful, result

(19.7) $$R_\lambda f(x) - R_\lambda^v f(x) = R_\lambda^v v R_\lambda f(x).$$

Martingale approach. Let us now give the 'right' treatment based on stochastic integral theory which is justified later. Though we now use the language of infinitesimal generators, the present approach is just as general as the Markov-process approach. The idea is to use the fact, for $f \in \mathscr{D}(\mathscr{G})$,

$$C_t^f := f(X_t) - \int_0^t \mathscr{G} f(X_s)\, ds$$

defines a martingale. Then

$$d[e^{-A(t)} f(X_t)] = e^{-A(t)}[\mathscr{G} f(X_t) - v(X_t) f(X_t)]\, dt + e^{-A(t)}\, dC_t^f.$$

(If you do not know that stochastic calculus obeys different rules from Newton's, you will not be surprised by this calculation.) Thus

(19.8) $$M_t f(X_t) - f(X_0) = \int_0^t M_s \mathscr{G}^v f(X_s)\, ds + N_t,$$

where

$$\mathscr{G}^v f(x) := \mathscr{G} f(x) - v(x) f(x), \qquad N_t := \int_0^t e^{-A(t)}\, dC_t^f.$$

(Continuity of v is not required in order for this to make sense.) The key fact (proved in Volume 2) is that N, as the stochastic integral of a bounded continuous adapted process relative to a martingale in L^2, is itself a martingale in L^2 (for each $\mathbf{P}^x$). Taking $\mathbf{P}^x$ expectations in (19.8),

$$P_t^v f(x) - f(x) = \int_0^t P_s^v \mathscr{G}^v f(x)\, ds.$$

This is essentially (19.7).

Exercise. How do you obtain (19.5) by the martingale method?

Note. By the *Feynman–Kac formula,* we mean any or all of (19.2), (19.5) and (19.7).

20. A Ciesielski–Taylor Theorem. The following strange result was discovered by Ciesielski and Taylor [1] by explicit calculation of the distributions involved. Only in the case when $n = 1$ is a simple non-computational explanation known (Williams [7]; see Section 49; but see also Getoor and Sharpe [2] and Biane [1].)

(*20.1*) THEOREM. *The hitting time of the sphere* $\{|x| = 1\}$ *by a* $BM_0(\mathbb{R}^n)$ *process has the same distribution as the total time spent in the ball* $\{|x| \leqslant 1\}$ *by a* $BM_0(\mathbb{R}^{n+2})$ *process.*

Let us now use the Feynman–Kac formula to obtain the distribution of the time spent in $\{|x| \leqslant 1\}$ by CBM ($\mathbb{R}^3$). The general case of CBM ($\mathbb{R}^n$) ($n \geqslant 3$) may be studied in exactly the same way, but we spare you (explicit use of) Bessel functions. Use of the Feynman–Kac formula is simpler than the Kac 'method of moments' technique employed by Ciesielski and Taylor.

So let X be CBM ($\mathbb{R}^3$). Let B be the unit ball in $\mathbb{R}^3$ and let

$$\varphi_t := \text{meas}\,\{s \leqslant t : X_s \in B\} = \int_0^t I_B(X_s)\,ds.$$

Fix $\alpha > 0$ for the moment and put

$$A_t := \alpha\varphi_t := \int_0^t v(X_s)ds, \quad v := \alpha I_B.$$

Introduce

$$R^v_\lambda 1(x) := \int_0^\infty e^{-\lambda t}\mathbf{E}^x[e^{-A(t)}]\,dt.$$

Put

$$A(\infty) := \uparrow \lim A(t) = \alpha\varphi(\infty).$$

Then

$$h(x) := \mathbf{E}^x[e^{-A(\infty)}] = \lim_{\lambda\downarrow\downarrow 0} \lambda R^v_\lambda 1(x).$$

By (19.5),

$$R_\lambda 1 - R^v_\lambda 1 = R_\lambda v R^v_\lambda 1,$$

so that, since $\lambda R_\lambda 1 = 1$,

$$1 - \lambda R^v_\lambda 1 = \lambda R_\lambda v R^v_\lambda 1.$$

On letting $\lambda \downarrow\downarrow 0$ (justification is completely trivial)

$$(20.2) \qquad 1 - h = R_0[vh] = \alpha \int_B g(x, y)h(y)\,dy$$

where g is the free-space Green function for $\frac{1}{2}\Delta$. See Section I.22. Spitzer [1] gives a neat treatment of (20.2), which we now follow. It is clear from the definition of h that h is spherically symmetric: $h(x) = f(|x|)$ for some f. The right-hand side of (20.2) is therefore the gravitational potential due to a spherically symmetric mass distribution. Gauss found the way to deal with such potentials. First, the potential outside a spherical *shell* due to a symmetric mass distribution on the shell is the same as if the whole mass of the shell were concentrated at its centre. Secondly, the potential inside a spherical shell due to a mass distribution† on the shell is constant, and therefore equal to the value at the shell's centre. We may therefore calculate, for $0 < |x| < 1$,

$$1 - h(x) = \alpha(2\pi|x|)^{-1} \int_{\{|y| < |x|\}} h(y)\,dy + \alpha \int_{\{|x| < |y| < 1\}} (2\pi|y|)^{-1} h(y)\,dy.$$

Thus, with $u(r) \equiv rf(r)$, where $h(x) = f(|x|)$, we have, for $0 < r < 1$,

$$r - u(r) = 2\alpha \int_0^r \rho u(\rho)\,d\rho + 2\alpha r \int_r^1 u(\rho)\,d\rho,$$

whence $u'' = 2\alpha u$ on $(0, 1)$. We now easily obtain

(20.3) $$\mathbf{E}^0 e^{-\alpha\varphi(\infty)} = \operatorname{sech} \delta, \qquad \delta := (2\alpha)^{1/2}.$$

Of course, the problem is really 1-dimensional, and we could have transformed the Gauss results into statements about BES (3); but it would not have been such fun.

(20.4) *Exercise. Let* X be CBM$(\mathbb{R})$. Let

$$T := \inf\{t : |X_t| = 1\} = H_1 \wedge H_{-1}.$$

Let $\alpha > 0$ and $\delta := (2\alpha)^{1/2}$. Explain why

$$\mathbf{E}^0[\exp(-\alpha H_1)] = \mathbf{E}^0[\exp(-\alpha H_1);\ H_1 < H_{-1}] + \mathbf{E}^0[\exp(-\alpha H_{-1});\ H_{-1} < H_1]e^{-2\delta},$$

and deduce that

$$\mathbf{E}^0 e^{-\alpha T} = \operatorname{sech} \delta.$$

Comparison of this result with (20.3) clinches the CT theorem in the case $n = 1$.

One of the classic applications of the Feynman–Kac formulae is to the arcsine law (see Itô and McKean [1]). However, we prove the arcsine law by using local-time theory, both in Section 24 and, by a better method, in Section VI.63 of Volume 2.

---

†Still assumed symmetric.

21. Time-substitution. Suppose that A is a PCHAF of X and that a null set has been weeded out so that properties (16.3)(i)–(iii) hold for all ω. Set

$$\tau_t(\omega) := \tau(t,\omega) := \inf\{s : A_s > t\},$$

so that τ is the right-continuous function inverse to A. Since

$$\{\tau_s < t\} = \{A_t > s\}$$

and $\{\mathscr{F}_t\}$ is right-continuous, for each s, τ_s is an $\{\mathscr{F}_t\}$ stopping time. Set $\tilde{X}_t := X_{\tau(t)}, \tilde{\mathscr{F}}_t := \mathscr{F}_{\tau(t)}$ and $\tilde{\theta}_t := \theta_{\tau(t)}$. Then

$$\tilde{X} = (\Omega, \mathscr{F}, \tilde{\mathscr{F}}_t, \tilde{X}_t, \tilde{\theta}_t, \mathbf{P}^x)$$

is a strong Markov process on $(E, \mathscr{E}^*)$. The point is that if $\tilde{T}$ is an $\{\tilde{\mathscr{F}}_t\}$ stopping time then $T := \tau(\tilde{T})$ is an $\{\mathscr{F}_t\}$ stopping time and $\tilde{\mathscr{F}}_t(\tilde{T}) = \mathscr{F}(T) := \mathscr{F}_T$; further, $\tau(\tilde{T}+t) = T + \tau_t \circ \theta_T$. Thus $\tilde{X}$ inherits the strong Markov property from X. As an exercise, write out all the details of the present argument—see Section X.5 of Dynkin [2]. This result is due to Volkonskii.

If A is strictly increasing, in which case τ is continuous, then it is clear that X and $\tilde{X}$ have the same hitting distributions:

(21.1) for every compact K, $\forall x$, $\forall B \in \mathscr{B}(K)$,

$$\mathbf{P}^x[X \circ H_K \in B] = \mathbf{P}^x[\tilde{X} \circ \tilde{H}_K \in B].$$

Indeed $X \circ H_K = \tilde{X} \circ \tilde{H}_K$. Of course, $\tilde{H}_K := \inf\{t : \tilde{X}_t \in K\}$.

The converse theorem is very deep. (See Theorem V.5.1 of Blumenthal and Getoor [1].)

(*21.2*) THEOREM (Blumenthal–Getoor–McKean). *If Y is a 'standard' process with the same hitting distributions as X then, for some strictly increasing PCHAF A of X, Y has the same laws as the $\tilde{X}$ process associated with A.*

For the definition of 'standard process', see Blumenthal and Getoor [1]. (FD processes are certainly standard.) Incidentally, the fact that *if X is standard and A is strictly increasing then $\tilde{X}$ is standard* emphasises the need to axiomatise processes by *probabilistic* axioms (as is done for standard processes and, better, *right* processes) instead of axiomatising via analytical properties like the FD property.

(*21.3*) *The generator of $\tilde{X}$.* The preceding sentence stands as a piece of pure mathematics. However, when we wish to talk about the generator of $\tilde{X}$, we should consider for example how the FD property behaves under time-substitution.

(*21.4*) *Volkonskii's formula.* Suppose that X is an FD process and that

$$A(t) = \int_0^t v(X_s)\,ds,$$

where v is a positive continuous function on E bounded away from 0. Then, for some K, $\tau_t \leqslant Kt$, $\forall t$. For $f \in \mathcal{D}(\mathcal{G})$,

$$C_t^f := f(X_t) - \int_0^t \mathcal{G}f(X_s)\,ds \quad \text{is an } R\text{-martingale}$$

relative to $(\Omega, \{\mathcal{F}_t^\mu\}, \mathbf{P}^\mu)$ for every μ. See (10.13). If we consider a fixed interval $[0, a]$ for t then $\tau_t \in [0, Ka]$, and we can apply the Optional Stopping Theorem II.77.5 to $C^f(t \wedge Ka)$. In this way, we deduce that

$$C_{\tau(t)}^f = f(\tilde{X}_t) - \int_0^t v(\tilde{X}_s)^{-1} \mathcal{G}f(\tilde{X}_s)\,ds \tag{21.5}$$

(t running over the whole half-line $[0, \infty)$) is a martingale relative to each $(\Omega, \{\tilde{\mathcal{F}}_t^\mu\}, \mathbf{P}^\mu)$. For $f \in \mathcal{D}(\mathcal{G})$, write

(21.6)(i) $$\tilde{G}f(x) := v(x)^{-1}\mathcal{G}f(x)$$

provided $f \in \mathcal{D}(\mathcal{G})$; then $\tilde{G}f \in C_0$. Taking $\mathbf{P}^x$ expectations of the martingale $C_{\tau(t)}^f$, we find that

$$\tilde{P}_t f(x) - f(x) = \int_0^t \tilde{P}_s \tilde{G} f(x)\,ds.$$

If we *assume* that (in the cases we study) the transition semigroup $\{\tilde{P}_t\}$ of $\tilde{X}$ is an FD semigroup with generator $\tilde{\mathcal{G}}$ (say), it is now clear that

(21.6)(ii) $$\tilde{\mathcal{G}} \supseteq \tilde{G}.$$

The results (21.6) comprise Volkonskii's formula. For Volkonskii's proof via Dynkin's formula, see X.10.24 in Dynkin [2].

Of course, if v is also bounded away from ∞, we can reverse the roles of X and $\tilde{X}$ to show that $\tilde{\mathcal{G}} = \tilde{G}$.

(21.7) *Finite Markov chains.* The time change of a finite Markov chain with Q-matrix Q by the additive functional $A_t = \int_0^t v(X_s)\,ds$ is, according to (21.6)(i), the Markov chain with Q-matrix $\tilde{Q}$, where

$$\tilde{q}_{ij} := v(i)^{-1} q_{ij}.$$

In terms of the jump–hold description of the chain (Section 2), the effect of the time change is easy to specify; when the time-changed chain visits i, it resides there for an exponential amount of time with mean $v(i)/q(i)$, compared with the mean $1/q(i)$ for the original chain.

22. Reflecting Brownian motion. Let X be CBM ($\mathbb{R}$). Let

$$A_t := \int_0^t I_{[0,\infty)}(X_s)\,ds = \text{meas}\,\{s \leqslant t : X_s \geqslant 0\}.$$

Let τ be the right-continuous inverse of A and let $\tilde{X}_t := X \circ \tau_t$. It is intuitively obvious that $\tilde{X}_t$ must be reflecting Brownian motion on $[0, \infty)$. Assume for now (see Note below) that $\tilde{X}$ is an FD process on $[0, \infty)$. Dynkin's formula $\tilde{\mathscr{G}} = \tilde{\mathscr{C}}$ for the generator of $\tilde{X}$ shows that

$$\tilde{\mathscr{G}}f(0) = \lim_{\varepsilon \downarrow\downarrow 0} \frac{f(\varepsilon) - f(0)}{\mathbf{E}^0[A(H_\varepsilon)]}.$$

Now, by the formula (I.13.7) for the transition-density function of Brownian motion killed at 0,

$$\mathbf{E}^0[A(H_\varepsilon)] = \int_{t=0}^{\infty} \int_{y=0}^{\varepsilon} [p(t; 0, \varepsilon - y) - p(t; 0, \varepsilon + y)]\, dy\, dt,$$

where p is the transition-density function of CBM ($\mathbb{R}$). But, for $\lambda > 0$ and with $\gamma := (2\lambda)^{1/2}$, formula (3.13) gives

$$\int_0^{\infty} e^{-\lambda t}[p(t; 0, \varepsilon - y) - p(t; 0, \varepsilon + y)]\, dt = 2\gamma^{-1} e^{-\gamma\varepsilon} \sinh \gamma y.$$

On letting $\lambda \downarrow\downarrow 0$, we obtain

$$\int_0^{\infty} [p(t; 0, \varepsilon - y) - p(t; 0, \varepsilon + y)]\, dt = 2y,$$

so that

$$\mathbf{E}^0[A(H_\varepsilon)] = \varepsilon^2.$$

Hence, for f in $C_0[0, \infty)$, we have $f \in \mathscr{D}(\tilde{\mathscr{G}})$ if and only if the formulae

(22.1) (i) $$\tilde{\mathscr{G}}f(x) = \tfrac{1}{2} f''(x) \quad (x > 0)$$

(22.1)(ii) $$\tilde{\mathscr{G}}f(0) = \lim_{\varepsilon \downarrow\downarrow 0} \varepsilon^{-2}[f(\varepsilon) - f(0)]$$

make sense and define a function $\tilde{\mathscr{G}}f$ in $C_0[0, \infty)$. Note that f in $\mathscr{D}(\tilde{\mathscr{G}})$ satisfies

$$f^+(0) := \lim_{\varepsilon \downarrow\downarrow 0} \varepsilon^{-1}[f(\varepsilon) - f(0)] = 0,$$

and observe how the formula for $\tilde{\mathscr{G}}$ checks with l'Hôpital's rule.

We saw in Section I.41.1 that reflecting Brownian motion $|X|$ is Markovian. From the explicit formula (I.14) for the transition-density function for $|X|$, it is clear that $|X|$ is an FD process. That $|X|$ has generator $\tilde{\mathscr{G}}$ as in (22.1) will be immediate once we show that

$$\mathbf{E}^0[H_\varepsilon \wedge H_{-\varepsilon}] = \varepsilon^2.$$

This follows from the more precise result that for $\lambda > 0$ and with $\gamma := (2\lambda)^{1/2}$ and $T := H_\varepsilon \wedge H_{-\varepsilon}$, we have

$$\mathbf{E}^0[\exp(-\lambda T)] = \operatorname{sech} \gamma\varepsilon.$$

See Exercise 20.4. The proof that $\tilde{X}$ and $|X|$ have the same generator $\tilde{\mathscr{G}}$, and hence the same transition function, is complete.

Important note on preservation of FD property. Deciding whether or not the FD property is preserved under probabilistic operations such as time-substitution is generally a very difficult problem. In the 1-dimensional case, special arguments apply that allow easy settlement of the matter, so that the FD property for simple transformations of CBM ($\mathbb{R}$) is usually taken for granted in the literature.

Let us work through the details for this example. So let X be CBM ($\mathbb{R}$), and define

$$A_t := \text{meas}\{s \leqslant t : X_s \geqslant 0\}, \qquad \tau_t := \inf\{u : A_u > t\}, \qquad \tilde{X}_t := X(\tau_t).$$

Our intuitive idea is this: if $\tilde{X}$ starts at a point x of $[0, \infty)$ then, with high probability, $\tilde{X}$ will very soon hit y.† We use this idea to show that $\tilde{X}$ has FD *resolvent*, whence, by the Hille–Yosida Theorem, $\tilde{X}$ has FD transition function $\{\tilde{P}_t\}$. (*Warning*: if you try to express the same intuitive idea directly in terms of the transition function, you will encounter some very awkward technical problems.)

For $y \in [0, \infty)$, put $\tilde{H}_y := \inf\{t : \tilde{X}_t = y\}$. By applying the strong Markov property of $\tilde{X}$ at time $\tilde{H}_y$—which we know amounts to applying the strong Markov property of X at time $\tau(H_y)$—we find that (with obvious notation) for $f \in C_0[0, \infty)$, $\lambda > 0$, and $x, y \in [0, \infty)$,

$$\tilde{R}_\lambda f(x) = \mathbf{E}^x \int_0^{\tilde{H}_y} e^{-\lambda t} f(\tilde{X}_t)\, dt + \mathbf{E}^x[\exp(-\lambda \tilde{H}_y) \tilde{R}_\lambda f(y)].$$

(You recognise that this is just Dynkin's formula for $\tilde{X}$.) Hence

$$\begin{aligned} (22.2) \qquad |\tilde{R}_\lambda f(x) - \tilde{R}_\lambda f(y)| &\leqslant \|f\| \mathbf{E}^x \int_0^{\tilde{H}_y} e^{-\lambda t}\, dt + |\tilde{R}_\lambda f(y)| \mathbf{E}^x[1 - \exp(-\lambda \tilde{H}_y)] \\ &\leqslant (\lambda^{-1} \|f\| + \|\tilde{R}_\lambda f\|) \mathbf{E}^x[1 - \exp(-\lambda \tilde{H}_y)] \\ &\leqslant 2\lambda^{-1} \|f\| \mathbf{E}^x[1 - \exp(-\lambda \tilde{H}_y)]. \end{aligned}$$

That $\tilde{R}_\lambda f$ is continuous will therefore follow once we show that

$$(22.3) \qquad \mathbf{E}^x[\exp(-\lambda \tilde{H}_y)] \to 1 \quad \text{as } x \to y.$$

But $\tilde{H}_y = A(H_y) \leqslant H_y$, a.s., so (22.4) follows from the corresponding property for X. It is trivial to show that, for $f \in C_0[0, \infty)$, $\tilde{R}_\lambda f(x) \to 0$ as $x \to \infty$. Hence

$$\tilde{R}_\lambda : C_0[0, \infty) \to C_0[0, \infty).$$

By Lemma 6.7, you will see that it is enough now to show that, for $f \in C_0(0, \infty)$,

$$(22.5) \qquad \tilde{P}_t f(x) \to f(x) \quad (t \downarrow 0).$$

(It will then follow that as $\lambda \uparrow\uparrow \infty$, $\lambda \tilde{R}_\lambda f \to f$ not only pointwise but also in the

† Here, y is a point of $[0, \infty)$ very close to x.

supremum norm.) However, since $\tilde{X}$ is right-continuous and

$$\tilde{P}_t f(x) = \mathbf{E}^x[f(\tilde{X}_t)],$$

the result (22.5) is obvious.

The same argument covers the case of $\tilde{X}^\lambda$ and the case of elastic Brownian motion in Section 24. The Feller–McKean example (now to be discussed) would require a little more thought, but whether or not *it* is FD (in fact, it *is*!) is irrelevant.

23. The Feller–McKean chain. Again let X be CBM ($\mathbb{R}$). We assume Trotter's Theorem I.5.9. Let $l(t, x)$ be a jointly continuous local time for X and set

$$A_t := \int_{\mathbb{R}} l(t, x) m(dx) = \sum_{\mathbb{Q}} l(t, q) m\{q\}$$

where m is a probability measure concentrated on the rationals with $m\{q\} > 0$, $\forall q \in \mathbb{Q}$. Then A is a strictly increasing PCHAF of X. (We suppress 'almost surely' qualifying phrases here.) Let $\tilde{X}$ be the corresponding time-transformation of X. Then $\tilde{X}$ is a continuous process that spends almost all its time in $\mathbb{Q}$ and can be regarded as a Markov chain with 'minimal' state-space $\mathbb{Q}$. The generator of $\tilde{X}$ (considered as an FD process on $\mathbb{R}$) is a highly singular second-order 'elliptic' operator

$$\frac{1}{2}\frac{d}{dm}\frac{d}{dx}.$$

Itô and McKean [1] contains the definitive account of such operators. Breiman [1] and Freedman [1] contain good easy introductions.

The Feller–McKean chain $\tilde{X}$ is historically important as the first chain with all states instantaneous:

$$q_i := -q_{ii} := \lim_{t \downarrow\downarrow 0} t^{-1}[1 - p_{ii}(t)] = \infty, \quad \forall i,$$

where

$$p_{ij}(t) := \mathbf{P}^i[\tilde{X}_t = j] \quad (i, j \in \mathbb{Q}).$$

(Some basic chain theory is recalled later in this chapter.) Since $\tilde{X}$ moves continuously and therefore does not jump, we have

$$q_{ij} := \lim_{t \downarrow\downarrow 0} t^{-1} p_{ij}(t) = 0 \quad (\forall i, j \in \mathbb{Q} : i \neq j).$$

Thus the Q-matrix Q of $\tilde{X}$ satisfies

$$Q = \begin{pmatrix} -\infty & 0 & 0 & \cdots \\ 0 & -\infty & 0 & \cdots \\ 0 & 0 & -\infty & \cdots \\ \vdots & \vdots & \vdots & \end{pmatrix}.$$

In Theorem 55.1, we describe all possible totally instantaneous Q-matrices.

24. Elastic Brownian motion; the arcsine law. Let X continue to denote CBM ($\mathbb{R}$) and let A continue to denote time spent by X above 0. Let ξ be an exponentially distributed variable of rate $\lambda > 0$ independent of the X process. Let

$$X_t^\lambda := \begin{cases} X_t & (t < \xi), \\ \partial & (t \geqslant \xi). \end{cases}$$

In other words, we have the 'killing at constant rate' situation of (18.14). Set

$$A_t^\lambda := \int_0^t I_{[0,\infty)}(X_s^\lambda)\, ds = A_{t \wedge \xi},$$

and let τ^λ be the right-continuous inverse to A^λ. Let $\tilde{\mathscr{G}}^\lambda$ be the generator of $\tilde{X}_t^\lambda := X^\lambda \circ \tau_t^\lambda$. Then

(24.1) (i) $$\tilde{\mathscr{G}}^\lambda f(x) = \tfrac{1}{2} f''(x) - \lambda f(x) \quad (x > 0)$$

(24.1)(ii) $$\tilde{\mathscr{G}}^\lambda f(0) = \lim_{\varepsilon \downarrow\downarrow 0} \varepsilon^{-2}[e^{-\gamma\varepsilon} f(\varepsilon) - f(0)], \quad \gamma := (2\lambda)^{1/2}$$

because

$$\mathbf{E}^0[A(H_\varepsilon) \wedge \xi] \sim \varepsilon^2 \quad (\varepsilon \downarrow\downarrow 0),$$
$$\mathbf{P}^0[H_\varepsilon < \xi] = \mathbf{E}^0[\exp(-\lambda H_\varepsilon)] = e^{-\gamma\varepsilon}.$$

Note the *elastic boundary condition*

(24.1)(iii) $$f^+(0) = \gamma f(0)$$

satisfied by elements f of $\mathscr{D}(\tilde{\mathscr{G}}^\lambda)$. Again note how the formula for $\tilde{\mathscr{G}}^\lambda$ checks with l'Hôpital's rule.

Now let us explain why this example is interesting. The lifetime of $\tilde{X}^\lambda$ is $A(\xi)$. If $X(\xi) > 0$ then $\tilde{X}^\lambda$ dies at position $X(\xi)$. The exciting thing is that if $X(\xi) < 0$, which happens for example with $\mathbf{P}^0$ probability $\frac{1}{2}$, then $\tilde{X}^\lambda$ *dies at* 0. In other words, $\tilde{X}^\lambda$ must be obtained by killing $\tilde{X}$ at rate λ while $\tilde{X}$ is away from 0, but in a way depending on the local time at 0 when $\tilde{X}$ is at 0. Indeed, if $\tilde{l}_t = l_{\tau(t)}$ denotes the local time at 0 for $\tilde{X}$ then

(24.2) $$\mathbf{P}^x[A(\xi) > t \mid \tilde{X}] = \exp(-\lambda t - \tfrac{1}{2}\gamma \tilde{l}_t), \quad \forall x \geqslant 0.$$

This formula, taken from Williams [6], is one way of introducing local time from *global* considerations! We prove (24.2) below. Note that it may be reformulated in a way that obviates the need for introducing ξ:

(24.3) $$\mathbf{E}^x[\exp(-\lambda\tau_t) \mid \tilde{X}] = \exp(-\lambda t - \tfrac{1}{2}\gamma l_{\tau(t)}), \quad \forall x \geqslant 0.$$

(24.4) *Arcsine law.* Lévy's arcsine law

$$\mathbf{P}^0[A(u) \leqslant t] = \frac{2}{\pi} \arcsin\left[\left(\frac{t}{u}\right)^{1/2}\right] \quad (t \leqslant u)$$

is an immediate consequence of (24.2). Take $x = 0$ in (24.2) and then take $\mathbf{P}^0$

expectations to get

$$\int_0^\infty \lambda e^{-\lambda u}\mathbf{P}^0[A(u)>t]\,du = e^{-\lambda t}\mathbf{E}^0[\exp(-\tfrac{1}{2}\gamma\tilde{l}_t)].$$

All that remains is to identify the law of $\frac{1}{2}\tilde{l}_t$. In fact, $\frac{1}{2}\tilde{l}_t$ has the same law as $S_t := \sup_{u\leq t} X_u$; you may consider this to be completely obvious in view of the construction of local time from upcrossings given in Section I.14. On the other hand, it may take Exercise 24.5 below to convince you. However, given the equality in law of $\frac{1}{2}\tilde{l}_t$ and S_t, the arcsine law now follows by consulting Laplace transform tables, or, better, by calculating a few integrals.

(24.5) *Exercise*

(i) Let $U_+(t,\varepsilon)$ be the number of upcrossings of $[0,\varepsilon]$ by X before time t, and let $U_-(t,\varepsilon)$ be the number of downcrossings of $[-\varepsilon,0]$ by X before time t. By the strong Markov property of X, argue that, conditional on $U(t,\varepsilon) := U_+(t,\varepsilon)+U_-(t,\varepsilon)=n$, $U_+(t,\varepsilon)$ has a $B(n,\frac{1}{2})$ distribution.

(ii) Observing that $U(t,\varepsilon)$ is the number of upcrossings of $[0,\varepsilon]$ by $|X|$ before time t, show, as in Lévy's Theorem I.14.7, that

$$\lim_{n\to\infty} 2^{-n}U_+(t,2^{-n}) = \tfrac{1}{2}l_t.$$

(iii) From the characterisation in Section 22 of reflecting Brownian motion, we have $\tilde{X}(t) \equiv X(\tau_t) \overset{\mathscr{D}}{=} |X_t|$. Thus

$$\lim_{n\to\infty} 2^{-n}U_+(\tau_t,2^{-n}) = \tfrac{1}{2}\tilde{l}_t \overset{\mathscr{D}}{=} \lim_{n\to\infty} 2^{-n}U(t,2^{-n}) = l_t.$$

(24.6) *Elastic Brownian motion.* Let X still denote Brownian motion and l_t the local time at 0 for X. Fix $\gamma>0$. Let $\hat{X}$ denote $|X|$ killed at rate $\gamma\,dl_t$, so that $\hat{X}$ has transition semigroup $\{\hat{P}_t\}$ with

$$\hat{P}_t f(x) = \mathbf{E}^x[\exp(-\gamma l_t) f(|X_t|)] = E^x[\exp(-\tfrac{1}{2}\gamma\tilde{l}_t) f(\tilde{X}_t)].$$

Obviously $\hat{\mathscr{G}} f = \frac{1}{2}f''$ away from 0. The situation at 0 is interesting.

$$\hat{\mathscr{G}} f(0) = \lim_{\varepsilon\downarrow\downarrow 0} \frac{f(\varepsilon)\mathbf{E}^0[\exp(-\gamma l\circ H_\varepsilon)] - f(0)}{\hat{\mathbf{E}}^0[\hat{H}_\varepsilon \wedge \hat{\zeta}]}$$

where H_ε and $\hat{H}_\varepsilon$ are of course the hitting times of ε by $|X|$ and $\hat{X}$ respectively. During the course of our proof of Lévy's Downcrossing Theorem (Section I.14), we showed that the $\mathbf{P}^0$ distribution of $l(H_\varepsilon)$ is exactly exponential with mean ε, so that

$$\mathbf{E}^0[\exp(-\gamma l\circ H_\varepsilon)] = (1+\gamma\varepsilon)^{-1}.$$

Further (why?),

$$\hat{\mathbf{E}}^0[\hat{H}_\varepsilon \wedge \hat{\zeta}] \sim \varepsilon^2 \quad (\varepsilon\downarrow\downarrow 0).$$

Hence, for $f \in \mathscr{D}(\hat{\mathscr{C}})$,

$$f^+(0) = \gamma f(0), \quad \mathscr{C} f(0) = \tfrac{1}{2} f'(0+).$$

The formula (24.2), which involves an extra killing at constant rate λ, is now obvious.

4. APPROACH TO RAY PROCESSES: THE MARTIN BOUNDARY

25. Ray processes and Markov chains. We now move on from the familiar FD semigroups and processes to Ray semigroups and processes. Quite rightly, you first want certain questions answered:

(25.1) (i) *Are there important examples of Ray processes that are not FD?*

(25.1) (ii) *Does the theory of Ray processes provide new information on FD processes?*

(25.1)(iii) *Do we have to move on again from Ray processes to still more general objects?*

The answers must be (i) Yes, (ii) Yes, (iii) No—or we would have not asked the questions. You will of course realise in regard to (iii) that 'No, never' must yield to 'Hardly ever' if the point is pressed.

For Ray processes, both the analysis and the probability theory are much richer than for the FD situation; and, for the pure mathematicians among you, this may be justification enough for studying the Ray theory. However, motivation never harmed anyone (least of all, pure mathematicians), and we propose to answer questions (i), (ii) and (iii) in some detail before we develop the theory.

In 1966, Chung made some shrewd and prophetic comments in his book [1] on Markov chains:

> The second edition of this book appears at a time when boundary theory (envisaged in this book as a study in depth of the behaviour of sample functions in relation to the 'infinities') has just begun to take shape. This vital theme, already announced in the preface to the first edition, will no doubt be the most challenging part of the theory to come. I have chosen not to enter into it in detail in the belief that such a development needs more time to mature. In this regard, it may be a timely observation that the theory of Markov processes in general-state-space, which flourished in recent-years and has built up a powerful machinery, has had to date little impact on the denumerable [chain] case. This is because the prevailing assumptions would allow the sample paths of chains virtually no other discontinuities than jumps—a situation which would make a trite object of a chain. On the other hand, the special theory of Markov chains has

yet to adapt its methodology to a broader context suitable for the general state-space. Thus there exists at the moment a state of mutual detachment which surely must not be allowed to continue. Future progress in the field looks to a meaningful fusion of these two aspects of the Markovian phenomenon.

The meaningful fusion is achieved in the theory of Ray processes. The benefits to chain theory are enormous. In the other direction, last-exit theory provides one of many examples where the general theory has benefited from adapting the methodology of chain theory. The prophetic character of Chung's comments will be seen throughout this book.

Anyone familiar with Chung's philosophy (one with which we are in full agreement) will know the stress to be laid on *sample functions* in the quotation. It is on the behaviour of sample functions rather than on that of (for example) excessive functions that we, as *probabilists*, must concentrate. It is therefore satisfying that the general theory will allow us to treat chains in a manner that greatly clarifies the probabilistic significance of $q_i, q_{ij}, q_{ij}(t)$ etc. and suppresses much of the analysis on which many previous treatments have relied. To be sure, Ray's theorem makes very heavy use of analysis in its early stages. The point is that, once it gets underway, the probability theory is more or less self-sufficient; and, by then, it trounces analysis at its own game.

In connection with question (25.1)(i), it is necessary to understand Chung's statement that the then-prevailing assumptions covered only 'trite' chains. This relates to the following fact (discussed briefly and illustrated in a moment, and explained fully later in this chapter). *If a transition matrix function* $\{p_{ij}(t)\}$ *on a countable set I has the FD property relative to the discrete topology of I, then the associated chain X is totally stable and Feller minimal.* Though nearly all chains that can serve as models for real-world phenomena are totally stable and Feller minimal, such chains are 'trite' from a pure-mathematical standpoint. (We shall attempt to provide a strong justification for the study of 'non-trite' chains later!)

The statement that X is *totally stable* means of course that every state i in I is stable:

$$q_i := \lim_{t \downarrow\downarrow 0} t^{-1}[1 - p_{ii}(t)] < \infty, \quad \forall i.$$

You can see why this has to be. Since X is FD, X is right-continuous in the discrete topology of I, so that if X starts in state i then X must stay at i throughout a time interval. The well-known fact that this time interval is exponentially distributed with rate q_i is proved (along with the existence of q_i) for general chains in Section 82. The *Feller-minimal* property refers to the fact that the behaviour of X is completely determined by Q in that if X explodes then X dies at its explosion time. In the next section we clarify these points in regard to a special example.

(25.2) *The Ray–Knight topology.* It is hard to believe now that Ray's tremendous paper [1] was published as long ago as 1959. Ray's choice of axioms for what is now called a Ray process was astonishingly perceptive. There was unfortunately an error in Ray's attempt to show that every 'acceptable' Markov process (and, in particular, *every* chain) could be made to accord with these axioms by introducing a suitable topology on, and compactification of, the state space. This extraordinary claim is however essentially true.

The gap in Ray's paper was corrected (for different particular situations and in different ways) by several people including Ray himself. It was Knight [1] who got things just right, and, especially after the appearance of the 1967 Kunita and T. Watanabe paper [1], the Ray–Knight compactification was firmly established. In particular it was known that *all* chains are Ray processes, which gives a strong 'Yes' to question (2.5.1)(i).

You can see the problem involved if we consider the Feller–McKean example (Section 23). The Feller–McKean process X is a continuous process on $\mathbb{R}$ that can be regarded as a chain on $\mathbb{Q}$. Now suppose that someone relabels $\mathbb{Q}$ as, say, $\mathbb{N}$ and presents us with the Feller–McKean transition matrix function $\{p_{ij}(t)\}$. Could be recognize from $\{p_{ij}(t)\}$ that we should 'unscramble' the situation by imbedding $\mathbb{N}$ as $\mathbb{Q}$ in $\mathbb{R}$? Yes; the Ray–Knight compactification will do the unscrambling for us.

26. Important example: birth process. The example we now discuss is extremely simple, but it serves well to illustrate points in the theory of Ray processes and (later) in the classification of stopping times and in the theory of jumps of martingales. We are sure that you know enough elementary chain theory (from Volume 1 of Feller [1], which discusses this example) to follow this account without difficulty. We are only concerned with intuitive understanding, and skip some details of rigour (along with some 'a.s.' phrases).

Let $I = \{1, 2, 3, \ldots\}$ and let Q be the $I \times I$ matrix

$$Q := \begin{pmatrix} -q_1 & q_1 & 0 & 0 & \cdots \\ 0 & -q_2 & q_2 & 0 & \cdots \\ 0 & 0 & -q_3 & q_3 & \cdots \\ \vdots & \vdots & \vdots & \vdots & \end{pmatrix}, \qquad \text{where } 0 < q_i < \infty, \forall i.$$

Let X be a right-continuous chain with Q-matrix Q, let

$$H_n := \inf\{t : X_t = n\}$$

and let $\eta := \lim_n H_n \leqslant \infty$ be the first explosion time of X. Up to time η, the paths of X are non-decreasing. Under the $\mathbf{P}^i$ law, the variables

$$H_{i+1} - H_i, H_{i+2} - H_{i+1}, \ldots$$

are independent and are exponentially distributed with rates $q_i, q_{i+1}, \ldots$ respectively. Thus

(26.1)
$$\mathbf{E}^i[e^{-\lambda\eta}] = \prod_{k\geqslant i}(1+\lambda q_k^{-1})^{-1} \quad (\lambda>0),$$

so that (as we see by letting $\lambda\downarrow\downarrow 0$)

$$\begin{aligned}&\text{if } \textstyle\sum q_k^{-1}=\infty \quad \text{then} \quad \eta=\infty, \quad \text{a.s.,}\\ &\text{if } \textstyle\sum q_k^{-1}<\infty \quad \text{then} \quad \eta<\infty, \quad \text{a.s.}\end{aligned}$$

In the case when $\eta=\infty$, a.s., there is nothing more to say: X is the unique chain with Q-matrix Q and X is FD because $P_t f(i)\leqslant \min_{j\geqslant i} f(j)$ and because (6.6)(iv)* is automatic.

We now devote our attention to the case when $\eta<\infty$, a.s.. If X is FD (for the discrete topology on I) then, by Theorem 11.1, X is quasi-left-continuous on $[0,\infty)$, so that

$$X(\eta)=\lim_n X(H_n)=\infty=\partial;$$

by the coffin condition, $X(t)=\partial=\infty$, $\forall t\geqslant\eta$. Thus there is only one FD chain, namely the 'Feller-minimal' chain killed at time η.

(26.2) *Exercise.* Show that the Feller-minimal chain $X^{\min}$ has resolvent

$$r_{ij}^{\min}(\lambda):=\int_0^\infty e^{-\lambda t}p_{ij}^{\min}(t)\,dt \quad (\lambda>0),$$

satisfying

$$r_{ij}^{\min}(\lambda)=\begin{cases}0 & (j<i),\\ (\lambda+q_j)^{-1}\displaystyle\prod_{i\leqslant k<j}[(1+\lambda q_k^{-1})^{-1}] & (j>i),\end{cases}$$

$$r_{jj}^{\min}(\lambda)=(\lambda+q_j)^{-1}.$$ □

When $\eta<\infty$ (a.s.), there are other (non-FD) chains X with Q-matrix Q. We now write $E:=\{1,2,3,\ldots;\infty\}$ for the one-point compactification of I and ∂ for a point isolated from E. Let μ be a probability measure on $I\cup\{\partial\}$ with $\mu\{\partial\}<1$. For each such μ, we can construct a chain X with Q-matrix Q by Doob's *immediate-return* procedure: we choose $X(\eta)\in I\cup\{\partial\}$ with distribution μ; if $X(\eta)=\partial$ then $X(t)=\partial$, $\forall t>\eta$; if $X(\eta)\in I$ then we run X according to the 'old' rules until the next explosion time η_2 (say); we choose $X(\eta_2)$ with distribution μ; etc., etc.

For this chain X, quasi-left-continuity breaks down and takes the modified form

(26.3)
$$\mathbf{P}^i[X(\lim H_n)=j\mid \vee\mathscr{F}(H_n)]=\mu_j:=\mu(\{j\}) \quad (j\in I\cup\partial),$$

where $\vee\mathscr{F}(H_n)$ is the smallest σ-algebra containing every $\mathscr{F}(H_n)$. We say that

∞ is a *branch-point with branching measure* μ and write

$$P_0(\infty, \{j\}) = \mu_j.$$

We note that X *never visits the branch-point* ∞. (It approaches ∞ but branches at the last moment.) The only sensible interpretation of the $\mathbf{P}^\infty$ law is as

$$\mathbf{P}^\infty = \sum_{k \in I \cup \partial} \mu_k \mathbf{P}^k.$$

Note that now we generally have $\mathscr{F}_\eta \neq \bigvee \mathscr{F}(H_n)$ (in contrast to the 'FD' situation in (11.6)).

(*26.4*) *Exercise.* For $k \in I \cup \partial$, put $x_k(\lambda) := \mathbf{E}^k[e^{-\lambda\eta}]$, so that $x_\partial(\lambda) = 0$. Show that, for $i, j \in I$,

$$\begin{aligned} r_{ij}(\lambda) &= r_{ij}^{\min}(\lambda) + x_i(\lambda) \sum_{k \in I} \mu_k r_{kj}(\lambda) \\ &= r_{ij}^{\min}(\lambda) + \frac{x_i(\lambda) \sum_{k \in I} \mu_k r_{kj}^{\min}(\lambda)}{\sum_{k \in I \cup \partial} \mu_k [1 - x_k(\lambda)]}. \end{aligned}$$

(Substitute the first equation into itself.)

(*26.5*) *Warning.* More complex ways of return from infinity are possible even for this example.

(*26.6*) *Note.* We must mention one further point in connection with this example. Consider the immediate-return process with $\mu_1 = 1$, so that X returns to 1 after each explosion. Then X *will* have the FD property (and so will be quasi-left-continuous) if I is topologised as the compact metric space with 1 as the unique limit-point of the sequence $2, 3, 4, \ldots$ Check this by using (26.4) and the HY theorem. The Ray–Knight topology will automatically make the point 1 an accumulation-point of I, as it should be.

In short, the Ray–Knight compactification will detect Feller properties when these are present. (More is true. It can happen that when a process X on (say) a compact metric space is constructed by complicated probabilistic methods, we are unable to prove directly that X has the FD property, but can prove the equivalent result that X is a Ray process without branch-points.) But the value of the Ray–Knight compactification is of course that it *always* works.

27. Excessive functions, the Martin kernel and Choquet theory. Let us do more than answer 'Yes' to question (25.1)(ii) by telling you that Ray theory yields a much simpler and more intuitive account of Martin (–Doob–Hunt) boundary theory, even in the case of discrete-parameter chains! To be honest, this most elementary case of Martin boundary theory is much the most interesting: it has many delightful and important applications. It will be helpful to run through some of the basic *analysis* for this case now. For one thing, it will help us understand the original Martin boundary of R. S. Martin. Later in this chapter, we shall derive both the analysis and the more interesting probability theory for the chain case from Ray's Theorem.

The 'Ray' treatment will be independent of Choquet theory, but we now explain how Choquet's famous theorem on integral representation of elements of simplexes shows that a Martin representation of harmonic (more generally, excessive) functions must hold. Meyer [2] and Phelps [1] have fine accounts of Choquet theory. Choquet's Theorem has been useful in establishing the existence and/or uniqueness of integral representations in many areas of probability theory. Inevitably, its very generality prevents its being useful in pinning down the explicit form of extremal elements.

Let I be a countable set and let Π be a substochastic $I \times I$ matrix. Define the *Green kernel* Γ of Π as the $I \times I$ matrix with

$$\Gamma(i, j) := \sum_{n=0}^{\infty} \Pi^n(i, j) \leqslant \infty, \quad \forall i, j,$$

so that, formally, $\Gamma = (I - \Pi)^{-1}$. Compare (I.22.1).

The probabilistic interpretation is obvious. Let $X = (X_n : n = 0, 1, 2, \ldots)$ be a Markov chain on I (with coffin state ∂ adjoined) with 1-step transition matrix Π, so that, for, $n \in \mathbf{N}$ and $i_0, i_1, i_2, \ldots, i_n \in I$,

$$\mathbf{P}^{i_0}[X_1 = i_1; \ldots; X_n = i_n] = \Pi(i_0, i_1)\Pi(i_1, i_2) \ldots \Pi(i_{n-1}, i_n).$$

(The Daniell–Kolmogorov Theorem immediately gives an appropriate X.) Then

$$\Gamma(i, j) = \mathbf{E}^i[\text{time}(\geqslant 0) \text{ spent by } X \text{ in } j].$$

(27*A*) ASSUMPTION. *There exists a reference point b in I such that*

$$0 < \Gamma(b, j) < \infty, \quad \forall j \in I.$$

This assumption is made throughout the remainder of Section 27. It says:

(*i*) *state b can feed into j (ultimately)* ($\forall j \neq b$);
(*ii*) *every state is transient.*

The easily-established strong Markov property of X shows that

(27.1) $$\Gamma(i, j) = \mathbf{P}^i[D_j < \infty]\Gamma(j, j) \leqslant \Gamma(j, j), \quad \forall i, j,$$

where

$$D_j := \inf\{n \geqslant 0 : X_n = j\}.$$

Now define the Martin kernel κ on $I \times I$ as

(27.2) $$\kappa(i, j) := \Gamma(i, j)/\Gamma(b, j).$$

It follows from (27A) and (27.1) that

(27.3) $$\kappa(i, j) \leqslant \kappa(j, j) < \infty, \quad \forall i, j.$$

It is another easy consequence of the strong Markov property that

$$\kappa(i, j) \leqslant \kappa(i, i) < \infty, \quad \forall i, j. \tag{27.4}$$

Exercise. Prove (27.4) probabilistically. Can you give a neat algebraic proof?

A function f from I to $\mathbb{R}$ is called *excessive* (respectively, *regular*) (for Π) if

(i) $0 \leqslant f < \infty$;
(ii) $\Pi f \leqslant f$ (respectively $\Pi f = f$).

The set of excessive functions forms a cone C in $\mathbb{R}^I$. For the topology of C, we take that of $\mathbb{R}^I$, that is, the topology of pointwise convergence.

Because of Assumption 27A,

$$\theta(j) := \sup_n \Pi^n(b, j) > 0,$$

and since a function f in C satisfies

$$f \geqslant \Pi f \geqslant \Pi^2 f \geqslant \cdots,$$

we have

$$f(b) \geqslant \theta(j) f(j), \quad \forall j. \tag{27.5}$$

In particular, every f in C may be written as

$$f = f(b) f^*, \quad f^* \in S := \{f \in C : f(b) = 1\}.$$

The study of the cone C thus reduces to the study of its section S.

(27.6) PROPOSITION. *The set S is a compact convex metrisable subset of the locally convex linear topological space $\mathbb{R}^I$.*

This proposition, which is an immediate consequence of (27.5) and Fatou's Lemma, states exactly that S satisfies the hypothesis of the metrisable case of Choquet's Theorem on the existence of integral representations. Recall that an element f of S is called *extremal* if the equation

$$f = \tfrac{1}{2} f_1 + \tfrac{1}{2} f_2 \quad (f_1, f_2 \in S)$$

implies that $f = f_1 = f_2$.

For our special situation, Choquet's Existence Theorem takes the following form. See Meyer [2] and Phelps [1].

(27.7) THEOREM. *The set S_e of extremal elements of S is a G_δ in S. If $f \in S$ then there exists a probability measure ν on $\mathcal{B}(S_e)$ such that*

$$f(i) = \int_{S_e} \xi(i) \nu(d\xi), \quad \forall i. \tag{27.8}$$

(Note that the map $\xi \mapsto \xi(i)$ is continuous on S.)

We wish to add the following theorems

(27.9) THEOREM. *Further, v is uniquely determined by f.*

Choquet's Uniqueness Theorem states that Theorem 27.9 is equivalent to the following lemma.

(27.10) LEMMA. *The cone C is a lattice in its intrinsic order.*

Note. The *intrinsic order* $\ll$ on C is defined as follows: for $x, z \in C$, we write $x \ll z$ if $\exists u \in C$ with $x + u = z$.

How to prove Lemma 27.10 will be explained in a moment. (As an exercise—not quite as easy as it may look!—try proving it now.)

The key technique for studying C is provided by the Riesz Decomposition Theorem 27.14. Let μ be a (non-negative) measure on I such that

$$\Gamma\mu(i) := \sum_{j \in I} \Gamma(i, j)\mu(j) < \infty, \quad \forall i.$$

Then $\Gamma\mu$ is called the *potential* (due to the *charge* μ). Since

$$\Pi\Gamma\mu = \Gamma\mu - \mu \leqslant \Gamma\mu, \tag{27.11}$$

the function $\Gamma\mu$ is excessive. Note that the equation

$$\mu = \Gamma\mu - \Pi\Gamma\mu \tag{27.12}$$

determines μ from $\Gamma\mu$, and that

$$\Pi^n\Gamma\mu = \sum_{k \geqslant n} \Pi^k\mu \downarrow 0 \quad (n \uparrow\uparrow \infty). \tag{27.13}$$

(27.14) THEOREM (Riesz Decomposition Theorem). *If f is excessive then f has a unique decomposition*

$$f = v + \Gamma\mu, \tag{27.15}$$

where v is regular and μ is a measure on I. Indeed,

$$v = \lim_n \Pi^n f, \tag{27.16}$$

$$\mu = f - \Pi f. \tag{27.17}$$

Proof. Define μ by (27.17). Then $\mu(i) \geqslant 0$, $\forall i$, and

$$(I + \Pi + \cdots + \Pi^n)\mu = f - \Pi^{n+1} f.$$

The Monotone-Convergence Theorem yields (27.15) with v as is (27.16). Properties (27.12) and/or (27.13) make the uniqueness assertion obvious. □

(27.18) *Exercise.* Now prove Lemma 27.10 (if you did not do so earlier) by showing that if

$$f_1 = v_1 + \Gamma\mu_1, \quad f_2 = v_2 + \Gamma\mu_2$$

then the lattice structure of C in its intrinsic order is exhibited by the equations

$$f_1 \wedge\wedge f_2 = \lim_n \Pi^n(v_1 \wedge v_2) + \Gamma(\mu_1 \wedge \mu_2),$$

$$f_1 \vee\vee f_2 = f_1 + f_2 - f_1 \wedge\wedge f_2,$$

where

$$(v_1 \wedge v_2)(i) := v_1(i) \wedge v_2(i), \qquad (\mu_1 \wedge \mu_2)(i) := \mu_1(i) \wedge \mu_2(i).$$

In this connection, we must mention Feller's historic paper [2], the impressive first attempt to define an appropriate 'boundary' for C by lattice methods.

Hints for exercise. First prove that if $v < \Gamma\mu(<\infty)$ then $v = 0$. Next deal separately with the cases (i) $v_1 = v_2 = 0$ and (ii) $\mu_1 = \mu_2 = 0$.

As an immediate consequence of the Riesz decomposition, we have the following proposition.

(27.19) PROPOSITION. *For each j in I, the function $\kappa(\cdot, j)$ is a (non-regular) extremal element of S. Every extremal element of S that is not of the form $\kappa(\cdot, j)$ for some j in I is regular.*

28. The Martin compactification. (We continue to assume 27A.) Since potential determines charge, the map

$$\varphi: I \to S \subset \mathbf{R}^I, \qquad \varphi(j) := \kappa(\cdot, j)$$

is one–one. *We now identify I with $\varphi(I)$ and let F be the compact closure of $I(=\varphi(I))$ in S.* The set F is called the Martin compactification of I, though, in this context, the theory is due to Doob and Hunt.

Since the topology of F is inherited from that of $\mathbf{R}^I$,

(28.1) *for each i, the map $\kappa(i, \cdot)$ extends continuously to F; then $\kappa: I \times F \to \mathbf{R}$, and, for $\xi \in F \backslash I$, we have the alternative notations: $\kappa(i, \xi) = \xi(i)$.*

(28.2) THOREM (Doob, Hunt). *Every extremal element of S is of the form $\kappa(\cdot, \xi)$ for some ξ in F. The following result therefore holds. Let F_e be the set of ξ in F for which $\kappa(\cdot, \xi)$ is extremal. Then each f in S can be written uniquely as*

$$f = \int_{F_e} \kappa(\cdot, \xi)\nu(d\xi) = v + \Gamma\mu,$$

where ν is a probability measure on $\mathscr{B}(F_e)$,

$$v := \int_{F_e \backslash I} \kappa(\cdot, \xi)\nu(d\xi) \quad \textit{is regular,}$$

and

$$\mu(j) := [\Gamma(b, j)]^{-1}\nu(j).$$

Once we establish the first sentence of Theorem 28.2, the remainder of the theorem follows from the Choquet results (27.7) and (27.9); and we then have $S_e = F_e \subset F$. (We are presently assuming the Choquet results, but recall that we later–see Section 44—give a full independent proof of Theorems 27.7 and 27.9.)

We now argue that it is enough to prove that

(28.3) *every element f of S may be written as*

$$f = \int_F \kappa(\cdot, \xi)\nu(d\xi)$$

for some (not necessarily unique) probability measure ν on $\mathscr{B}(F)$.

First argument. From (28.3), it follows that S is the closed convex hull of F. By a standard result (much more elementary than Choquet theory)—see Theorem V.8.5 of Dunford and Schwartz [1]—the extremal elements of S are contained in F. □

Second argument. First, we clarify notation by (temporarily) writing α for a typical element of S_e (not yet known to belong to F) and ξ for a typical element of F.

Let β be an element of S_e. Its unique Choquet representing measure on $\mathscr{B}(S_e)$ is of course the unit mass at β, denoted by $\varepsilon_\beta(\cdot)$. However, by (28.3), there is at least one probability measure ν_β on $\mathscr{B}(F)$ such that

$$\beta(i) = \int_{\xi \in F} \kappa(i, \xi)\nu_\beta(d\xi) = \int_{\xi \in F} \nu_\beta(d\xi) \int_{\alpha \in S_e} \alpha(i)\mu_\xi(d\alpha), \quad \forall i,$$

where μ_ξ on $\mathscr{B}(S_e)$ is the Choquet representing measure for $\kappa(\cdot, \xi)$. Hence

$$\varepsilon_\beta(\cdot) = \int_{\xi \in F} \nu_\beta(d\xi)\mu_\xi(\cdot) \quad \text{on } \mathscr{B}(S_e),$$

and so, for (ν_β)-almost-all ξ in F, $\mu_\xi = \varepsilon_\beta$. But, for any ξ in F for which $\mu_\xi = \varepsilon_\beta$, we have

$$\kappa(\cdot, \xi) = \beta(i),$$

so that $\beta = \xi \in F$. □

We have now reduced the problem of proving Theorem 28.2 to that of proving the statement (28.3).

Proof of (28.3). Fix f in S. Choose a measure β such that

$$0 < \Gamma\beta(i) < \infty, \quad \forall i.$$

(By (27.3), it is enough to choose β so that $\beta(j) > 0, \forall j$, and $\sum \Gamma(j, j)\beta(j) < \infty$.) Let

$$f_n(i) := \min(f(i), n\Gamma\beta(i)).$$

Then f_n is excessive, and since f_n is dominated by the potential $n\Gamma\beta$, it follows from (27.13) and the Riesz theorem that f_n is a potential:

$$f_n(i) = \sum \Gamma(i, j)\mu_n(j) = \sum \kappa(i, j)\nu_n(j) \tag{28.4}$$

where $\nu_n(j) = \Gamma(b, j)\mu_n(j)$. Since $f_n(b) = f(b) = 1$ for large n, and $\kappa(b, j) = 1, \forall j$, it follows that (for large n) ν_n is a probability measure on F with $\nu_n(I) = 1$. Since F is compact metrisable, $\Pr(F)$ is compact metrisable in the weak topology. Let ν be a subsequential limit of (ν_n) in $\Pr(F)$. Then (28.3) follows from (28.4) and (28.1). □

The following analytical problem remains: *how can we determine the 'extremal' part F_e of F?*

A striking probabilistic solution is provided as one part of the Doob–Hunt probabilistic theory of the Martin boundary for this case. (See Section 29.)

(28.5) *Example.* Let X be a simple random walk on $\mathbf{Z}^d$ such that

$$\Pi(i, j) = \begin{cases} (2d)^{-1} & \text{if } |j - i| = 1, \\ 0 & \text{otherwise.} \end{cases}$$

We now prove that *every regular function* (that is, every solution of $\Pi f = f \geqslant 0$) *is constant.*

Case 1: $d = 1$ or 2. In this case, it is well known that X is recurrent. If f is regular then (under each $\mathbf{P}^i$) $f(X_n)$ is a non-negative martingale, so that $\lim f(X_n)$ exists. Since X visits every point of $\mathbf{Z}^d$ infinitely often, the only possible explanation is that f is constant on $\mathbf{Z}^d$. This proof is (of course) due to Doob.

Case 2: $d \geqslant 3$. In this case, X is transient and Assumption 27A holds with $b = 0$ (say). It is well known (see Spitzer [1]) that

$$\Gamma(i, j) \sim \text{constant}\, |j - i|^{2-d} \quad (|j - i| \to \infty),$$

as one might expect by analogy with the Brownian-motion results. Since

$$\kappa(i, j) \sim |j - i|^{2-d}/|j|^{2-d},$$

it is clear that F is the one-point compactification $I \cup \{\infty\}$ of I and that $\kappa(i, \infty) = 1$, $\forall i$. The desired result follows. It is worth mentioning that no particularly simple proof is known. See Spitzer [1] for a specialisation of the Martin-boundary argument to this case and for an Itô–McKean proof.

(28.6) *Example.* We now consider 'space–time coin-tossing'. Think of X_n as (H_n, n), where H_n represents the number of heads in n tosses. We put

$$I := \{(m, n) \in \mathbf{Z}^2 : 0 \leqslant m \leqslant n\}.$$

$$\Pi((m, n); (m + 1, n + 1)) = 1 - \Pi((m, n); (m, n + 1)) = \tfrac{1}{2}.$$

Then, for (m,n) and (r,s) in I,

$$\Gamma((m,n);(r,s)) = \begin{cases} \binom{s-n}{r-m} 2^{-(s-n)} & \text{if } 0 \leqslant r-m \leqslant s-n, \\ 0 & \text{otherwise.} \end{cases}$$

Taking $b=(0,0)$, we find from Stirling's formula that if $s \to \infty$ and $r/s \to t \in [0,1]$ then

$$\kappa((m,n);(r,s)) \to h_t(m,n) := 2^n t^m (1-t)^{n-m}.$$

It is now clear (why?) that the Martin topology can be regarded as identifying (m,n) in I with $(1+n)^{-1}(m,n) \in \mathbb{R}^2$, with $F \backslash I = [0,1] \times \{1\}$ and

$$h_t = \kappa(\cdot, \xi) \quad (\xi = (t,1) \in F \backslash I).$$

Thus f is a regular element of S if and only if there exists a probability measure ν on $\mathscr{B}[0,1]$ such that

$$f(m,n) = \int_0^1 2^n t^m (1-t)^{n-m} \nu(dt). \tag{28.7}$$

This result yields an immediate solution of the *Hausdorff moment problem.* (See Spitzer [1].)

The Weierstrass Approximation Theorem makes it obvious that ν is uniquely determined by f in (28.7). Hence, for every $t \in [0,1]$, h_t is extremal in S.

29. The Martin representation: Doob–Hunt explanation. We retain the notation of Section 28. In particular, we have

$$\zeta := \inf\{n : X_n = \partial\}.$$

(29.1) THEOREM (Doob, Hunt). *Almost surely on* $\{\zeta = \infty\}$,

$$X_{\zeta-} := \lim_n X_n$$

exists in the topology of F, *and* $X_{\zeta-} \in F_e$.

(Recall that the probabilistic results of this section and the analytical results of the last section will all soon (Section 43) be exhibited as consequences of Ray's Theorem.) As explained in Section 8.5 of Itô and McKean [1], the point is that, while F_e is large enough to allow representation of regular functions, it is also small enough (think!) to describe the exits of X.

Let us agree to define

$$X_{\zeta-} := X_{\zeta-1} \quad \text{on } \{\zeta < \infty\}.$$

(We are not interested in the case when X starts at ∂.) Recall that b is our 'reference' point in terms of which the Martin kernel is defined.

(29.2) THEOREM (Doob, Hunt). *Let*

$$1 = \int_{F_e} \kappa(\cdot, \xi)\nu_1(d\xi)$$

be the Martin representation of the (excessive) constant function 1. *Then*

$$\mathbf{P}^b[X_{\zeta-} \in B] = \nu_1(B), \quad \forall B \in \mathscr{B}(F_e).$$

Example. To get $f(\cdot,\cdot) = 1$ in (28.7), you have to choose ν to be the unit mass at $\frac{1}{2}$. Thus Theorems 29.1 and 29.2 contain the strong law for tossing a fair coin.

Now let $h \in S$. To avoid trivial nuisances, we assume that h is *strictly* positive on I. The *Doob h-transform* Π_h of Π is defined as follows:

$$\Pi_h(i, j) := h(i)^{-1}\Pi(i, j)h(j).$$

Then Π_h is substochastic. We have, with obvious notation,

$$\kappa_h(i, j) = \frac{h(b)}{h(i)}\kappa(i, j),$$

and $f \in S_h$ if and only if $hf \in S$. Thus F and F_e are unaffected if we change from Π to Π_h. (You should formulate this a little more carefully.) Hence if $X^{(h)}$ is a chain with one-step transition matrix Π_h then

$$X^{(h)}(\zeta^{(h)}-) \textit{ exists in } F_e \quad \textit{almost surely.}$$

Here $\zeta^{(h)}$ denotes $\inf\{n : X^{(h)}(n) = \partial\}$.

(29.3) THEOREM (Doob). *A strictly positive function h in S is extremal in S if and only if for some single point ξ of F, we have*

$$X^{(h)}(\zeta^{(h)}-) = \xi, \quad \textit{almost surely.}$$

Then $\xi \in F_e$ and $h = \kappa(\cdot, \xi)$. We then say that $X^{(h)}$ represents X conditioned to converge to ξ.

(29.4) *Notes*

(a) If $h = \kappa(\cdot, c)$, where $c \in I$, then $X^{(h)}$ has the same laws as $\{X_n : n \leqslant \sigma_c\}$, where σ_c is the time of the *last* visit by X to state c. (Prove this as an exercise.) This gives the correct interpretation of X *conditioned by* $\{X(\zeta-) = c\}$ for $c \in I$.
(b) Theorems 29.2 and 29.3 are obviously closely related. (Investigate the connection.)

(29.2) *Example.* Return to Example 28.6 and take $h = h_t$ for some fixed $t \in [0, 1]$. Then $h = \kappa(\cdot, \xi)$, where $\xi = (t, 1) \in F \backslash I$. Then

$$\Pi_h((m, n); (m + 1, n + 1)) = 1 - \Pi_h((m, n); (m, n + 1)) = t.$$

Thus

(29.6) *the h_t transform corresponds to the case of space–time coin-tossing for a coin with probability t of heads.*

By the strong law of large numbers, $X^{(h)} \to \xi$, so that h_t is extremal.

(29.7) *A statistical interpretation.* The property (19.6) illustrates the attractive idea that the Doob *h*-transforms correspond to the set of 'appropriate' *alternative hypotheses* in hypothesis-testing contexts. This idea has been developed to cover maximum-likelihood estimation, sufficiency etc. by members of the Copenhagen school.

30. R. S. Martin's boundary. The original Martin boundary (Martin [1]) was introduced to describe the non-negative harmonic functions on a Greenian domain D in $\mathbb{R}^n$. Let us run quickly (and in heuristic fashion) through some of the basic ideas. We take $n = 3$, which case provides the most familiar potential theory. Every domain D in $\mathbb{R}^3$ is Greenian, so we do not need to explain what 'Greenian' means.

Let g be the free-space Green function for $\mathbb{R}^3$ for $\frac{1}{2}\Delta$:

$$g(x, y) = (2\pi|y - x|)^{-1}.$$

Then the Green function g_D for D is the smallest non-negative function on $D \times D$ such that

(30.1)(i) $g - g_D$ is bounded in the neighbourhood of each 'diagonal' point (x, x) of $D \times D$;

(30.1)(ii) $\frac{1}{2}\Delta_x(g - g_D) = \frac{1}{2}\Delta_y(g - g_D) = 0$ on D;

(30.1)(iii) $g_D(x, y) = g_D(y, x)$.

Such a function g_D is known to exist. See Section 7.4 of Itô and McKean [1].

Let $V := \mathbb{R}^3 \backslash D$. The *physical* significance of $g_D(x, y)$ is as the potential at x due to a unit charge (we are using 'probabilistic' units—see Section I.22) placed at y when V is earthed. The *probabilistic* significance of g_D is that if X is a Brownian motion on $\mathbb{R}^3$ then, for $x, y \in D$,

(30.2) $$\mathbf{E}^x[\text{time spent by } X \text{ in } dy \text{ before time } H_V] = g_D(x, y)\, dy.$$

The strong Markov theorem yields the probabilistic formula

(30.3) $$g_D(x, y) = g(x, y) - \mathbf{E}^x g(X \circ H_V, y),$$

or, in better notation,

(30.3′) $$g_D(\cdot, y) = g(\cdot, y) - P_V g(\cdot, y).$$

(Of course there are all sorts of details of rigour to be chased up, but, for the moment, who cares?) Classical Martin boundary theory reduces to the analytical part of the Martin–Doob–Hunt theory of the process $\{X_t : t < H_V\}$ killed on leaving D.

Pick a reference point b in D, introduce the Martin kernel

$$\kappa(x, y) := \frac{g_D(x, y)}{g_D(b, y)} \in [0, \infty]$$

and take the Gelfand–Stone–Čech (GSČ) compactification F of D determined by the separable class of functions

$$\{\arctan \kappa(x, \cdot) : x \in D\}.$$

Full details of such GSČ compactifications are recalled later in this chapter. The point is that each function $\kappa(x, \cdot)$ extends continuously to F. Martin's Theorem is that every non-negative harmonic function f on D with $f(b) = 1$ has a unique representation

$$f(x) = \int_{F_e \backslash D} \kappa(x, \xi) \nu(d\xi)$$

where F_e is what you expect, and ν is a probability measure on $\mathscr{B}(F_e \backslash D)$.

(30.4) *Example.* Let D be the open ball $\{x \in \mathbf{R}^3 : |x| < 1\}$. Then Kelvin's method of images shows that, for $y \in D$,

$$g_D(x, y) = \begin{cases} g(x, y) - |x|^{-1} g(x^*, y) & (x \in D \backslash \{0\}), \\ g(0, y) - (2\pi)^{-1} & (x = 0), \end{cases}$$

where $x^* := |x|^{-2} x$ is the point inverse to x in the sphere $\partial D = \{\xi \in \mathbf{R}^3 : |\xi| = 1\}$. Taking $b = 0$, you find that as $y \to \xi \in \partial D$ (in the Euclidean topology),

$$\kappa(x, y) \to \kappa(x, \xi) := \frac{1 - |x|^2}{|\xi - x|^3}.$$

Hence (why?) $F = D \cup \partial D$ with the Euclidean topology, and every positive harmonic function f on D with $f(0) = 1$ may be written as

$$f(x) = \int_{\partial D} \kappa(x, \xi) \nu(d\xi). \tag{30.5}$$

Invariance under SO(3) implies that every point of ∂D is in F_e, so that the representation (30.5) of f is unique.

We investigate this example further in Section 31.

31. Doob–Hunt theory for Brownian motion. It is easy to guess the basic form of the Doob–Hunt theorems for Brownian motion or indeed for a general Markov process. However, *we need to be careful about the precise formulation* *of the concepts of harmonic function and excessive function in the context of continuous-parameter processes.* In Example 30.4, 'f is harmonic' means of course that $\frac{1}{2}\Delta f = 0$ (on D). The correct probabilistic formulation relies on the *mean-value property*.

Here are some guidelines.

Let $\{P_t\}$ be a transition function (in the 'abstract' sense of (1.1)) on a measurable space $(E, \mathscr{E})$. An $\mathscr{E}^*$-measurable function f from E to $[0, \infty]$ is called *excessive* (for $\{P_t\}$) if

(31.1)(i) $\qquad f$ is supermedian: $\quad P_t f \leqslant f, \quad \forall t,$

(31.1)(ii) $$\lim_{t \downarrow 0} P_t f(x) = f(x) \quad \forall x.$$

Of course, if x is excessive, then $P_t f(x) \uparrow f(x)$ $(t \downarrow 0)$. Do note that, for excessive functions, the 'smoothness' condition (31.1)(ii) is imposed in addition to the 'supermedian' condition (31.1)(i). In Ray theory the difference between supermedian and excessive becomes critically important.

Now let X be a 'nice' process on a nice space E. Let $\{P_t\}$ be the transition function of X. If f is an $\mathscr{E}^*$-measurable function from E to $[0, \infty]$ satisfying the smoothness condition (31.1)(ii), then f is excessive if and only if f is *superharmonic* in the following sense:

$$P_{E \setminus \bar{A}} f \leqslant f,$$

whenever A is an open subset of E with compact closure $\bar{A}$. We call an $\mathscr{E}^*$-measurable function from E to $[0, \infty]$ satisfying (31.1)(ii) *harmonic* if the *mean-value property*

$$P_{E \setminus \bar{A}} f = f$$

holds whenever A is an open subset of E with compact closure $\bar{A}$.

It is important to notice that, in general, a harmonic function will *not* satisfy $P_t f = f, \forall t$. Example 31.4 will clarify this matter.

After all that, the analogues of Theorems 29.1–29.3 for the situation discussed in Section 30 are obvious (in form!). Let D be our domain in $\mathbb{R}^3$. Let $V := \mathbb{R}^3 \setminus D$ and let F be the Martin compactification of D. Then, for Brownian motion X started inside D, it is a.s. true that

$$X(H_V -) := \lim_{t \uparrow\uparrow H_V} X(t) \quad \text{exists in } F$$

and $X(H_V -) \in F_e \setminus D$. This should set you wondering about the relation between regular points for the Dirichlet problem and extremal points of the Martin boundary. The following exercise should make you think further about Dirichlet–Martin connections.

(*31.2*) *Exercise.* Return to the case where D is the open ball $\{|x| < 1\}$ for which

$F_e = D \cup \partial D$ with the Euclidean topology. By considering the effect of changing the reference point b from 0 to a point c of D, show that

$$\mathbf{P}^c[X(H_{\partial D}) \in d\xi] = \kappa(c, \xi)\mu(d\xi),$$

where μ is a normalised surface area measure on ∂D. If g is a continuous function on ∂D then the unique continuous function h on $\bar{D}$ with $\Delta h = 0$ on D and $h = g$ on ∂D is therefore given by the Poisson formula

$$h(x) = \int_{\partial D} \kappa(x, \xi) g(\xi) \mu(d\xi). \tag{31.3}$$

For fixed $\xi \in \partial D$, calculate explicitly the differential generator

$$\tfrac{1}{2}\kappa(x, \xi)^{-1} \Delta \kappa(x, \xi) = \tfrac{1}{2}\Delta + \kappa(x, \xi)^{-1} \operatorname{grad} \kappa(x, \xi) \cdot \operatorname{grad}$$

of Brownian motion conditioned to hit ∂D at ξ, and convince yourself that the 'extra' term behaves as it should.

(31.4) *The Helms–Johnson example revisited* (see EII.79.77). Take $D := \mathbf{R}^3 \backslash \{0\}$. Let X be Brownian motion in $\mathbf{R}^3$ started in D. Since X never visits 0, the Green function for D is just the restriction to $D \times D$ of the free-space Green function

$$g_D(x, y) = (2\pi |y - x|)^{-1} \quad (x, y \in D).$$

The Martin compactification adjoins two points 0 and ∞ to D, producing the expected one-point compactification of $\mathbf{R}^3$. If the reference point b is chosen with $|b| = 1$ then

$$\kappa(x, 0) = |x|^{-1}, \quad \kappa(x, \infty) = 1, \quad \forall x \in D.$$

The function f on D with $f(x) \equiv |x|^{-1}$ is harmonic in D, but $P_t f(x) \neq f(x)$, $\forall t > 0$, $\forall x \in D$. Indeed, you can check that

$$P_t f(x) = [\Phi(x[2t]^{-1/2}) - \Phi(-x[2t]^{-1/2})] f(x), \quad \forall t > 0, x \in D, \tag{31.5}$$

where Φ is the normal distribution function:

$$\Phi(y) = (2\pi)^{-1/2} \int_{-\infty}^{y} \exp(-\tfrac{1}{2}u^2)\, du.$$

Equation (31.5) makes it clear that f is excessive for $\{P_t\}$.

The fact that f is supermedian implies that, $\forall x \in D$, $f(X_t)$ is a *super*martingale relative to $(\{\mathscr{F}_t^\circ\}, \mathbf{P}^x)$. The fact that f is not invariant under $\{P_t\}$ implies that $x \in D$, $f(X_t)$ is *not* a martingale relative to $(\{\mathscr{F}_t^\circ\}, \mathbf{P}^x)$. The 'intermediate' fact that f is harmonic corresponds to the statement that f is a *local martingale.* It is in local-martingale theory that the main illustrative importance of the present example lies.

Let

$$T_n := \inf\{t : X_t = n^{-1}\} = \inf\{t : f(X_t) = n\} \leqslant \infty.$$

We can find g_n in $\mathscr{D}(\frac{1}{2}\Delta)$ with $g_n = f$ on $\{|y| \geqslant n^{-1}\}$. For $|x| \geqslant n^{-1}$, we have, by Dynkin's formula (10.12),

$$\begin{aligned} P_{t\wedge T_n}f(x) - f(x) &= \mathbf{E}^x f\circ X(t\wedge T_n) - f(x) \\ &= \mathbf{E}^x g\circ X(t\wedge T_n) - g(x) \\ &= \mathbf{E}^x \int_0^{t\wedge T_n} \tfrac{1}{2}\Delta g(X_s)\,ds = 0, \end{aligned}$$

since $\frac{1}{2}\Delta g = \frac{1}{2}\Delta f = 0$ on $\{|y| \geqslant n^{-1}\}$. Hence, for each fixed x, we have

(31.6) $$P_{t\wedge T_n}f(x) = f(x), \quad \forall t, \forall n > |x|^{-1}.$$

Since $T_n \uparrow \infty$, this property can be regarded as an appropriate ('local') correction of the false statement that $P_t f(x) = f(x)$, $\forall x\,\forall t$. Note further that, on letting $n \uparrow\uparrow \infty$ in (31.6), we obtain

$$P_t f(x) \leqslant f(x), \quad \forall t, \forall x \in D,$$

by Fatou's Lemma. Since $P_t f(x) \neq f(x)$, $\forall t > 0$, $\forall x \in D$, it follows that

(31.7) *for each t and each $x \in D$, the sequence $\{f(X_{t\wedge T_n}) : n \in \mathbf{N}\}$ is not uniformly integrable relative to $\mathbf{P}^x$.*

We shall appreciate the full significance of (31.7) only when we come to study local-martingale theory.

It is worth noting that, since $|X_t|^{-1}$ is a continuous supermartingale relative to each $\mathbf{P}^x$ $(x \in D)$, $\lim_{t\to\infty} |X_t|^{-1}$ exists a.s.$(\mathbf{P}^x)$.

Exercise. Explain why the limit must be 0, a.s.$(\mathbf{P}^x)$.

Thus a 3-dimensional Brownian motion started away from 0 will never hit 0 and will drift to ∞. The corresponding result is true all the more in dimension $n > 3$. Of course, we have known and used these properties since Section I.18.

Next consider 'X Doob-conditioned to converge to 0' with generator $f^{-1}(\frac{1}{2}\Delta)f$. The radial part of this process is nothing other than a 1-dimensional Brownian motion absorbed at 0. (Check this!)

We can see this working in reverse. Let B be a 1-dimensional Brownian motion on $(0, \infty)$ with 0 as killing boundary (see Section I.13). The Martin compactification is of course $[0, \infty]$, and if we take $b = 1$, we obtain

$$\kappa(x, 0) = 1, \quad \kappa(x, \infty) = x.$$

Thus 'Brownian motion on $(0, \infty)$ Doob-conditioned to converge to ∞' has generator

$$x^{-1}\left(\frac{1}{2}\frac{d^2}{dx^2}\right)x = \frac{1}{2}\frac{d^2}{dx^2} + x^{-1}\frac{d}{dx};$$

in other words, it is a BES (3) process.

(*31.8*) *Comments.* Martin–Doob–Hunt–Ray–Kunita–Watanabe boundary theory may be developed under assumptions of extreme generality. We present proofs only for the discrete-parameter chain case. But we do so by the most modern *continuous* (!)-parameter methods. We then give a list of references in which you may chase up the general theory.

The long account of Martin–Doob–Hunt theory that we have already given has been a gentle introduction to compactifications. It helps prepare the way for the much more sophisticated Ray–Knight compactification. Further, it will be reassuring after working through the Ray–Knight theory to find that in the case of discrete-parameter chains we obtain *exactly* (there are many boundaries!) the familiar results of Section 27–29.

After all this talk of Martin boundary theory, let us remind you that it is but one application of Ray theory. Indeed, it is not the one that chiefly concerns us in this book!

32. Ray processes and right processes. It remains to answer question (25.1)(iii). Blumenthal and Getoor [1] present a theory based, not on assumptions like the FD property that rely on analytical properties of the transition function, but on *probabilistic* axioms for a setup

$$X = (\Omega, \mathscr{F}, \mathscr{F}_t, \theta_t, X_t, \mathbf{P}^x).$$

The fundamental concept in Blumenthal and Getoor [1] is that of a *standard process*. Now there are obvious advantages in using probabilistic axioms: for example, the property of being standard is preserved under various probabilistic operations that do not preserve the FD property.

Briefly, the situation is this. Some generalisation of standard process was needed in order to cope with branch-points. Meyer introduced the concept of a *right process* (process satisfying *les hypothèses droites*) as the 'natural' concept for Markov-process theory. Ray's hypotheses remained the most general analytical hypotheses on transition functions. Ray, Knight, Meyer, Shih, Walsh, Getoor and Sharpe all participated in work that culminated in proving that the concepts of Ray process and right process are essentially the same. This is explained very clearly in the books by Getoor [1], Sharpe [1] and Dellacherie and Meyer [1]. Let us quote part of Getoor's book (perhaps you will not mind if we emphasise the obvious fact that the set D of non-branch-points has no connection with our domain D in the classical Martin-boundary case!):

> One has the following inclusions among the various classes of processes:
>
> $$(\text{Feller}) \subset (\text{Hunt}) \subset (\text{special standard}) \subset (\text{standard}) \subset (\text{right}).$$
>
> ...These different types of processes were introduced at various stages during the development of the modern theory of Markov processes. In view of the theory to be developed in the sections it seems to me that [except for right processes] they are now mainly of historic interest.... A

Ray process Y on a compact metric space E is not necessarily a right process since $P_0 \neq I$ [in general]. However, if one restricts Y to the set of non-branch points D, then it is a Borel right process. Since $Y_t \in D$ for all $t \geq 0$, this amounts to considering initial measures that are carried by D. In what follows we shall see that the converse is true in the sense that if X is a right process with state-space E, then by changing the topology on E one can essentially regard X as a Ray process restricted to its set of non-branch points.

There is thus a perfect match between the probabilistic and analytical parts of the theory. However, at least in special circumstances, one can achieve such a match in other ways, as is illustrated by the important books by Fukushima [1], Silverstein [1, 2] on symmetrisable processes. But the moral is that the subject's masters regard its theoretical foundations as being fully achieved by the essentially equivalent theories of Ray processes and right processes. We content ourselves with introducing you to the more concrete Ray processes; the masters can then set you right.

5. RAY PROCESSES

33. Orientation. Our basic datum will be a resolvent $\{R_\lambda : \lambda > 0\}$ on a space I. In general, there will *not* exist a nice (right) process with resolvent $\{R_\lambda\}$ and taking all its values in I. In order to construct a nice (Ray) process with resolvent $\{R_\lambda\}$, we must generally allow the process to take values in a suitable compactification F of I. (If this process starts in I, and if I is a Borel subset of F, then the set of times spent in $F \setminus I$ before the death-time of the process will be of measure zero.) First we must construct the Ray–Knight compactification F of I determined by $\{R_\lambda\}$, extend $\{R_\lambda\}$ to a 'Ray' resolvent on F, and construct the 'Ray' transition function $\{P_t\}$ on F with resolvent $\{R_\lambda\}$. Then we shall construct the Ray process on F with transition function $\{P_t\}$.

The main case in which we are interested is that when we begin with a 'standard' transition function on a countable set I; and our notation for the general case reflects that for chains.

Here (*as a guide*—you are not expected to understand about RK compactifications, branch-points etc. now!) is a list of our notations for chains:

I: countable 'minimal' state space.
$\{P_t\}$: ('standard') transition function on I with resolvent $\{R_\lambda\}$;
F: Ray-Knight compactification of I determined by $\{R_\lambda\}$;
F_e: set of non-branch- (or extremal) points in F;
F_{br}: set of branch-points in F;
$E := \{\xi \in F : P_t(\xi, F \setminus I) = 0, \forall t > 0\}$.

In the definition of E, we have used the notation $\{P_t\}$ to denote the Ray extension on $(F, \mathscr{B}_F)$ of the original transition function on I.

(*33.1*) *Important note.* Let I (for reasons explained in Section 32, we abandon the notation 'D for domain') be a bounded domain in $\mathbb{R}^3$ and let $V := \mathbb{R}^3 \backslash I$. To obtain the Martin–Doob–Hunt results for Brownian motion on I, we apply Ray theory not to $\{X_t : t < H_V; \mathbf{P}^b\}$ but to its time-reversal

$$\hat{X} := \{X(H_V - t) : 0 < t < H_V; \mathbf{P}^b\},$$

which is also *Markovian* with *stationary* transition probabilities and which has Green function

$$\hat{g}_I(y, x) = g_I(0, x) g_I(x, y)/g_I(0, y) = g_I(0, x)\kappa(x, y).$$

Note the order in which y and x appear, and also the appearance of the Martin kernel. Time-reversal will be studied in Part 6 of this chapter.

Ray theory is an entrance-boundary theory, and time reversal has to be invoked to apply it to theory of Martin exit boundary. Where X goes to is where $\hat{X}$ comes from.

If I is an arbitrary (possible unbounded) domain in $\mathbb{R}^3$, we first speed up X via a time substitution of the type described in Section 21, so as to ensure that X exits I (or else dies) within a finite time. *Then* we reverse X to produce $\hat{X}$. The faster the speeding up, the smoother will be the analytic properties of $\hat{X}$. As we have already mentioned, you will see all the tricks during our 'Ray' treatment of the discrete-parameter chain case.

(*33.2*) *Our plan.* We first describe the 'good' situation, that in which we have a Ray resolvent on a compact metric space F. Then we explain how a resolvent on a measurable space $(I, \mathscr{I})$ may, under minimal conditions, be extended to a Ray resolvent on a Ray–Knight compactification F of I. We then construct the Ray semigroup and Ray process on F.

34. Ray resolvents. *Let F now denote an arbitrary compact metric space. Let $\{R_\lambda : \lambda > 0\}$ be an honest Feller resolvent on $C(F)$:*

$$R_\lambda : C(F) \to C(F), \quad 0 \leqslant f \leqslant 1 \Rightarrow 0 \leqslant \lambda R_\lambda f \leqslant 1,$$
$$\lambda R_\lambda 1 = 1, \quad R_\lambda - R_\mu + (\lambda - \mu) R_\lambda R_\mu = 0.$$

(*34.1*) DEFINITION (continuous α-supermedian function). *For $\alpha \geqslant 0$, an element f of $C(F)$ is called a* (continuous) α-supermedian function relative to $\{R_\lambda\}$, *and we write $f \in \mathrm{CSM}^\alpha$, if*

$$0 \leqslant \lambda R_{\lambda+\alpha} f \leqslant f \quad (\forall \lambda > 0).$$

For $\beta > \alpha$,

$$\lambda R_{\lambda+\beta} f = \lambda R_{\lambda+\alpha} f - \lambda(\beta - \alpha) R_{\lambda+\alpha} R_{\lambda+\beta} f, \tag{34.2}$$

so that

$$\mathrm{CSM}^\alpha = \bigcap_{\beta > \alpha} \mathrm{CSM}^\beta.$$

(34.3) **DEFINITION** (Ray resolvent). *Our honest Feller resolvent on F is called a* Ray resolvent *if*

(34.4) $$\bigcup_{\alpha \geqslant 0} \mathrm{CSM}^{\alpha} \text{ separates points of } F.$$

(34.5) **LEMMA.** *$\{R_\lambda\}$ is a Ray resolvent if and only if* CSM^{α} *separates points of F for each fixed strictly positive α.*

Lemma 34.5 obviously follows from the following result.

(34.6) **LEMMA.** *The vector space*

$$\mathscr{L} := \mathrm{CSM}^{\alpha} - \mathrm{CSM}^{\alpha} \quad (\alpha > 0)$$

is independent of $\alpha > 0$.

Proof. Let $0 < \alpha < \beta$. Since $\mathrm{CSM}^{\alpha} \subseteq \mathrm{CSM}^{\beta}$, it is enough to prove that if (as we now assume) $f \in \mathrm{CSM}^{\beta}$ then f may be written as the difference of two elements of CSM^{α}. But

$$f = [f + (\beta - \alpha)R_\alpha f] - (\beta - \alpha)R_\alpha f,$$

and, since $(\beta - \alpha)R_\alpha f \in \mathrm{CSM}^{\alpha}$ because of the resolvent equation, it is enough to show that $f + (\beta - \alpha)R_\alpha f \in \mathrm{CSM}^{\alpha}$. Now, since $f \in \mathrm{CSM}^{\beta}$, it follows from (34.2) that

$$\lambda R_{\lambda+\alpha} f \leqslant f + (\beta - \alpha)R_{\lambda+\alpha} f.$$

Using this fact and the resolvent equation again, we obtain

$$\begin{aligned}\lambda R_{\lambda+\alpha}[f + (\beta-\alpha)R_\alpha f] &\leqslant f + (\beta-\alpha)R_{\lambda+\alpha} f + (\beta-\alpha)(R_\alpha f - R_{\lambda+\alpha} f)\\ &\leqslant f + (\beta-\alpha)R_\alpha f,\end{aligned}$$

so that $f \in \mathrm{CSM}^{\alpha}$ as required. □

(34.7) **LEMMA.** *If $\{R_\lambda\}$ is a Ray resolvent then $\mathscr{L}$ is a dense subspace of the Banach space C(F).*

Proof. For each α, CSM^{α} is obviously closed under the operation $\wedge$. For f, g, h, $k \in \mathrm{CSM}^{\alpha}$, we have

$$\begin{aligned}(f - g) \wedge (h - k) &= [(f + k) - (g + k)] \wedge [(h + g) - (g + k)]\\ &= [(f + k) \wedge (h + g)] - (g + k) \in \mathscr{L}.\end{aligned}$$

Hence $\mathscr{L}$ is closed under $\wedge$, and, since $\mathscr{L}$ is a vector space, $\mathscr{L}$ is a *lattice*. Since $\mathscr{L}$ contains constant functions and $\mathscr{L}$ separates points of F because of the Ray hypothesis, the lattice form of the Stone–Weierstrass Theorem gives the result. □

35. The Ray–Knight compactification. We make the following hypothesis.

(*35.1*) GENERAL HYPOTHESIS. *Suppose given*

(*i*) *a measurable space* $(I, \mathscr{I})$;
(*ii*) *an honest resolvent* $\{R_\lambda : \lambda > 0\}$ *on* $(I, \mathscr{I})$, *so that in particular* $R_\lambda : \mathrm{b}\mathscr{I} \to \mathrm{b}\mathscr{I}$;
(*iii*) *a sequence* $S = (f_k : k \in \mathbf{N})$ *of elements of* $\mathrm{b}\mathscr{I}$ *such that, for* $x, y \in I$ *with* $x \neq y$, *there exist* $\lambda > 0$ *and* $k \in \mathbf{N}$ *such that* $R_\lambda f_k(x) \neq R_\lambda f_k(y)$.

Heuristic comment. If this last property failed for some x and y then the process started at x would be identical to that started from y on the time-parameter set $(0, \infty)$; so we would identify x and y.

We want to view $\{R_\lambda\}$ as a Ray resolvent on a certain compactification F on I. For this purpose, we introduce certain Banach subalgebras of $\mathrm{b}\mathscr{I}$. Let $A(\cdot)$ denote 'Banach algebra generated by'. Define inductively

(35.2) $$Z_1 := A\left(1, \bigcup_{\lambda>0} R_\lambda S\right), \quad Z_{n+1} := A\left(Z_n, \bigcup_{\lambda>0} R_\lambda Z_n\right).$$

It is an immediate consequence of the resolvent equation that the family $\{R_\lambda : \lambda > 0\}$ of bounded operators on the Banach space $\mathrm{b}\mathscr{I}$ is separable in the uniform operator topology; and it follows easily that each Z_n is separable. Put

(35.3) $$Z := \text{closure}\left(\bigcup_n Z_n\right).$$

Then it is easily verified that the *separable Banach algebra* Z *is the smallest Banach subalgebra of* $\mathrm{b}\mathscr{I}$ *that contains constant functions, contains* $R_\lambda S$ *for each* $\lambda > 0$, *and satisfies* $R_\lambda : Z \to Z$ *for each* $\lambda > 0$. Further, it is immediate from Hypothesis 35.1 that Z *separates points of* I.

Let $\{g_n : n \in \mathbf{N}\}$ be a countable dense subset of Z. Define the map $\varphi : I \to \mathbf{R}^{\mathbf{N}}$ as follows:

$$\varphi(x) := (g_1(x), g_2(x), \ldots) \in \mathbf{R}^{\mathbf{N}}.$$

Since Z separates points of I, φ is one–one. *We now identify* I *with* $\varphi(I)$. Since

$$\varphi(I) \subseteq \prod_n [-\|g_n\|, \|g_n\|],$$

the closure F of $\varphi(I)$ is compact in $\mathbf{R}^{\mathbf{N}}$. We call F the *Ray–Knight compactification of* $I = \varphi(I)$ *induced by* S. We skip discussion of the influence of S: it has no practical importance for us.

Since every g_n has a unique continuous extension from I to F, it follows that every g in Z has a continuous extension to F. Thus $Z \subseteq C(F)$ in an obvious sense. However, the closed algebra Z contains all constant functions, and, by construction of F, separates points of F. Hence, by the Stone–Weierstrass

Theorem,

$$Z = C(F). \tag{35.4}$$

(35.5) LEMMA. *$\{R_\lambda : \lambda > 0\}$ is a Ray resolvent on $C(F)$.*

Proof. Since $R_\lambda : Z \to Z$ for each $\lambda > 0$, $\{R_\lambda : \lambda > 0\}$ is a Feller resolvent on F. Now let CSM^α denote the set of continuous α-supermedian functions for $\{R_\lambda : \lambda > 0\}$ on F, and let $\mathscr{L} := \mathrm{CSM}^\alpha - \mathrm{CSM}^\alpha$ (independently of α) as in Lemma 34.6. Suppose that there exist two distinct points ξ and η of F that are not separated by $\mathscr{L}$. Now, for $h \in C(F) \cap S$, $R_\lambda h = R_\lambda(h^+) - R_\lambda(h^-) \in \mathscr{L}$. We can therefore show inductively that ξ and η are not separated by Z_1, or by Z_2, etc. But this leads to the ridiculous conclusion that ξ and η are not separated by $Z = C(F)$. □

Here is an important situation in which we may use the above construction.

(35.6) SPECIAL HYPOTHESIS. *Suppose that*

(i) I is an LCCB;
(ii) $\{R_\lambda : \lambda > 0\}$ is a Feller resolvent on $C_b(I)$;
(iii) $\lambda R_\lambda 1 = 1$;
(iv) $\lim_{\lambda \to \infty} \lambda R_\lambda f(x) = f(x)$, $\forall f \in C_b(I)$, $\forall x \in I$.

One case in which this special hypothesis holds in that in which we have a Markov chain on a countable set I with standard transition function and we begin with the discrete topology on I, taking S to be the set of indicator function of elements of I.

Suppose now that the special hypothesis (35.6) obtains. Then we may take S to be a countable dense subset of $C_0(I)$. Then the general hypothesis (35.1) holds. Since every function g_n will be continuous on I, the map φ will here be continuous: the topology on I induced by F will be at least as coarse as the original topology. We emphasize that in this situation, it can happen that

$$\lim_{\lambda \to \infty} \lambda R f(\xi) \neq f(\xi)$$

for some $\xi \in F$.

Note that if the special hypothesis (35.6) holds, and I is compact, then, by the proof of Lemma 6.7, $\{R_\lambda : \lambda > 0\}$ will be a strongly continuous resolvent on $C(I)$, and we shall be back in the FD situation.

(35.7) *The Feller-McKean chain* (see Section 23). If we are given the Feller–McKean chain viewed as a chain on $\mathbb{Q}$ with its discrete topology, and take S to be the collection of indicator functions $(I_q : q \in \mathbb{Q})$, then the Ray–Knight compactification of $\mathbb{Q}$ will be the usual one-point compactification of $\mathbb{R}$. Hence the Ray–Knight topology of $\mathbb{Q}$—that induced from the topology of F—will be coarser than the original topology. (The set $\{0\}$ was open in the discrete topology). The proof of the statements just made follows from results on one-dimensional diffusions X in Section V.50 in Volume 2, the essential point being that, by Dynkin's formula,

$$|R_\lambda f(x) - R_\lambda f(y)| \leqslant \left| \mathbb{E}^x \int_0^{H_y} e^{-\lambda s} f(X_s)\, ds + [\mathbb{E}^x(e^{-\lambda H_y}) - 1] R_\lambda f(y) \right|$$
$$\leqslant 2\lambda^{-1} \| f \| [1 - \mathbb{E}^x(e^{-\lambda H_y})].$$

(35.8) *Example.* We now look at a famous example which illustrates a case where our Ray hypotheses do not hold. Let $I := [0, \infty)$. Our process X stays at 0 for an exponential holding time of rate 1 and then drifts towards ∞ at rate 1. Thus

$$P_t f(x) = \begin{cases} f(x+t) & \text{if } x \neq 0, \\ e^{-t} f(0) + \int_0^t e^{-s} f(t-s)\, ds & \text{if } x = 0. \end{cases}$$

This time, R_λ does *not* map $C_b(I)$ to $C_b(I)$.

Let us concentrate on the situation when X starts at 0. Put $T := \inf\{t : X_t \neq 0\}$. Then T is an $\{\mathscr{F}^\circ_{t+}\}$ stopping time, but, since $X(T) = 0$ and $X(T+\varepsilon) \neq 0$, $\forall \varepsilon > 0$, the process X does *not* start afresh at time T. (This example is the standard example of a simple Markov process that is not strong Markov relative to the $\{\mathscr{F}^\circ_{t+}\}$ filtration.)

For this example, what we need the Ray–Knight compactification to do is to enforce the strong Markov property by tearing $[0, \infty)$ apart at 0, producing a 'corrected' state space $F := 0 \cup [0_+, \infty)$, where $[0_+, \infty)$ is homeomorphic to the usual half-line $[0, \infty)$ and where 0 is now a point isolated from $[0_+, \infty)$. Our process (started at 0) is then modified by setting

$$X_t := 0 \in F \quad (t < T), \qquad X_T := 0_+, \qquad X_t := T \quad (t > T);$$

and it is now strong Markov, as required.

In this example, the mapping $I \mapsto \varphi(I)$ is not continuous: the Ray–Knight topology on I is now *bigger* than the original topology because $\{0\}$ is open in the RK topology.

(35.9) *Lévy's diagonal Q-matrix.* Lévy 'jazzed up' Example 35.8 to produce a remarkable illustrative example for Markov chain theory. Let $I := \mathbb{Q} \cap [0, 1]$. Let q be a strictly positive function on $I \setminus \{1\}$ such that

$$\sum_{i \in I \setminus \{1\}} q_i^{-1} < \infty.$$

Put $q_1 := 0$. Our process takes values in $[0, 1]$. Its sample paths are continuous increasing functions which spend almost all their time in I. (The paths are 'random Cantor functions'.) We now describe the law of the process *started at* 0. Let $\{T_j : j \in I\}$ be independent exponentially distributed variables, T_j having rate q_j, so that, in particular, $T_1 = \infty$. Define

$$X_t := j \quad \text{if} \sum_{k<j} T_k \leqslant t < \sum_{k \leqslant j} T_k$$

and interpolate X by continuity (or monotonicity). Then X is a simple Markov process with resolvent satisfying

$$r_{ij}(\lambda) = (\lambda + q_j)^{-1} \prod_{i \leqslant k < j} [(1 + \lambda q_k^{-1})^{-1}] \quad (i \leqslant j)$$

as for the pure-birth process in Section 26.

The discussion of Example 35.8 makes it clear that the Ray–Knight compactification should tear the set $[0,1]$ apart at each point i of $I \backslash \{1\}$, so that the process will spend an exponentially distributed time at i, then jump to i_+ and leave i_+ immediately. The RK topology on I arises from a metric d, which, in this example, is given by

$$d(i,j) = \mathbf{E}^i(H_j) = \sum_{i \leqslant k < j} q_k^{-1}.$$

Whether or not we deal in compactifications, the process X cannot jump from a point i of I to a point j of I. (The point i_+ is of course *not* in I; it is a 'fictitious' state from the point of view of chain theory.) Hence the Q-matrix Q of X satisfies $q_{ij} = 0$ whenever $i \neq j$. Thus Q is the *diagonal* $I \times I$ matrix

$$Q = \operatorname{diag}(-q_i) = (-q_i \delta_{ij}).$$

Ray's Theorem: analytical part

36. From semigroup to resolvent. Here is Part 1 of Ray's amazing achievement.

(*36.1*) THEOREM (Ray's Theorem: Part 1). *Let* $\{R_\lambda\}$ *be a Ray resolvent on a compact metric space. Thus* $\{R_\lambda\}$ *is an honest Feller resolvent, and the space* $\mathscr{L}$ *in Lemma 34.6 is dense in* $C(F)$. *Then there exists a unique honest measurable transition function* $\{P_t\}$ *on* $(F, \mathscr{B}(F))$ *(see Section 3) such that*

(i) $t \mapsto P_t f(x)$ *is right-continuous on* $[0, \infty)$ *for* $x \in F$ *and* $f \in C(F)$;

(ii) $$R_\lambda f(x) = \int_0^\infty e^{-\lambda t} P_t f(x)\, dt \quad (f \in C(F), x \in F, \lambda > 0).$$

Points to note

(a) It is *not* true in general that $P_t : C(F) \to C(F)$.
(b) It is *not* true in general that $P_0 = I$, though, of course,

$$P_0 P_t = P_t P_0 = P_t, \quad \forall t \geqslant 0.$$

(c) It is *not* true in general that $t \mapsto P_t f(x)$ is continuous for $f \in C(F)$ and $x \in F$.

An example illustrating these points will be given at the end of the next section.

Getoor [1] and Dellacherie and Meyer [1, Chapter XII] prove the theorem from first principles, that is, without using the Hille–Yosida Theorem. Since we already have the HY Theorem, we give a proof based on it, guided in part by Meyer [2].

Proof of Theorem 36.1. Because of the importance of the theorem, we go carefully through the proof. *Throughout the proof, α denotes a fixed positive number.* Though α is used in its construction, the final transition semigroup $\{P_t\}$ will, of course, be independent of α.

We make use of the well-known fact that if (x_{mn}) is a double sequence such that both $m \mapsto x_{mn}$ and $n \mapsto x_{mn}$ are monotone non-decreasing then

$$\lim_m \lim_n x_{mn} = \lim_n \lim_m x_{mn}.$$

(This is of course what underlies the Monotone-Convergence Theorem.) As a consequence, we see that

(36.2) *the limit of a non-decreasing sequence of right-continuous non-increasing functions mapping $[0, \infty)$ into $[0, \infty)$ is right-continuous.*

Strategy. The idea of the proof is to begin by using the Hille–Yosida Theorem to establish results on the domain Z_0 of strong continuity of the resolvent; to extend these results by monotonicity to functions in CSM^α; then by linearity to functions in $\mathscr{L}^\alpha := \mathrm{CSM}^\alpha - \mathrm{CSM}^\alpha$; then by continuity to functions in $C(F)$, since $\mathscr{L}^\alpha$ is dense in $C(F)$; then by the Monotone-Class Theorem to functions in $b\mathscr{B}(F)$. But we have to be careful of the order in which this strategy is applied.

Step 1: Write

(36.3) $$\mathscr{R} := R_\lambda C(F) \quad (\text{independently of } \lambda > 0), \qquad Z_0 := \bar{\mathscr{R}},$$

the closure being in $C(F)$ or $b\mathscr{B}(F)$. By the Hille–Yosida Theorem, there exists a strongly continuous contraction semigroup $\{Q_t : t \geqslant 0\}$ on Z_0 with resolvent the restriction of $\{R_\lambda\}$ to Z_0. It is clear from (5.4) and (5.5) that each Q_t is positive on Z_0 in that if $f \in Z_0^+$ (that is, $f \in Z_0$ and $f \geqslant 0$) then $Q_t f \geqslant 0$. The constant function 1 is in Z_0, and $Q_t 1 = 1$ for all t. Finally, for $f \in Z_0$, $\lambda R_{\lambda+\alpha} f \to f$ as $\lambda \to \infty$, whence

(36.4) for $f \in Z_0$ and $t \geqslant 0$, $Q_t(\lambda R_{\lambda+\alpha} f) \to Q_t f$ (strong topology).

Step 2: Let $t \geqslant 0$ and let $f \in \mathrm{CSM}^\alpha$. The map $\lambda \mapsto \lambda R_{\lambda+\alpha} f$ is non-decreasing because, for $0 < \lambda < \nu$,

$$\nu R_{\lambda+\alpha} f - \lambda R_{\lambda+\alpha} f = (\nu - \lambda) R_{\nu+\alpha}[f - \lambda R_{\lambda+\alpha} f] \geqslant 0.$$

Since $\lambda R_{\lambda+\alpha} f \in \mathscr{R}$, so that $Q_t(\lambda R_{\lambda+\alpha} f)$ is defined, we may (as we must!) define

(36.5) for $f \in \mathrm{CSM}^\alpha$ and $x \in F$, $P_t f(x) := \uparrow \lim_{\lambda \uparrow\uparrow \infty} \{Q_t(\lambda R_{\lambda+\alpha} f)\}(x)$.

The formula (36.5) and the linearity and positivity of $\{Q_t\}$ make it clear that if $f_1, f_2 \in \mathrm{CSM}^\alpha$ and $c_1, c_2 \in [0, \infty)$ then

$$P_t(c_1 f_1 + c_2 f_2) = c_1 P_t f_1 + c_2 P_2 f_2,$$

and if $f, g \in \mathrm{CSM}^\alpha$ with $f \geqslant g$ then

$$P_t f \geqslant P_t g.$$

It is clear on comparing (36.5) with (36.4) that

$$P_t = Q_t \quad \text{on} \quad R_\alpha(C(F)^+).$$

The map P_t extends uniquely by linearity to a positive map

$$P_t \colon \mathscr{L}^\alpha \to \mathrm{b}\mathscr{B}(F), \quad \text{where} \quad \mathscr{L}^\alpha := \mathrm{CSM}^\alpha - \mathrm{CSM}^\alpha,$$

and P_t agrees with Q_t on $\mathscr{R}$. Because P_t is positive and linear and $P_t 1 = 1$, and since $\mathscr{L}^\alpha$ is dense in $C(F)$ (by Lemma 34.6),

(36.6) P_t has a unique extension to a bounded linear operator $P_t \colon C(F) \to \mathrm{b}\mathscr{B}(F)$.

We know from Theorem 6.2 that there exists a unique kernel $P_t \colon F \times \mathscr{B}(F) \to [0, 1]$ such that

$x \mapsto P_t(x, B)$ is $\mathscr{B}(F)$-measurable for each $B \in \mathscr{B}(F)$,

$B \mapsto P_t(x, B)$ is a probability measure on $(F, \mathscr{B}(F))$ for each $x \in F$,

$$P_t f(x) = \int_F P_t(x, dy) f(y), \quad \forall f \in C(F).$$

The map $f \mapsto P_t f$ from $C(F)$ to $\mathrm{b}\mathscr{B}(F)$ therefore extends canonically to a map from $\mathrm{b}\mathscr{B}(F)$ to $\mathrm{b}\mathscr{B}(F)$.

Step 3: We claim that the linear operators $\{P_t : t \geqslant 0\}$ on $\mathrm{b}\mathscr{B}(F)$ have the semigroup property:

(36.7) $$P_s P_t = P_{s+t} \quad (s \geqslant 0, t \geqslant 0).$$

Proof of (36.7). We know from the Hille–Yosida Theorem that $P_s P_t f = P_{s+t} f$ for $f \in Z_0$. Now let $f \in \mathrm{CSM}^\alpha$. Then $\lambda R_{\lambda+\alpha} f \in Z_0$, and

$$P_s P_t(\lambda R_{\lambda+\alpha} f) = P_{s+t}(\lambda R_{\lambda+\alpha} f).$$

We have $P_t(\lambda R_{\lambda+\alpha} f) \uparrow P_t f$ by (36.5); and, *since P_s arises from a kernel*, the Monotone-Convergence Theorem yields

$$(P_s P_t f)(x) = (P_{s+t} f)(x) \quad (f \in \mathrm{CSM}^\alpha, x \in F).$$

The result extends according to the remainder of our strategy: by linearity to $f \in \mathscr{L}^\alpha$, thence by continuity to $f \in C(F)$, and thence by the Monotone-Class Theorem to $f \in b\mathscr{B}(F)$.

Step 4: We now show that

(36.8) $\quad$ *for* $g \in C(F)^+$ *and* $x \in F, e^{-\alpha t}P_tR_\alpha g(x) \leqslant R_\alpha g(x)$,

and the map $t \mapsto e^{-\alpha t}P_tR_\alpha g(x)$ *is right-continuous and non-increasing.*

Proof of (36.8). For $\lambda > 0$, let $h_\lambda := \lambda R_{\lambda+\alpha} g \in Z_0$. By the Hille–Yosida Theorem, we have

$$e^{-\alpha t}P_t(\lambda R_{\lambda+\alpha}R_\alpha g) = e^{-\alpha t}P_tR_\alpha h_\lambda = \int_t^\infty e^{-\alpha r}P_r h_\lambda \, dr$$
$$\leqslant R_\alpha h_\lambda = R_\alpha g - R_{\lambda+\alpha} g \leqslant R_\alpha g.$$

But $R_\alpha g \in \mathrm{CSM}^\alpha$, so that, by (36.5),

$$(e^{-\alpha t}P_tR_\alpha g)(x) \leqslant (R_\alpha g)(x) \quad (x \in F).$$

That

$$e^{-\alpha(s+t)}P_{s+t}R_\alpha g(x) \leqslant e^{-\alpha s}P_sR_\alpha g(x)$$

is now clear from the semigroup property.

By the principle (36.2), the map $t \mapsto e^{-\alpha t}P_tR_\alpha g(x)$, the non-decreasing limit of the right-continuous non-increasing maps $t \mapsto e^{-\alpha t}P_t(\lambda R_{\lambda+\alpha}R_\alpha g)(x)$, is right-continuous.

Step 5: Next we prove the following.

(36.9) LEMMA. *For* $f \in \mathrm{CSM}^\alpha$ *and* $x \in F$, *the map* $t \mapsto e^{-\alpha t}P_tf(x)$ *is non-increasing and right-continuous.*

Remarks. This fact is of central importance in the probabilistic theory.

Proof of Lemma 36.9. We have

$$\lambda R_{\lambda+\alpha}f = R_\alpha \lambda(f - \lambda R_{\lambda+\alpha}f),$$

and $\lambda(f - R_{\lambda+\alpha}f) \in C(F)^+$, whence, by (36.8),

$$t \mapsto e^{-\alpha t}P_t\lambda R_{\lambda+\alpha}f(x) \text{ is non-increasing and right-continuous.}$$

The desired result now follows from (36.5) and the principle (36.2).

Step 6: It is now clear, since $\mathscr{L}^\alpha$ is dense in $C(F)$, that,

(36.10) $\quad$ for $f \in C(F)$, the map $t \mapsto P_tf(x)$ is right-continuous.

The Monotone-Class Theorem shows that P_t defines a measurable transition function on $(F, \mathscr{B}(F))$, and all that remains is to confirm that, for $f \in C(F)$ (or for $f \in b\mathscr{B}(F)$), for $x \in F$ and $\lambda > 0$, we have

$$\int_0^\infty e^{-\lambda t} P_t f(x)\, dt = R_\lambda f(x). \tag{36.11}$$

Proof of (36.11). *If* $f \in Z_0$ then (36.11) holds by the Hille–Yosida Theorem. The argument is now completed by the machinery in our strategy. □

(*36.12*) *Exercise.* Prove that

$$Z_0 = \{f \in C(F) : P_0 f = f\}.$$

Hint. Compare Lemma 6.7.

37. Branch-points. We continue with the notation and hypotheses of Section 36. The set F_e of *non-branch-points* of F (for $\{P_t\}$) is defined as follows:

$$\begin{aligned} F_e &:= \{x \in F : P_0 f(x) = f(x), \quad \forall f \in C(F)\} \\ &= \{x \in F : P_0(x, \cdot) = \varepsilon_x\}. \end{aligned}$$

The set F_{br} of *branch-points* is defined as $F_{br} := F \backslash F_e$.

The proper explanation of the rôle of branch-points escapes the analysis, and has to wait for the probability: you need to see the paths of the process.

(*37.1*) LEMMA. *A semigroup on a compact metric space F is FD if and only if it is a Ray semigroup without branch-points.*

(Of course, a 'Ray semigroup' is a semigroup derived from a Ray resolvent as described in Theorem 36.1.) Lemma 37.1 is an immediate consequence of the proof of Lemma 6.7.

Now let (h_m) be a dense sequence in CSM^1. Then

$$P_0 h_m = \uparrow \lim \lambda R_{\lambda+1} h_m \leqslant h_m, \quad \forall m.$$

Since $\mathscr{L} = \mathrm{CSM}^{-1} - \mathrm{CSM}^1$ is dense in $C(F)$,

$$\begin{aligned} F_e &= \bigcap_m \{P_0 h_m = h_m\} \\ &= \bigcap_m \bigcap_n \{P_0 h_m > h_m - n^{-1}\} \\ &= \bigcap_m \bigcap_n \left(\bigcup_\lambda \{\lambda R_{\lambda+1} h_m > h_m - n^{-1}\} \right), \end{aligned}$$

so that, since the set in large parentheses is open,

$$F_e \text{ is a } G_\delta \text{ in } F. \tag{37.2}$$

In particular, $F_e \in \mathscr{B}_F$.

Let μ be a probability measure on $(F, \mathscr{B}_F)$. We write μP_t for the measure

$$\mu P_t(\cdot) := \int_F \mu(dx) P_t(x, \cdot).$$

However, it is better to think in terms of the functional notation

$$(\mu, g) := \mu(g) := \int g \, d\mu$$

for measures, and define μP_t via

$$(\mu P_t, f) = (\mu, P_t f).$$

(37.3) LEMMA. *For* $\mu \in \Pr(F)$,

$$\mu P_0 = \mu \quad \textit{if and only if} \quad \mu(F_{\text{br}}) = 0.$$

Proof. With (h_m) as above, we have

$$\begin{aligned} \mu P_0 = \mu &\Leftrightarrow (\mu P_0, h_m) = (\mu, h_m), \quad \forall m, \\ &\Leftrightarrow \mu\{P_0 h_m > h_m - n^{-1}\} = 1, \quad \forall m, n. \\ &\Leftrightarrow \mu(F_e) = 1. \end{aligned}$$

□

(37.4) COROLLARY. $P_t(x, F_{\text{br}}) = 0, \forall x \in F, \forall t \geqslant 0.$

Proof. Since $P_t P_0 = P_t$, the measure $\mu := \varepsilon_x P_t$ satisfies $\mu P_0 = \mu$. □

(37.5) *Example.* Here is a rather artificial example to illustrate points (a), (b) and (c) made after the statement of Theorem 36. Let $F := [-1, 1]$. We take a process that while away from 0 drifts towards 0 at rate 1, and which on approaching 0 jumps (or *branches*) to $+1$ or -1 with probability $\frac{1}{2}$ each. For $x \geqslant 0$,

$$P_t f(x) = \begin{cases} f(x-t) & (t < x), \\ \frac{1}{2} f(1 - \langle t - x \rangle) + \frac{1}{2} f(-1 + \langle t - x \rangle) & (t \geqslant x), \end{cases}$$

where $\langle \cdot \rangle$ denotes 'fractional part of'. For $x \leqslant 0$,

$$P_t f(x) = \begin{cases} f(x+t) & (t < |x|), \\ P_t f(-x) & (t \geqslant |x|). \end{cases}$$

You can see that P_t does *not* map $C(F)$ to $C(F)$ and that $t \mapsto P_t f(x)$ has points of discontinuity.

Now, for $x > 0$,

$$R_\lambda f(x) = \int_0^x e^{-\lambda t} f(x-t)\, dt + e^{-\lambda x} R_\lambda f(0),$$

with a similar 'Dynkin formula' for $x<0$. It follows that $R_\lambda : C(F) \to C(F)$. The result $Z_0 = \{f \in C(F) : P_0 f = f\}$ of Ray theory leads us to believe that

$$Z_0 := \text{closure}\, R_\lambda C(F) = \{f \in C(F) : f(0) = \tfrac{1}{2} f(1) + \tfrac{1}{2} f(-1)\},$$

and this result is easily verified directly. The fact that $\{R_\lambda\}$ is indeed a Ray resolvent now follows immediately from the fact that Z_0 separates points of F.

38. Choquet representation of 1-excessive probability measures. We continue with the notation of Sections 36 and 37. For $\alpha \geqslant 0$, a probability measure μ on $(F, \mathscr{B}_F)$ is called *α-supermedian* (relative to $\{R_\lambda\}$) if

$$\lambda \mu R_{\lambda+\alpha} \leqslant \mu, \quad \forall \lambda > 0,$$

that is, if

$$\mu(\lambda R_{\lambda+\alpha} f) \leqslant \mu(f), \quad \forall f \in C(F)^+, \quad \forall \lambda.$$

For every μ in $\Pr(F)$ and every $\alpha \geqslant 0$, we have (as $\lambda \to \infty$)

$$\mu(\lambda R_{\lambda+\alpha} f) \to \mu(P_0 f), \quad \forall f \in C(F),$$

so that, in the weak topology of $\Pr(F)$,

$$\lambda \mu R_{\lambda+\alpha} \to \mu P_0.$$

Hence a probability measure μ on $(F, \mathscr{B}_F)$ will be called *α-excessive* (relative to $\{R_\lambda\}$) if

(38.1) (i) μ is α-supermedian;
(38.1)(ii) $\mu = \mu P_0$, or equivalently (Lemma 37.3) $\mu(F_{br}) = 0$.

Fix $\alpha = 1$ for convenience. The set of 1-excessive probability measures on F is easily shown to satisfy the hypotheses of both the existence and uniqueness parts of Choquet's representation theorem. However, we can now obtain the explicit form of the representation by simple direct methods.

(38.2) THEOREM (Ray's theorem: Part 2). *Let μ be a 1-excessive element of* $\Pr(F)$. *Then there exists a unique element ν of* $\Pr(F)$ *such that $\nu(F_{br}) = 0$ and*

$$\mu = \nu R_1 = \int_{F_0} \nu(d\xi) R_1(\xi, \cdot).$$

Proof. For $\lambda > 0$, set

$$\tilde{\nu}_\lambda := (\lambda + 1)(\mu - \lambda \mu R_{\lambda+1}).$$

Because $\{R_\lambda\}$ is honest, $\tilde{\nu}_\lambda \in \Pr(F)$. Recall that $\Pr(F)$ is compact and let $\tilde{\nu}$ be any limit point of $\tilde{\nu}_\lambda$ as $\lambda \to \infty$. The resolvent equation shows that

$$\tilde{\nu}_\lambda R_1 = (\lambda + 1) \mu R_{\lambda+1},$$

so that on letting $\lambda \to \infty$ through a suitable sequence,

$$\tilde{\nu} R_1 = \mu P_0 = \mu.$$

Put $\nu = \tilde{\nu} P_0$. Since $\nu = \nu P_0$, we have $\nu(F_{br}) = 0$; and, since $P_0 R_1 = R_1$, we have $\nu R_1 = \mu$.

Now if $\nu^* R_1 = \mu$ and $\nu^* P_0 = \nu^*$ for some $\nu^* \in \Pr(F)$ then (as $\lambda \to \infty$)

$$\tilde{\nu}_\lambda = (\lambda + 1)\nu^*(R_1 - \lambda R_1 R_{\lambda+1}) = (\lambda + 1)\nu^* R_{\lambda+1} \to \nu^* P_0 = \nu^*,$$

so that $\nu^* = \nu$. □

Ray's Theorem: Probabilistic part

39. The Ray process associated with a given entrance law. Our treatment of FD processes was designed to make the transition to Ray processes as easy as possible; and we shall not dwell on those 'FD' arguments which apply in the present context. We shall highlight those places where the Ray theory is more subtle.

For the Martin-boundary application (and for other purposes), we need to deal with processes with time-parameter set $(0, \infty)$ *open at* 0. (The $\hat{X}$ process in (33.1) is not defined at time 0.)

Let $\rho = \{\rho_t : t > 0\}$ be a *probability entrance law* for $\{P_t\}$, that is, a family of elements of $\Pr(F)$ satisfying

(39.1) $$\rho_{s+t} = \rho_s P_t, \quad \forall s, t > 0.$$

Note that, for $0 < \varepsilon < t$,

$$\rho_t P_0 = \rho_\varepsilon P_{t-\varepsilon} P_0 = \rho_\varepsilon P_{t-\varepsilon} = \rho_t,$$

so that

(39.2) $$\rho_t(F_e) = 1, \quad \forall t > 0.$$

Step 1: *use of Daniell–Kolmogorov Theorem*. For $\omega \in F^{(0,\infty)}$, write $Y_t(\omega) = \omega(t)$; and put $\mathscr{B}_F^{(0,\infty)} := \sigma\{Y_t : t > 0\}$. The Daniell–Kolmogorov theorem implies the existence of unique measure $\mathbf{P}^\rho$ on $(F^{(0,\infty)}, \mathscr{B}_F^{(0,\infty)})$ such that, for $0 < t_1 < t_2 < \cdots < t_n$ and $x_1, x_2, \ldots, x_n \in F$,

(39.3) $$\mathbf{P}^\rho[Y_{t_1} \in dx_1; Y_{t_2} \in dx_2; \ldots; Y_{t_n} \in dx_n] = \rho_{t_1}(dx_1) P_{t_2 - t_1}(x_1, dx_2) \ldots P_{t_n - t_{n-1}}(x_{n-1}, dx_n).$$

In particular, (39.2) shows that

(39.4) $$\mathbf{P}^\rho[Y_t \in F_e] = \rho_t(F_e) = 1, \quad \forall t > 0.$$

Step 2: *regularising Y to produce X*. If $h \in \mathrm{CSM}^1$ then (Lemma 36.9) $e^{-t} P_t h \leqslant h, \forall t,$

so that

$$\{e^{-t}h(Y_t):0<t<\infty\} \tag{39.5}$$

is a $\mathbf{P}^\rho$ *supermartingale* relative to the filtration induced by the Y-process. The crucial property that CSM[1] *separates points* now allows us to repeat the argument of (7.9) to show that, a.s.($\mathbf{P}^\rho$), the limit

$$X_t := \lim_{\mathbb{Q}\ni q\downarrow\downarrow t} Y_q$$

exists for all $t>0$ and $t\mapsto X_t$ is Skorokhod from $(0,\infty)$ to F. The proof that

$$\mathbf{P}^\rho[X_t = Y_t] = 1, \quad \forall t>0,$$

proceeds as did the proof of (7.12), but you find that you now need to use (39.4).

Step 3: *extending* X *to* $t=0$. By applying (II.69.4) to supermartingales of the form (39.5), we see that

$$X_0 := \lim_{t\downarrow\downarrow 0} X_t \text{ exists in } F \tag{39.6}$$

and the process X is an R-process in $[0,\infty)$.

Set $\rho_0 := \mathbf{P}^\rho\circ X_0^{-1}$. By right-continuity of paths,

$$\rho_0 = \lim_{t\downarrow\downarrow 0} \rho_t \quad \text{in } \Pr(F). \tag{39.7}$$

We must now check that

$$\rho_0 P_t = \rho_t, \quad \forall t, \tag{39.8}$$

so that (39.3), with X replacing Y, extends to the case when $0\leqslant t_1\leqslant t_2\leqslant\cdots\leqslant t_n$. It is tempting to write $(\rho_\varepsilon, P_t f) = (\rho_{t+\varepsilon}, f)$, $f\in C(F)$, and to try to deduce (39.8) by letting $\varepsilon\downarrow\downarrow 0$. However, $P_t f$ need not be continuous.

We *can* prove (38.8)—with important extra information—as follows. We have, for $\lambda>0$, $\varepsilon>0$ and $f\in C(F)$,

$$(\rho_\varepsilon, R_\lambda f) = \int_0^\infty e^{-\lambda t}(\rho_{t+\varepsilon}, f)\,dt = \mathbf{E}^\rho\int_0^\infty e^{-\lambda t} f(X_{t+\varepsilon})\,dt;$$

and we can let $\varepsilon\downarrow\downarrow 0$, using the facts that $R_\lambda f$ is continuous on F and that X is right-continuous, to obtain

$$(\rho_0, R_\lambda f) = \mathbf{E}^\rho\int_0^\infty e^{-\lambda t} f(X_t)\,dt = \int_0^\infty e^{-\lambda t}(\rho_t, f)\,dt.$$

Now $\rho_0 R_1$ is clearly 1-supermedian, and, because of (39.2), it is 1-excessive. Hence, by Theorem 38.2, $\rho_0 R_1 = \nu R_1$ for a unique ν carried by F_e. From the resolvent equation, $\rho_0 R_\lambda = \nu R_\lambda$ for every λ. We now know that for $f\in C(F)$, we have, for almost all t,

$$(\nu, P_t f) = (\rho_0, P_t f) = \mathbf{E}^\rho f(X_t) = (\rho_t, f). \tag{39.9}$$

But $t \mapsto P_t f(x)$ is right-continuous for each x, and X is an R-process. Hence, by right-continuity, equality holds in (39.9) for all $t \geq 0$ and all $f \in C(F)$. Thus, (38.8) holds. Since $\nu P_0 = \nu$, we see that $\rho_0 = \mathbf{P}^\rho \circ X_0^{-1} = \nu$.

Since $\rho_0 = \nu$, so that $\rho_0(F_e) = 1$, we obtain the following improvement of (39.6):

(39.10) $$X_0 := \lim_{t \downarrow\downarrow 0} X_t \in F_e, \quad \text{a.s.}(\mathbf{P}^\rho).$$

It is (39.10), applied in reversed time, that gives the Doob–Hunt Convergence Theorem for the Martin boundary.

40. Strong Markov property of Ray processes. *Let Ω now denote the space of R-paths ω from* $[0, \infty)$ *to F.* (For $\omega \in \Omega$, we define $\omega(\infty) = \partial$, where ∂ is either a 'new' coffin state or the 'old' coffin state already adjoined to arrange honesty—it does not matter.) Let X now denote the coordinate process, $X_t(\omega) := \omega(t)$, let $\mathscr{F}^\circ := \sigma\{X_t : t \geq 0\}$ and let $\mathscr{F}_t^\circ := \sigma\{X_s : s \leq t\}$.

For any $\mu \in \Pr(F)$,

(40.1) $$\rho_t := \mu P_t$$

defines an entrance law ρ. The results of Section 3 show that we can define a corresponding law $\mathbf{P}^\rho$ on $(\Omega, \mathscr{F}^0)$. Our new $\mathbf{P}^\rho$ is what we called $\mathbf{P}^\rho \circ X^{-1}$ in Section 39. Now, with ρ as in (40.1), it is natural to do what we did in FD theory and write $\mathbf{P}^\mu$ for the measure $\mathbf{P}^\rho$ on $(\Omega, \mathscr{F}^0)$. The significance of μ as 'initial law' is no longer accurate, however, since

$$\mathbf{P}^\mu \circ X_0^{-1} = \mu P_0.$$

(We shall see shortly that we can think of μ as the $\mathbf{P}^\mu$ law of X instantaneously before time 0.) As usual, we shall write $\mathbf{P}^x$ $(x \in F)$ for $\mathbf{P}^{\varepsilon_x}$. Then

$$\mathbf{P}^x[X_0 \in \Gamma] = P_0(x, \Gamma) \quad (\Gamma \in \mathscr{B}_F),$$

and, for $0 \leq t_1 \leq t_2 \leq \cdots \leq t_n$,

$$\begin{aligned}\mathbf{P}^x&[X_0 \in dx_0; X_{t_1} \in dx_1; \ldots; X_{t_n} \in dx_n] \\ &= P_0(x, dx_0) P_{t_1}(x_0, dx_1) \cdots P_{t_n - t_{n-1}}(x_{n-1}, dx_n).\end{aligned}$$

Of course

$$\mathbf{P}^\mu(\cdot) = \int_F \mathbf{P}^x(\cdot)\mu(dx) \quad \text{on } \mathscr{F}^0.$$

The setup

$$(\Omega, \mathscr{F}^\circ, \mathscr{F}_t^\circ, \mathbf{P}^x, \theta_t, X_t)$$

is the *Ray process* with transition function $\{P_t\}$. It has the simple Markov property.

Because P_t need not map $C(F)$ to $C(F)$, we have to use a 'Laplace-transformed' version of the proof of Theorem 8.3 for FD processes. Here are the necessary modifications to that proof. For T an a.s. finite $\{\mathscr{F}^\circ_{t+}\}$ stopping time, $T^{(n)}$ its nth dyadic approximation, $\Lambda \in \mathscr{F}^\circ_{T+}$ and $f \in C(F)$, we have

$$\int_0^\infty e^{-\lambda s}\mathbf{E}^\mu[f\circ X(T^{(n)}+s);\Lambda]\,ds = \mathbf{E}^\mu[(R_\lambda f)\circ X(T^{(n)});\Lambda].$$

Letting $n \to \infty$ and using right-continuity of paths and the fact that $R_\lambda f \in C(F)$, we find that

$$\int_0^\infty e^{-\lambda s}\mathbf{E}^\mu[f\circ X(T+s);\Lambda]\,ds = \mathbf{E}^\mu[(R_\lambda f)\circ X(T);\Lambda]$$

$$= \int_0^\infty e^{-\lambda s}\mathbf{E}^\mu[(P_s f)\circ X(T);\Lambda]\,ds.$$

Hence

(40.2) $$\mathbf{E}^\mu[f\circ X(T+s);\Lambda] = \mathbf{E}^\mu[P_s f\circ X(T);\Lambda]$$

for almost all s. But (same old argument!) each side of (40.2) is right-continuous in s, so that (40.2) holds for all s.

We can now follow exactly the same course as we did in Section 9, introducing various completions etc., and establish that the 'Ray process proper'

$$(\Omega, \mathscr{F}, \mathscr{F}_t, \mathbf{P}^x, \theta_t, X_t)$$

is strong Markov. Assume this done.

41. The role of branch-points. We already know that *for fixed* $t \geqslant 0$, $X_t \in F_e$ a.s. We now prove the much stronger result that

(41.1) *almost surely, X never visits F_{br}:* $\mathbf{P}^x[X_t \in F_e, \forall t \geqslant 0] = 1, \quad \forall x.$

Proof of (41.1). The 'right' way to prove this is by using Meyer's section theorem (Theorem II.76.2), but there is an elementary proof avoiding capacitability theory—see the exercise below. The Section Theorem shows that it is enough to prove that if T is a finite stopping time and $x \in F$ then

(41.2) $$\mathbf{P}^x[X(T) \in F_{br}] = 0.$$

But $\rho_t := \mathbf{P}^x \circ X_{T+t}^{-1}$ defines a probability entrance law because of the Strong Markov Theorem, and (41.2) now follows immediately from Step 3 of Section 39.

Exercise. Deduce (41.1) from (41.2) by using Lemma II.75.1 and the fact (37.2) that F_{br} is K_σ (countable union of compact sets).

The 'exploding birth process' example in Section 26 shows that X can have left limits in F_{br}. Thus X may be able to approach F_{br}, but it will branch at the moment of approach. The following theorem, which is the 'Ray' analogue of Blumenthal's Quasi-Left-Continuity Theorem (Theorem 11.1), is the true explanation of the role of branch-points.

(41.3) THEOREM. *If $(T_n : n \in \mathbf{N})$ is a strictly increasing sequence of $\{\mathscr{F}_t\}$ stopping times with $T_n \uparrow\uparrow T < \infty$ then, for $x \in F$ and $\Gamma \in \mathscr{B}_F$,*

$$\mathbf{P}^x\left[X_T \in \Gamma \,\middle|\, \bigvee_n \mathscr{F}_{T(n)}\right] = P_0(X_{T-}, \Gamma),$$

where $\bigvee \mathscr{F}_{T(n)} := \sigma\{\mathscr{F}_{T(n)} : n \in \mathbf{N}\}$.

Proof. Let $\Lambda \in \mathscr{F}_{T(k)}$ for some k. Then, for $f \in C(F)$, we have, for $n \geqslant k$,

$$\int_0^\infty e^{-\lambda t}\mathbf{E}^x[f\circ X(T_n+t); \Lambda]\,dt = \mathbf{E}^x[R_\lambda f\circ X(T_n); \Lambda].$$

Since the set of discontinuities of X is countable (why?), and hence of measure 0, we can let $n \to \infty$ to obtain

$$\int_0^\infty e^{-\lambda t}\mathbf{E}^x[f\circ X(T+t); \Lambda]\,dt = \mathbf{E}^x[R_\lambda f\circ X(T-); \Lambda]$$

$$= \int_0^\infty e^{-\lambda t}\mathbf{E}^x[P_t f\circ X(T-); \Lambda]\,dt.$$

Hence, by the old right-continuity argument yet again,

$$\mathbf{E}^x[f\circ X(T+t); \Lambda] = \mathbf{E}^x[P_t f\circ X(T-); \Lambda]$$

for *all* $t \geqslant 0$. In particular,

$$\mathbf{E}^x[f(X_T); \Lambda] = E^x[P_0 f(X_{T-}); \Lambda]. \tag{41.4}$$

Monotone-class arguments now show that (41.4) is true for all $\Lambda \in \bigvee \mathscr{F}_{T(k)}$ and all $f \in b\mathscr{B}_F$. On taking $f = I_\Gamma$, we obtain the desired result. □

(41.5) *Note.* If we set $X_{0-} := X_0$ then Theorem 41.3 and its proof apply to the case when '$T_n \uparrow\uparrow T$' is interpreted in the slightly wider sense that

(*i*) when $T(\omega) = 0$, $T_n(\omega) = 0$, $\forall n$;
(*ii*) when $T(\omega) > 0$, $T_n(\omega) \to T(\omega)$ and $T_n(\omega) \leqslant T_{n+1}(\omega) < T(\omega)$, $\forall n$:

From now on, we always allow this wider interpretation of '$T_n \uparrow\uparrow T$' for stopping times.

We have at last finished Ray's Theorem (though Ray's amazing 1959 paper goes further in certain directions). Recall that we have utilised ideas of Knight and Kunita and Watanabe as well as those of Ray.

6. APPLICATIONS

Martin boundary theory in retrospect

42. From discrete to continuous time. The idea of combining time reversal with Ray theory to produce Martin-boundary results goes back to the very important Kunita and T. Watanabe papers [2, 3]. A fine account appears in Meyer [4]. The optimal way to handle the discrete-parameter chain case has been known in folklore for some time, and the extremely effective device of using time transformation to make the resolvents into *compact* operators finds general expression in the paper [1] by Garcia Alvarez and Meyer.

As in Section 27, let I be a countable set, let Π be a substochastic $I \times I$ matrix with Green matrix $\Gamma := \Sigma \Pi^n$ and make the following assumption:

there exists a (reference) point b in I such that $0 < \Gamma(b, j) < \infty, \quad \forall j.$

Define the Martin kernel κ on $I \times I$:

$$\kappa(i, j) := \Gamma(i, j)/\Gamma(b, j).$$

Let J (rather than X)—J stands for 'jump chain'—now denote a discrete-parameter chain with one-step transition matrix Π. By observing J with a (discrete) clock that 'ticks' only when J *changes* position (and which therefore 'ignores' times n when $J(n) = J(n-1)$), we produce a new chain $\tilde{J}$. You can easily check that

$$\tilde{\Pi}(i, j) = [1 - \delta(i, j)][1 - \Pi(i, i)]^{-1}\Pi(i, j)$$

and that the change from Π to $\tilde{\Pi}$ preserves excessive and regular functions and also preserves the Martin kernel. Since, further, results of Doob–Hunt type are invariant under the change from J to $\tilde{J}$, we may as well make the assumption

$$\Pi(i, i) = 0, \quad \forall i.$$

Now let $q(\cdot)$ be a strictly positive function on I such that

$$\sum_j \frac{\Gamma(j, j)}{q(j)} < \infty. \tag{42.1}$$

We shall often write q for the diagonal $I \times I$ matrix diag$\{q(i)\}$. Introduce the $I \times I$ matrix

$$Q := -q + q\Pi.$$

We now *let X be a (right-continuous, continuous-parameter) chain on I with Q-matrix Q*. It is well known that each visit by X to state i in I is exponentially distributed with rate $q(i)$ and is independent of the behaviour of X prior to that particular visit. It is also well known that the jumps of X are made in accordance with the law of J. The lifetime ζ of X is a.s. finite. Indeed, for each i,

$$\mathbf{E}^i[\zeta] = \sum_j \frac{\Gamma(i,j)}{q(j)} \leqslant \sum_j \frac{\Gamma(j,j)}{q(j)} < \infty.$$

The $\mathbf{P}^i$ probability that X is at j at time t having made the n ($\geqslant 0$) jumps $i = i_0$ to i_1, i_1 to $i_2, \ldots, i_{n-1}$ to $i_n = j$ has (as a function of t) Laplace transform

$$\Pi_\lambda(i_0, i_1)\Pi_\lambda(i_1, i_2)\cdots\Pi_\lambda(i_{n-1}, i_n)[\lambda + q(j)]^{-1},$$

where Π_λ denotes the substochastic matrix:

$$\Pi_\lambda(i,j) := [\lambda + q(i)]^{-1} q(i)\Pi(i,j).$$

thus the resolvent $\{R_\lambda\}$ of X satisfies

$$(42.2) \qquad R_\lambda(i,j) = \frac{\Gamma_\lambda(i,j)}{\lambda + q(j)},$$

where Γ_λ is the Green matrix of Π_λ:

$$\Gamma_\lambda := \sum_{n \geqslant 0} \Pi_\lambda^n.$$

The formula (42.2) is intuitively obvious from the *formal* calculation

$$R_\lambda = (\lambda - Q)^{-1} = (\lambda + q - q\Pi)^{-1} = [(\lambda + q)(I - \Pi_\lambda)]^{-1}.$$

All of these formulae are due to Feller [3].

We write $\{P_t\}$ for the transition (matrix) function of X,

$$P_t(i,j) := \mathbf{P}^i\{X_t = j\},$$

and G for the Green matrix of X,

$$G(i,j) = R_0(i,j) = \Gamma(i,j)/q(j).$$

Note that the Martin kernel for X is again κ.

(*42.3*) *Time-reversal.* We now insist that X *starts at* b, so we work with the $\mathbf{P}^b$ law.

Define

$$\hat{Y}(t) := \begin{cases} X(\zeta - t) & (0 < t \leqslant \zeta), \\ \hat{\partial} & (t > \zeta), \end{cases}$$

where $\hat{\partial}$ is a new coffin state isolated from I. Since $\hat{Y}$ is left-continuous and we wish to work with right-continuous processes, we put

$$\hat{X}(t) := \hat{Y}(t+) \quad (t > 0).$$

We do *not* define $\hat{X}(0)$. It is easy to believe that $\{\hat{X}(t) : t > 0\}$ is a right-continuous ($\mathbf{P}^b$) *modification* of $\{\hat{Y}(t) : t > 0\}$.

For $t > 0$, define

$$\hat{\rho}_t(j) := \mathbf{P}^b[\hat{X}(t) = j] \quad (j \in I \cup \hat{\partial}).$$

Then, as we shall see in Section 45, we have the following.

(42.4) *Nagasawa's formula.* For $0 < t_1 < t_2 < \cdots < t_n$ and $j_1, j_2, \ldots, j_n \in I \cup \hat{\partial}$,

$$\mathbf{P}^b[\hat{X}(t_1) = j_1, \ldots, \hat{X}(t_n) = j_n] = \hat{\rho}_{t_1}(j_1)\hat{P}_{t_2 - t_1}(j_1, j_2) \cdots \mathbf{P}_{t_n - t_{n-1}}(j_{n-1}, j_n)$$

where

(42.5)
$$\hat{P}_t(i, j) := G(b, j)P_t(j, i)/G(b, i) \quad (i, j \in I),$$
$$\hat{P}_t(i, \hat{\partial}) := 1 - \hat{P}_t(i, I), \quad \hat{P}_t(\hat{\partial}, \cdot) = \text{unit mass at } \hat{\partial}.$$

You can easily check that $\{\hat{P}_t\}$ is an honest transition function on $I \cup \hat{\partial}$. It follows from (42.4) that $\{\hat{\rho}_t : t > 0\}$ *is a probability entrance law for* $\{\hat{P}_t\}$ *on* $I \cup \hat{\partial}$.

The resolvent $\{\hat{R}_\lambda\}$ of $\hat{X}$ is obtained by taking Laplace transforms in (42.5). Thus

(42.6)
$$\hat{R}_\lambda(i, j) = G(b, j)R_\lambda(j, i)/G(b, i) \quad (i, j \in I),$$
$$\hat{R}_\lambda(i, \hat{\partial}) = \lambda^{-1} - \hat{R}_\lambda(i, I), \quad \text{etc.}$$

In particular, $\hat{X}$ has Green function $\hat{G}$ on $I \times I$ satisfying

$$\hat{G}(i, j) = \hat{R}_0(i, j) = G(b, j)\kappa(j, i).$$

(42.7) *Exercise.* Explain (intuitively—we do not yet have a strong Markov theorem for $\hat{X}$) why the fact that $\hat{X}$ can reach $\hat{\partial}$ from a point i of I only via b corresponds precisely to the result

$$\hat{R}_\lambda(i, \hat{\partial}) = \hat{R}_\lambda(i, b)[\hat{R}_\lambda(b, b)]^{-1}\hat{R}_\lambda(b, \hat{\partial}) \leqslant \hat{R}_\lambda(b, \hat{\partial}),$$

and prove this result analytically.

43. Proof of the Doob–Hunt Convergence Theorem. The plan should now be obvious to you.

Take the discrete topology on $I \cup \hat{\partial}$. Then $I \cup \hat{\partial}$ is an LCCB and $\{\hat{R}_\lambda\}$ is a resolvent on $I \cup \hat{\partial}$ satisfying the special 'pre-Ray' hypotheses (35.6). We can therefore build the Ray–Knight compactification $F \cup \hat{\partial}$ (say) of $I \cup \hat{\partial}$ determined by $\{\hat{R}_\lambda\}$. Since the indicator function of the set $\{\hat{\partial}\}$ is continuous and of compact support on $I \cup \hat{\partial}$, it follows from the definition of Z_1 in (35.2) that

(43.1) $\quad i \mapsto \hat{R}_\lambda(i, \hat{\partial})$ *extends continuously from* $I \cup \hat{\partial}$ *to* $F \cup \hat{\partial}$.

But since, by (42.7),

$$\hat{R}_\lambda(i, \hat{\partial}) \leqslant \hat{R}_\lambda(b, \hat{\partial}) < \hat{R}_\lambda(\hat{\partial}, \hat{\partial}), \quad \forall i \in I,$$

it is clear that $\hat{\partial}$ is isolated in $F \cup \hat{\partial}$. Hence F is a compactification of I.

(43.2) THEOREM. *The Ray–Knight compactification F of I is identical to the Martin compactification F_M (say) of I based on κ.*

That is, there exists a homeomorphism of F to F_M that leaves points of I invariant.

Proof. Let $l(I)$ be the usual Banach space of absolutely convergent series $(\mu_i : i \in I)$ with norm

$$\|u\| := \sum |u_i| < \infty.$$

Usually, but not always, we think of $l(I)$ as the space of signed measures u on I with $\|u\|$ as total-variation norm.

The whole of the present proof could be based on the fact that each $\hat{R}_\lambda$ $(\lambda \geqslant 0)$, acting on the *right* by multiplication $u \mapsto u\hat{R}_\lambda$, is a *compact* operator on $l(I)$. (This follows immediately from (43.3) below and the well-known Cohen–Dunford criterion.) However, we shall phrase the argument in terms of *uniform integrability*, which is the concept underlying the Cohen–Dunford result.

For $\lambda \geqslant 0$, we have

$$\hat{R}_\lambda(i,j) \leqslant \hat{R}_0(i,j) \leqslant \hat{R}_0(j,j) = \frac{\Gamma(j,j)}{q(j)}, \tag{43.3}$$

and, since $\sum \Gamma(j,j)/q(j) < \infty$, *the functions* $\{\hat{R}_\lambda(i,\cdot) : \lambda \geqslant 0, i \in I\}$ *on I are uniformly integrable with respect to counting measure on I.* Hence

(43.4) *if (i_n) is a sequence of I and $\lambda \geqslant 0$ is fixed then* $\lim_n \hat{R}_\lambda(i_n, \cdot)$ *exists in* $l_1(I)$ *if and only if* $\lim_n \hat{R}_\lambda(i_n, j)$ *exists* $\forall j$.

Indeed, this follows directly from (43.3) and the Dominated-Convergence Theorem.

It is immediate from (43.3) that each $\hat{R}_\lambda$ $(\lambda \geqslant 0)$ acting on the right by multiplication on $l(I)$ is a *bounded* operator, and it is easy to see that the resolvent equation extends to give

$$\hat{R}_0(I - \lambda\hat{R}_\lambda) = \hat{R}_\lambda, \qquad \hat{R}_0 = \hat{R}_\lambda(I + \lambda\hat{R}_0)$$

in the language of bounded operators on $l(I)$. Since, for example,

$$\hat{R}_0(i_n, \cdot) = \hat{R}_\lambda(i_n, \cdot)(I + \lambda\hat{R}_0),$$

we can improve (43.4) to the following form:

(43.5) $\lim_n \hat{R}_\lambda(i_n, \cdot)$ *exists in $l(I)$ for some (and then all) $\lambda \geqslant 0$ if and only if* $\lim_n \hat{R}_0(i_n, j) = \lim_n G(b, j)\kappa(j, i_n)$ *exists* $\forall j$.

The argument leading to (43.1) with j replacing ∂ shows that if (i_n) is a sequence in I converging to ζ_{RK} in the Ray–Knight topology of F, then

$$\hat{R}_\lambda(\zeta_{RK}, j) = \lim \hat{R}_\lambda(i_n, j) \textit{ exists } \forall j \in I.$$

Hence $\lim \kappa(j, i_n)$ exists $\forall j$, so that (i_n) converges to a point ξ_M of F_M in the Martin topology and

$$\kappa(j, \xi_M) = G(b, j)^{-1} \sum_{k \in I} \hat{R}_\lambda(\xi_{RK}, k)(I + \lambda \hat{R}_0)(k, j), \quad \forall j \in I, \tag{43.6}$$

for every $\lambda > 0$.

Conversely, suppose that a sequence (i_n) in I converges to a point ξ_M of F_M (in the topology of F_M). Then, from (43.5) (you can check that the extra $\hat{\partial}$ term causes no trouble),

$$\lim_n \sum_{j \in I \cup \hat{\partial}} |\hat{R}_\lambda(i_n, j) - \hat{R}_\lambda(\xi_M, j)| = 0,$$

where

$$\hat{R}_\lambda(\xi_M, j) := \begin{cases} \sum_{k \in I \cup \hat{\partial}} G(b, k)\kappa(k, \xi_M)(I - \lambda \hat{R}_\lambda)(k, j) & (j \in I), \\ \lambda^{-1} - \hat{R}_\lambda(\xi_M, I), & (j = \hat{\partial}). \end{cases} \tag{43.7}$$

Hence, for every bounded function f on I,

$$\lim \hat{R}_\lambda f(i_n) = \hat{R}_\lambda f(\xi_M) := \sum_{j \in I \cup \hat{\partial}} \hat{R}_\lambda(\xi_M, j) f(j).$$

By the construction (Section 35) of the Knight algebra $C(F \cup \hat{\partial})$, it now follows that $\lim g(i_n)$ exists for every g in $C(F \cup \hat{\partial})$. Hence $\xi_{RK} := \lim i_n$ exists in F, and

$$\hat{R}_\lambda(\xi_{RK}, \cdot) = \hat{R}_\lambda(\xi_M, \cdot).$$

You can round off the argument (that $\xi_M = \xi_{RK}$ etc.) to your own satisfaction, but (43.6) and (43.7) really say it all. In particular, these results show that F_e and $(F_M)_e$ agree. □

The Doob–Hunt Convergence Theorem 29.1 now follows immediately from (39.10) because $X(\zeta -) = \hat{X}(0+)$ □

44. The Choquet representation of Π-excessive functions. Suppose that $0 \leqslant \Pi f \leqslant f$ and $f(b) = 1$; thus, in the notation of Section 27, $f \in S$. It is an elementary fact (used in the proof of Theorem 28.2) that

$$f(i) = \uparrow \lim \sum \Gamma(i, j)\beta_n(j)$$

for some non-negative 'charges' β_n on I. Thus

$$f(i) = \uparrow \lim \sum G(i, j)q(j)\beta_n(j),$$

and, since $\lambda R_\lambda G = G - R_\lambda \leqslant G$, it follows that f is supermedian for $\{R_\lambda\}$: $\lambda R_\lambda f \leqslant f$ on I.

Recall from (27.5) that $f(j) \leqslant \theta(j)^{-1}$. It is now convenient to assume (and we *do* assume) $q(\cdot)$ chosen so that (in addition to (42.1)) we have

$$\sum G(j, j)\theta(j)^{-1} = \sum \Gamma(j, j)q(j)^{-1}\theta(j)^{-1} < \infty.$$

Then

$$R_0 f = \sum G(i, j) f(j) < \sum G(j, j) \theta(j)^{-1},$$

so that $R_0 f$ is bounded on I (indeed, uniformly over f in S). If

$$f^1 := f - R_1 f (\geqslant 0) \quad \text{on } I$$

then f^1 satisfies $\lambda R_{\lambda+1} f^1 \leqslant f^1$ and also

$$f = f^1 + R_0 f^1.$$

The resolvents $\{R_\lambda\}$ and $\{\hat{R}_\lambda\}$ are 'in duality relative to the measure $G(b, \cdot)$ on I' in the sense that

$$\langle h_1, R_\lambda h_2 \rangle_{G(b,\cdot)} = \langle \hat{R}_\lambda h_1, h_2 \rangle_{G(b,\cdot)},$$

where h_1 and h_2 are non-negative functions on I and

$$\langle h_1, h_2 \rangle_{G(b,\cdot)} := \sum_i h_1(i) h_2(i) G(b, i).$$

It is a general principle of potential theory that *a* 1-*excessive function for* $\{R_\lambda\}$ *is the density relative to* $G(b, \cdot)$ *of a* 1-*excessive measure for* $\{\hat{R}_\lambda\}$.

We can use (respectively, understand) the *idea* behind this general principle in (via) our simple situation. Put

$$\hat{\mu}^1(j) := f^1(j) G(b, j) \quad (j \in I).$$

Then, for $j \in I$,

(44.1) $$\lambda \hat{\mu}^1 \hat{R}_{\lambda+1}(j) = G(b, j) \lambda R_{\lambda+1} f^1(j) \leqslant \hat{\mu}^1(j),$$

so that $\hat{\mu}^1$ is (at least) 1-supermedian for $\{\hat{R}_\lambda\}$ on I. If we ignore technical details, the general principle is 'trivial', but it is extremely useful, as we shall soon see.

Let us now make the cunning definition

$$\hat{\mu}^1(\hat{\partial}) := f^1(b).$$

By several applications of the resolvent equation, you can check that

$$\hat{\mu}^1(\hat{\partial}) - \lambda \hat{\mu}^1 \hat{R}_{\lambda+1}(\hat{\partial}) = (\lambda + 1)^{-1} [f(b) - (\lambda + 1) R_{\lambda+1} f(b)] \geqslant 0,$$

so that $\hat{\mu}^1$ is 1-supermedian for $\{\hat{R}_\lambda\}$ on $I \cup \hat{\partial}$. Note that

$$\hat{\mu}^1(I \cup \hat{\partial}) = f^1(b) + G f^1(b) = f(b) = 1,$$

so that $\hat{\mu}^1$ is a probability measure on $I \cup \hat{\partial}$. Now every point of $I \cup \hat{\partial}$ is obviously a non-branch-point of $\{\hat{R}_\lambda\}$ now considered as extended to $F \cup \hat{\partial}$, so $\hat{\mu}^1$ can be considered as a 1-excessive probability measure on $F \cup \hat{\partial}$.

By Theorem 38.2,

$$\hat{\mu}^1(j) = \int_{F_* \cup \hat{\partial}} \nu(d\xi) \hat{R}_1(\xi, j)$$

for some probability measure ν on $\mathscr{B}(F_e \cup \hat{\partial})$. Define

$$\hat{\mu}(j) := G(b, j)f(j) \quad (j \in I).$$

Then, *on* I (not $I \cup \hat{\partial}$), we have

$$\hat{\mu} = \hat{\mu}^1[I + \hat{R}_0] = \int_{F_e} \nu(d\xi)\hat{R}_0(\xi, \cdot),$$

whence

$$f(i) = \int_{F_e} \kappa(i, \xi)\nu(d\xi) \tag{44.2}$$

and $\nu(F_e) = f(b) = 1$. That the 'Choquet' representation (44.2) is unique follows easily from the 'uniqueness' part of Theorem 38.2. Our direct proof (avoiding Choquet theory) of the analytical results in Section 27 is now complete.

45. Doob's *h*-transforms. Now the 'Ray' approach to Martin boundary theory really pays dividends.

Let ν be a probability measure on F_e and let

$$h(i) := \int \kappa(i, \xi)\nu(d\xi)$$

denote the excessive function (with $h(b) = 1$) that ν represents. We can define the $\hat{\mathbf{P}}^\nu$ law of '$\hat{X}$ started according to the law ν' in the usual way. It is automatic from (42.7) that $\hat{X}$ dies at b:

$$\hat{\mathbf{P}}^\nu[\hat{X}(\hat{\zeta}-) = b] = 1, \quad \text{where} \quad \hat{\zeta} := \inf\{t : \hat{X}(t) = \hat{\partial}\}.$$

Now write Y_h for the time-reversal of $(\hat{X}, \hat{\mathbf{P}}^\nu)$:

$$Y_h(t) := \begin{cases} \hat{X}(\hat{\zeta} - t) & (0 < t \leqslant \hat{\zeta}), \\ \partial & (t > \hat{\zeta}), \end{cases}$$

where ∂ is a 'forward' coffin state. Put $X_h(t) := Y_h(t+)$, so that, for $t = 0$, $X_h(0) = b$ (a.s.($\hat{\mathbf{P}}$)). By Nagasawa's formula, X_h (starts at b and) has transition probabilities

$$\begin{aligned} P_h(t; i, j) &= \frac{\int \nu(d\xi)\hat{G}(\xi, j)}{\int \nu(d\xi)\hat{G}(\xi, i)} \hat{P}(t; j, i) \\ &= h(i)^{-1}P(t; i, j)h(j). \end{aligned}$$

In particular, the Q-matrix Q_h of X_h satisfies

$$Q_h(i, j) = h(i)^{-1}Q(i, j)h(j),$$

so that the 'jump chain' J_h of successive states visited by X has one-step transition matrix Π_h, where

$$\Pi_h(i,j) = h(i)^{-1}\Pi(i,j)h(j).$$

We have proved the following theorem.

(*45.1*) THEOREM (Doob). *The probability measure ν that represents an element h of S is the distribution of $J_h(\zeta-)$, where J_h is a discrete-parameter chain with transition matrix Π_h and started at b.*

Theorems 29.2 and 29.3 follow immediately.

We have presented a full acount of Martin-boundary theory in its simplest setting, but by the most powerful methods. You will wish to follow up the general theory in Meyer [4]. To get a first idea of the scope of applications, see Blackwell and Kendall [1] on Polya's urn and population growth, Revuz [1] for the culmination of work of Kesten, Spitzer, Brunel and Revuz on random walks on groups, and Dynkin [3] for an extraordinary and deep result on random deformations of ellipsoids. Norris, Rogers and Williams [1] gives a simpler proof of Dynkin's result.

Time reversal and related topics

It is something of a mystery that a number of results are much clearer in reversed time than in forward time. Martin-boundary theory has already illustrated this; and we shall shortly see other examples.

Our concern is with how to use the *idea* of time reversal, without full discussion of certain technical difficulties, which tend to blur the subject. Chung and Walsh [1] is a very important paper on identifying and resolving those difficulties. The last two volumes of Dellacherie and Meyer [1] have the latest news. Time reversal is deeply connected with *duality*, a major topic from the work of Hunt on. See Blumenthal and Getoor [1]. Joanna Mitro's papers [1, 2] greatly clarify duality.

46. Nagasawa's formula for chains. Let I be a countable set and let $\{p_{ij}(t)\}$ be a transition matrix function on I (satisfying the usual continuity condition: $p_{ij}(0+) = \delta_{ij}$). Let $\{\rho(t)\}$ be a probability entrance law for the usual extension $\{P_t^{+\partial}\}$ of $\{P_t\}$ to $I \cup \partial$. But let us write P_t instead of the clumsy $P_t^{+\partial}$. Let $\{X_t : t > 0;\ \Omega, \mathscr{F}^\circ, \mathbf{P}^\rho\}$ be a process on $I \cup \partial$ with

$$\mathbf{P}^\rho[X_s = i; X_{s+t} = j; X_{s+t+u} = k; \ldots] = \rho_i(s)p_{ij}(t)p_{jk}(u)\ldots$$

whenever $s, t, u, \ldots > 0$ and $i, j, k, \ldots \in I \cup \partial$. If you like, you can imagine X to be the Ray process taking values in the Ray–Knight compactification of $I \cup \partial$ determined by $\{R_\lambda\}$. (The precise details of this are clarified in Section 50.) But let us not be too specific about the technicalities—assume that X is 'sufficiently smooth'. It is well known (see Section 52 for proof) that, for $k \in I$, there exists a continuous function $f_{k\partial}(\cdot)$ on $(0, \infty)$ such that

$$\mathbf{P}^k[\zeta \in dt] = f_{k\partial}(t)\, dt, \quad \zeta := \inf\{t : X_t = \partial\}.$$

(Strictly speaking, we should be concentrating on a particular triple $(\Omega, \mathscr{F}^\circ, \mathbf{P}^\rho)$, but we assume that $\mathbf{P}^k$ can be defined on $(\Omega, \mathscr{F}^\circ)$, as will happen when $(\Omega, \mathscr{F}^\circ)$ is the usual path-space. As far as we are concerned, this point is rather academic. When you read the Chung–Walsh paper, you will see why we have mentioned it, and you will recall comments on regular conditional probabilities made before Theorem II.90.11.

Let us suppose that $\mathbf{P}^\rho(0 < \zeta < \infty) = 1$. We now consider the process $\{\hat{Y}(t), \mathbf{P}^\rho\}$, where

$$\hat{Y}(t) := \begin{cases} X(\zeta - t) & (0 < t \leqslant \zeta), \\ \hat{\partial} & (t > \zeta). \end{cases}$$

It is plausible that, for $k \in I$,

(46.1) $$\mathbf{P}^\rho[\hat{Y}(t) = k] = \int_0^\infty \mathbf{P}^\rho[X(\zeta - t) = k; \zeta \in t + dv]$$

$$= \int_{v=0}^\infty \mathbf{P}^\rho[X(v) = k](\mathbf{P}^k[\zeta \in dt]/dt)\, dv = \xi_k f_{k\partial}(t),$$

where

(46.2) $$\xi_k := \int_0^\infty \rho_k(v)\, dv.$$

We do not need to tell you that (46.1) is not a rigorous calculation. One way (Chung–Walsh) in which to make it rigorous involves approximating the integrals by Riemann sums, using right-continuity of X to push things through. Another way (Meyer [4]) is to justify directly the result obtained when both sides of (46.1) are multiplied by an arbitrary measurable non-negative function of t and integrated over $(0, \infty)$. In either case, we can use left-continuity of $\hat{Y}$ to show that (46.1) must hold for all (not merely almost all) t.

Without any *further* difficulty, we can calculate for $i, j, k \in I$ and $s, t, u > 0$,

(46.3) $$\mathbf{P}^\rho[\hat{Y}_s = i; \hat{Y}_{s+t} = j; \hat{Y}_{s+t+u} = k]$$

$$= \int_{v=0}^\infty \mathbf{P}^\rho[X_{\zeta-s} = i; X_{\zeta-s-t} = j; X_{\zeta-s-t-u} = k; \zeta \in s + t + u + dv],$$

$$= \int_{v=0}^\infty \mathbf{P}^\rho[X_v = k; X_{v+u} = j; X_{v+u+t} = i] f_{i\partial}(s)\, dv$$

$$= \xi_k p_{kj}(u) p_{ji}(t) f_{i\partial}(s) = \mathbf{P}^\rho[\hat{Y}_s = i] \hat{p}_{ij}(t) \hat{p}_{jk}(u),$$

where

$$\hat{p}_{ij}(t) := \xi_j p_{ji}(t)/\xi_i$$

(with some arbitrary conventions for 0/0). The extension of (46.3) 'to n terms' is obvious. Thus

(46.4) $\{\hat{Y}_t : t > 0; \mathbf{P}^\rho\}$ *is Markovian with stationary transition probabilities* $\{\hat{p}_{ij}(t)\}$ *and with probability entrance law on* $I \cup \hat{\partial}$ *determined by (46.1).*

47. Strong Markov property under time reversal. Consider a process X that starts at 1, drift towards 0 at constant rate 1, stays at 0 for an exponentially distributed time S of rate 1, and then dies. Thus

$$X_t = 1 - t \quad (0 \leqslant t < 1); \qquad X_t = 0 \quad (1 \leqslant t < 1 + S); \qquad X_t = \partial \quad (t \geqslant 1 + S).$$

It is easy to see that X has FD transition function, so that X is strong Markov. The time-reversal $\hat{Y}$ of X satisfies

$$\hat{Y}_t = 0 \quad (0 < t \leqslant S); \qquad \hat{Y}_t = t - S \quad (S < t \leqslant S + 1); \qquad \hat{Y}_t = \hat{\partial} \quad (t > S + 1).$$

The right-continuous modification $\hat{X}$ of $\hat{Y}$ satisfies

$$\hat{X}_t = 0 \quad (0 \leqslant t < S); \qquad \hat{X}_t = t - S \quad (S \leqslant t < S + 1); \qquad \hat{X}_t = \hat{\partial} \quad (t \geqslant S + 1).$$

Note that 1 acts as a branch-point for $\hat{X}$, with $\hat{P}_0(1, \hat{\partial}) = 1$. The process $\hat{X}$ is very similar to the process in Example 35.8. In particular, $\hat{X}$ is not strong Markov relative to its natural $\{\mathscr{F}^\circ_{t+}\}$-algebras, because $\hat{X}$ does not start afresh at time S. Thus *time reversal can destroy the strong Markov property.*

Chung and Walsh [1] show that the time-reversal (made right-continuous) of a strong Markov process has a Markov property intermediate between the simple and strong Markov properties: the so-called *moderately strong Markov property*. This interesting concept is 'correct' from the point of view of the theory of previsible processes.

There is however another way out of the 'difficulty', which is more satisfactory: *Doob's simultaneous compactification.* The problem with time reversal is that the Ray–Knight compactifications associated with X and $\hat{X}$ may be totally different and may induce different topologies on the initially given state-space. For the example that we have just been discussing, we know from (35.8) that, in order to 'force' the strong Markov property of $\hat{X}$, we must tear $[0, 1]$ apart at 0, producing a state-space $\{0\} \cup [0_+, 1]$. (Recall that $[0_+, 1]$ is the same as the conventional $[0, 1]$ and that 0 now denotes a point isolated from $[0_+, 1]$.) The important thing is that *both* X *and* $\hat{X}$ have good (right-continuous, strong Markov) modifications with state-space $\{0\} \cup [0_+, 1]$, with 0_+ made a branch-point for X from which X branches to 0, and that these modifications are properly related: each is the 'time-reversal made right-continuous' of the other. Here you have a clue to Doob's idea (see Doob [2]) of constructing an *entrance–exit* space for a general process X by using a 'simultaneous' compactification based on both X and a time-reversal $\hat{X}$. We leave aside discussion of the worries you are beginning to have about whether we will now have to abandon the Ray property.

48. Equilibrium charge. We now give the promised intuitive explanation (expanded from Williams [1]) of Hunt's Theorem I.22.7 on equilibrium potential and the Chung–Getoor–Sharpe Theorem I.22.13 on equilibrium charge. Though we skip rigour for interest's sake, it is not too hard to supply it even in much more general contexts.

Take $n \geqslant 3$ and let g be the free-space Green function for Δ (in $\mathbb{R}^n$). Let $b \in \mathbb{R}^n$ and let B be canonical Brownian motion in $\mathbb{R}^+$ starting at b. Let v be a strictly positive C^∞ function on $\mathbb{R}^n$ such that

$$\mathbf{E}^b A_\infty = 2\int g(b,y)v(y)\,dy < \infty$$

where

$$A_t := \int_0^t v(B_s)\,ds.$$

Put $\tau_t := \inf\{s : A_s > t\}$. Then, by Volkonskii's Theorem, $X_t := B \circ \tau_t$ defines a continuous strong Markov process with finite lifetime A_∞. The Green function of X *with respect to the measure* $2v(y)\,dy$ is just g.

Now let $\hat{X}$ be the time-reversal of X with Green function $\hat{g}$ relative to the measure $2v(y)\,dy$ given by the Nagasawa formula:

(48.1) $$\hat{g}(y,z) = g(b,z)g(z,y)/g(b,y).$$

Define

$$H_K := \inf\{s > 0 : X_s \in K\}, \qquad \hat{H}_K := \inf\{s > 0 : \hat{X}_s \in K\},$$
$$L_K^B := \sup\{s > 0 : B_s \in K\}, \qquad L_K^X := \sup\{s > 0 : X_s \in K\}.$$

For $\Gamma \in \mathcal{B}(\partial K)$, put

$$\lambda(\Gamma) := \mathbf{P}^b[B(L_K^B) \in \Gamma] = \mathbf{P}^b[X(L_K^X) \in \Gamma] = \mathbf{P}^b[\hat{X}(\hat{H}_K) \in \Gamma\}.$$

Then, using the (unproved but 'obvious') strong Markov property of $\hat{X}$, we find that

$$2v(z)\,dz \int_{\partial K} \lambda(dy)\hat{g}(y,z) = \mathbf{E}^b[\text{time spent by } X \text{ in } dz \text{ before } L_K^X]$$
$$= \mathbf{E}^b[\text{time spent by } X \text{ in } dz]\mathbf{P}^z[H_K < \infty]$$
$$= 2g(b,z)v(z)\,dz\,\mathbf{P}^z[H_K < \infty].$$

On substituting the formula (48.1) for $\hat{g}(y,z)$, we see that, for almost all z.

(48.2) $$\mathbf{P}^z[H_K < \infty] = \int_{\partial K} g(z,y)e_K(dy),$$

where

(48.3) $$e_K(dy) := g(b,y)^{-1}\mathbf{P}^b[B(L_K^B)\in dy].$$

Since $\mathbf{P}^z[H_K<\infty]$ has the same significance for B as for X, the main results of Theorems I.22.7 and I.22.13 are more or less proved. Rounding off Theorem I.22.7 is largely a case of repeating arguments used in connection with the Dirichlet problem. Let us emphasise one thing, however. The two sides of (48.2) are discontinuous at sufficiently singular points of ∂K, as Lebesgue's thorn shows. To obtain (48.2) for *all z*, which is important because of such singularities, we use the fact that both sides of (48.2) are *excessive* (for B or for X).

Exercise. Explain the 'excessive' property and why it implies equality for all z in (48.2).

You should now be convinced that time reversal provides the natural approach to many problems.

49. BM (ℝ) and BES(3); splitting times. Many relations exist between 1-dimensional Brownian motion and the 3-dimensional Bessel process. Here is a first one. (A $\mathrm{BM}_b(\mathbb{R})$ has starting position b, etc.)

(*49.1*) THEOREM. *Let B be a $\mathrm{BM}_0(\mathbb{R})$ and define $H_1^B := \inf\{t : B_t = 1\}$. Let R be a $\mathrm{BES}_0(3)$ and define $L_1^R := \sup\{t : R_t = 1\}$. Then the processes*

$$\{1 - B(H_1^B - t) : 0 \leqslant t < H_1^B\} \quad \textit{and} \quad \{R(t) : 0 \leqslant t < L_1^R\}$$

are identical in law.

After the discussion at the end of Section 31, it should be clear to you that Martin-boundary theory makes this result extremely plausible. It is not difficult (*Exercise*!) to prove it directly by bare-hand computation—see Williams [4], where the result is applied to local-time theory.

For $0 \leqslant t < \infty$, put

$$A_t := \mathrm{meas}\{s \leqslant t : B_s \geqslant 0\}, \qquad \tau_t := \inf\{s : A_s > t\}.$$

Then (see Section 22) $Y_t := B\circ\tau_t$ defines a reflecting Brownian motion Y. In other words, Y is a $\mathrm{BES}_0(1)$ process. Put $H_1^Y := \inf\{t : Y_t = 1\}$. Then

$$H_1^Y = A(H_1^B) \sim \mathrm{meas}\{t \leqslant L_1^R : R(t) \leqslant 1\} = \mathrm{meas}\{t < \infty : R(t) \leqslant 1\},$$

'$\sim$' signifying equality in law. This is the special case '$n = 1$' of the Ciesielski–Taylor Theorem that if R_n is a $\mathrm{BES}_0(n)$ and R_{n+2} is a $\mathrm{BES}_0(n+2)$ then

(49.2) $$\inf\{t : R_n(t) = 1\} \sim \mathrm{meas}\{t < \infty : R_{n+2}(t) \leqslant 1\}.$$

See (II.20) and Ciesielski and Taylor [1]. Many proofs of (49.2) are now known. None seems to provide a clear geometrical explanation, and one conjectures that no such explanation exists.

The following result (Williams [4, 7]) was needed for certain applications to excursion theory and local-time theory.

(49.3) **THEOREM.** *Fix b in* $(0, \infty)$. *On a suitable probability triple* $(\Omega, \mathcal{F}, \mathrm{P})$, *set up three independent random elements* (see Fig. III.1):

a random variable γ *uniformly distributed on* $(0, b)$;
a $\mathrm{BM}_b(\mathbb{R})$ $\{B(t): t \geqslant 0\}$;
a $\mathrm{BES}_0(3)$ $\{R(t): t \geqslant 0\}$.

Define

$$\rho := \inf\{t : B(t) = \gamma\},$$

$$X(t) := \begin{cases} B(t) & (t < \rho), \\ R(t-\rho) + \gamma & (t \geqslant \rho). \end{cases}$$

Then $\{X(t): t \geqslant 0\}$ *is a* $\mathrm{BES}_b(3)$.

We regard this result as providing a *path decomposition* of the $\mathrm{BES}_b(3)$ X at the time ρ at which X attains its minimum value γ. Williams [4, 7] proved Theorem 49.3 by bare-hand calculation, but did try (unsuccessfully) to produce a theory of *splitting-times* as the 'natural' times at which one might expect to have path decompositions of Markov processes. His idea was to call a random time ρ an algebraic splitting-time (for a Markov process X) if, for every t, we can write

$$\{\rho = t\} = F_t \cap G_t,$$

where

$$F_t \in \sigma\{X_s : s \leqslant t\}, \qquad G_t \in \sigma\{X_u : u \geqslant t\},$$

the σ-algebras being *uncompleted.* The way in which this definition would be ruined if we allowed completion of algebras is a first indication of how much more difficult it must be to prove a splitting-time theorem than the Stopping-Time (that is, Strong Markov) Theorem.

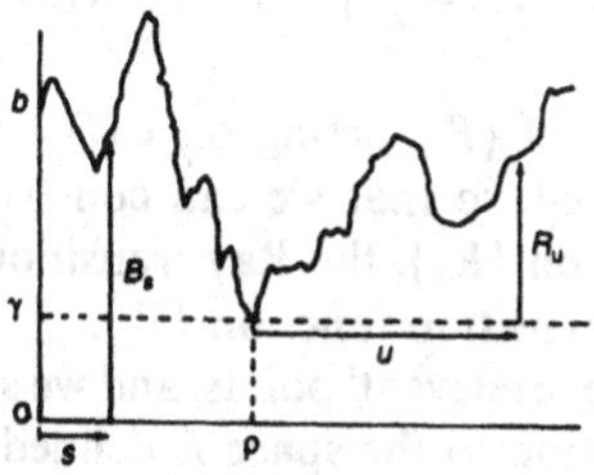

Figure III.1

Jacobsen [1] gave a nice formulation of 'splitting-time' in terms of the 'crossover' property, and also gave a more illuminating proof of Theorem 49.3. Pitman [1] then proved Theorem 49.3 by using a clever random-walk approximation, and in [2] he described other applications of the splitting-time idea. Millar has done some fine work in this area (see his survey [1] and papers referred to there) and has (Millar [2]) the definitive proof of results on path decomposition at times of minima for a wide class of Markov processes based-on work of Getoor and Sharpe.

New proofs of Theorem 49.3 appear regularly. See, for example, Walsh [1], le Gall [7], Ikeda and Watanabe [1], Revuz and Yor [1], and Section VI.55 in Volume 2. Non-standard analysis provides a promising approach to splitting times; see Cutland and Kendall [1].

Theorem 49.3 helped motivate some of the original work on *grossissements* (enlargements of filtrations). See Barlow [1,2], Jeulin [1], Jeulin and Yor [1]. Jeulin gives a martingale proof of Theorem 49.3.

A first look at Markov-chain theory

Many key ideas of modern Markov-process theory—last-exit decompositions, excursion laws, boundary theory etc.—first appeared in clear form in Markov-chain theory; and chain theory still seems to us the ideal vehicle for learning process theory and assessing its achievements. Like number theory, chain theory is at the same time 'concrete' and sufficiently rich to accommodate the most sophisticated ideas.

50. Chains as Ray processes. Let I be a countable set, so that I is an LCCB in its discrete topology. Let $\{P_t\} = \{p_{ij}(t)\}$ be a transition matrix function on I, assumed 'standard' in that

$$p_{ij}(0+) = p_{ij}(0) = \delta_{ij}, \quad \forall i, j \in I.$$

Without loss of generality, we can assume that $\{P_t\}$ is honest. So let us assume that

$$P_t(i, I) = \sum_{j \in I} p_{ij}(t) = 1, \quad \forall i \in I.$$

Let $\{R_\lambda\}$ be the resolvent of $\{P_t\}$ acting on $C(I) = B(I)$. Then the conditions (35.6) are obviously satisfied, so that we can construct the Ray–Knight compactification F of I based on $\{R_\lambda\}$, the Ray transition function $\{P_t\}$ on F, and the Ray process X with transition function $\{P_t\}$.

The space F may include 'irrelevant' points, and we shall see that we can—with advantage—restrict attention to the space E defined as follows:

$$E := \{x \in F : P_t(x, I) = 1, \forall t > 0\} \supseteq I. \tag{50.1}$$

For $x \in F$, the map $t \mapsto P_t(x, I)$ is obviously (why?) non-decreasing in t, so that

$$E = \{x \in F : R_1(x, I) = 1\}. \tag{50.2}$$

(If, for example, we have the Poisson chain on $\mathbf{Z}^+$ with

$$q_{ij} := \begin{cases} -1 & \text{if } j = i, \\ 1 & \text{if } j = i + 1, \\ 0 & \text{otherwise,} \end{cases}$$

then F is the one-point compactification of $\mathbf{Z}^+$, and $P_t(\infty, \{\infty\}) = 1$ for $t \geqslant 0$; here, $E = \mathbf{Z}^+$.)

The space E may be described very simply. Think of the point i of I as identified with the element

$$r_{i \cdot}(1) = R_1(i, \{\cdot\}) \in \ell_1(I).$$

Then the arguments in Section 43 show that (up to homeomorphism) E is nothing other than the closure of I in $\ell_1(I)$. The map

$$x \mapsto r_{x \cdot}(1) = R_1(x, \{\cdot\})$$

is therefore a homeomorphism of E *to* $\ell_1(I)$.

(50.3) THEOREM (Neveu [5]). *For* $t > 0$, *the map*

$$x \mapsto p_{x \cdot}(t) = P_t(x, \{\cdot\})$$

from E *to* $\ell_1(I)$ *is continuous. Hence, for every* f *in* $B(I)$ *and every* $t > 0$, *the map* $x \mapsto P_t f(x)$ *is continuous on* E.

Proof. We are guided by Neveu [5]. With slight (but allowable) misuse of notation, let $x \to y$ in E. Then

$$r_{x \cdot}(1) \to r_{y \cdot}(1) \quad (\text{in } \ell_1(I)).$$

Since, for $0 < u < v$,

$$\int_u^v e^{-s} p_{x \cdot}(s)\, ds = r_{x \cdot}(1)[e^{-u} P(u) - e^{-v} P(v)],$$

we have, for $0 < \delta < t$,

$$\delta^{-1} \int_{t-\delta}^{t} e^{t-s} p_{xj}(s)\, ds \to \delta^{-1} \int_{t-\delta}^{t} e^{t-s} p_{yj}(s)\, ds, \quad \forall j \in I. \tag{50.4}$$

Fix $j \in I$. Let $\varepsilon > 0$ be given, and let $\delta > 0$ be so small that $p_{jj}(u) \geqslant 1 - \varepsilon$ for $u \leqslant \delta$. Then, for $t - \delta < s < t$,

$$p_{xj}(t) \geqslant e^{t-s} p_{xj}(s) e^{-\delta}(1 - \varepsilon),$$

whence, from (50.4),

$$\liminf_{x\to y} p_{xj}(t) \geqslant (1-\varepsilon)e^{-\delta}\delta^{-1}\int_{t-\delta}^{t} e^{t-s}p_{yj}(s)\,ds.$$

Now let $\varepsilon\downarrow\downarrow 0$ (and insist that $\delta\downarrow\downarrow 0$) to obtain

$$\liminf_{x\to y} p_{xj}(t) \geqslant p_{yj}(t). \tag{50.5}$$

Fatou's Lemma combined with $\Sigma p_{yj}(t)=1$ shows that equality must hold in (50.5) and this implies (why?) that we can replace 'lim inf' in (50.5) by 'lim'. □

The point of Theorem 50.3 is that $\{P_t\}$ has nice analytical properties on E. We must now show that 'only E matters' in the probabilistic theory.

(*50.6*) THEOREM (Ray, Kunita–Watanabe, Meyer). *For every x in E,*

$$\mathbf{P}^x[X_t\in E, \forall t\geqslant 0]=1, \tag{50.7}$$

$$\mathbf{P}^x[X_{t-}\in E, \forall t> 0]=1. \tag{50.8}$$

Proof. The proof is a fine illustration of the need for Meyer's section theorems.

Let $x\in E$. Then, with χ_I denoting the characteristic (indicator) function of I, we have, for any stopping time T,

$$1=R_1\chi_I(x)=\mathbf{E}^x\int_0^T e^{-t}\chi_I(X_t)\,dt+\mathbf{E}^x[e^{-T}R_1\chi_I(X_T)]$$

$$\leqslant \mathbf{E}^x[1-e^{-T}]+\mathbf{E}^x[e^{-T}]=1.$$

You can see that, because of (50.2),

$$\mathbf{P}^x[T<\infty, X_T\in F\backslash E]=0,$$

and the result (50.7) now follows from the Section Theorem II.76.2.

To prove (50.8), we need some of the general theory of processes, which we study further in Chapter VI. Since the process $\{X_{t-}\}$ is left-continuous, it is 'previsible'. Hence if

$$\mathbf{P}^x[X_{t-}\in F\backslash E \quad \text{for some } t]>0$$

then (by Meyer's 'previsible' section theorem—see Dellacherie and Meyer [1]) there exists a sequence (T_n) of stopping times with $T_n\uparrow\uparrow T$ (a.s.($\mathbf{P}^x$)) such that

$$\mathbf{P}^x[T<\infty, X_{T-}\in F\backslash E]>0.$$

However, this contradicts the 'Quasi-Left-Continuity' Theorem 41.3, because (as we saw above) $X_T\in E$ (a.s.($\mathbf{P}^x$)), whereas

$$P_0(X_{T-}, E)<1 \quad (\text{if } X_{T-}\in F\backslash E).$$

Clarification. Since $R_1\chi_I \geqslant \chi_E$, we have, for $y \in F\backslash E$,

$$1 > R_1(y, I) = \varepsilon_y R_1 \chi_I = \varepsilon_y P_0 R_1 \chi_I \geqslant \varepsilon_y P_0 \chi_E = P_0(y, E). \qquad \square$$

We *may (and shall) now regard E as the state-space of our process (chain?!)* X. Of course, E need not be compact, but E is Polish. We write E_{br} for the set of branch-points in E, and E_e for the set of non-branch-points in E.

51. Significance of q_i. Let $i \in I$ and define

(51.1) $$S_i := \inf\{t : X_t \neq i\}.$$

(Then S_i is an $\{\mathscr{F}^0_{t+}\}$ stopping time.) By the simple Markov property, we have, for $s, t > 0$,

$$\mathbf{P}^i[S_i > s + t] = \mathbf{P}^i[S_i > s]\mathbf{P}^i[S_i > t],$$

so that

(51.2) $$\mathbf{P}^i[S_i > t] = e^{-q_i t} \quad \text{for some } q_i \in [0, \infty].$$

Now let $\varepsilon_n \downarrow\downarrow 0$. Then it is clear that

$$\mathbf{P}^i[S_i \geqslant t] = \lim_n p_{ii}(\varepsilon_n)^{[t/\varepsilon_n]+1}$$

so that

$$-q_i t = \lim_n \frac{t}{\varepsilon_n} \log p_{ii}(\varepsilon_n).$$

It is now trivial that

(51.3) $$q_i = \lim_{\varepsilon \downarrow\downarrow 0} \varepsilon^{-1}[1 - p_{ii}(\varepsilon)],$$

the existence of the limit being part of our conclusion. Of course, we have

(51.4) $$p_{ii}(t) \geqslant \mathbf{P}^i[S_i > t] = e^{-q_i t}.$$

We make the usual classification:

(51.5) *a state i of I is called stable if $q_i < \infty$ and instantaneous if $q_i = \infty$.*

Since paths are right-continuous, it is clear that

(51.6) *a point of I that is isolated in I in the Ray–Knight topology is a stable state.*

Of course (give an example!) a stable state may be a point of accumulation of I.

52. Taboo probabilities; first-entrance decomposition. We stick to DW's traditional use of 'b', the first letter in the alphabet to cause printers (and readers) no trouble, for a state in I on which we concentrate our attention. (Of course, b is not a branch-point of X.) We try to abide by Chung's terminology and notation for chains.

So fix b in I. Introduce Chung's *taboo transition probability*:

(52.1) $${}_bp_{ij}(t) := \mathbf{P}^i[X_t = j, H_b > t] \quad (i, j \in I \backslash b)$$

where H_b is the hitting-time of b. Then, because H_b has the *terminal-time property*

$$H_b = t + H_b \circ \theta_t \quad \text{on } \{H_b > t\},$$

it follows (*Exercise*—but see Section 54) that $\{{}_bp_{ij}(t)\}$ *is a 'standard' transition matrix function on* $I \backslash b$ and that, for $s, t > 0$,

(52.3) $$F_{ib}(t+s) - F_{ib}(t) = \sum_{j \in I \backslash b} {}_bp_{ij}(t) F_{jb}(s) \quad (i \in I \backslash b),$$

where

(52.4) $$F_{ib}(t) := \mathbf{P}^i[H_b \leqslant t].$$

Chung and Neveu independently showed that (52.3) may be differentiated with respect to s in the following precise sense: *for $i \neq b$, there exists a (finite) continuous function $f_{ib}(\cdot)$ on $[0, \infty)$ such that*

(52.5) $$F_{ib}(t) = \int_0^t f_{ib}(s)\, ds;$$

further, for $s > 0$, but not necessarily for $s = 0$,

(52.6) $$f_{ib}(t+s) = \sum_{j \in I \backslash b} {}_bp_{ij}(t) f_{jb}(s).$$

We skip the analytical derivation of (52.5) and (52.6) from (52.3). See Neveu [3, 4], where the idea is that we can write f_{ib} explicitly as

$$f_{ib}(t) = \sum_{j \in I \backslash b} t^{-1} \int_0^t {}_bp_{ij}(t-s)\, dF_{jb}(s),$$

or Chung [1; Theorem II.12.4] for proof by appeal to classical (Fubini) theorems on differentiation.

The important fact that f_{ib} extends continuously from $(0, \infty)$ to $[0, \infty)$ with $f_{ib}(0) < \infty$ needs special pleading. We reproduce the argument from Chung and Neveu. We have

$$f_{ib}(s) \geqslant {}_bp_{ii}(s-u) f_{ib}(u) \quad (0 < u < s).$$

Hence, as we see by letting $u \downarrow\downarrow 0$ (through a suitable subsequence),

(52.7) $$f_{ib}(s) \geqslant {}_bp_{ii}(s) \left[\limsup_{u \downarrow\downarrow 0} f_{ib}(u) \right];$$

and now letting $s \downarrow\downarrow 0$,

$$\liminf_{s \downarrow\downarrow 0} f_{ib}(s) \geqslant \limsup_{u \downarrow\downarrow 0} f_{ib}(u).$$

Put

(52.8) $$q_{ib} := f_{ib}(0) := f_{ib}(0+) \quad (i \in I\backslash b).$$

We see from (52.7) that

(52.9) $$0 \leqslant q_{ib} < \infty \quad (i \in I\backslash b).$$ □

If we let $s \downarrow\downarrow 0$ in (52.6) and apply Fatou's Lemma, we obtain

(52.10) $$f_{ib}(t) \geqslant \sum_{j \in I\backslash b} {}_b p_{ij}(t) q_{jb}.$$

Strict inequality may obtain here. The right-hand side of (52.10) represents the $\mathbf{P}^i$ probability density of first entering b *via a jump from* $I\backslash b$ at time t. (Accept this intuitively obvious fact for now.) Note how 'wrong' things go in the case of the Feller–McKean process.

Let us now agree to use the 'tilde' notation for Laplace transforms. So, for $\lambda > 0$, write

(52.11) $$\tilde{f}_{ib}(\lambda) := \int_0^\infty e^{-\lambda t} f_{ib}(t)\,dt = \mathbf{E}^i[\exp(-\lambda H_b)];$$

(52.12) $$\tilde{p}_{ij}(\lambda) := \int_0^\infty e^{-\lambda t} p_{ij}(t)\,dt = r_{ij}(\lambda).$$

Dynkin's formula gives

$$\tilde{p}_{ib}(\lambda) = R_\lambda \chi_b(i) = \mathbf{E}^i[\exp(-\lambda H_b)] R_\lambda \chi_b(b) = \tilde{f}_{ib}(\lambda) \tilde{p}_{bb}(\lambda),$$

and, on inverting the Laplace transform, we obtain the *first-entrance decomposition*:

(52.13) $$p_{ib}(t) = \int_0^t f_{ib}(s) p_{bb}(t-s)\,ds \quad (i \in I\backslash b)$$

which (by continuity) is valid for *all* $t \geqslant 0$.

(*52.14*) **Exercise.** Prove the more precise result

$$\mathbf{P}^i[H_b \in ds; X(t) = b] = f_{ib}(s) p_{bb}(t-s)\,ds \quad (0 < s < t).$$

53. The Q-matrix; DK conditions. We have already seen that

(53.1) $$q_{ii} := \lim_{\varepsilon \downarrow\downarrow 0} \varepsilon^{-1}[p_{ii}(\varepsilon) - 1] = -q_i \in [-\infty, 0].$$

Since $q_{ib} := f_{ib}(0+)$ exists in $[0, \infty)$, it follows immediately from (52.13) that

(53.2) $$\lim_{\varepsilon \downarrow\downarrow 0} \varepsilon^{-1} p_{ib}(\varepsilon) = q_{ib} \in [0, \infty).$$

The existence of the limits in (53.1) and (53.2) was first established by Kolmogorov and Doob. We write

$$Q = P'(0), \tag{53.3}$$

with the interpretation provided by (53.1) and (53.2). We call the conditions

(DK1) $$0 \leqslant q_{ij} < \infty \quad (i \neq j);$$

(DK2) $$\sum_{j \neq i} q_{ij} \leqslant q_i \leqslant \infty$$

the Doob–Kolmogorov conditions. The condition (DK2) is obtained by letting $\varepsilon \downarrow\downarrow 0$ in the equation

$$\sum_{j \neq i} \varepsilon^{-1} p_{ij}(\varepsilon) \leqslant \varepsilon^{-1}[1 - p_{ii}(\varepsilon)] \tag{53.4}$$

and applying Fatou's Lemma. (Under our assumption that $\{P_t\}$ is honest, equality holds in (53.4).) The Feller–McKean example shows that (even for an honest $\{P_t\}$) we can have the 'worst' possible situation:

$$\sum_{j \neq i} q_{ij} = 0, \quad q_i = \infty, \quad \forall i.$$

Actually this is the 'best' possible situation for chains with all states instantaneous. (See Theorem 55.1.)

The probabilistic interpretation of $q_{ij} (j \neq i)$ is given in Section 57.

54. Local-character condition for Q. Let G be an open subset of E. Define

$$p_{ij}^G(t) := \mathbf{P}^i[X(t) = j, H_{E \backslash G} > t] \quad (i, j \in I \cap G),$$

where, as usual $H_{E\backslash G}$ denotes the hitting-time of $E\backslash G$. Then

$$p_{ii}^G(t) \geqslant p_{ii}(t) - \mathbf{P}^i[H_{E\backslash G} \leqslant t] \quad (i \in I \cap G).$$

Since G is open and X is right-continuous, $\mathbf{P}^i[H_{E\backslash G} > 0] = 1$, so that

$$p_{ii}^G(t) \to 1 \quad (t \to 0) \tag{54.1}$$

for $i \in I \cap G$. That $\{P^G(t)\} := \{p_{ij}^G(t) : i, j \in I \cap G\}$ is a transition matrix function on $I \cap G$ is easily shown (recall Exercise 18.15). Note that (54.1) ensures that $\{P^G(t)\}$ is 'standard'. Extend $\{P^G(t)\}$ to $(I \cap G) \cap \partial$ in the usual way and observe that, for $b \in I \cap G$,

$$p_{b\partial}^G(\varepsilon) \geqslant \mathbf{P}^b[X_\varepsilon \in I\backslash G] = \sum_{j \in I\backslash G} p_{bj}(\varepsilon).$$

Multiply through by ε^{-1}, let $\varepsilon \downarrow\downarrow 0$, and apply Fatou's Lemma and (DK1) to obtain

$$Q(b, I\backslash G) := \sum_{j \in I\backslash G} q_{bj} \leqslant q_{b\partial}^G < \infty.$$

We have proved the following result (Williams [8, 9]).

(54.2) LEMMA (local-character condition). *Let $b \in I$, and let G be an open subset of E containing b. Then*

$$Q(b, I \backslash G) < \infty.$$

The result is interesting, not for its own sake, but for its 'concrete' applications. we follow Williams [9].

(54.3) COROLLARY 1. *If a and b are distinct points of I, then*

(N) $$\sum_{j \notin \{a,b\}} q_{aj} \wedge q_{bj} < \infty.$$

Note. The 'N' is in deference to Neveu, in whose work this condition is implicit but nowhere explicit.

Proof of Corollary 1. Since E is Hausdorff, there exist disjoint open subsets G_a, G_b of E with $a \in G_a$, $b \in G_b$. Then

$$\sum_{j \notin \{a,b\}} q_{aj} \wedge q_{bj} \leqslant Q(a, I \backslash G_a) + Q(b, I \backslash G_b) < \infty. \qquad \square$$

(54.4) COROLLARY 2. *Suppose that H is a finite subset of I such that*

(54.5) $$\liminf_j \sum_{h \in H} q_{hj} > 0.$$

Then every state i in $I \backslash H$ is stable.

Note. The meaning of 'lim inf' should be obvious: for some $\varepsilon > 0$,

$$\sum_{h \in H} q_{hj} < \varepsilon \quad \text{for only finitely many } j.$$

Proof of Corollary 2. It is clearly enough to suppose that I is infinite and H is *minimal* subject to the requirement (54.5). Then every state in H is instantaneous (by DK2)), and is therefore a point of accumulation of I. Let G be an open subset of E that contains H. Then Lemma 54.2 and the condition (54.5) imply that $E \backslash G$ contains only finitely many points of I. Thus I is homeomorphic to the disjoint union of $|H|$ copies of $\{1, 2, 3, \ldots; \infty\}$ and is already *compact*: $I = F = E$. (Thus X takes all its values in I.) Any point i of $I \backslash H$ is isolated in the Ray–Knight topology, and is therefore stable. $\square$

It is important that

(54.6) *under the hypotheses of (54.4), we can add that*

(54.7) $$\sum_{j \in I \backslash \{i\}} q_{ij} = q_i \quad (< \infty), \quad \forall i \in I \backslash H.$$

This follows because, since X takes all its values in I, X must exist the stable state i in $I\backslash H$ by jumping to another state of I. (Since $\{P_t\}$ is honest, X cannot jump to ∂; but the real point is that there are no 'fictitious' states in $E\backslash I$ to which X can jump from i.) We are sure that you already know that (54.7) is equivalent to the statement that X exits i by jumping to another state of I. (If not, wait for something very much better in Section 57.) □

55. Totally instantaneous Q-matrices. Recall that the conditions

(DK1) $$0 \leqslant q_{ij} < \infty, \quad \forall i, j : i \neq j,$$

(N) $$\sum_{j \notin \{a,b\}} q_{aj} \wedge q_{bj} < \infty, \quad \forall a, b : a \neq b,$$

hold.

Suppose now that every state of I is instantaneous:

(TI) $$q_i = -q_{ii} = \infty, \quad \forall i \in I.$$

Then we can argue that the following 'safety' condition holds:

(S): There exists an *infinite* subset J of I with

$$Q(i, J\backslash i) := \sum_{j \in J\backslash i} q_{ij} < \infty, \quad \forall i \in I.$$

'J is a large set (comparatively) safe from hits'.

Proof of (S) (Williams [9]). Label I as $\mathbb{N}$. Since (TI) holds, it follows from (54.4) that there is an infinite set $J = \{j(1), j(2), \ldots\} \subset I$ such that $j(n) > n$ and

$$\sum_{i \leqslant n} q_{i,j(n)} < 2^{-n}.$$

Then

$$Q(i, J\backslash i) < \infty, \quad \forall i.$$ □

It was known back in 1967 (see Williams [10]) that if Q is a Q-matrix that satisfies (TI) and (DK1) then the conditions (N) and (S) hold. Swayed by the then-prevalent belief that totally instantaneous chains are impossibly complicated, DW spent the next seven years trying to find additional necessary conditions. He was then somewhat annoyed to discover that there are none.

(55.1) THEOREM. *Let Q be an $I \times I$ matrix satisfying (TI) and (DK1). Then $Q = P'(0)$ for some transition function $\{P(t)\}$ on I if and only if the conditions (N) and (S) hold; and then $\{P(t)\}$ may be chosen to be honest.*

It is obvious that the 'if' part is proved by bare-hands construction of a suitable $\{P(t)\}$. (See Williams [9].)

Theorem 55.1 is all very well as an analytical result, but it does not probe deeply into the *probabilistic* structure of chains. For a much more challenging problem than that solved by Theorem 55.1, see Section 60.

Note that the easiest way to guarantee conditions (N) and (S) is to take $q_{ij}=0$ $(i \neq j)$. Thus the Feller–McKean Q-matrix is the *most* 'likely' candidate for a TI Q-matrix, not the least likely, as was once thought. This explains remarks at the end of Section 53.

56. Last exits. We now wish to prove the following result, which is 'dual' to (52.13): for $b, j \in I$ with $j \neq b$, there exists a continuous function $g_{bj}(\cdot)$ on $[0, \infty)$ such that

$$p_{bj}(t) = \int_0^t p_{bb}(s) g_{bj}(t-s)\, ds. \tag{56.1}$$

One interpretation is provided by the dual of (52.14):

$$\mathbf{P}^b[\sigma_b(t) \in ds;\ X(t) = j] = p_{bb}(s) g_{bj}(t-s)\, ds, \tag{56.2}$$

where

$$\sigma_b(t) := \sup\{s \leqslant t : X(s) = b\}. \tag{56.3}$$

The intuitively obvious (but very unfashionable!) thing to do is to derive these results from those in Section 5.2 by time reversal. We used the same idea in connection with the Chung–Getoor–Sharpe description of equilibrium charge in terms of (spatial) last-exit distribution.

Since the 'hat' notation is standard both for Laplace transforms and for time reversal, let us follow the notation of Doob [2], *using the 'tilde' notation for Laplace transforms (as we did in Section 52), and the 'star' notation for time-reversed processes.*

It is appropriate that we should follow Doob's paper [2], because we now make rather trivial use of a method fully developed there.

Let ξ be an exponentially distributed random variable of rate $\alpha > 0$ and independent of X. Let Y^* and X^* be $(E \cup \partial^*)$-valued processes defined as follows:

$$Y^*(t) := X(\xi - t) \quad (0 < t \leqslant \xi); \qquad Y^*(t) := \partial^* (t > \xi); \qquad X^*(t) := Y^*(t+).$$

By Nagasawa's formula, the process X^* under the $\mathbf{P}^b$ law is Markovian with transition matrix function $\{p^*_{ij}(t)\}$ satisfying

$$p^*_{ij}(t) := e^{-\alpha t} p_{ji}(t) \tilde{p}_{bj}(\alpha) / \tilde{p}_{bi}(\alpha). \tag{56.4}$$

Apply the first entrance decomposition to X^* to obtain

$$\tilde{p}^*_{jb}(\lambda) = \tilde{f}^*_{jb}(\lambda) \tilde{p}^*_{bb}(\lambda),$$

whence

$$\tilde{p}_{bj}(\lambda + \alpha) = \tilde{f}^*_{jb}(\lambda) \tilde{p}_{bb}(\lambda + \alpha) \tilde{p}_{bj}(\alpha) / \tilde{p}_{bb}(\alpha).$$

Inversion of the Laplace transform yields (56.1), with

$$g_{bj}(t) := e^{\alpha t} f^*_{jb}(t)\tilde{p}_{bj}(\alpha)/\tilde{p}_{bb}(\alpha).$$

It is formally obvious how (56.2) follows from (52.14). However, one has to be rather careful here because of the difficulty mentioned earlier than X^* need not be strong Markov. Recall that the Ray–Knight compactification induced by X^* may be quite different from that induced by X, that we need Doob's double (or simultaneous) compactification to do things properly, and so on....

Of course, our proof that (for $j \neq b$) there exists a continuous function g_{bj} on $[0, \infty)$ satisfying (56.1) is totally rigorous. We could have made the argument independent of the concept of time reversal by making (56.4) an analytical definition of a transition matrix function $\{p^*_{ij}(t)\}$.

Important exercise. Deduce from (56.1) that

(56.5) $$g_{bj}(0) = q_{bj}.$$

Next prove that, for $s > 0$ and $t \geqslant 0$,

(56.6) $$g_{bj}(s+t) = \sum_{i \neq b} g_{bi}(s)\,{}_bp_{ij}(t) \quad (j \neq b),$$

and deduce that

(56.7) $$g_{bj}(t) = \sum_{i \neq b} q_{bi}\,{}_bp_{ij}(t) \quad (j \neq b).$$

(*Hint for (56.6)*: compare (52.6).) Remembering that $\{P(t)\}$ is assumed honest, show that (56.1) implies that

(56.8) $$1 - p_{bb}(t) = \int_0^t p_{bb}(s) g_b(t-s)\, ds,$$

where

(56.9) $$g_b(t) := \sum_{j \neq b} g_{bj}(t) \quad (t > 0).$$

Deduce from (56.6) that g_b *is non-increasing*; then from (56.8) that g_b *is finite on* $(0, \infty)$; then from (56.6) that g_b *is continuous on* $(0, \infty)$. Use (56.8) to show that

(56.10) $$g_b(0) := g_b(0+) = q_b \leqslant \infty.$$

Note that (56.7) does not necessarily extend to $t = 0$. Let ν_b be the measure on $(0, \infty]$ defined by

(56.11) $$\nu_b(t, \infty] := g_b(t) \quad (t > 0).$$

Deduce from (56.8) that, for $\lambda > 0$,

(56.12) $$\tilde{p}_{bb}(\lambda) = \left[\lambda + \int_{(0,\infty]} (1 - e^{-\lambda l})\nu_b(dl) \right]^{-1}.$$

57. Excursions from b. Neveu's paper [2–5] are perhaps the finest written on chains. Our concern here is to describe what is in modern terminology Neveu's description of the Itô excursion law from the point b of I.

Let

(57.1) $$L_b(t) := \text{meas}\,\{s \leqslant t : X_s = b\},$$

and, for $\tau \geqslant 0$, write

(57.2) $$\rho(\tau) := \inf\{t : L_b(t) > \tau\} \leqslant \infty.$$

The strong Markov property shows that $\{\rho(\tau) : \tau \geqslant 0 : \mathbf{P}^b\}$ is a subordinator in the sense of Section II.37, but with a slight generalisation (which you can easily make) to allow ρ to take the value ∞. Exactly as in Section II.37, we find that, for $\lambda > 0$,

(57.3) $$\mathbf{E}^b[e^{-\lambda\rho(\tau)}] = e^{-\tau\Psi(\lambda)}$$

for some function Ψ. Hence

(57.4) $$\Psi(\lambda)^{-1} = \mathbf{E}^b \int_0^\infty e^{-\lambda\rho(\tau)}\,d\tau = \mathbf{E}^b \int_0^\infty e^{-\lambda t}\,dL_b(t) = \tilde{p}_{bb}(\lambda),$$

so that, from (56.12),

(57.5) $$\Psi(\lambda) = \lambda + \int_{(0,\infty]} (1 - e^{-\lambda t})\nu_b(dt).$$

The Lévy–Itô formula (II.37.4) now shows that we can write

(57.6) $$\rho(\tau) = \tau + \int_{(0,\infty]} l N((0,\tau] \times dl)$$

where N is a Poisson measure on $(0,\infty) \times (0,\infty]$ with expectation measure $d\tau\,\nu_b(dl)$. Of course, $\rho(\tau)$ is finite if and only if no atom of N lies in $(0,\tau] \times \{\infty\}$. It is therefore clear that

$$\mathbf{P}^b[\rho(\tau) < \infty] = \mathbf{P}^b[L_b(\infty) > \tau] = \exp(-\tau\nu_b\{\infty\}).$$

The jumps made by $\rho(\cdot)$ correspond to the lengths of the excursions made by X from b, so our description of the Itô excursion law of X at b must be consistent with (57.6).

(57.7) *Excursion space.* Let U be the *excursion space* of Skorokhod maps e from $[0,\infty)$ to $E \cup \partial$ such that

(i) if $e(s) = \partial$ then $e(t) = \partial$, $\forall t \geqslant s$;
(ii) $e(s) \neq b$ for $s > 0$.†

---

† An excursion of X from b is to be considered as killed (sent to ∂) at its lifetime, so it cannot return to b.

Define $\zeta_e := \inf\{t : e(t) = \partial\}$. Let $\mathscr{U}^\circ$ be the smallest σ-algebra on U measuring each projection $e \mapsto e(t)$.

(57.8) THEOREM (Neveu). *There exists a unique σ-finite measure n on $(U, \mathscr{U}^\circ)$ such that, for $0 < t_1 < t_2 < \cdots < t_m$ and $i_1, i_2, \ldots, i_m \in I \backslash b$,*

$$n\{e \in U : e(t_k) = i_k \ (1 \leqslant k \leqslant m)\}$$
$$= g_{bi_1}(t_1)\, {}_bp_{i_1 i_2}(t_2 - t_1) \cdots {}_bp_{i_{m-1} i_m}(t_m - t_{m-1}).$$

The following statements hold:

(57.9) $$n(U) = q_b, \quad n\{e : e(0) = j\} = q_{bj} \quad (j \in I \backslash b);$$

(57.10) $$n\{e : \zeta_e > t\} = g_b(t), \quad n\{e : e(t) = j\} = g_{bj}(t) \quad (j \in I \backslash b);$$

(57.11) $$n\{e : \zeta_e \in dt\} = \nu_b(dt) = \eta_b(t)\, dt \quad on \ (0, \infty),$$

where $\eta(t)$ $(t > 0)$ is defined independently of s in $(0, t)$ by

(57.12) $$\eta(t) := \sum_{j \in I \backslash b} g_{bj}(s) f_{jb}(t - s).$$

The measure n is the Itô excursion law of X at b in the sense explained in (57.13) below.

The business of assigning credit is terribly tortuous. What is certain is that Neveu's papers have received much less credit than they deserve. However, we must be careful not to overcompensate and thereby do injustice to later work. (See, for example, Freedman's interpretation of q_{bj} described below.) We are expressing our belief that *if* Itô excursion laws had been discovered when Neveu was writing on chains, then he would have expressed his ideas in the form of Theorem 57.8. (Recall what Gauss did with ideal-class groups!)

(57.13) *The role of n as Itô excursion law from b.* Part 8 of Chapter VI in Volume 2 is an extensive study of Itô excursion theory; and you will have to look there for proofs, and for much fuller explanation, of the statements now to be made. We describe how a process Z with the $\mathbf{P}^b$ law of X may be built out of excursions from b.

Construct, as in Section II.37, a Poisson random measure Λ on

$$((0, \infty) \times U, \mathscr{B}(0, \infty) \times \mathscr{U}^\circ)$$

with intensity measure Leb $\times\, n$. Let A be the set of atoms of Λ, the 'points' of the Poisson point process. A typical point of A is, of course, a pair (σ, e) where $\sigma \in (0, \infty)$ and e is an element of U with lifetime $\zeta_e \leqslant \infty$.

For $\tau \geqslant 0$, define

$$\gamma(\tau) = \sum \{\zeta_e : (\sigma, e) \in A : \sigma \leqslant \tau\}.$$

Because of (56.11), (56.12), (57.6), (57.10) and (57.11), γ is a subordinator identical in law to the 'inverse local time' process $(\rho, \mathbf{P}^b)$. The jumps of γ correspond to excursions of Z from b, and it remains only to interpolate within these excursions. For $t \geqslant 0$, we define

$$Z(t) := e(t - \gamma(\tau -))$$

if for some $(e, \tau) \in A$, we have $\gamma(\tau -) \leqslant t < \gamma(\tau)$; define $Z(t) := b$ otherwise. Then Z has the $\mathbf{P}^b$ law of X. In particular, if C is a measurable subset of U then the local time at b for X before the first excursion from b that lies in C has the exponential distribution with rate parameter $n(C)$.

The second equation in (57.9) shows that if we set T_{bj} for the time of the first jump made by X from b to j,

$$T_{bj} := \inf\{t > 0 : X(t-) = b, X(t) = j\},$$

then

(57.14)
$$\mathbf{P}^b[L_b(T_{bj}) > \tau \mid T_{bj} < \infty] = \exp(-\tau q_{bj}).$$

This is the interpretation of q_{bj} discovered by Freedman [2].

Another (and very closely related) interpretation of q_{bj} is provided by the theory of Lévy kernels. Let $J_{bj}(t)$ be the number of jumps made by X from b to j during time t. Then the idea of q_{bj} as 'jump intensity' is perfectly captured by the statement that

(57.15) $J_{bj}(t) - q_{bj} L_b(t)$ *is a martingale (relative to* $(\{\mathscr{F}_t^\circ\}, \mathbf{P}^\mu)$ *for every* μ*).*

In particular, for every probability measure μ on E,

(57.16)
$$\mathbf{E}^\mu J_{bj}(t) = \int_E \mu(dx) \int_0^t p_{xb}(s) q_{bj}\, ds.$$

The object of *Lévy kernel theory* is to describe (simultaneously) *all* the jumps of X. The Q-matrix Q, which describes the I-to-I jumps, is just the restriction to $I \times \mathscr{B}(I)$ of the Lévy kernel of X, which describes all possible E-to-E jumps. See Benveniste and Jacod [1] and Volumes IV and V of Dellacherie and Meyer [1].

58. Kingman's solution of the 'Markov characterisation problem'. Kingman calls a function p on $[0, \infty)$ a *Markov p-function* if there exists a ('standard') transition function on a countable set I and a state b in I such that $p_{bb}(t) = p(t)$, $\forall t$.

(58.1) THEOREM (Kingman). *A continuous function p on* $[0, \infty)$ *is a Markov p-function if and only if its Laplace transform may be written as*

$$\tilde{p}(\lambda) = \left[\lambda + \int_{(0,\infty)} (1 - e^{-\lambda t}) \eta(t)\, dt + \nu(\{\infty\}) \right]^{-1} \quad (\lambda > 0)$$

where $\nu(\{\infty\}) \geqslant 0$ *and where* η *is a lower-semicontinuous function on* $(0, \infty)$ *such that either*

(i) $\eta(t) = 0, \forall t$, *or*
(ii) for some a in $(0, \infty)$,

$$\eta(t) > 0 \quad (0 < t < 1), \qquad \eta(t) > e^{-at} \quad (t \geqslant 1).$$

You can see that the 'only if' part of the theorem is largely a consequence of Neveu's work. The 'if' part is surprising and very much more difficult, and Kingman's proof of it is a splendid *tour de force*. You will find this proof and much else of interest in Kingman's book [3].

59. Symmetrisable chains. Our paper, Rogers and Williams [2], was designed to advertise the power of *Dirichlet-form theory* initiated by Beurling and Deny and spectacularly developed for use in probability theory by Fukushima [1], Silverstein [1, 2]. We proved the following theorem by a finite-state approximation technique originally used by Reuter and Ledermann for birth-and-death processes.

(*59.1*) THEOREM. *Let Q be an* $I \times I$ *matrix such that*

(59.2) $$q_{ij} \geqslant 0 \quad (i \neq j), \qquad \sum_{k \neq i} q_{ik} = -q_{ii} \leqslant \infty, \quad \forall i,$$

and such that Q is m-symmetrisable in that

$$m_i q_{ij} = m_j q_{ji} \quad (i, j \in I)$$

for some strictly positive numbers $(m_i : i \in I)$. *Then there exists a standard transition function* $\{P(t)\}$ *on I with Q-matrix Q and m-symmetrisable in that*

$$m_i p_{ij}(t) = m_j p_{ji}(t) \quad (i, j \in I; t \geqslant 0)$$

if and only if

(59.3) $$\mathscr{D}(\mathscr{E}) := \{f \in \ell^2(m) : \mathscr{E}(f, f) < \infty\} \quad \text{is dense in } \ell^2(m)$$

where $\mathscr{E}$ *is the* Dirichlet form *or* energy norm *associated with Q*:

$$\mathscr{E}(f, f) := \sum_i \sum_j m_i q_{ij}(f_j - f_i)^2.$$

Though, in general, $\{P(t)\}$ *is by no means unique, there is a 'canonical'* $\{P(t)\}$ *with the properties described. If m is a finite measure, the canonical* $\{P(t)\}$ *is honest.*

It was too glibly stated in our paper that if every q_{ii} is finite then the canonical $\{P(t)\}$ process corresponds to the chain reflected off its Martin boundary. Ivor McGillivray has pointed out that we should have said 'reflected off its Kuramochi boundary', the Kuramochi boundary being a boundary analogous to the Martin boundary (and agreeing with it in the cases most frequently encountered) but specially tailored for symmetrisable processes.

60. An open problem. Here, to end with, is a problem we should like to solve.

Suppose that m is a probability measure on I, and that Q is an $I \times I$ matrix satisfying (59.2) and also

$$\sum_{j \neq i} m_j q_{ji} = -m_i q_{ii} \leqslant \infty, \quad \forall i. \tag{60.1}$$

When does there exist a (positive-recurrent) chain X with Q-matrix Q and with m as invariant measure?

The condition (59.3) is necessary—see Rogers and Williams [2].

References for Volumes 1 and 2

ABRAHAMS, R. and ROBBIN, J,
[1] *Transversal Mappings and Flows*, Benjamin, New York, Amsterdam, 1967.
ADLER, R. J.
[1] *The Geometry of Random Fields*, Wiley Chichester, 1981.
[2] *An Introduction to Continuity, Extrema, and Related Topics for General Gaussian Processes*, IMS Lecture Notes—Monograph Series Vol. 12, IMS, Hayward, Calif., 1990.
AIZENMANN, M. and SIMON, B.
[1] Brownian motion and the Harnack inequality for Schrödinger operators, *Comm. Pure and Appl. Math.*, **35**, 209–273 (1982).
ALBEVERIO, S., BLANCHARD, P. and HØEGH-KROHN, R.
[1] Newtonian diffusions and planets, with a remark on non-standard Dirichlet forms and polymers, *Stochastic Analysis and Applications: Lecture Notes in Mathematics 1095*, Springer, Berlin, 1984, pp. 1–24.
ALBEVERIO, S., FENSTAD, I.E., HØEGH-KROHN, R. and LINDSTRÖM, T.
[1] *Non-standard Methods in Probability and Mathematical Physics*, Academic Press, New York (1986).
ALDOUS, D. J.
[1] Stopping times and tightness, *Ann. Prob.*, **6**, 335–40 (1978).
ANCONA, A.
[1] Negatively curved manifolds, elliptic operators and Martin boundary *Ann. Math.*, **125**, 495–536 (1987).
ARNOLD, L. and WIHSTUTZ, V. (editors)
[1] *Lyapunov Exponents (Proceedings): Lecture Notes in Mathematics 1186*, Springer, Berlin, 1986.
AZÉMA, J.
[1] Sur les fermés aléatoires, *Séminaires de Probabilités XIX: Lecture Notes in Mathematics 1123*, Springer, Berlin, 1985, pp. 297–495.
AZÉMA, J. and YOR, M.
[1] Une solution simple au problème de Skorokhod, *Séminaire de probabilités XIII: Lecture Notes in Mathematics 721*, Springer, Berlin, 1979, pp. 90–115, 625–633.
[2] (editors) Temps locaux, *Astérisque* **52–53** Société Mathématique de France (1978).
[3] Etude d'une martingale remarquable, *Séminaire de Probabilités XXIII: Lecture Notes in Mathematics 1372*, Springer, Berlin, 1989, pp. 88–130.
AZENCOTT, R.
[1] Grandes déviations et applications, *Ecole d'Été de Probabilités de Saint-Flour VIII: Lecture Note in Mathematics 774*, Springer, Berlin, 1980.

BARLOW, M. T.
[1] Study of a filtration expanded to include an honest time, *Z. Wahrscheinlichkeitstheorie*, **44**, 307–323 (1978).
[2] Decomposition of a Markov process at an honest time (unpublished).
[3] One dimensional stochastic differential equation with no strong solution, *J. London Math. Soc.*, **26**, 335–347 (1982).
[4] On Brownian local time, *Séminaire de Probabilités XV: Lecture Notes in Mathematics 850*, Springer, Berlin, 1981, pp. 189–190.
[5] Necessary and sufficient conditions for the continuity of local time of Lévy processes, *Ann. Prob.* **16**, 1389–1427 (1988).
BARLOW, M. T. and HAWKES, J.
[1] Application d'entropie métrique à la continuité des temps locaux des processus de Lévy. *C.R. Acad. Sci. Paris Ser. I*, **301**, 237–239 (1985).
BARLOW, M. T., JACKA, S. and YOR, M.
[1] Inequalities for a pair of processes stopped at a random time, *Proc. London Math. Soc.*, **52**, 142–172 (1986).
[2] Inégalities pour un couple de processus arrêtes à un temps quelconque, *C.R. Acad. Sci.*, **299**, 351–354 (1984).
BARLOW, M. T. and PERKINS, E.
[1] One-dimensional stochastic differential equations involving a singular increasing process, *Stochastics*, **12**, 229–249 (1984).
[2] Strong existence, uniqueness and non-uniqueness in an equation involving local time, *Séminaire de Probabilités XVII: Lecture Notes in Mathematics 986*, Springer, Berlin, 1983, pp. 32–66.
BARLOW, M. T. and YOR, M.
[1] (Semi-) martingale inequalities and local times, *Z. Wahrscheinlichkeitstheorie* **55**, 237–254 (1981).
[2] Semi-martingale inequalities via the Garsia-Rodemich-Rumsey lemma and applications to local times, *J. Funct. Anal.*, **49**, 198–229 (1982).
BASS, R. and CRANSTON, M.
[1] The Malliavin calculus for pure jump processes and applications to local time, *Ann. Prob.*, **14**, 490–532 (1986).
BATCHELOR, G. K.
[1] Kolmogoroff's theory of locally isotropic turbulence, *Proc. Camb. Phil. Soc*, **43**, 553–559 (1947).
BAXENDALE, P.
[1] Asymptotic behaviour of stochastic flows of diffeomorphisms; two case studies, *Prob. Th. Rel. Fields*, **73**, 51–85 (1986).
[2] Moment stability and large deviations for linear stochastic differential equations, *Proc. Taniguchi Symposium on Probabilistic Methods in Mathematical Physics, Katata and Kyoto, 1985* (ed. N. Ikeda), Kinokuniya, Tokyo, 31–54 (1986).
[3] The Lyapunov spectrum of a stochastic flow of diffeomorphisms, in Arnold and Wihstutz [1], pp. 322–337 (1986).
[4] Brownian motions on the diffeomorphism group, I, *Compos. Math.*, **53**, 19–50 (1984).
BAXENDALE, P. and HARRIS, T. E.
[1] Isotropic stochastic flows. *Ann. Prob.*, **14**, 1155–1179 (1986).
BAXENDALE, P. and STROOCK, D. W.
[1] Large deviations and stochastic flows of diffeomorphisms, *Prob. Th. Rel. Fields*, **80**, 169–215 (1988).

BENSOUSSAN, A.
[1] Lectures on stochastic control, *Nonlinear Filtering and Stochastic Control: Lecture Notes in Mathematics 972*, Springer, Berlin, 1982, pp. 1–62.
BENEŠ, V. E., SHEPP, L. A. and WITSENHAUSEN, H. S.
[1] Some solvable stochastic control problems, *Stochastics*, **4**, 39–83 (1980).
BENVENISTE, A. and JACOD, J.
[1] Systèmes de Lévy des processus de Markov, *Invent. Math.*, **21**, 183–198 (1973).
BERMAN, S. M.
[1] Local times and sample function properties of stationary Gaussian processes, *Trans. Amer. Math. Soc.*, **137**, 277–300 (1969).
[2] Harmonic analysis of local times and sample functions of Gaussian processes, *Trans. Amer. Math. Soc.*, **143**, 269–281 (1969).
[3] Gaussian processes with stationary increments: local times and sample function properties, *Ann. Math. Statist.*, **41**, 1260–1272 (1970).
BIANE, P.
[1] Comparaison entre temps d'atteinte et temps de séjour de certaines diffusions réelles, *Seminaire de Probabilités XIX*, *Lecture Notes in Mathematics 1123*, Springer, Berlin, 1985, pp. 291–296.
BICHTELER, K.
[1] Stochastic integration and L^p-theory of semi-martingales, *Ann. Prob.*, **9**, 49–89 (1981).
BICHTELER, K. and FONKEN, D.
[1] A simple version of the Malliavin calculus in dimension one, *Martingale Theory in Harmonic Analysis and Banach Spaces: Lecture Notes in Mathematics 939*, Springer, Berlin, 1982, pp. 6–12.
BICHTELER, K. and JACOD, J.
[1] Calcul de Malliavin pour les diffusions avec sauts: Existence d'une densité dans le cas unidimensionnel, *Séminaire de Probabilités XVII: Lecture Notes in Mathematics 986*, Springer, Berlin, 1983, pp. 132–157.
BILLINGSLEY, P.
[1] *Ergodic Theory and Information*, Wiley, New York, 1965.
[2] *Convergence of Probability Measures*, Wiley, New York, 1968.
[3] Conditional distributions and tightness, *Ann. Prob.*, **2**, 480–485 (1974).
BINGHAM, N. H.
[1] Fluctuation theory in continuous time, *Adv. Appl. Prob.*, **7**, 705–766 (1975).
BINGHAM, N. H. and DONEY, R. A.
[1] On higher-dimensional analogues of the arc-sine law, *J. Appl. Prob.* **25**, 120–131 (1988).
BISHOP, R. and CRITTENDEN, R. J.
[1] *Geometry of Manifolds*, Academic Press, New York, 1964.
BISMUT, J.-M.
[1] *Méchanique Aléatoire: Lecture Notes in Mathematics 866*, Springer, Berlin, 1981.
[2] Martingales, the Malliavin calculus and hypoellipticity under general Hörmander's conditions, *Z. Wahrscheinlichkeitstheorie*, **56**, 469–505 (1981).
[3] Calcul de variations stochastiques et processus de sauts, *Z. Wahrscheinlichkeitstheorie* **56**, 469–505 (1983).
[4] Large deviations and the Malliavin calculus, *Progress in Mathematics*, Birkhäuser, Boston, 1984.
[5] The Atiyah–Singer theorems; a probabilistic approach: I, The index theorem, *J. Funct. Anal.*, **57**, 56–98 (1984); II, The Lefschetz fixed-point formulas, *ibid*, 329–348.

BISMUT, J.-M. and MICHEL, D.

[1] Diffusions conditionnelles, I, II, *J. Funct. Anal.*, **44**, 174–211 (1981), **45**, 274–292 (1981).

BLACKWELL, D. and KENDALL, D. G.

[1] The Martin boundary for Polya's urn scheme and an application to stochastic population growth, *J. Appl. Prob.* **1**, 284–296 (1964).

BLUMENTHAL, R. M. and GETOOR, R. K.

[1] *Markov Processes and Potential theory*, Academic Press, New York, 1968.

[2] Local times for Markov processes. *Z. Wahrscheinlichkeitstheorie verw. Geb.*, **3**, 50–74 (1964).

BONDESSON, L.

[1] Classes of infinitely divisible distributions and densities. *Z. Wahrscheinlichkeitstheorie verw Geb.*, **57**, 39–71 (1981).

BOUGEROL, P. and LACROIX, J.

[1] *Products of Random Matrices with Applications to Schrödinger Operators*, Birkhauser, Boston, 1985.

BOURBAKI, N.

[1] Topologie générale, in *Eléments de Mathématique*, Hermann, Paris, 1958, Chap. IX, 2nd edition.

BREIMAN, L.

[1] *Probability*, Addison-Wesley, Reading, Mass., 1968.

BRÉMAUD, P.

[1] *Point Processes and Queues: Martingale Dynamics*, Springer, New York, 1981.

BRETAGNOLLE, J.

[1] Résultats de Kesten sur les processus à accroissements indépendantes, *Séminaire de Probabilités V, Lecture Notes in Mathematics 191*, Springer, Berlin, 1971, pp. 21–36.

BRYDGES, D., FRÖHLICH, J. and SPENCER, T.

[1] The random walk representation of classical spin systems and correlation inequalities. *Comm. Math. Phys.*, **83**, 123–150 (1982).

BURDZY, K.

[1] On nonincrease of Brownian motion. *Ann. Prob.* **18**, 978–980 (1990).

[2] Brownian paths and cones, *Ann. Prob.* **13**, 1006–1010 (1985).

[3] Cut points on Brownian paths. *Ann. Prob.* **17**, 1012–1036 (1989).

BURKHOLDER, D.

[1] Distribution function inequalities for martingales, *Ann. Prob.*, **1**, 19–42 (1973).

CARLEN, E. A.

[1] Conservative diffusions, *Comm. Math. Phy.*, **94**, 293–315 (1984).

[2] Potential scattering in quantum mechanics, *Ann. Inst. H. Poincaré*, **42**, 407–428 (1985).

CARVERHILL, A. P.

[1] Flows of stochastic dynamical systems: ergodic theory, *Stochastics*, **14**, 273–318 (1985).

[2] A formula for the Lyapunov exponents of a stochastic flow. Application to a perturbation theorem, *Stochastics*, **14**, 209–226 (1985).

[3] A nonrandom Lyapunov spectrum for nonlinear stochastic dynamical systems, *Stochastics*, **17**, 209–226, 1986.

CARVERHILL, A. P., CHAPPELL, M. J. and ELWORTHY, K. D.

[1] Characteristic exponents for stochastic flows, *Proceedings, BIBOS I: Stochastic Processes.*

CARVERHILL, A. P. and ELWORTHY, K. D.

[1] Flows of stochastic dynamical systems: the functional analytic approach, *Z. Wahrscheinlichkeitstheorie*, **65**, 245–268 (1983).

CHALEYAT-MAUREL, M.
[1] La condition d'hypoellipticité d'Hörmander, *Astérisque*, **84–85**, 189–202 (1981).
CHALEYAT-MAUREL, M. and EL KAROUI, N.
[1] Un problème de réflexion et ses applications au temps local et aux équations différentielles stochastiques sur **R**, case continu. In Azema and Yor [2], pp. 117–144.
CHEEGER, J. and EBIN, D. G.
[1] *Comparison Theorems in Riemannian Geometry*, North-Holland, Amsterdam, 1975.
CHUNG, K. L.
[1] *Markov Chains with Stationary Transition Probabilities*, 2nd edition, Springer, Berlin, 1967.
[2] Probabilistic approach in potential theory to the equilibrium problem, *Ann. Inst. Fourier, Grenoble*, **23**, 313–322 (1973).
[3] Excursions in Brownian motion, *Ark. Mat.*, **14**, 155–177 (1976).
CHUNG, K. L. and GETOOR, R. K.
[1] The condenser problem, *Ann. Prob.*, **5**, 82–86 (1977).
CHUNG, K. L. and WALSH, J. B.
[1] To reverse a Markov process, *Acta Math.*, **123**, 225–251 (1969).
[2] Meyer's theorem on previsibility, *Z. Wahrscheinlichkeitstheorie*, **29**, 253–256 (1974).
CHUNG, K. L. and WLLIAMS, R. J.
[1] *Introduction to Stochastic Integration* Birkhäuser, Boston, 1983.
CIESIELSKI, Z.
[1] Hölder conditions for realisations of Gaussian processes. *Trans. Amer. Math. Soc.*, **99**, 403–413 (1961).
CIESIELSLKI, Z. and TAYLOR, S. J.
[1] First passage times and sojourn times for Brownian motion in space and the exact Hausdorff measure of the sample path, *Trans. Amer. Math. Soc.*, **103**, 434–450 (1962).
ÇINLAR, E., CHUNG, K. L. and GETOOR, R. K. (editors)
[1] *Seminars on Stochastic Processes 1981, 1982; 1983, 1984* (four volumes), Birkhäuser, Boston, 1982, 1983, 1984, 1985.
ÇINLAR, E, CHUNG, K. L., GETOOR, R. K. and GLOVER, J. (editors)
[1] *Seminar on Stochastic Processes 1986*, Birkhäuser, Boston, 1987.
ÇINLAR, E., JACOD, J., PROTTER, P. and SHARPE, M. J.
[1] Semimartingales and Markov processes, *Z. Wahrscheinlichkeitstheorie*, **54**, 161–220 (1980).
CLARK, J. M. C.
[1] The representation of functionals of Brownian motion by stochastic integrals, *Ann. Math. Stat.*, **41**, 1282–1295 (1970); **42**, 1778 (1971).
[2] An introduction to stochastic differential equations on manifolds, *Geometric Methods in Systems Theory* (eds. D. Q. Mayne and R. W. Brockett), Reidel, Dordrecht, 1973.
[3] The design of robust approximations to the stochastic differential equations of nonlinear filtering, *Communications Systems and Random Process Theory* (ed. J. Skwirzynski), Sijthoff and Noordhoff, Alphen aan den Rijn, 1978.
CLARKSON, B. (editor)
[1] *Stochastic Problems in Dynamics*, Pitman, London, 1977.
COCOZZA, C. and YOR, M.
[1] Démonstration simplifiée d'un théoreme de Knight, *Séminaire de Probabilités XIV: Lecture Notes in Mathematics 721*, Springer, Berlin, 1980, pp. 496–499.
CRANK, J.
[1] *The Mathematics of Diffusion*, 2nd ed. Oxford University Press, Oxford (1975).

CRANSTON, M.
[1] On the means of approach of Brownian motion *Ann. Probab.*, **15**, 1009–1013 (1987).
CUTLAND, N.
[1] Non-standard measure theory and its applications, *Bull. London. Math. Soc.*, **15**, 529–589 (1983).
CUTLAND, N. and KENDALL, W. S.
[1] A non-standard proof of one of David Williams' splitting-time theorems, in D. G. Kendall [5], pp. 37–48.
DARLING, R. W. R.
[1] Martingales in manifolds—definition, examples, and behaviour under maps, *Séminaire de Probabilités XVI Supplement: Lecture Notes in Mathematics 921*, Springer, Berlin, 1982, pp. 217–236.
DAVIES, E. B. and SIMON, B.
[1] Ultracontractivity and the heat kernel for Schrödinger operators and Dirichlet Laplacians, *J. Funct. Anal.* **59**, 335–395 (1984).
DAVIS, B.
[1] Picard's theorem and Brownian motion, *Trans. Amer. Math. Soc.*, **213**, 353–362 (1975).
[2] Applications of the conformal invariance of Brownian motion, *Harmonic analysis in Euclidean Space. Proc. Symp. Pure Math. XXXV*, Amer. Math Soc., 303–310.
DAVIS, M. H. A.
[1] On a multiplicative functional transformation arising in non-linear filtering theory, *Z. Wahrscheinlichkeitstheorie*, **54**, 125–139 (1980).
[2] Pathwise non-linear filtering, *Stochastic Systems: the Mathematics of Filtering and Identification and Applications* (eds. M. Hazewinkel and J. C. Willems), Reidel, Dordrecht, 1981.
[3] Some current issues in stochastic control theory, *Stochastics*.
[4] *Markov Models and Optimization*, Chapman & Hall, London, 1993.
DAVIS, M. H. A. and VARAIYA, P.
[1] Dynamic programming conditions for partially observed stochastic systems, *SIAM J. Control*, **11**, 226–261 (1973).
DAWSON, D. A.
[1] Measure-valued Markov processes, *Ecole d'Eté de Probabilités de Saint-Flour XXI, 1993* (ed. P. L. Hennequin), *Lecture Notes in Mathematics 1541*, 1993.
DAWSON, D. A. and GÄRTNER, J.
[1] Large deviations from the McKean–Vlasov limit for weakly-interacting diffusions, *Stochastics*, **20**, 247–308 (1987).
DELLACHERIE, C.
[1] *Capacités et Processus Stochastiques*, Springer, Berlin, 1972.
[2] Quelques exemples familiers en probabilités d'ensembles analytiques non-Boréliens, *Séminaire de Probabilités XII: Lecture Notes in Mathematics*, Springer, Berlin, 1978, pp. 742–745.
[3] Un survoi de la theorie de l'intégrale stochastique, *Stoch. Proc. Appl.*, **10**, 115–144 (1980).
DELLACHERIE, C., DOLÉANS(-DADE), CATHERINE, LETTA, G. and MEYER, P. A.
[1] Diffusions à coefficients continus d'après D. W. Stroock et S. R. S. Varadhan, *Séminaire de Probabilités IV: Lecture Notes in Mathematics 124*, Springer, Berlin, 1970, pp. 241–282.

DELLACHERIE, C. and MEYER, P. A.
[1] *Probabilités et Potentiel*, Chaps. I–VI, Hermann, Paris, 1975; Chaps. V–VIII, Hermann, Paris, 1980; Chaps. IX–XI, Hermann, Paris, 1983; Chapters XII–XVI, Hermann, Paris, 1987; Chaps XVII–XXIV, Hermann, Paris, 1993.
DEUSCHEL, J.-D. and STROOCK, D. W.
[1] *Large Deviations*. Academic Press, Boston, 1989.
DE WITT-MORETTE, C. and ELWORTHY, K. D. (editors)
[1] New stochastic methods in physics, *Phys. Rep.*, **77**, 121–382 (1981).
DOLÉANS(-DADE), C.
[1] Existence du processus croissant natural associé à un potentiel de la classe (D), *Z. Wahrscheinlichkeitstheorie* **9**, 309–314 (1968).
[2] Quelques applications de la formule de changement de variables pour les semimartingales, *Z. Wahrescheinlichkeitstheorie*, **16**, 181–194 (1970).
DOLÉANS-DADE, C. and MEYER, P. A.
[1] Equations différentielles stochastiques, *Séminaires de Probabilités XI: Lecture Notes in Mathematics 581*, Springer, Berlin, 1977, pp. 376–382.
DONEY, R. A.
[1] On the maxima of random walks and stable processes and the arc-sine law, *Bull. London Math. Soc.*, **19**, 177–182 (1987).
[2] A path decomposition for Lévy processes, *Stoch. Proc. Appl.* **47**, 167–181 (1993).
DOOB, J. L.
[1] *Stochastic Processes*, Wiley, New York, 1953.
[2] State-spaces for Markov chains, *Trans. Amer. Math. Soc.* **149**, 279–305 (1970).
[3] *Classical Potential Theory and its Probabilistic Counterpart*, Springer, New York, 1981.
DOSS, H.
[1] Liens entre équations différentielles stochastiques et ordinaires, *Ann. Inst. Henri Poincaré B*, **13**, 99–126 (1977).
DUBINS, L. and SCHWARZ, G.
[1] On continuous martingales, *Proc. Natl. Acad. Sci. USA*, **53**, 913–916 (1965).
DUNFORD, N. and SCHWARTZ, J. T.
[1] *Linear Operators: Part I, General Theory*, Interscience, New York, 1958.
DURRETT, R.
[1] *Brownian Motion and Martingales in Analysis*, Wadsworth, Belmont, Calif. 1984.
[2] (editor) Particle systems, random media, large deviations, *Contemp. Math.* **41**, Amer. Math. Soc., Providence, RI, 1985.
[3] *Probability: Theory and Examples*, Wadsworth & Brooks Cole, Pacific Grove, Calif., 1991.
DVORETSKY, A., ERDÖS, P. and KAKUTANI, S.
[1] Double points of paths of Brownian motion in n-space, *Acta. Sci. Math. (Szeged)*, **12**, 64–81 (1950).
[2] Multiple points of paths of Brownian motion in the plane, *Bull. Res. Council Isr. Sect. F*, **3**, 364–371 (1954).
[3] Points of multiplicity c of plane Brownian paths, *Bull. Res. Council Isr. Sect. F*, **7**, 175–180 (1958).
DVORETSKY, A., ERDÖS, P., KAKUTANI, S. and TAYLOR, S. J.
[1] Triple points of Brownian motion in 3-space, *Proc. Camb. Phil. Soc.*, **53**, 856–862 (1957).
DYNKIN, E. B.
[1] *Theory of Markov Processes*, Pergamon Press, Oxford, 1960.

[2] *Markov Processes* (two volumes), Springer, Berlin, 1965.
[3] Non-negative eigenfunctions of the Laplace–Beltrami operator and Brownian motion in certain symmetric spaces (in Russian), *Dokl. Akad. Naud SSSR*, **141**, 288–291 (1961).
[4] Diffusion of tensors, *Dokl. Akad. Nauk. SSSR*, **179**, 1264–1267 (1968).
[5] Local times and quantum fields, in Çinlar, Chung and Getoor [1, 1983].
[6] Gaussian and non-Gaussian random fields associated with Markov processes, *J. Func. Anal.*, **55**, 344–376 (1984).
[7] Self-intersection local times, occupation fields and stochastic integrals, *Adv. App. Math.*, **65**, 254–271 (1987).
[8] Random fields associated with multiple points of the Brownian motion, *J. Funct. Anal.*, **62**, 397–434 (1985).
[9] Local times and quantum fields, in Çinlar, Chung and Getoor [1, 1984].

ELLIOTT, R. J.
[1] *Stochastic Calculus and Applications*, Springer, Berlin, 1982.

ELLIOTT, R. J. and ANDERSON, B. D. O.
[1] Reverse time diffusions, *Stochastic Processes and their Applications*, **19**, 327–339 (1985).

ELWORTHY, K. D.
[1] *Stochastic Differential Equations on Manifolds*, London Mathematical Society Lecture Note Series 20, Cambridge University Press, Cambridge, 1982.
[2] (editor) *From Local Time to Global Geometry, Control and Physics, Proceedings, Warwick Symposium 1984/85*, Longman, Harlow/Wiley, New York, 1986.

ELWORTHY, K. D. and STROOCK, D. W.
[1] Large deviation theory for mean exponents of stochastic flows, Appendix to Carverhill, Chappell and Elworthy [1].

ELWORTHY, K. D. and TRUMAN, A.
[1] Classical mechanics, the diffusion (heat) equation and the Schrödinger equation on a Riemannian manifold, *J. Math. Phys.*, **22**, 2144–2166 (1981).
[2] The diffusion equation and classical mechanics: an elementary formula, *Stochastic processes in quantum theory and statistical physics* (ed. S. Albeverio *et al.*), *Lecture Notes in Physics 173*, Springer, Berlin, 1982, pp. 136–146.

EMÉRY, M.
[1] Annoncabilité des temps prévisibles: deux contre-exemples, *Séminaire de Probabilités IV: Lecture Notes in Mathematics 784*, Springer, Berlin, 1980, pp. 318–323.
[2] On the Azéma martingales, *Séminaire de Probabilitiés XXIII: Lecture Notes in Mathematics 1372*, Springer, Berlin 1989 pp. 66–88.

ETHIER, S. N. and KURTZ, T. G.
[1] *Markov Processes: Characterization and Convergence*, Wiley, New York, 1986.

EVANS, S. N.
[1] On the Hausdorff dimension of Brownian cone points, *Math. Proc. Camb. Phil. Soc.*, **98**, 343–353 (1985).
[2] Multiple points in the sample paths of a Lévy process, *Prob. Th. Rel. Fields*, **76**, 359–367 (1987).

FELLER, W.
[1] *Introduction to Probability Theory and its Applications*, Vol. 1, 2nd edition Wiley, New York, 1957; Vol. 2, Wiley, New York, 1966.
[2] Boundaries induced by non-negative matrices, *Trans. Amer Math. Soc.*, **83**, 19–54 (1956).
[3] On boundaries and lateral conditions for the Kolmogorov equations, *Ann. Math., Ser. II*, **65**, 527–570 (1957).

[4] Generalized second-order differential operators and their lateral conditions, *Illinois J. Math.*, **1**, 459–504 (1957).

FLEMING, W. H. and RISHEL, R. W.

[1] *Deterministic and Stochastic Optimal Control*, Springer, Berlin, 1975.

FÖLLMER, H.

[1] Calcul d'Itô sans probabilités, *Seminaire de Probabilités XV: Lecture Notes in Mathematics 850*, Springer, Berlin, 1981, pp. 143–150.

FREEDMAN, D.

[1] *Brownian Motion and Diffusion*, Holden-Day, San Francisco, 1971.

[2] *Approximating Countable Markov Chains*, Holden-Day, San Francisco, 1972.

FRIEDMAN, A.

[1] *Stochastic Differential Equations and Applications* (two volumes), Academic Press, New York, 1975.

FRISTEDT, B.

[1] Sample functions of stochastic processes with stationary independent increments, *Adv. Prob.*, **3**, 241–396 (1973).

FUJISAKI, M., KALLIANPUR, G. and KUNITA, H.

[1] Stochastic differential equations for the non-linear filtering problem, *Osaka J. Math.*, **9**, 19–40 (1972).

FUKUSHIMA, M.

[1] *Dirichlet Forms and Markov Processes*, Kodansha, Tokyo, 1980.

[2] Basic properties of Brownian motion and a capacity on the Wiener space, *J. Math. Soc. Japan*, **36**, 161–176 (1984).

GARCIA ALVAREZ, M. A. and MEYER, P. A.

[1] Une théorie de la dualité à un ensemble polaire près: I, *Ann. Prob.*, **1**, 207–222 (1973).

GARSIA, A.

[1] *Martingale Inequalities: Seminar Notes on Recent Progress*, Benjamin, Reading, Mass, 1973.

GARSIA, A., RODEMICH, E. and RUMSEY, H. Jr

[1] A real variable lemma and the continuity of paths of some Gaussian processes. *Indiana Univ. Math. J.*, **20**, 565–578 (1970).

GEMAN, D. and HOROWITZ, J.

[1] Occupation densities, *Ann. Prob.*, **8**, 1–67 (1980).

GEMAN, D. HOROWITZ, J. and ROSEN, J.

[1] A local time analysis of intersections of Brownian paths in the plane, *Ann. Prob.*, **12**, 86–107 (1984).

GETOOR, R. K.

[1] *Markov processes: Ray Processes and Right Processes: Lecture Notes in Mathematics 440*, Springer, Berlin, 1975.

[2] Excursions of a Markov process, *Ann. Prob.*, **8**, 244–266 (1979).

[3] Splitting times and shift functionals, *Z. Wahrscheinlichkeitstheorie*, **47**, 69–81 (1979).

GETOOR, R. K. and SHARPE, M. J.

[1] Last exit times and additive functionals, *Ann. Prob.*, **1**, 550–569 (1973).

[2] Excursions of Brownian motion and Bessel process, *Z. Wahrscheinlichkeitstheorie*, **47**, 83–106 (1979).

[3] Last exit decompositions and distributions, *Indiana Univ. Math. J.*, **23**, 377–404 (1973).

[4] Excursions of dual processes, *Adv. Math.*, **45**, 259–309 (1982).

[5] Conformal martingales, *Invent Math.*, **16**, 271–308 (1972).

GIKHMAN, I. I. and SKOROKHOD, A. V.

[1] *The Theory of Stochastic Processes* (three volumes), Springer, Berlin, 1979.

GRAY, A., KARP, L. and PINSKY, M. A.
[1] The mean exit time from a tube in a Riemannian manifold, *Probability and Harmonic Analysis* (eds. J. Chao and W. Woyczynski), Dekker, 1986, pp. 113–137.
GRAY, A. and PINSKY, M. A.
[1] The mean exit time from a small geodesic ball in a Riemannian manifold, *Bull. Sci Math.*, **107**, 345–370 (1983).
GREENWOOD, P. and PERKINS, E.
[1] A conditional limit theorem for random walk and Brownian local time on square root boundaries, *Ann. Prob.* **11**, 227–261 (1982).
[2] Limit theorems for excursions from a moving boundary. *Th. Prob. Appl.* **29**, 703–714 (1984).
GREENWOOD, P. and PITMAN, J. W.
[1] Construction of local time and Poisson point processes from nested arrays, *J. London Math. Soc.* (2), **22**, 182–192 (1980).
[2] Fluctuation identities for Lévy processes and splitting at the maximum, *Adv. Appl. Prob.*, **12**, 893–902 (1980).
GRENANDER, U.
[1] *Probabilities on Algebraic Structures*, Wiley, New York, 1963.
GRIFFEATH, D.
[1] Coupling methods for Markov processes, *Advances in Mathematics Supplementary Studies: Studies in Probability and Ergodic Theory*, Vol. 2, Academic Press, New York, 1978, pp. 1–43.
GROMOV, M. and ROHLIN, V. A.
[1] *Russian Math. Surveys*, **25**, 1–57 (1970).
GROSSWALD, E.
[1] The Student *t*-distribution of any degree of freedom is infinitely divisible, *Z. Wahrsheinlichkeitscheorie verw. Geb.*, **36**, 103–109 (1976).
HALMOS, P.
[1] *Measure Theory*, Van Nostrand, Princeton, NJ, 1959.
HARRIS, T. E.
[1] Brownian motions on the homeomorphisms of the plane, *Ann. Prob.*, **9**, 232–254 (1981).
HAUSSMANN, U.
[1] On the integral representation of Itô processes, *Stochastics*, **3**, 17–7 (1979).
[2] *A Stochastic Maximum Principle for Optimal Control of Diffusions*, Longman, Harlow, 1986.
HAWKES, J.
[1] Multiple points for symmetric Lévy processes, *Math. Proc. Camb. Phil.*, **83**, 83–90 (1978).
[2] The measure of the range of a subordinator, *Bull. London Math. Soc.*, **5**, 21–28 (1973).
[3] Local times as stationary processes, *From Local to Global Geometry, Control and Physics, Research Notes in Math. 150*, Pitman, Harlow, 1986, pp. 111–120.
HAZEWINKEL, M. and WILLEMS, J. C. (editors)
[1] *Stochastic Systems: The Mathematics of Filtering and Identification and Applications*, Reidel, Dordrecht, 1981.
HELGASON, S.
[1] *Differential Geometry and Symmetric Spaces*, Academic Press, New York, 1962.
HELMS, L. L.
[1] *Introduction to Potential Theory*, Robert E. Krieger, Huntington, NY, 1975.

HILLE, E. and PHILLIPS, R. S.
[1] *Functional Analysis and Semigroups*, Amer. Math. Soc., Providence, RI, 1957.
HOLLEY, R., STROOCK, D. W. and WILLIAMS, D.
[1] Applications of dual processes to diffusion theory, *Proc. Amer. Math. Soc. Prob. Symp., Urbana*, 1976, pp. 23–36.
HÖRMANDER, L.
[1] Hypoelliptic second-order differential equations, *Acta Math.*, **117**, 147–171 (1967).
HSU, P.
[1] On excursions of reflecting Brownian motion, *Trans. Math. Soc.*, **296**, 239–264 (1986).
[2] Brownian motion and the index theorem (to appear).
HUNT, G. A.
[1] Markoff processes and potentials: I, II, III, *Illinois J. Math.*, **1**, 44–93; 316–369 (1957); **2**, 151–213 (1958).
IKEDA, N. and WATANABE, S.
[1] *Stochastic Differential Equations and Diffusion Processes*, North Holland–Kodansha, Amsterdam and Tokyo, 1981.
[2] Malliavin calculus of Wiener functionals and its applications, in Elworthy [2], pp. 132–178.
ISMAIL, M. E. and KELKER, D. H.
[1] The Bessel polynomials and the Student t-distribution, *SIAM J. Math. Anal.*, **7**, 82–91 (1976).
ITÔ, K.
[1] Stochastic integral, *Proc. Imp. Acad. Tokyo*, **20**, 519–524 (1944).
[2] On a stochastic integral equation, *Proc. Imp. Acad. Tokyo*, **22**, 32–35 (1946).
[3] Stochastic differential equations in a differential manifold, *Nagoya Math. J.*, **1**, 35–47 (1950).
[4] The Brownian motion and tensor fields on a Riemannian manifold, *Proc. Int. Congr. Math, Stockholm*, 1963, pp. 536–539.
[5] Stochastic parallel displacement, *Probabilistic Methods in Differential Equations: Lecture Notes in Mathematics 451*, Springer, Berlin, 1975, pp. 1–7.
[6] Poisson point processes attached to Markov processes, *Proc. 6th Berkeley Symp. Math. Statist. Prob.*, Vol. 3, University of California Press, Berkeley, 1971, pp. 225–240.
[7] (editor) *Proceedings of the 1982 Taniguchi Int. Symp. on Stochastic Analysis*, Kinokuniya–Wiley, 1984.
[8] Stationary random distributions. *Mem Coll. Sci. Kyoto Univ. Ser. A*, **28**, 209–223 (1954).
ITÔ, K. and MCKEAN, H. P.
[1] *Diffusion Processes and their Sample Paths*, Springer, Berlin, 1965.
JACKA, S.
[1] A finite fuel stochastic control problem, *Stochastics*, **10**, 103–113 (1983).
[2] A local time inequality for martingales, *Séminaires de Probabilités XVII: Lecture Notes in Mathematics 986*, Springer, Berlin, 1983.
JACOBSEN, M.
[1] Splitting times for Markov processes and a generalised Markov property for diffusions, *Z. Wahrscheinlichkeitstheorie*, **30**, 27–43 (1974).
[2] *Statistical Analysis of Counting Processes: Lecture Notes in Mathematics 12*, Springer, New York, 1982.
JACOD, J.
[1] A general theorem of representation for martingales, *Proc. Amer. Math. Soc. Prob. Symp.*, Urbana, 1976, 37–53.

[2] *Calcul Stochastique et Problèmes de Martingales: Lecture Notes in Mathematics 714*, Springer, Berlin, 1979.

JACOD, J. and YOR, M.

[1] Etude des solutions extrémales et représentation intégrale des solutions pour certains problèmes de martingales, *Z. Wahrscheinlichkeitstheorie*, **38**, 83–125 (1977).

JEULIN, T.

[1] *Semimartingales et Grossissement d'une Filtration: Lecture Notes in Mathematics 833*, Springer, Berlin, 1980.

JEULIN, T. and YOR, M.

[1] Grossissement d'une filtration et semi-martingales: formules explicites, *Séminaire de Probabilités XII: Lecture Notes in Mathematics 649*, Springer, Berlin, 1978, pp. 78–97.

[2] (editors) *Grossissements de Filtrations: Exemples et Applications: Lecture Notes in Mathematics 1118*, Springer, Berlin, 1985.

JOHNSON, G. and HELMS, L. L.

[1] Class (D) supermartingales, *Bull. Amer. Math. Soc.*, **69**, 59–62 (1963).

KAILATH, T.

[1] An innovations approach to least squares estimation, Part I: Linear filtering with additive white noise, *IEEE Trans. Autom. Control.* **13**, 646–655 (1968).

KALLIANPUR, G.

[1] *Stochastic Filtering Theory*, Springer, Berlin, 1980.

KARATZAS, I. SHREVE, S. E.

[1] *Brownian Motion and Stochastic Calculus*, Springer, Berlin, 1988.

KELLOGG, O. D.

[1] *Foundations of Potential Theory*, Dover, New York, 1953.

KENDALL, D. G.

[1] Pole-seeking Brownian motion and bird navigation (with discussion), *J. Roy. Statist. Soc.* **B**, **36**, 365–417 (1974).

[2] The diffusion of shape, *Adv. Appl. Prob.*, **9**, 428–430 (1979).

[3] Shape manifolds, Procrustean metrics, and complex projective spaces, *Bull. London Math. Soc.*, **16**, 81–121 (1984).

[4] A totally unstable Markov process, *Quart. J. Math. Oxford*, **9**, 149–160 (1958).

[5] (editor) *Analytic and Geometric Stochastics* (special supplement to *Adv. Appl. Prob.* to honour G. E. H. Reuter), Appl. Prob. Trust, 1986.

KENDALL, D. G. and REUTER, G. E. H.

[1] Some pathological Markov processes with a denumerable infinity of states and the associated contraction semigroups of operators on ℓ, *Proc. Int. Congr. Math. 1954 (Amsterdam)*, *3*, 377–415 (1956).

KENDALL, W. S.

[1] Knotting of Brownian motion in 3-space, *J. London Math. Soc.* (2), **19**, 378–384 (1979).

[2] Brownian motion, negative curvature, and harmonic maps, *Stochastic Integrals: Lecture Notes in Mathematics 851*, Springer, Berlin, 1981, pp. 479–491.

[3] Brownian motion on a surface of negative curvature, *Séminaire de Probabilités XVIII: Lecture Notes in Mathematics 1059*, Springer, Berlin, 1984, pp. 70–76.

[4] Survey article on stochastic differential geometry (to appear).

KENT, J.

[1] Some probabilistic properties of Bessel functions, *Ann. Prob.*, **6**, 760–770 (1978).

[2] The infinite divisibility of the von Mises–Fisher distribution for all values of the parameter in all dimensions, *Proc. London Math. Soc.*, **3**, 359–384 (1977).

[3] Continuity properties for random fields. *Ann. Prob.* **17**, 1432–1440 (1989).

KESTEN, H.
[1] Hitting probabilities of single points for processes with stationary independent increments, *Mem. Amer. Math. Soc.*, **93** (1969).
KHASMINSKII, R. Z.
[1] Ergodic properties of recurrent diffusion processes and stabilization of the solution of the Cauchy problem for parabolic equations, *Th. Prob. Appl.*, **5**, 179–196 (1960).
[2] *Stochastic Stability of Differential Equations*, Sijthoff and Noordhoff, Alphen aan den Rijn, 1980.
KIFER, Y.
[1] Brownian motion and positive harmonic functions on complete manifolds of non-positive curvature, in Elworthy [2], pp. 187–232.
KINGMAN, J. F. C.
[1] Subadditive ergodic theory, *Ann. Prob.*, **1**, 883–909 (1973).
[2] Completely random measures, *Pacific J. Math.*, **21**, 59–78 (1967).
[3] *Regenerative Phenomena*, Wiley, New York, 1972.
[4] *Poisson Processes*, Oxford University Press, Oxford, 1993.
KNIGHT, F. B.
[1] Note on regularisation of Markov processes, *Illinois, J. Math.*, **9**, 548–552 (1965).
[2] A reduction of continuous square-integrable martingales to Brownian motion, *Martingales: A Report on a Meeting at Oberwolfach* (ed. H. Dinges): *Lecture Notes in Mathematics 190*, Springer, Berlin, 1971, pp. 19–31.
[3] Random walks and the sojourn density process of Brownian motion, *Trans. Amer. Math. Soc.*, **107**, 56–86 (1963).
KNIGHT, F. B. and PITTENGER, A.O.
[1] Excision of a strong Markov process, *Z. Wahrscheinlichkeitstheorie*, **23**, 114–120 (1972).
KOBAYASHI, S. and NOMIZU, K.
[1] *Foundations of Differential Geometry* (two volumes) Wiley-Interscience, New York, 1963, 1969.
KOLMOGOROV, A. N.
[1] The local structure of turbulence in an incompressible fluid at very large Reynolds numbers, *Dokl. Akad. Nauk SSSR*, **30**, 229–303 (1941).
[2] The distribution of energy in locally isotropic turbulence. *Dokl. Akad. Nauk SSSR*, **32**, 19–21 (1941).
KOZIN, F. and PRODROMOU, S.
[1] Necessary and sufficient conditions for almost sure sample stability of linear Itô equations, *SIAM J. Appl. Math.*, **21**, 413–425 (1971).
KRYLOV, N. V.
[1] *Controlled Diffusion Processes*, Springer, New York, 1980.
KUELBS, J.
[1] The law of the iterated logarithm for Banach space valued random variables, *Probability in Banach Spaces: Lecture Notes in Mathematics 526*, Springer, Berlin, 1976, pp. 131–142.
KUNITA, H.
[1] On the decomposition of the solutions of stochastic differential equations, *Stochastic Integrals: Lecture Notes in Mathematics 851*, Springer, Berlin, 1981, pp. 213–255.
[2] On backward stochastic differential equations, *Stochastics*, **6**, 293–313 (1982).
[3] Stochastic differential equations and stochastic flows of homeomorphisms, *Stochastic Analysis and Applications, Adv. Probab. Related Topics*, **7**, Dekker, New York, 1984, pp. 269–291.
[4] Stochastic partial differential equations connected with nonlinear filtering, in Mitter and Moro [1].

[5] *Stochastic Flows and Stochastic Differential Equations*, Cambridge University Press, Cambridge, 1990.

KUNITA, H. and WATANABE, S.

[1] On square integrable martingales, *Nagoya Math. J.*, **30**, 209–245 (1967).

KUNITA, H. and WATANABE, T.

[1] Some theorems concerning resolvents over locally compact spaces, *Proc. 5th Berkeley Symp. Math. Statist. Prob.*, Vol. 2, Part 2, University of California Press, Berkeley 1967, pp. 131–164.

[2] Markov processes and Martin boundaries, I, *Illinois J. Math.*, **9**, 485–526 (1965).

[3] On certain reversed processes and their application to potential theory and boundary theory, *J. Math. Mech.*, **15**, 393–434 (1966).

KUSUOKA, S. and STROOCK, D.

[1] Applications of the Malliavin calculus, Part I, *Proceedings of the 1982 Taniguchi Int. Symp. on Stochastic Analysis* (ed. K. Itô), Kinokuniya–Wiley, 1984, 271–306.

[2] Applications of the Malliavin calculus, Part II, *J. Fac. Sci. Univ. Tokyo (IA)*, **32**, 1–76 (1985).

LE GALL, J.-F.

[1] Applications du temps local aux equations différentielles stochastiques unidimensionelles, *Séminaire de Probabilités XVII: Lecture Notes in Mathematics 986*, Springer, Berlin, 1983, pp. 15–31.

[2] Sur la saucisse de Wiener et les points multiples du mouvement Brownien plan at la méthode de renormalization de Varadhan, *Séminaire de Probabilités XIX: Lecture Notes in Mathematics 1123*, Springer, Berlin, 1985, pp. 314–331.

[4] Fluctuation results for the Wiener sausage, *Ann. Prob.*, **16**, 991–1018 (1988).

[5] The exact Hausdorff measure of Brownian multiple points, in Çinlar, Chung, and Getoor and Glover [1], pp. 107–137.

[6] Planar Brownian motion, cones and stable processes, *C. R. Acad. Sci. Paris Ser. I*, **302**, 641–643 (1986).

[7] Une approche élementaire des théorèmes de decomposition de Williams, *Séminaire de Probabilités, XX, Lecture Notes in Mathematics 1204*, Springer, Berlin, 1986, pp. 447–464.

LE GALL, J.-F., ROSEN, J. and SHIEH, N. R.

[1] Multiple points of Lévy processes, *Ann. Prob.*, **17**, 503–515 (1989).

LE GALL, J.-F. and YOR, M.

[1] Etude asymptotique de certains mouvements browniens complexes avec drift, *Prob. Th. Rel. Fields*, **71**, 183–229 (1986).

[2] Etude asymptotique des enlacements due mouvement brownien autour des droites de l'espace, *Prob. Th. Rel. Fields*, **74**, 617–635 (1987).

LE JAN, Y.

[1] Flots de diffusion dans $\mathbb{R}^d$, *C.R. Acad. Sci. Paris Ser. I*, **294**, 697–699 (1982).

[2] Equilibre et exposants de Lyapounov de certains flots Browniens, *C.R. Acad. Sci. Paris Ser. I*, **298**, 361–364 (1984).

[3] Exposants de Lyapounov pour les mouvements Browniens isotropes, *C. R. Acad. Sci. Paris Ser. I*, **299**, 947–949 (1984).

[4] On isotropic Brownian motions, *Z. Wahrscheinlichkeitstheorie verw. Geb.*, **70**, 609–620 (1985).

LE JAN, Y. and WATANABE, S.

[1] Stochastic flows of diffeomorphisms, *Proceedings of the 1982 Taniguchi Int. Symp. on Stochastic Analysis*, 1984, pp. 307–332.

LENGLART, E., LEPINGLE, D. and PRATELLI, M.

[1] Présentation unifiée de certaines inégalités de la théorie des martingales, *Séminaire*

de Probabilités XIV: Lecture Notes in Mathematics 784, Springer, Berlin, 1980.
LÉVY, P.
[1] *Théorie de l'Addition des Variables Aléatoires*, Gauthier Villars, Paris, 1954.
[2] *Processus Stochastiques et Mouvement Brownien*, Gauthier Villars, Paris, 1965.
[3] Systèmes markoviens et stationnaires. Cas dènombrable, *Ann. Ecole Norm. Sup. (3)*, **68**, 327–381 (1951); **69**, 203–212 (1952).
[4] Processus markoviens et stationnaires du cinquième type (infinité dénombrable des états possibles, parametre continu), *C. R. Acad. Sci. Paris*, **236**, 1630–1632, (1953).
[5] Processus markoviens et stationnaires. Cas dènombrable, *Ann. Inst. H. Poincaré*, **16**, 7–25 (1958).
LEWIS, J. T.
[1] Brownian motion on a submanifold of Euclidean space, *Bull. London Math. Soc.*, **18**, 616–620 (1986).
LIGGETT, T.
[1] *Interacting Particle Systems*, Springer, New York, 1985.
LINDVALL, T.
[1] On coupling of diffusion processes, *J. Appl. Prob.*, **20**, 82–93 (1983).
LIPSTER, R. S. and SHIRYAYEV, A. N.
[1] *Statistics of Random Processes*, I, Springer, Berlin, 1977.
LONDON, R. R., MCKEAN, H. P., ROGERS, L. C. G. and WILLIAMS, D.
[1] A martingale approach to some Wiener–Hopf problems, I, *Séminaire de Probabilités XVI: Lecture Notes in Mathematics 920*, Springer, Berlin, 1982, pp. 41–67.
LYONS, T. J.
[1] Finely holomorphic functions, *J. Funct. Anal.*, **37**, 1–18 (1980).
[2] Instability of the Liouville property for quasi-isometric Riemannian manifolds and reversible Markov chains, *J. Diff. Geom.* **26**, 33–66 (1987).
[3] The critical dimension at which quasi-every path is self-avoiding, in D. G. Kendall [5], pp. 87–100.
LYONS, T. J. and MCKEAN, H. P.
[1] Windings of the plane Brownian motion, *Adv. Math.*, **51**, 212–225 (1984).
MCGILL, P.
[1] Calculation of some conditional excursion formulae, *Z. Wahrscheinlichkeitstheorie*, **61**, 255–260 (1982).
[2] Markov properties of diffusion local time: a martingale approach, *Adv. Appl. Prob.*, **14**, 789–810 (1980).
[3] Integral representation of martingales in the Brownian excursion filtration, *Séminaire de Probabilités XX: Lecture Notes in Mathematics 1204*, Springer, Berlin, 1986, pp. 465–502.
MCKEAN, H. P.
[1] *Stochastic Integrals*, Academic Press, New York, 1969.
[2] Excursions of a non-singular diffusion, *Z. Wahrscheinlichkeitstheorie*, **1**, 230–239 (1963).
[3] Brownian local times, *Adv. Math.*, **16**, 91–111 (1975).
[4] Brownian motion with a several-dimensional time, *Teor. Veroyatnost.*, **4**(4), 357–378 (1963).
MCNAMARA, J. M.
[1] A regularity condition on the transition probability measure of a diffusion process. *Stochastics*, **15**, 161–182 (1985).
MAISONNEUVE, B.
[1] Systèmes régeneratifs, *Astérisque*, Soc. Mathématique de France, **15** (1974).

MAISONNEUVE, B. and MEYER, P. -A.
[1] Ensembles aléatoires markoviens homogènes, *Séminaire de Probabilités VIII: Lecture Notes in Mathematics 381*, Springer, Berlin, 1974, pp. 172–261.
MALLIAVIN, M.P. and MALLIAVIN, P.
[1] Factorisations et lois limites de la diffusion horizontale au dessus d'un espace riemannien symmetrique, *Lecture Notes in Mathematics 404*, Springer, Berlin, 1974, pp. 166–217.
MALLIAVIN, P.
[1] Stochastic calculus of variation and hypo-elliptic operators, *Proc. Int. Symp. Stoch. Diff. Equations*, Kyoto, 1976 (ed. K. Itô), Kinokuniya–Wiley, 1978, pp. 195–263.
[2] C^k-hypoellipticity with degeneracy, *Stochastic Analysis* (eds. A. Friedman and M. Pinksy), Academic Press, New York, 1978, pp. 199–214.
[3] Formula de la moyenne, calcul de perturbations et théorèmes d'annulation pour les formes harmoniques, *J. Funct. Anal.*, **17**, 274–291 (1974).
MARCUS, M.B. and ROSEN, J.
[1] Sample path properties of the local times of strongly symmetric Markov processes via Gaussian processes. *Ann. Prob.*, **20**, 1603–1684 (1992).
MANDL, P.
[1] *Analytic Treatment of One-Dimensional Markov Processes*, Springer, Berlin, 1968.
MELÉARD, S.
[1] Application du calcul stochastique à l'étude de processus de Markov réguliers sur [0, 1], *Stochastics*, **19**, 41–82 (1986).
MESSULAM, P. and YOR, M.
[1] On D. Williams' 'pinching method' and some applications, *J. London Math. Soc.*, **26**, 348–364 (1982).
METIVIER, M. and PELLAUMAIL, J.
[1] *Stochastic Integration*, Academic Press, New York, 1979.
MEYER, P. A.
[1] Un cours sur les intégrales stochastiques, *Séminaire de Probabilités X: Lecture Notes in Mathematics 511*, Springer, Berlin, 1976, pp. 245–400.
[2] *Probability and Potential*, Blaisdell, Waltham, Mass., 1966.
[3] *Processus de Markov: Lecture Notes in Mathematics 26*, Springer, Berlin, 1967.
[4] *Processus de Markov: la Frontière de Martin: Lecture Notes in Mathematics 77*, Springer, Berlin, 1970.
[5] Démonstration simplifiée d'un théorème de Knight, *Séminaire de Probabilités V: Lecture Notes, in Mathematics 191*, Springer, Berlin, 1971, pp. 191–195.
[6] Démonstration probabiliste de certaines inégalités de Littlewood-Paley, *Séminaire de Probabilités X: Lecture Notes in Mathematics 511*, Springer, Berlin, 1976, pp. 125–183.
[7] Flot d'un équation différentielle stochastique, *Séminaire de Probabilités XV: Lecture Notes in Mathematics 850*, Springer, Berlin, 1981, pp. 103–117.
[8] Sur la démonstration de prévisibilité de Chung and Walsh, *Séminaire de Probabilites IX: Lecture Notes in Mathematics 465*, Springer, Berlin, 1975, pp. 530–533.
[9] Géometrie stochastique sans larmes, *Séminaire de Probabilités XV: Lecture Notes in Mathematics 850*, Springer, Berlin, 1981, pp. 44–102.
[10] Géometrie stochastique sans larmes (bis), *Séminaire de Probabilités XVI: Supplément, Lecture Notes in Mathematics 921*, Springer, Berlin, 1982, pp. 165–207.
[11] Eléments de probabilités quantiques, *Séminaire de Probabilités XX: Lecture Notes in Mathematics 1204*, Springer, Berlin, 1986, pp. 186–312.

[12] *Quantum Theory for Probabilists, Lecture Notes in Mathematics 1538*, Springer, Berlin, 1993.

MIHLSTEIN, G. N.

[1] Approximate integration of stochastic differential equations, *Th. Prob. Appl.*, **19**, 557–562 (1974).

MILLAR, P. W.

[1] Random times and decomposition theorems, in *Probability: Proc. Symp. Pure Math. XXXI*, Amer. Math. Soc., Providence, RI, 1977, pp. 91–103.

[2] A path decomposition for Markov processes, *Ann. Prob.*, **6**, 345–348 (1978).

MILLAR, P. W. and TRAN, L. T.

[1] Unbounded local times, *Z. Wahrscheinlichkeitstheorie verw. Geb.*, **30**, 87–92 (1974).

MITRO, J.

[1] Dual Markov processes: construction of a useful auxiliary process, *Z. Wahrscheinlichkeitstheorie*, **47**, 139–156 (1979).

[2] Dual Markov functions: applications of a useful auxiliary process, *Z. Wahrscheinlichkeitstheorie*, **48**, 97–114 (1979).

MITTER, S. K.

[1] Lectures on non-linear filtering and stochastic control, in Mitter and Moro [1], pp. 170–207.

MITTER, S. K. and MORO, A. (editors)

[1] *Non-linear Filtering and Stochastic Control: Lecture Notes in Mathematics 972*, Springer, Berlin, 1982.

MOTOO, M.

[1] Application of additive functionals to the boundary problem of Markov processes (Lévy's system of U-processes), *Proc. 5th Berkeley Symp. Math. Statist. Prob.*, Vol. 2, Part 2, Univ. of California Press, Berkeley, 1967, pp. 75–110.

[2] Proof of the law of iterated logarithm through diffusion equation, *Ann. Inst. Statist. Math.*, **10**, 21–28 (1959).

MOTOO, M. WATANABE, S.

[1] On a class of additive functionals of Markov processes, *J. Math. Kyoto Univ.*, **4**, 429–469 (1965).

NAKAO, S.

[1] On the pathwise uniqueness of solutions of one-dimensional stochastic differential equations, *Osaka J. Math.*, **9**, 513–518 (1972).

NASH, J. F.

[1] The imbedding problem for Riemannian manifolds, *Ann. Math.*, **63**, 20–63 (1956).

NELSON, E.

[1] *Dynamical Theories of Brownian Motion*, Princeton University Press, 1967.

[2] *Quantum Fluctuations*, Princeton University Press, 1984.

NEVEU, J.

[1] *Bases Mathématiques du Calcul des Probabilités*, Masson, Paris, 1964.

[2] Sur les états d'entrée et les états fictifs d'un processus de Markov, *Ann. Inst. Henri Poincaré*, **17**, 323–337 (1962).

[3] Lattice methods and submarkovian processes, *Proc. 4th Berkeley Symp. Math. Statist. Prob.*, Vol. 2, University of California Press, Berkeley, 1960, pp. 347–391.

[4] Une généralisation des processus à accroissements positifs indépendants, *Abh. Math. Sem. Univ. Hamburg*, **25**, 36–61 (1961).

[5] Entrance, exit and fictitious states for Markov chains, *Proc. Aarhus Colloq. Combin Prob.*, 1962, pp. 64–68.

NORRIS, J. R.
[1] Simplified Malliavin calculus, *Séminaire de Probabilités XX: Lecture Notes in Mathematics 1204*, Springer, Berlin, 1986, pp. 101–130.
NORRIS, J. R., ROGERS, L. C. G. and WILLIAMS, D.
[1] Brownian motion of ellipsoids, *Trans. Amer. Math. Soc.*, **294**, 757–765 (1986).
[2] Self-avoiding random walk: a Brownian motion model with local time drift, *Prob. Th. Rel. Fields*, **74**, 271–287 (1987).
OCONE, D.
[1] Malliavin's calculus and stochastic integral: representation of functionals of diffusion processes, *Stochastics*, **12**, 161–185 (1984).
ORIHARA, A.
[1] On random ellipsoid, *J. Fac. Sci. Univ. Tokyo, Sect. IA Math.*, **17**, 73–85 (1970).
PARDOUX, E.
[1] Stochastic differential equations and filtering of diffusion processes, *Stochastics*, **3**, 127–167 (1979).
[2] Grossissement d'une filtration et retournement du temps d'une diffusion, *Séminaire de Probabilités XX: Lecture Notes in Mathematics 1204*, Springer, Berlin, 1986, pp. 48–55.
[3] Equations of non-linear filtering, and applications to stochastic control with partial observations, in Mitter and Moro [1], pp. 208–248.
PARDOUX, E. and TALAY, D.
[1] Discretization and simulation of stochastic differential equations, *Acta Appl. Math.*, 3, 23–47 (1985).
PARTHASARATHY, K. R.
[1] *Probability Measures on Metric Spaces*, Academic Press, New York, 1967.
PAUWELS, E. and ROGERS, L. C. G.
[1] Skew-product decompositions of Brownian motions, *Contemp. Math.* **73**, 237–262 (1988).
PERKINS, E.
[1] Local time and pathwise uniqueness for stochastic differential equations, *Séminaire de Probabilités XVI: Lecture Notes in Mathematics 920*, Springer, Berlin, 1982, pp. 201–208.
[2] Local time is a semimartingale, *Z. Wahrscheinlichkeitstheorie*, **60**, 79–117 (1982).
PHELPS, R. R.
[1] *Lectures on Choquet's Theorem*, Van Nostrand, Princeton, NJ, 1966.
PINSKY, M. A.
[1] Homogenization and stochastic parallel displacement, in Williams [13], pp. 271–284.
[2] Stochastic Riemannian geometry, *Probabilistic Analysis and Related Topics*, 1 (ed. A. T. Bharucha-Reid), Academic Press, New York, 1978.
PITMAN, J. W.
[1] One-dimensional Brownian motion and the three-dimensional Bessel process, *J. Appl. Prob.*, **7**, 511–526 (1975).
[2] Path decomposition for conditional Brownian motion, *Inst. Math. Statist. Univ. Copenhagen*, Preprint No. 11 (1974).
[3] Lévy systems and path decompositions, in Çinlar, Chung and Getoor [1, 1981].
PITMAN, J. W. and YOR, M.
[1] Bessel processes and infinitely divisible laws, *Stochastic Integrals* (ed. D. Williams), *Lecture Notes in Mathematics 851*, Springer, Berlin, 1981, pp. 285–370.
[2] A decomposition of Bessel bridges. *Z. Wahrscheinlichkeitstheorie*, **59**, 425–457 (1982).

[3] The asymptotic joint distribution of windings of planar Brownian motion, *Bull. Amer. Math. Soc.*, **10**, 109–111 (1984).
[4] Asymptotic laws of planar Brownian motion, *Ann. Prob.*, **14**, 733–779 (1986).
PITTENGER, A. O. and SHIH, C. T.
[1] Coterminal families and the strong Markov property, *Trans. Amer. Math. Soc.*, **182**, 1–42 (1973).
POOR, W. A.
[1] *Differential Geometric Structures*, McGraw-Hill, New York, 1981.
PORT, S. C. and STONE, C. J.
[1] Classical potential theory and Brownian motion, *Proc. 6th Berkeley Symp. Math. Statist. Prob.*, Vol. 3, University of California Press, Berkeley, 1972, pp. 143–176.
[2] Logarithmic potentials and planar Brownian motion, *Proc. 6th Berkeley Symp. Math. Statist. Prob.*, Vol. 3, University of California Press, Berkeley 1972, pp. 177–192.
[3] *Brownian Motion and Classical Potential Theory*, Academic Press, New York, 1978.
PRICE, G. C. and WILLIAMS, D.
[1] Rolling with 'slipping': I, *Séminaire de Probabilités XVII: Lecture Notes in Mathematics 986*, Springer, Berlin, 1983, pp. 194–297.
PROHOROV, YU, V.
[1] Convergence of random processes and limit theorems in probability, *Th. Prob. Appl.*, **1**, 157–214 (1956).
PROTTER, P.
[1] On the existence, uniqueness, convergence and explosions of solutions of stochastic differential equations, *Ann. Prob.*, **5**, 243–261 (1977).
RAO, K. M.
[1] On decomposition theorems of Meyer, *Math. Scand.*, **24**, 66–78 (1969).
[2] Quasimartingales, *Math. Scand.*, **24**, 79–92 (1969).
RAY, D. B.
[1] Resolvents, transition functions and strongly Markovian processes, *Ann. Math.*, **70**, 43–72 (1959).
[2] Sojourn times of a diffusion process, *Illinois J. Math.*, **7**, 615–630 (1963).
REUTER, G. E. H.
[1] Denumerable Markov processes, II, *J. London Math. Soc.*, **34**, 81–91 (1959).
REVUZ, D.
[1] The Martin boundary of a recurrent random walk has one or two points, *Probability: Proc. Symp. Pure Math. XXXI*, Amer. Math. Soc., Providence, RI, 1977, pp. 125–130.
REVUZ, D. and YOR, M.
[1] *Continuous Martingales and Brownian Motion*, Springer, Berlin, 1991.
ROGERS, L. C. G.
[1] Williams' characterization of the Brownian excursion law: proof and applications, *Séminaire de Probabilités XV: Lecture Notes in Mathematics 850*, Springer, Berlin, 1981, pp. 227–250.
[2] Itô excursion theory via resolvents, *Z. Wahrscheinlichkeitstheorie*, **63**, 237–255 (1983).
[3] Smooth transition densities for one-dimensional diffusions, *Bull. London Math. Soc.*, **17**, 157–161 (1985).
[4] Continuity of martingales in the Brownian excursion filtration, *Prob. Th. Rel. Fields* **76**, 291–298 (1987).
[5] Multiple points of Markov processes in a complete metric space, *Séminaire de Probabilités XXIII: Lecture Notes in Mathematics 1372*, Springer, Berlin, 1989, pp. 186–197.
[6] A new identity for real Lévy processes. *Ann. Inst. Henri Poincaré*, **20**, 21–34 (1984).

ROGERS, L. C. G. and PITMAN, J. W.
[1] Markov functions, *Ann. Prob.* **9**, 573–582 (1981).
ROGERS, L. C. G. and WILLIAMS, D.
[1] *Diffusions, Markov Process, and Martingales: Volume 2: Itô Calculus*, Wiley, Chichester, 1987.
[2] Construction and approximation of transition matrix functions, in D. G. Kendall [5], pp. 133–160.
ROGOZIN, B. A.
[1] On the distribution of functionals related to boundary problems for processes with independent increments, *Th. Prob. Appl.*, **11**, 580–591 (1966).
ROSEN, J.
[1] A local time approach to self-intersections of Brownian paths in space, *Comm. Math. Phys.*, **88**, 327–338 (1983).
SCHWARTZ, L.
[1] Géometrie différentielle du 2ième ordre, semimartingales et équations différentielles stochastiques sur une variéte différentielle, *Séminaire de Probabilités XVI, Supplément: Lecture Notes in Mathematics 921*, Springer, Berlin, 1982, pp. 1–148.
SHARPE, M. J.
[1] *General Theory of Markov Processes*, Academic Press, New York, 1988.
SHEPPARD, P.
[1] On the Ray–Knight property of local times, *J. London Math. Soc.*, **31**, 377–384 (1985).
SHIGA, T. and WATANABE, S.
[1] Bessel diffusions as a one-parameter family of diffusion processes, *Z. Wahrscheinlichkeitstheorie*, **27**, 37–46 (1973).
SHIGEKAWA, I.
[1] Derivatives of Wiener functionals and absolute continuity of induced measure, *J. Math. Kyoto Univ.*, **20**, 263–289 (1980).
SHIMURA, M.
[1] Excursions in a cone for two-dimensional Brownian motion, *J. Math. Kyoto Univ.*, **25**, 433–443 (1985).
SILVERSTEIN, M. L.
[1] *Symmetric Markov Processes: Lecture Notes in Mathematics 426*, Springer, Berlin, 1974.
[2] *Boundary Theory for Symmetric Markov Processes: Lecture Notes in Mathematics 516*, Springer, Berlin, 1976.
SIMON, B.
[1] *Functional Integration and Quantum Physics*, Academic Press, New York, 1979.
[2] Semiclassical analysis of low-lying eigenvalues, II. Tunneling, *Ann. Math.* **120**, 89–118 (1984).
SKOROKHOD, A. V.
[1] Limit theorems for stochastic processes, *Th. Prob. Appl.* **1**, 261–290 (1956).
[2] Limit theorems for Markov processes, *Th. Prob. Appl.* **3**, 202–246 (1958).
SPITZER, F.
[1] *Principles of Random Walk*, Van Nostrand, Princeton, NJ, 1964.
[2] Some theorems concerning two-dimensional Brownian motion, *Trans. Amer. Math. Soc.*, **87**, 187–197 (1958).
STRASSEN, V.
[1] An invariance principle for the law of the iterated logarithm, *Z. Wahrscheinlichkeitstheorie*, **3**, 211–226 (1964).
[2] Almost sure behaviour of sums of independent random variables and martingales,

Proc. 5th Berkeley Symp. Math. Statist. Prob., Vol. 2, Part 1, University of California Press, Berkeley, 1966, pp. 315–343.

STROOCK, D. W.

[1] The Malliavin calculus and its applications to second-order parabolic differential operators I, II, *Math. System Theory*, **14**, 25–65, 141–171 (1981).

[2] The Malliavin calculus; a functional analytical approach, *J. Funct. Anal.*, **44**, 217–257 (1981).

[3] Diffusion processes associated with Lévy generators, *Z. Wahrscheinlichkeitstheorie*, **32**, 209–244 (1975).

[4] *An Introduction to the Theory of Large Deviations*, Springer, Berlin, New York, 1984.

STROOCK, D. W. and VARADHAN, S. R. S.

[1] *Multidimensional Diffusion Processes*, Springer, New York, 1979.

[2] On the support of diffusion processes with applications to the strong maximum principle, *Proc. 6th Berkeley Symp. Math. Statist. Prob.*, Vol. 3, University of California Press, Berkeley, 1972, pp. 333–359.

[3] Diffusion processes with boundary conditions, *Comm. Pure Appl. Math.*, **24**, 147–225 (1971).

STROOCK, D. W. and YOR, M.

[1] Some remarkable martingales, *Seminaire de Probabilités XV: Lecture Notes in Mathematics 850*, Springer, Berlin, 1981, pp. 590–603.

SUSSMANN, H. J.

[1] On the gap between deterministic and stochastic ordinary differential equations, *Ann. Prob.*, **6**, 19–41 (1978).

SYMANZIK, K.

[1] Euclidean quantum field theory, *Local Quantum Theory* (ed. R. Jost), Academic Press, New York, 1969.

TALAGRAND, M.

[1] Regularity of Gaussian processes, *Acta Math.*, **159**, 99–149 (1987).

TAYLOR, G. I.

[1] Statistical theory of turbulence, *Proc. Roy. Soc. London A*, **151**, 421–478 (1935).

TAYLOR, H. M.

[1] A stopped Brownian motion formula, *Ann. Prob.*, **3**, 234–246 (1975)

TAYLOR, S. J.

[1] Sample path properties of processes with stationary independent increments, *Stochastic Analysis* (eds. D. G. Kendall and E. F. Harding), Wiley, New York, 1973, pp. 387–414.

THORIN, O.

[1] On the infinite divisibility of the lognormal distribution, *Scand. Actuarial J.*, 121–148 (1977).

TSIREL'SON, B. S.

[1] An example of the stochastic equation having no strong solution, *Teoria Verojatn. i Primenen.*, **20**, 427–430 (1975).

VAN DEN BERG, M. and LEWIS, J. T.

[1] Brownian motion on a hypersurface, *Bull. London Math. Soc.*, **17**, 144–150 (1985).

VARADHAN, S. R. S.

[1] *Large Deviations and Applications*, SIAM, Philadelphia, 1984.

VARADHAN, S. R. S. and WILLIAMS, R. J.

[1] Brownian motion in a wedge with oblique reflection, *Comm. Pure Appl. Math.*, **38**, 405–443 (1985).

WALSH, J. B.
[1] Excursions and local time, in Azema and Yor [2], pp. 159–192.
[2] Stochastic integration with respect to local time, in Çinlar, Chung and Getoor [1, 1983].
[3] An introduction to stochastic partial differential equations, *Ecole d'Eté de Probabilités de St Flour XIV–1984, Lecture Notes in Mathematics 1180*, Springer, Berlin, 1986.
WARNER, F. W.
[1] *Foundations of Differentiable Manifolds and Lie Groups*, Springer, Berlin 1983.
WATANABE, S.
[1] On discontinuous additive functionals and Lévy measures of a Markov process, *Jap. J. Math.*, **34**, 53–79 (1964).
WATSON, G. N.
[1] *A Treatise on the Theory of Bessel Functions*, Cambridge University Press, Cambridge, 1966.
WHITNEY, H.
[1] *Geometric Integration Theory*, Princeton University Press, Princeton, NJ, 1957.
WHITTLE, P.
[1] *Optimization over Time* (two volumes), Wiley, Chichester, 1982, 1983.
WILLIAMS, D.
[1] Brownian motions and diffusions as Markov processes, *Bull. London Math. Soc.*, **6**, 257–303 (1974).
[2] Some basic theorems on harnesses, *Stochastic Analysis* (eds. D. G. Kendall and E. F. Harding), Wiley, New York, 1973, pp. 349–366.
[3] On Lévy's downcrossing theorem, *Z. Wahrscheinlichkeitstheorie*, **40**, 157–158 (1977).
[4] Path decomposition and continuity of local time for one-dimensional diffusions, I, *Proc. London Math. Soc., Ser. 3*, **28**, 738–768 (1974).
[5] On the stopped Brownian motion formula of H. M. Taylor, *Séminaire de Probabilités X: Lecture Notes in Mathematics 511*, Springer, Berlin, 1976, pp. 235–239.
[6] Markov properties of Brownian local time, *Bull. Amer. Math. Soc.*, **75**, 1035–1036 (1969).
[7] Decomposing the Brownian path, *Bull. Amer. Math. Soc.*, **76**, 871–873 (1970).
[8] The Q-matrix problem for Markov chains, *Bull. Amer. Math. Soc.*, **81**, 1115–1118 (1975).
[9] The Q-matrix problem, *Séminaire de Probabilités X: Lecutre Notes in Mathematics 511*, Springer, Berlin, 1976, pp. 216–234.
[10] A note on the Q-matrices of Markov chains, *Z. Wahrscheinlichkeitstheorie*, **7**, 116–121 (1967).
[11] Some Q-matrix problems, *Probability: Proc. Symp. Pure Math. XXXI*, Amer. Math. Soc., Providence, RI, 1977, pp. 165–169.
[12] *Diffusions, Markov Processes, and Martingales, Volume 1: Foundations*, Wiley, Chichester, 1979.
[13] (editor) *Stochastic Integrals: Proceedings, LMS Durham Symposium, Lecture Notes in Mathematics 851*, Springer, Berlin, 1981.
[14] Conditional excursion theory, *Séminaire de Probabilités XIII: Lecture Notes in Mathematics 721*, Springer, Berlin, 1979, pp. 490–494.
[15] (=[W]) *Probability with Martingales*, Cambridge University Press, Cambridge, 1991.

YAGLOM, A. M.
[1] Some classes of random fields in n-dimensional space, related to stationary random processes, *Th. Prob. Appl.*, **2**, 273–319 (1957).
YAMADA, T.
[1] On a comparison theorem for solutions of stochastic differential equations and its applications, *J. Math. Kyoto Univ.*, **13**, 497–512 (1973).
YAMADA, T. and OGURA, Y.
[1] On the strong comparison theorems for solutions of stochastic differential equations, *Z. Wahrscheinlichkeitstheorie*, **56**, 3–19 (1981).
YAMADA, T. and WATANABE, S.
[1] On the uniqueness of solutions of stochastic differential equations, *J. Math. Kyoto Univ.*, **11**, 155–167 (1971).
YOR, M.
[1] Sur certains commutateurs d'une filtration, *Séminaires de Probabilités XV: Lecture Notes in Mathematics 850*, Springer, Berlin, 1981, pp. 526–528.
[2] Sur la continuité des temps locaux associès à certaines semimartingales, in Azéma and Yor [2], pp. 23–35.
[3] Rappel et préliminaires généraux, in Azéma and Yor [2], pp. 17–22.
[4] Précisions sur l'existence et la continuité des temps locaux d'intersection du mouvement Brownien dans $\mathbf{R}^2$, *Séminaire de Probabilités XX: Lecture Notes in Mathematics 1204*, Springer, Berlin, 1986, pp. 532–542.
[5] Sur la répresentation comme intégrales stochastique des temps d'occupation du mouvement Brownien dans $\mathbf{R}^d$, *ibid*, pp. 543–552.
YAMADA, T.
[1] *Functional Analysis*, Springer, Berlin, 1965.
[2] Brownian motion in homogeneous Riemannian space, *Pacific J. Math.*, **2**, 263–296. (1952).
ZAKAI, M.
[1] The Malliavin calculus, *Acta Appl. Math.*, **3**, 175–207 (1985).
ZHENG, W. A. and MEYER, P.-A.
[1] Quelques résultats de 'méchanique stochastique', *Séminaire de Probabilités XVIII: Lecture Notes in Mathematics 1059*, Springer, Berlin, 1984, pp. 223–244.
ZVONKIN, A. K.
[1] A transformation of the phase space of a diffusion process that removes the drift, *Math. USSR Sbornik*, **22**, 129–149 (1974).

Index to Volumes 1 and 2

Absolute continuity: II.9.
Absorbing state: III.12.
Accessible stopping time: VI.13-14.
Adapted process: II.45.
Additive functional: III.16; construction from λ-potential, III.16, I.17.
Affine group: V.35.
Algebra: II.1.
Almost surely: II.14, III.9.
Announceable time: VI.12.
Approximation to compensators: VI.31.
Arcsine law: II.34, III.24, V.53.
Arzelà–Ascoli theorem: II.85.
Atlas: V.34.
Atom: II.88.
Awaiting the almost inevitable: II.57.
Azéma's martingale: II.37.

Backward equation: I.4.
Barlow's example: V.41.
Basic integrands: IV.5.
Bessel process: IV.35, V.48; time reversal, III.49.
Bi-invariant metric: V.35.
Birth process: III.26, VI.14.
Blumenthal's 0–1 Law: for Brownian motion, I.12; for FD processes, III.9.
Bochner's theorem: I.24.
Bochner's horizontal Laplacian: V.34.
Borel–Cantelli Lemmas: II.15.
Boundary points of one-dimensional diffusions: V.47, V.51.
Boundary theory: see Martin–Doob–Hunt theory.
Branch points: definition, III.37; illustrative example, III.37; probabilistic significance III.41.
Brownian motion: definition, I.1; on affine group, V.35; arcsine law, II.34, III.24, VI.53; Brownian bridge, I.25, II.91, IV.40; canonical, II.90; complex, see complex Brownian motion; and continuous martingales, I.2, IV.34; Dirichlet problem, I.22; elastic Brownian motion, III.24; of ellipses, V.36, V.37; energy of charge, I.22; excursion law, VI.50; exponential martingales, I.2, I.9; Feller Brownian motions, VI.57; on filtered probability space, I.2, II.72; first-passage distribution, I.9, I.14, III.10; Gaussian description, I.3; generator, I.4, III.6; Green function, I.22; iterated-logarithm laws, I.16; Kolmogorov's

backward and forward equations, I.4; Kolmogorov's test, I.13; Lévy's characterization, IV.33; on Lie groups, V.35; local time, I.5, I.14, III.16; on a manifold, V.30, V.31; martingales of, I.17; martingale characterisations, I.2; martingale representation, IV.36, IV.41; modulus of continuity, I.10; no-increase property, I.10; nowhere-differentiability, I.12; on the orthonormal frame bundle, V.30, V.33, V.34; path decomposition, VI.55; potential theory, I.22; quadratic variation, I.11, IV.2; Ray–Knight Theorems, VI.52; recurrence, I.3; reflecting Brownian motion, I.14, III.22, V.6; reflection principle, I.13; resolvent, III.3; rotational invariance, I.18; scaled Brownian excursion, IV.40; scaling, I.3; skew-product representation, V.31; on SO(3), V.35; Skorokhod embedding, I.7, VI.51; slow points, I.10; strong Markov property, I.12; on a surface, V.4, V.31; time-reversed, II.38; transition density, I.4; unbounded variation, I.11, IV.2; wandering to infinity, I.18.
Brownian sheet: I.25.
Burkholder–Davis–Gundy inequalities: IV.42.

Càdlàg maps: see *R*-paths.
Cameron–Martin–Girsanov change of measure: IV.38-41, V.27.
Canonical decomposition of a special semimartingale: VI.40.
Canonical process: II.28, II.71, III.7.
Capacity: I.22.
Carathéodory's Extension Theorem: II.5.
Carverhill's noisy North–South flow: V.14.
Cauchy law: I.20.
Cauchy process: I.28, VI.2, VI.28.
Change of time scale: see time substitution.
Chapman–Kolmogorov equations: I.4, III.1.
Characteristic exponent: I.28.
Characteristic operator: III.12.
Charge: I.22, III.27; see also equilibrium charge.
Chart: V.34.
Choquet capacitability theory: III.76.
Choquet representation of λ-excessive functions: III.44.
Choquet representation of 1-excessive probabilities: III.38.
Choquet's theorem on integral representations: III.27.
Christoffel symbols: V.31, V.34.
Cieselski–Taylor Theorem: III.20, III.49.
Clark's Theorem on Brownian martingale representation: IV.41.
Coffin state: III.3.
Comparison theorem: V.43.
Compensated Poisson process: II.64.
Compensator: VI.29, VI.31; see also dual previsible projection.
Completions: II.75.
Complex Brownian motion: I.19; cone point, I.21; cut point, I.21; multiple points, I.21; Spitzer's theorem, I.20; windings of, I.20.
Compound Poisson process: I.28.
Condition N: III.54.
Condition S: III.55.
Conditional expectations and probabilities: II.41, II.44; regular, II.42, II.43.
Conditional independence: II.60.
Cone point: I.21.

Conformal martingales: IV.34.
Connection: V.32, V.34.
Continuous Lévy processes, characterization of: I.28, III.14.
Continuous local martingale: pure local martingales, IV.34; quadratic-variation process IV.30; as time change of Brownian motion, IV.34.
Continuous mapping principle: II.84.
Continuous semimartingale: canonical decomposition, IV.30, VI.24; Itô's formula, IV.32; local time, IV.43.
Contraction resolvent: III.4; strongly continuous (SCCR), III.4.
Contraction semigroup: III.4; strongly continuous (SCCSG), III.4.
Control problems: see stochastic control.
Controlled variance problem: V.6, V.42.
Convergence of random variables: II.19.
Coupling inequality: V.54.
Coupling of one-dimensional diffusions: V.54.
Covariance of a diffusion: V.1.
Covariant differentiation: V.32, V.34.
Cumulative risk: II.64, VI.22.
Curvature: V.38.
Cut point: I.21.
Cylinder: II.25.

d-system: II.1.
Daniell–Kolmogorov Theorem: II.30, II.31; limitations of, II.34.
Début: of open set for R-process, II.74; of compact set of R-process, II.75; of progressive set, VI.3.
Début Theorem II.76, III.9, VI.3.
De Finetti's Theorem: II.51.
Diffeomorphism: V.34.
Diffeomorphism Theorem: V.13.
Diffusion equation:I.4.
Diffusion: III.13, V.1, V.2; diffusion SDE, V.8; in one dimension, see one-dimensional diffusions; physical, I.23.
Directed set: II.80.
Dirichlet form: I.23; for Markov chains, III.59.
Dirichlet problem: I.22.
Distribution function: II.16.
Doléans' characterization of FV processes: VI.20, VI.25-27.
Doléans exponential: IV.19.
Doléans' proof of the Meyer Decomposition Theorem: VI.30.
Dominated-Convergence Theorem: II.8.
Donsker's Invariance Principle: I.8.
Doob decomposition of a submartingale: II.54.
Doob *h*-transform: III.29, III.45, IV.39.
Doss–Sussmann method: V.28.
Downcrossing Theorem (Lévy): I.14.
Drift of a diffusion: V.1.
Dual previsible projection: VI.1, VI.21, VI.23.
Dynkin's formula: III.10.
Dynkin's Local-Maximum Principle: III.13.

Dynkin's Isomorphism Theorem: I.27.
Dynkin's Maximum Principle: III.6.

Elastic boundary: III.24.
Elementary process: IV.6, IV.25.
Elworthy's example: V.13.
Empirical distribution: II.91.
Entrance laws: III.39.
Equilibrium charge and potential: I.22, III.48, VI.35.
Ergodic Theorem for one-dimensional diffusions: V.53.
Evanescent process: IV.13.
Excessive functions: III.27; representation, see Martin–Doob–Hunt theory; Riesz decomposition, III.27; uniformly λ-excessive, III.16.
Excessive measures: III.38.
Excursion intervals: VI.42.
Excursion law: VI.47, VI.50; for Brownian motion, VI.50, VI.55; for Markov chain, VI.43, VI.50.
Excursion theory: Ch.VI; censoring and reweighting of excursion laws, VI.58; characteristic measure, VI.47; excursion filtration, VI.59; for a finite Markov chain, VI.43; lifetime, VI.47; marked excursions, VI.49; for a Markov chain, III.57; Markovian character of excursion law, VI.48; path decomposition for Brownian excursions, VI.55; from a point which is not regular extremal, VI.50; Poisson point process, VI.43, VI.47; starred excursion, VI.49; by stochastic calculus, VI.59.
Excursion space: VI.43, VI.47.
Expectation: II.17.
Exponential map: V.34.
Exponential semimartingale: IV.19, IV.22, IV.37.
Extending the generator: III.4.

FD (Feller–Dynkin) diffusions: III.13, V.22; martingale representation, V.25.
FD processes: existence, III.7; strong Markov property, III.8, III.9.
FD semigroups: III.6.
FV: see finite-variation.
Fair stopping time: VI.12.
Fatou Lemma: II.8; for non-negative supermartingales, IV.14.
Feller Brownian motions: VI.57.
Feller property: III.6.
Feller–McKean chain: III.23, III.35.
Feynman–Kac formula: III.19; for Markov chains, IV.22.
Field: see algebra.
Fick's Law: I.23.
Filtering: VI.8; Bayesian approach, VI.10; change-detection filter, V.10, V.22; Kalman–Bucy filter, VI.9; robust filtering, VI.11.
Filtration: II.45, II.63; natural, II.45.
Finite-dimensional distributions: II.29, II.87.
Finite fuel control problem: V.7, V.15.
Finite-variation functions: II.13.
Finite-variation processes: IV.7; Doléans' characterization, VI.20.
First-approach times: II.74.

First-entrance decomposition: III.52.
First-entrance times: see début.
First-hitting times: II.74.
Forward equation: I.4, I.23.
Freedman's interpretation of q_{bj}: III.57.
Fubini's Theorem: II.12.
Fundamental Theorem of Algebra: I.20.

Gamma process: I.28.
Gaussian process: definition, I.3.
Gaussian random fields, isotropic: I.26.
Generator: see infinitesimal generator.
Geodesic: V.32, V.34.
Girsanov SDE: V.26.
Good λ inequality: IV.42.
Green function: I.22, III.27, III.30.
Gronwall's lemma: V.11.

Harmonic function: III.31.
Hausdorff moment problem: III.28.
Hazard function: II.64.
Heat equation: I.4.
Helms–Johnson example: II.79, III.31, IV.14, VI.33.
Hermite polynomials: I.2.
Hewitt–Savage 0–1 law: II.51.
Hille–Yosida Theorem: III.5.
Hölder inequality: II.10.
Honest transition function: III.3.
Horizontal lift: V.34.
Horizontal vector field: V.34.
Hörmander's Theorem: V.38.
Hunt's Theorem: VI.35.
Hyperbolic plane: V.34, V.35, V.36.
Hyperboloid sheet: V.36.
Hypothèses droites: VI.46.

Identical hitting-distributions: III.21.
Imbedding: V.34.
Independence: II.21, II.23.
Indistinguishable processes: II.36, IV.13.
Infinite divisibility: I.28.
Infinitesimal generator: III.2, III.4; Brownian motion, I.4, III.6; one-dimensional diffusion V.47, V.50.
Inner measure: II.6.
Inner regularity of measures: II.80.
Innovations process: VI.8.
Instantaneous state of a Markov chain: III.51.
Integral: II.7.
Integrable-variation processes: IV.7.

Integral curve: V.34.
Integration by parts: IV.2, VI.38; for continuous semimartingales, IV.32; for finite-variation processes, IV.18.
Integrators: IV.16.
Isometric imbedding: V.34.
Isotropic Gaussian random fields: I.26.
Itô's formula: IV.3, VI.39; for continuous semimartingales, IV.32; for convex functions IV.45, V.47; for FV processes, IV.18.
Itô integral: see stochastic integral.

Jensen's inequality: II.18, II.41, II.52.
Joint law: II.16.

Kalman–Bucy filter: VI.9.
Khasminskii's method for stability: V.37.
Khasminskii's test for explosion: V.52.
Killing: III.18
Kingman's Markov Characterization Theorem: III.58.
Knight's Theorem on continuous local martingales: IV.34.
Kolmogorov's backward equations: I.4.
Kolmogorov's forward equations: I.4.
Kolmogorov's lemma: I.25, II.85, IV.44.
Kolmogorov's test for Brownian motion: I.13.
Kolmogorov's 0–1 Law: II.50.
Krylov's example: V.29.
Kunita–Watanabe inequalities: IV.28.

L-process, L-path: IV, Introduction.
λ-potential operator: see resolvent.
Laplace exponent: I.28, I.37.
Laplace–Beltrami operator: V.30, V.34.
Last-exit decomposition: III.56, VI.43, VI.48.
Last-exit distribution for Brownian motion: VI.35.
Law of the Iterated Logarithm: I.16.
Law of process: II.27.
Law of a random variable: II.16.
LCCB: locally compact Hausdorff space with countable base, II.6.
Lebesgue measure: II.5.
Lebesgue's thorn: III.9.
Left-invariant vector field: V.35.
Lévy Brownian motion: I.24.
Lévy's characterization of Brownian motion: I.2, IV.33.
Lévy–Doob 'Downward' Theorem: II.51.
Lévy kernels: III.57, IV.21.
Lévy–Hincin formula: I.28, VI.2.
Lévy measure: I.28, II.37, VI.2; Lévy system: VI.28.
Lévy process: I.28, I.29, I.30, II.37, VI.2.
Lévy's 'Upward' Theorem: II.50.
Lie algebra: V.35.
Lie bracket: V.34, 38.

Lie group: V.35.
Lifetime: III.7.
Likelihood ratio: II.79, IV.17; for Markov chains, IV.22.
Lipschitz square root: V.12.
Local martingale: IV.1, IV.14; on a manifold, V.30, V.33.
Local time: for Brownian motion, I.5, I.14; for continuous semimartingales, IV.43-4; growth set, VI.45; for Lévy processes, I.30; Markovian local time, IV.43; as an occupation density, IV.45; for one-dimensional diffusions, V.49; at regular extreme point of a Ray process, VI.45; from upcrossings of Brownian motion, I.14, II.79.
Localization: IV.9.
Locally bounded previsible process: IV.10.
Lusin space: II.31, II.82.
Lyapunov exponent: V.37.

Malliavin–Bismut integration-by-parts: V.38.
Malliavin calculus: V.38.
Manifold: V.34.
Marked excursions: VI.49.
Markov chains: III.2; birth process, IV.26; Dirichlet form, III.59; Feller–McKean chain, III.23, III.35; Lévy's diagonal Q-matrix, III.35, IV.35; Martin boundary, III.48; martingale problem, IV.20-22; as Ray processes, III.50; stable and instantaneous states, III.51; see also Q-matrices, standard transition functions.
Markov p-function: III.58.
Markov inequality: II.18.
Markov processes: III.1; see also FD processes, Ray processes.
Martin compactification: III.28.
Martin kernel: III.27; for Brownian motion in the unit ball, III.30.
Martin–Doob–Hunt theory: for discrete-parameter chains, III.28, III.29, III.42; for Brownian motion, III.30, III.31.
Martingales: definitions, II.46, II.63; for Brownian motion, I.17; convergence theorems, II.49, II.50, II.51, II.69; in L^p, II.53; regularity of paths, II.65, II.66, II.67; for FD processes, III.10; for Brownian motion, I.17.
Martingale inequalities: Burkholder–Davis–Gundy inequality, IV.42; Doob's L^p inequality, II.52, II.70; Doob's submartingale inequality, II.52, II.54, II.70; Doob's Upcrossing Lemma, II.48.
Martingale problem: V.19; existence of solutions, V.23; for Markov chains, IV.20; Markov property of solution, V.21; relationship to weak solutions of SDEs, V.19-20; Stroock–Varadhan Theorem, V.24; well-posed, V.19.
Martingale representation: for Brownian motion, IV.36, 41; for FD diffusion, V.25; for Markov chains, IV.21.
Maximum-Modulus Theorem: I.20.
Maximum Principle: III.13.
McGill's Lemma: VI.59.
Mean curvature: V.4.
Measurable function: II.2.
Measurable space: II.1.
Measurable transition function: III.3.
Measure space: II.4.
Meyer decomposition: III.17, VI.29, VI.32, VI.46.
Meyer's Previsibility Theorem: VI.15.

Minkowski inequality: II.10.
Moderate function: IV.42.
Modification: II.36.
Monotone-Class Theorems: II.3.
Monotone Convergence Theorem: II.8.
Multiple points of Brownian motion: I.21.
Multiplicative functional: see PCHMF.

Nagasawa's formula: III.42, III.46.
Natural scale: V.46.
Net: II.80.
Normal coordinates: V.34.
Normal transition function: III.3.

Observation process: VI.8
Occupation density formula: I.5, IV.45.
One-dimensional diffusions: I.5, V.44-54, absorbing, inaccessible, reflecting end-points, V.47, V.51; exit, entrance boundary points, V.51; infinitesimal generator, V.47, V.50; natural scale, V.46; regular diffusion, V.45; resolvent, V.50, VI.54; scale function, V.46; speed measure, V.47; time substitution, V.47.
One-parameter subgroup: V.35.
Optional processes, σ-algebra: VI.4.
Optional projection: VI.7.
Optional-Sampling Theorem: II.59, II.77.
Optional Section Theorem: VI.5.
Optional-Stopping Theorem: II.57.
Optional time: VI.4; see also stopping time.
Orthonormal frame bundle: V.30, V.33, V.34.
Ornstein–Uhlenbeck process: I.23, V. 5; spectral measure, I.24.
Outer measure: I.6, II.35.

π-system: II.1.
Parallel-displacement: V.32, 34.
Parallel transport: V.32.
Path decomposition: III.49, VI.55.
Path regularization: II.67, III.7.
Path-space: II.28, V.8.
Pathwise-exact SDE: see SDE.
Pathwise uniqueness: V.9, V.17; Nakao theorem, V.41; Yamada–Watanabe Theorem, V.40.
PCHAF (perfect, continuous, homogeneous, additive functional): III.16.
PCHMF (perfect, continuous, homogeneous, multiplicative functional): III.18.
PFA theorem: VI.12, VI.16.
Picard's Theorem: I.20.
Polish space: II.82.
Poisson measures: II.37.
Potential (supermartingale): II.59.
Potential theory: I.22; see also Dirichlet problem, Martin–Doob–Hunt theory.
Poisson kernel for the half plane: I.19.
Poisson kernel for the unit ball: III.30.

Poisson measure, process: II.37, VI.2.
Polish space: II.82; characterization of, II.82.
Probability triple: II.14.
Pre-Brownian motion: II.32, II.68.
Pre-Poisson set function: II.33.
Pre-T σ-algebra: II.58, II.73, VI.17.
Previsible: II.47; path functionals, V.8; processes, σ-algebra, IV.6; Section Theorem, VI.19; stopping time, VI.12.
Previsible projection: VI.19.
Product σ-algebras: II.ll; product measures, II.12, II.22.
Progressive process, σ-algebra: II.73, VI.3.
Prohorov's Theorem: II.83.
Pseudo-Riemannian metric: V.36.
Pure local martingale: IV.34, IV.35, V.28.
Purely discontinuous martingales: IV.24.

Q-matrices: III.1; DK conditions, III.53; local-character condition, III.54; probabilistic significance, III.57, IV.21; of symmetric chains, III.59; of totally instantaneous chains, III.55.
Quadratic-covariation process: IV.26.
Quadratic variation: I.11.
Quadratic-variation process: IV.26, VI.36; for continuous local martingales, IV.30; previsible angle-bracket process, VI.34.
Quantum fluctuations: V.5.
Quasi-left-continuity: for FD processes, III.11; of filtrations, VI.18; for Ray processes III.41, III.50.
Quasimartingales: VI.41.
Quaternions: V.35.

R-filtered space: II.67.
R-path, R-process: II.62, II.63, Introduction to Chapter IV.
R-regularisation: II.67.
R-supermartingale convergence theorem: II.69.
Radon–Nikodým Theorem: II.9.
Random field: I.24.
Random walk, Martin boundary of: III.28.
Ray–Knight compactification: III.35.
Ray–Knight Theorem on local times: VI.52.
Ray processes: III.36; application to chains, III.50.
Ray resolvent: III.34.
Ray's Theorem: III.36, III.38.
Reducing sequence: IV.11.
Reduction: IV.11, IV.29.
Reflecting Brownian motion: see Brownian motion.
Reflection principle: I.13.
Regular conditional probabilities: existence theorem, II.89; counterexample, II.43.
Regular class (D) submartingale: VI.31.
Regular diffusion: V.45.
Regular increasing process: VI.21.
Regular point: I.22.

Regular function: III.27.
Regularizable path: II.62.
Resolvent: III.2, III.3; of Brownian motion, III.3.
Resolvent equation: III.2, III.3.
Reuter's Theorem on drifting Brownian motion: IV.39.
Reversed Martingale Convergence Theorem: II.51.
Riemannian connection: V.32, V.34.
Riemannian manifold: V.31.
Riemann mapping theorem: I.19.
Riemannian metric: V.34.
Riemannian structure induced by non-singular diffusion: V.34.
Riesz decomposition of excessive functions: III.27.
Riesz decomposition of a UI supermartingale: II.59.
Riesz representation Theorem: II.80, III.6.
Rolling without slipping: V.33.

σ-additivity: II.4.
σ-algebra: II.1; countably-generated, II.88.
σ-field: see σ-algebra.
SDE: of diffusion type, V.8; exact, V.9, V.17; Itô's Theorem on existence and uniqueness of solutions, V.11; links with martingale problem, V.19-20; with (locally) Lipschitz coefficients, V.11-13; Markov property of solutions, V.13; pathwise uniqueness, V.9, V.17; strong solution, V.10; Tanaka's SDE, V.16; time-reversal, V.13; Tsirel'son's SDE, V.18; uniqueness in law, V.16; weak solution, V.16.
Scale function: V.28, V.46.
Scheffé's Lemma: II.8.
Section Theorem: II.76; Optional Section Theorem, VI.5; Previsible Section Theorem, VI.19.
Semimartingale: IV.15; continuous, see continuous semimartingale; as integrator, IV.16; local time, IV.43; in a manifold, IV.15.
Signal process: VI.8.
Skew product of Brownian motion: IV.35.
Skorokhod embedding: I.7, VI.51.
Skorokhod's equation: V.6.
Special semimartingale: VI.40.
Spectral measure of stationary Gaussian process: I.24.
Spitzer–Rogozin identity for Lévy processes: I.29.
Splitting time: III.49.
Stable process: I.28.
Stable subspace: IV.24.
Standard process: III.49.
'Standard' transition matrix function: III.2.
Stochastic control, optimality principle: V.15.
Stochastic development: V.33.
Stochastic differential equation: see SDE.
Stochastic differentials: IV.32, V.1.
Stochastic flows: V.13.
Stochastic integral: IV.27, VI.36-38; Riemann-sum approximation, IV.47.
Stochastic partial differential equations: VI.11.
Stochastic process: II.27.

Stone–Weierstrass Theorem: II.80.
Stopping time: II.56, II.73.
Strassen's Law: I.16; invariance principle, I.16.
Stratonovich calculus: IV.46; switch to Itô, V.30.
Strong Law of Large Numbers: II.51.
Strong Markov property: for Brownian motion, I.12; for FD processes, III.8, III.9; for Ray processes, III.40; under time reversal, III.47.
Strong reduction: VI.37.
Structural constants for Lie groups: V.35.
Structural equations: V.34.
Subadditive Ergodic Theorem: I.22.
Submanifold: V.34; regular submanifold, V.31.
Sub-Markov semigroup: III.3.
Submartingale: II.46, II.63.
Subordinator: I.28, II.37, VI.43.
Summation convention: V.1.
Superharmonic function: III.31.
Supermartingale: convergence theorem, II.49; definition, II.46, II.63; sup of a sequence of, II.78.
Supermedian function: III.34.
Symmetrisable Q-matrix, transition matrix function: III.59.

Taboo probabilities: III.52.
Tanaka's formula: IV.43.
Tanaka's SDE: V.16.
Tangent bundle: V.34.
Tangent vector: V.30, V.34.
Tchebychev inequality: II.18.
Terminal time: III.18.
Tightness: II.83, II.85.
Time change: see time substitution.
Time reversal: III.42, III.47, III.49.
Time-reversed Brownian motion: II.38.
Time substitution: III.21, IV.30, V.26.
Torsion: V.34.
Totally inaccessible stopping time: V.21, VI.13-14.
Tower property of conditional expectation: II.41.
Transition function: III.1; measurable, III.3.
Trotter's Theorem: I.5.
Tsirel'son's SDE: V.18.

Uniform asymptotic negligibility: I.28.
UI: see uniform integrability.
Uniform integrability II.20, II.29, II.21 II.44.
Uniqueness in law: V.16.
Universal completion: III.9.
Upcrossings: II.62.
Upcrossing Lemma: II.48.
Usual augmentation: II.67, II.75.
Usual conditions: II.67, IV, Introduction.

Volkonskii's formula: III.21.
Volkonskii–Šur–Meyer Theorem: III.16, III.17.
Volume element: V.34.
Von Mises distribution: IV.39.

Weak convergence: II.83; Prohorov's Theorem, II.83; in *W*, II.85; Skorokhod's interpretation, II.84, II.86.
Weak* topology: II.80.
Weyl's Lemma: V.38.
Whittle's flypaper example: V.7, V.15.
Wiener–Hopf factorisation of Lévy processes: I.29 .
Wiener measure: I.6.
Wiener process: see Brownian motion.
Wiener's Theorem: I.6; proofs, I.6, II.71.

Yor's addition formula: IV.19.
Yor's Theorem on semimartingale local time: IV.44.

Zvonkin's observation: V.18, V.28.

Printed in the United States
By Bookmasters